Molecular Biology of the Female Reproductive System

Molecular Biology of the Female Reproductive System

Edited by

J. K. Findlay

Prince Henry's Institute of Medical Research
Clayton, Victoria
Australia

ACADEMIC PRESS

A Division of Harcourt Brace & Company

San Diego New York Boston London Sydney Tokyo Toronto

This book is printed on acid-free paper. ∞

Academic Press, Inc.
525 B Street, Suite 1900, San Diego, California 92101-4495

United Kingdom Edition published by
Academic Press Limited
24–28 Oval Road, London NW1 7DX

Library of Congress Cataloging-in-Publication Data

Molecular biology of the female reproductive system / edited by
J.K. Findlay.
p. cm.
Includes bibliographical references and index.
ISBN 0-12-256365-4 (hardcover)
1. Generative organs, Female--Molecular aspects. 2. Generative organs, Female--Cytology. 3. Human reproduction--Molecular aspects.
I. Findlay, Jock K.
[DNLM: 1. Reproduction--physiology. 2. Sex Hormones--physiology.
3. Placental Hormones--physiology. WQ 205 C393 1994]
QP259.C46 1994
612.6'2--dc20
DNLM/DLC
for Library of Congress 93-30113
CIP

PRINTED IN THE UNITED STATES OF AMERICA
94 95 96 97 98 99 BC 9 8 7 6 5 4 3 2 1

Contents

1

Hormonal Control of Folliculogenesis and Luteinization

Stephen G. Hillier

2

Control Systems of Theca-Interstitial Cells

Denis A. Magoffin & Gregory F. Erickson

3

Growth Factor-Mediated Cell–Cell Interactions in the Ovary

Michael K. Skinner & Jeff A. Parrott

4

Potential Relevance of Cytokines to Ovarian Physiology

Eli Y. Adashi, Ehud Kokia, & Arye Hurwitz

5

Role of Insulin-like Growth Factors (IGF) and IGF Binding Proteins in Folliculogenesis

Gregory F. Erickson, Akira Nakatani, X.-J. Liu, Shunichi Shimasaki, & Nicholas Ling

6

Expression of Genes Encoding Steroidogenic Enzymes in the Ovary

Margaret M. Hinshelwood, Michelle Demeter-Arlotto, Gary D. Means, & Evan R. Simpson

7

Molecular Biology of Endocrine Receptors in the Ovary

Ok-Kyong Park-Sarge & Kelly E. Mayo

8

Molecular Mechanisms in Ovulation

Alex Tsafriri & Nava Dekel

9

Hormonal Control of Luteal Function

Timothy D. Braden, Carol J. Belfiore, & Gordon D. Niswender

10

Cell Biology of the Oviduct and Endometrium

Lois A. Salamonsen & Colin D. Nancarrow

11

Cell Biology of Migratory Cells in the Endometrium

David A. Clark & Ewa Muzikova

12

Growth Factor and Cytokine Expression in the Endometrium

Liam J. Murphy & Gustavo Ballejo

13

Paracrine Interactions in Endometrial Function

Markku Seppälä & Eeva-Marja Rutanen

14

Molecular Biology of Trophectoderm and Placental Hormones

R. Michael Roberts & Russell V. Anthony

Contributors

Numbers in parentheses indicate the pages on which the authors' contributions begin.

Eli Y. Adashi (83), Departments of Obstetrics/Gynecology and Physiology, University of Maryland School of Medicine, Baltimore, Maryland 21201

Russell V. Anthony (395), Departments of Animal Sciences and Biochemistry, University of Missouri–Columbia, Columbia, Missouri 65211

Gustavo Ballejo (345), Department of Pharmacology, Faculdade de Medicina, Universidade de São Paulo, Ribreirao Preto, Brazil

Carol J. Belfiore (259), Department of Molecular Biology, University of Wyoming, Laramie, Wyoming 82071

Timothy D. Braden (259), Department of Zoology, Oklahoma State University, Stillwater, Oklahoma 74078

David A. Clark (329), Departments of Medicine, Obstetrics and Gynecology, and Pathology, Molecular Immunology–Virology Program, McMaster University, Faculty of Health Sciences, Hamilton, Ontario, Canada L8N 3Z5

Nava Dekel (207), The Department of Hormone Research, The Weizmann Institute of Science, Rehovot 76100, Israel

Michelle Demeter-Arlotto (129), The Cecil H. and Ida Green Center for Reproductive Biology Sciences, and the Departments of Obstetrics/Gynecology and Biochemistry, The University of Texas Southwestern Medical Center, Dallas, Texas 75235

Gregory F. Erickson (39, 101), Department of Reproductive Medicine, University of California at San Diego, School of Medicine, La Jolla, California 92093

Stephen G. Hillier (1), Reproductive Endocrinology Laboratory, Department of Obstetrics and Gynaecology, University of Edinburgh Centre for Reproductive Biology, Edinburgh EH3 9EW, United Kingdom

Margaret M. Hinshelwood (129), The Cecil H. and Ida Green Center for Reproductive Biology Sciences, and the Departments of Obstetrics/Gynecology and Biochemistry, The University of Texas Southwestern Medical Center, Dallas, Texas 75235

Arye Hurwitz (83), Departments of Obstetrics/Gynecology and Physiology, University of Maryland School of Medicine, Baltimore, Maryland 21201

Ehud Kokia (83), Departments of Obstetrics/Gynecology and Physiology, University of Maryland School of Medicine, Baltimore, Maryland 21201

Nicholas Ling (101), Neuroendocrine Biosciences, Inc., La Jolla, California 92037

X.-J. Liu (101), Department of Molecular Endocrinology, Whittier Institute for Diabetes and Endocrinology, La Jolla, California 92037

Denis A. Magoffin (39), Department of Obstetrics and Gynecology, Cedars-Sinai Medical Center, Los Angeles, California 90048

Kelly E. Mayo (153), Department of Biochemistry, Molecular Biology, and Cell Biology, Northwestern University, Evanston, Illinois 60208

Gary D. Means (129), Fred Hutchinson Cancer Research Center, Basic Sciences Division, Seattle, Washington 98104

Liam J. Murphy (345), Departments of Internal Medicine and Physiology, Faculty of Medicine, University of Manitoba, Winnipeg, Manitoba, Canada R3E 0W3

Ewa Muzikova (329), Molecular Immunology–Virology Program, McMaster University, Faculty of Health Sciences, Hamilton, Ontario, Canada L8N 3Z5

Akira Nakatani (101), Department of Pathology, Nagasaki University School of Medicine, Nagasaki 852, Japan

Colin D. Nancarrow (289), Division of Animal Production, CSIRO, Prospect, New South Wales 2148, Australia

Gordon D. Niswender (259), Animal Reproduction and Biotechnology Laboratory, Department of Physiology, Colorado State University, Fort Collins, Colorado 80523

Ok-Kyong Park-Sarge (153), Department of Physiology, University of Kentucky Medical School, Lexington, Kentucky 40536

Jeff A. Parrott (67), Reproductive Endocrinology Center, University of California at San Francisco, San Francisco, California 94143

R. Michael Roberts (395), Departments of Animal Sciences and Biochemistry, University of Missouri–Columbia, Columbia, Missouri 65211

Eeva-Marja Rutanen (379), Department of Obstetrics and Gynecology, Helsinki University Central Hospital, SF-00290 Helsinki, Finland

Lois A. Salamonsen (289), Prince Henry's Institute of Medical Research, Clayton, Victoria 3168, Australia

Markku Seppälä (379), Department of Obstetrics and Gynecology, Helsinki University Central Hospital, SF-00290 Helsinki, Finland

Shunichi Shimasaki (101), Department of Molecular Endocrinology, Whittier Institute for Diabetes and Endocrinology, La Jolla, California 92037

Evan R. Simpson (129), The Cecil H. and Ida Green Center for Reproductive Biology Sciences, and the Departments of Obstetrics/Gynecology and Biochemistry, The University of Texas Southwestern Medical Center, Dallas, Texas 75235

Michael K. Skinner (67), Reproductive Endocrinology Center, University of California at San Francisco, San Francisco, California 94143

Alex Tsafriri (207), The Bernhard Zondek Hormone Research Laboratory, The Department of Hormone Research, The Weizmann Institute of Science, Rehovot 76100, Israel

Preface

Our understanding of the control of female reproduction has advanced significantly in the past decade. Thus, the purpose of this book is twofold: First, to show how the techniques of cellular and molecular biology have led to this increase in knowledge and second, to review the autocrine, paracrine, and endocrine systems controlling reproductive processes in the female. With chapters by recognized leaders in the field, this book will be of primary interest to students and researchers working in biomedical and animal research. It will also be of interest to specialist medical and veterinary practitioners, toxicologists, biologists, and teachers as a way of updating their knowledge.

The chapters focus on the ovary and uterus, including the interaction between the developing trophectoderm of the blastocyst and the endometrium. Chapters on the ovary include hormonal control of folliculogenesis and luteal function, cell–cell interactions in the follicle and the roles of cytokines in regulating steroid and protein hormone production, and an update on endocrine receptors and the mechanisms involved in ovulation. The cell biology of the oviduct and uterus is reviewed, and there are chapters on migratory cells and paracrine regulation in the endometrium. The final chapter discusses the recent data on hormones of the trophectoderm and placenta. The authors pay particular attention to the application of cellular and molecular approaches to understanding control of the reproductive processes and have provided extensive reference lists at the end of each chapter for follow up.

This book differs from others in this field in both its focus and content and, together with the previously published volume *Molecular Biology of the Male Reproductive System*, edited by David de Kretser, will provide the reader with an extensive catalog of knowledge of the control of reproductive processes at the cellular and molecular levels.

I extend my thanks to the authors for giving their valuable time to write the excellent contributions, to my secretary, Faye Coates, and to Academic Press for helping to produce the book.

Jock Findlay

1

Hormonal Control of Folliculogenesis and Luteinization

STEPHEN G. HILLIER

I. Introduction

During each menstrual cycle, human ovaries sequentially synthesize and secrete estradiol-17β (estradiol) and progesterone–steroid hormones that coordinate the function of the entire female reproductive system. Increased estradiol secretion characterizes the follicular phase of the cycle, reflecting the specialized endocrine function of the preovulatory follicle. After ovulation, the follicle luteinizes and secretes mainly progesterone at rates mirroring the development and regression of the corpus luteum during the luteal phase. This chapter surveys the major endocrine and paracrine mechanisms that give rise to these processes and discusses the practical implications for more efficient and effective use of exogenous gonadotropins to stimulate ovarian function in infertile women.

II. Control of Follicular Growth

A. The Ovarian Cycle

The human ovarian cycle lasts approximately 28 days and encompasses preovulatory follicular development (10–14 days), ovulation, and the functional lifespan of the corpus luteum (12–15 days). Peripheral plasma sex steroid, gonadotropin, and inhibin levels throughout the menstrual cycle are illustrated in Figure 1.

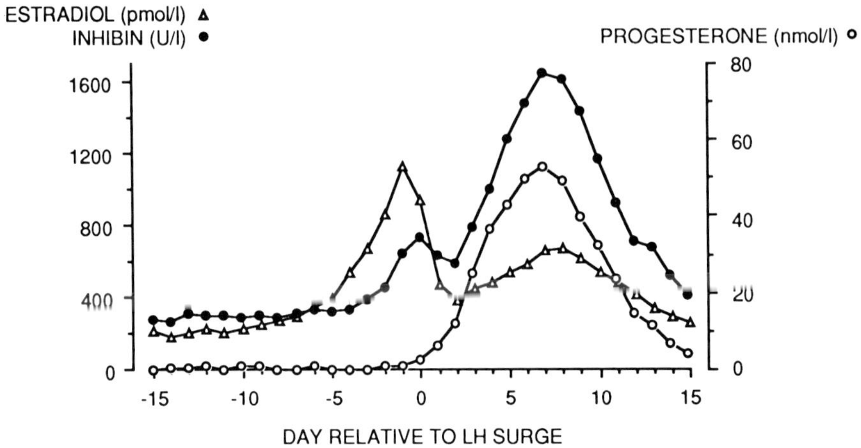

FIGURE 1 Serum levels of estradiol, progesterone, and inhibin during the human menstrual cycle. Data redrawn from McLachlan *et al.* (1990) with permission.

B. Preovulatory Follicular Development

The follicle destined to ovulate grows during the follicular phase of the cycle, increasing in diameter from ~5 mm to ≥20 mm and becoming the major ovarian source of secreted estrogen. This estrogen secretory stage follows a programmed sequence of cell growth and differentiation in the follicle wall that terminates in ovulation and transformation of the follicle into a corpus luteum. The entire sequence of events depends upon primary (endocrine) stimulation of the ovaries by the gonadotropins, follicle-stimulating hormone (FSH) and luteinizing hormone (LH), underpinned by local (paracrine and autocrine) levels of control emanating from within the follicle itself.

Multiple immature antral follicles that are potentially able to undergo preovulatory development are usually present in the ovaries at the beginning of each normal menstrual cycle. These incipient preovulatory follicles comprise a fluid-filled antral cavity surrounded by inner granulosa (to which the cumulus-enclosed secondary oocyte is attached) and outer thecal cell layers separated by a lamina basalis. Tonic stimulation of the ovaries by both FSH and LH is necessary for follicles to achieve this intermediate stage of development, but "recruitment" to the preovulatory stage of development requires additional stimulation by FSH (Figure 2). By the midfollicular phase, the follicle destined to ovulate will have grown to about 10 mm in diameter and will be recognizable as the largest healthy follicle in either ovary (Gougeon and Lefévre, 1983). Thereafter, during the second half of the follicular phase, estrogen synthesis in this dominant follicle becomes

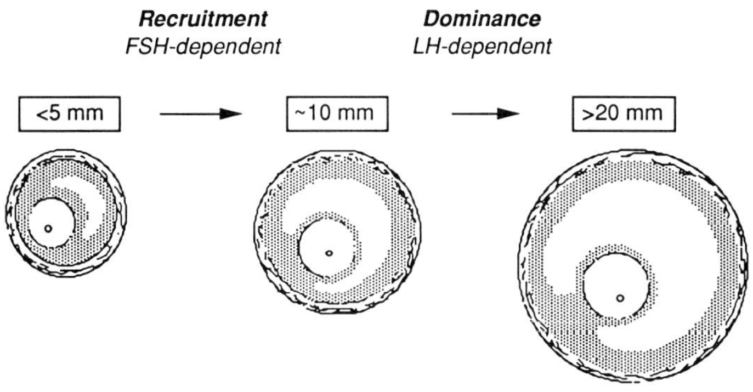

FIGURE 2 Development-dependent stages in selection of the preovulatory follicle in the human menstrual cycle. Recruitment: Blood FSH levels rise and stimulate the proliferation and functional differentiation of granulosa cells in multiple immature follicles, including induction of LH receptors and aromatase activity. Tonic stimulation by LH maintains thecal androgen synthesis in these follicles. The follicle that will next ovulate is that which is most responsive to FSH (i.e., that with the lowest FSH "threshold"). Dominance: By midfollicular phase, the preovulatory follicle can be recognized as the largest (≥10 mm diameter) healthy follicle in either ovary. Because its granulosa cells express LH receptors coupled to aromatase, this follicle continues to grow and secrete estradiol in the face of declining levels of FSH. Basal levels of LH are sufficient to sustain preovulatory follicular growth and estrogen secretion. Development-related paracrine signaling sustains the dominance of this follicle, amplifying its responsiveness to FSH and LH. Modified from Hillier (1990).

increasingly responsive to LH (Figure 2). The individual contributions of FSH and LH to this process are described in the following paragraphs.

C. Function of FSH

The primary signal for starting preovulatory follicular development is increased secretion of FSH by the pituitary gland. This occurs in response to the withdrawal of estrogen-mediated inhibition of pituitary FSH release as the corpus luteum of the previous cycle regresses (Le Nestour *et al.*, 1993). It takes 10–12 days of sustained stimulation by FSH for a ~5 mm follicle to attain a full preovulatory diameter of ≥20 mm; during this time the number of its granulosa cells doubles 5 or 6 times to reach over 50 million (McNatty, 1981). The FSH directly stimulates proliferation of granulosa cells and induces the LH-responsive mechanisms in these cells that sustain steroid secretion by the preovulatory follicle and, after ovulation, the corpus luteum (Richards *et al.*, 1987).

Granulosa cells are the only cells in the female body known to possess FSH receptors. When FSH is bound to its receptor on the cell surface, adenylyl cyclase and cyclic adenosine monophosphate (cAMP)-dependent protein kinase(s) are activated, leading to increased expression of the diverse messenger ribonucleic acids (mRNAs) that encode the proteins crucial to cytoproliferation and differentiation. The list of FSH-responsive genes is extensive and at present includes aromatase (P450arom), the steroidogenic cytochrome P450 crucial to estrogen synthesis (Hickey *et al.*, 1988; Steinkampf *et al.*, 1988); cholesterol side chain cleavage (P450scc), which is rate-limiting in progesterone synthesis (Richards *et al.*, 1987); the LH receptor (Segaloff *et al.*, 1990); several polypeptide growth factors and their binding proteins, such as the insulin-like growth factors (IGFs) and IGFBPs (Oliver *et al.*, 1987; Voutilainen and Miller, 1987; Hernandez *et al.*, 1992); proteolytic enzymes and inhibitors implicated in the mechanism of follicular rupture at ovulation, such as tissue plasminogen activator (TPA) (O'Connell *et al.*, 1987) and plasminogen activator inhibitor (PAI) (Ny *et al.*, 1985); local regulatory peptides such as inhibin, activin, and follistatin [see

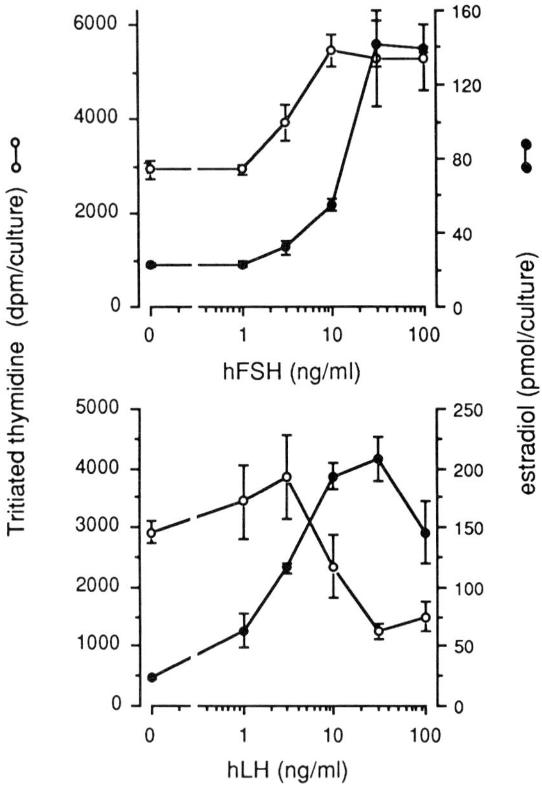

FIGURE 3 Dose-dependent effects of FSH (top) and LH (bottom) on thymidine uptake (open circles) and estradiol production (closed circles) by human granulosa cells obtained from a preovulatory follicle before onset of the LH surge. The cells were incubated for 4 days with hFSH or hLH and 1.0 μ*M* testosterone (aromatase substrate). Data are means ± SE. Reproduced with permission from Yong *et al.* (1992b). Copyright © 1992 The Endocrine Society.

Ying (1988) for review]; and the heat-shock protein hsp90 (Ben-Ze'ev and Amsterdam, 1989).

In vitro treatment with FSH stimulates deoxyribonucleic acid (DNA) synthesis in cultured human granulosa cells (Figure 3). The mitogenic action of FSH is greatly enhanced if IGF-I is also present in the culture medium, indicating that locally produced growth factors facilitate FSH-stimulated follicular growth *in vivo* (Yong *et al.*, 1992b).

D. Function of LH

The secretion of sex steroids by the preovulatory follicle and corpus luteum is critically dependent on LH. Estrogen secretion by the follicles increases during the second half of the follicular phase when the frequency of pulsatile LH discharge by the pituitary increases and FSH secretion declines. During this stage of development, granulosa cells express LH receptors, and tonic levels of LH are able to directly stimulate steroidogenic enzyme systems preinduced by FSH (Hsueh *et al.*, 1984). Luteinizing hormone receptors are also located on thecal/interstitial cells, and steroidogenesis at these sites is under direct LH control throughout the menstrual cycle (Erickson *et al.*, 1985). The hormone acts through its receptors on mature granulosa cells and thecal/interstitial cells to activate adenylyl cyclase signaling, which increases precursor-cholesterol uptake and sustains characteristically high steroidogenic enzyme activity in these cells (Gwynne and Strauss, 1982). Luteinizing hormone also activates the transmission of intracellular signals in thecal (Hofeditz *et al.*, 1988) and luteal (Davis *et al.*, 1989; Wheeler and Veldhuis, 1989) cells through inositol lipid hydrolysis and the diacylglycerol/protein kinase C pathway (see Leung and Steele [1992] for review).

An ovulation-inducing LH bolus is released by the pituitary gland at midcycle. This terminates proliferation of granulosa cells (Yong *et al.*, 1992a,b); temporarily inhibits estrogen secretion; and simultaneously initiates rupture of the follicle formation of the corpus luteum, and onset of luteal progesterone secretion [see Hillier (1991a) for review]. Thereafter, tonic stimulation by LH transiently sustains steroidogenesis in the corpus luteum [see Zeleznik (1991) for review]. If pregnancy occurs, the functional lifespan of the corpus luteum is extended by chorionic gonadotropin (CG) acting through luteal cell LH receptors (see Section VII).

III. Endocrine Regulation of Follicular Estrogen Secretion

Major tenets of the "two-cell, two-gonadotropin" model of estrogen synthesis are that LH stimulates precursor androgen synthesis in the theca interna and that FSH stimulates granulosa cell aromatase activity. Hence

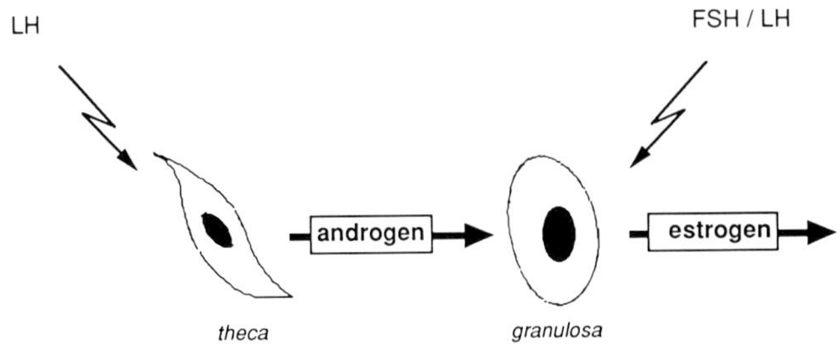

FIGURE 4 Contemporary "two-cell, two-gonadotropin" model of follicular estradiol biosynthesis. Androgen synthesis occurs in the theca interna regulated by LH; the rate-limiting metabolism of C_{21} precursor to C_{19} androgen is catalyzed by 17-hydroxylase/C17-20 lyase (P450c17). Aromatase (P450arom), the enzyme that converts androstenedione to C_{18} estrogen (estradiol), is induced in granulosa cells by FSH. FSH also induces granulosa cell LH receptors functionally coupled to aromatase. Therefore, in the preovulatory follicle, both the synthesis of androgen (in thecal cells) and its aromatization to estradiol (in granulosa cells) are directly regulated by LH. Based on Armstrong and Dorrington (1979).

both cell types and both gonadotropins are crucial to estrogen synthesis (Armstrong and Dorrington, 1979) (Figure 4).

A. Thecal Androgen Synthesis

Thecal cells minimally express P450arom and are therefore unable to synthesize estrogens *de novo* (Tamura *et al.*, 1992; Whitelaw *et al.*, 1992). However, thecal cells do abundantly express 17-hydroxylase/C-17,20-lyase (P450c17), the cytochrome P450 that catalyzes rate-limiting steps in androgen synthesis, and which is positively regulated by LH (Erickson *et al.*, 1985). The vascularized theca interna has direct access to the blood-borne precursor cholesterol, and thecal cells contain all the steroidogenic enzymes necessary to synthesize androgens from acetate and cholesterol (Gwynne and Strauss, 1982). IGF-I receptors functionally coupled to androgen synthesis are also present on thecal cells (Bergh *et al.*, 1993), and picomolar amounts of IGF-I augment basal and LH-stimulated androgen synthesis in rat (Cara and Rosenfield, 1988) and human thecal cells *in vitro* (Hillier *et al.*, 1991c). Insulin-like growth factors of granulosa and/or hepatic origin may therefore fulfill a cogonadotropic role in augmenting LH-stimulated androgen synthesis *in vivo*.

The major androgen secreted by the human preovulatory follicle is androstenedione. During the follicular phase of the menstrual cycle, the ovary contributes about 30% of the total blood androstenedione and the adrenal accounts for the rest. Toward midcycle, the ovarian contribution rises to about 60% as a result of increased synthesis and secretion of the steroid by the LH-stimulated theca of the preovulatory follicle, reflecting its role as a precursor for estradiol (Baird, 1977). The mechanism by which androgen synthesis is selectively upregulated in this follicle appears to depend on paracrine control; this is discussed later in Section V.

Thecal androgen synthesis in the preovulatory follicle is suppressed at midcycle. The mechanism is poorly understood but seems to involve a form of LH-induced desensitization due to a temporary but extreme elevation in the circulating LH level. Thecal 17-hydroxylase/C-17,20 lyase activity is attenuated at this time, possibly due to locally produced estradiol (Erickson *et al.*, 1985).

B. Estrogen Synthesis by Granulosa Cells

Granulosa cells are unable to synthesize androgens *de novo* since they do not express P450c17. They are, however, intrafollicular sites of androgen metabolism, and express 17-oxoreductase, 3β-hydroxysteroid dehydrogenase/$\Delta^{5\text{-}4}$ isomerase (3β-HSD), as well as aromatase (Tamura *et al.*, 1992). These steroidogenic enzymes are positively regulated by FSH during preovulatory follicular development [see Hillier (1985) for review]. Granulosa cells also undertake 5α reduction (McNatty, 1981) and conjugation (Lischinsky *et al.*, 1983) of androgens, but it is not clear if the enzymes that catalyze these processes are developmentally regulated.

Within the preovulatory follicle, the avascular granulosa cell layer is exposed to high concentrations of aromatizable androgen that reach it by diffusion from the theca interna. Practically the entire aromatizing capacity of the preovulatory follicle resides in its granulosa cells, judging from comparisons of aromatase enzyme activity in isolated granulosa and thecal cells (Hillier *et al.*, 1981), and the cellular distribution of P450arom protein (Sasano *et al.*, 1989; Tamura *et al.*, 1992) and P450arom mRNA (Whitelaw *et al.*, 1992) in the follicle wall. Cells in the outer (mural) granulosa cell layer are thought to be particularly active sites of aromatization, since they express more P450arom (Zoller and Weisz, 1978; Whitelaw *et al.*, 1992) and have more LH receptors (Amsterdam *et al.*, 1975) than cells distal to the lamina basalis. Mural granulosa cells are in more or less direct contact with the thecal vasculature, being separated from blood vessels only by the lamina basalis in many regions of the follicle wall. They are therefore in a favorable position to respond immediately to changes in the circulating LH level. Moreover, the estrogen they produce can be discharged directly into the venous effluent of the preovulatory follicle (Zeleznik *et al.*, 1981).

C. Development-Dependent LH Action

Estradiol is released into the bloodstream in greatest amounts as the follicle destined to ovulate completes its development in the late follicular phase. At this time the preovulatory follicle secretes up to 400 μg (~1.5

μmol) of estradiol a day and is the source of more than 90% of secreted estrogen (Baird, 1977). At this stage of follicular development, granulosa cells express LH receptors that are functionally coupled to aromatase activity so that both the synthesis of androgen in the theca interna and its conversion to estrogen in granulosa cells are directly regulated by LH (Figure 4). Paracrine signaling within the follicle wall presumably reinforces this development-related function of LH, as discussed in the next section.

IV. Paracrine Regulation of Follicular Estrogen Secretion

Fundamental to the endocrine control of follicular estrogen secretion is that LH promotes the synthesis of androgen in thecal cells while FSH induces its metabolism (aromatization) to estrogen in granulosa cells. The realization that androgen functions in this way as a biochemical link between adjacent cell types was the first step toward our current understanding that paracrine signaling ultimately underpins all aspects of gonadotropin action on the ovaries. Aspects of the follicular paracrine system that directly affect estrogen synthesis are surveyed in this section.

A. Theca-Derived Paracrine Control

In addition to androgens, the theca interna produces many polypeptide regulatory factors (Green, 1989) that are potentially able to influence granulosa cell function. These last include epidermal growth factor (EGF) and the transforming growth factors TGFα and TGFβ.

1. Androgens

Besides serving as an aromatase substrate, theca-derived androstenedione is metabolized in granulosa cells to testosterone and other androgens that act locally to modulate cytodifferentiative responses to FSH. The regulatory action of androgen in granulosa cells is mediated through specific androgen receptors (Horie *et al.*, 1992) and entails amplification of cyclic AMP-mediated intracellular signaling [see Hillier (1985) for review]. Aromatase activity (Daniel and Armstrong, 1986; Harlow *et al.*, 1988) and inhibin production (Hillier *et al.*, 1991a) are conspicuous among the FSH-inducible functions known to be augmented by androgens *in vitro*. Conversely, and most intriguingly, inhibin is able to promote LH-dependent thecal androgen synthesis *in vitro* (Hsueh *et al.*, 1987; Hillier *et al.*, 1991c). Thus the potential exists for a reciprocal interaction between granulosa-derived inhibin and theca-derived androgen that could contribute to the local regulation of preovulatory follicular estrogen secretion, as discussed in Section V.

2. EGF and TGFα

Epidermal growth factor and TGFα are closely related gene products with similar properties and which bind to the same receptors (Massagué, 1983; Derynck, 1986). The active factors are polypeptides consisting of about 50 amino acids which have been shown to modulate the development of epidermis, breast, and gut. They also act as angiogenic factors and may mediate hypercalcemia (Gill *et al.*, 1987; Waterfield, 1989). Granulosa cells express hormonally regulated receptors for EGF/TGFα (Feng *et al.*, 1987). Treatment with EGF/TGFα *in vitro* promotes the proliferation of granulosa cells (Gospodarowicz and Bialecki, 1979) and interferes with FSH induction of cytodifferentiation (expression of LH receptors, aromatase activity, and progesterone and inhibin synthesis) (Jones *et al.*, 1982; Knecht and Catt, 1983; Franchimont *et al.*, 1986; Steinkampf *et al.*, 1988). Intrafollicular expression of TGFα has been localized to the theca interna (Kudlow *et al.*, 1987). Theca-derived EGF/TGFα could therefore be involved in the paracrine control of granulosa cell responsiveness to FSH, serving to enhance cell proliferation and inhibit steroidogenesis.

3. TGFβs

TGFβs are 25-kDa homodimeric proteins usually expressed in regions of epithelio–mesenchymal interaction, notably during embryonic organogenesis (Slack, 1989). They are multifunctional regulators of cell growth, differentiation, and function that either stimulate or inhibit cellular proliferation *in vitro*, depending on the cells, growth conditions, and presence of other growth factors (Roberts *et al.*, 1988).

Ovarian thecal (Skinner *et al.*, 1987; Bendell and Dorrington, 1988) and granulosa cells (Kim and Schomberg, 1989) have been identified as sites of TGFβ synthesis, and steroid synthesis in both cell types is influenced by treatment with TGFβ *in vitro*. In rat granulosa cell cultures, treatment with TGFβ modulates the stimulatory effects of FSH and the inhibitory effects of EGF/TGFα on LH receptor induction (Knecht *et al.*, 1986) as well as steroidogenesis (Adashi and Resnick, 1986; Feng *et al.*, 1986). TGFβ also promotes the proliferation of granulosa cells *in vitro* (Dorrington *et al.*, 1988). In rat thecal/interstitial cell cultures, TGFβ inhibits androgen synthesis and stimulates progesterone accumulation (Magoffin *et al.*, 1989). TGFβ is therefore a potential autocrine regulator of steroid synthesis in both cell types, as well as being a possible mediator of paracrine theca–granulosa cell interaction in the follicle wall. In cultured rat granulosa cells, TGFβ was shown to augment cell responsiveness (cAMP production, LH receptor induction, and steroidogenesis) to low doses of FSH but to attenuate responsiveness to high-dose FSH (Knecht *et al.*, 1987). This important finding suggests a mechanism by which locally produced TGFβs might contribute to interfollicular variations in FSH thresholds in the human menstrual cycle (see later).

Many mesenchymal and epithelial cell types whose growth and differentiation are affected by TGFβs respond to these factors with elevated deposition of cell adhesion proteins such as fibronectin, production of protease inhibitors, and expression of integrin receptors (Roberts *et al.*, 1988). Accordingly, TGFβs are thought to affect tissue growth and differentiation by influencing the abundance and architecture of the extracellular matrix as well as the ability of cells to interact with it (Massagué, 1987). It is noteworthy that granulosa cell differentiation *in vitro* is associated with reduced deposition of fibronectin (Skinner *et al.*, 1985), and that extracellular matrix is known to influence granulosa cell responsiveness to gonadotropins (Amsterdam *et al.*, 1989). Follicular development, which consists of interactions between epithelial granulosa cells layer and mesenchyme-derived thecal cells separated by a basement membrane, may therefore constitute an adult form of tissue morphogenesis that also involves locally produced TGFβs (Slack, 1989).

B. Granulosa-Derived Paracrine Control

FSH-induced granulosa cell differentiation is associated with increased production of various steroidal and nonsteroidal factors that have potential regulatory functions within the follicle wall (Hsueh *et al.*, 1984). Estrogen, IGFs, inhibin/activin, and follistatin will be discussed here.

1. Estrogens

Estradiol is generally assumed to contribute to the local mechanism by which a single preovulatory follicle is selected to ovulate in the human menstrual cycle. However, this assumption is based mainly on evidence that estrogens augment FSH action in rat granulosa cells (Goldenberg *et al.*, 1973; Hillier *et al.*, 1980; Richards *et al.*, 1987). There is no such evidenceto support a regulatory action by estrogen in either human (Hillier *et al.*, 1991a) or nonhuman primate (Hild-Petito *et al.*, 1988) granulosa cells. An assessment of the action of estrogen in isolated granulosa cells is complicated by the fact that preovulatory granulosa cells have an increased capacity to metabolize estradiol to catecholestrogens (Hammond *et al.*, 1986). Catecholestrogens have been identified in ovarian follicular fluids from various species, including man (Dehennin *et al.*, 1984), and 2-hydroxyestradiol augments FSH-induced steroid hormone synthesis in cultured granulosa cells (Hudson and Hillier, 1985), pointing to intrafollicular regulatory functions for these estrogen metabolites.

Granulosa-derived estrogen is implicated as a negative regulator of thecal cell function and may be involved in the suppression of thecal androgen synthesis by the midcycle LH surge (Leung and Armstrong, 1980; Erickson *et al.*, 1985).

2. IGFs

There is an increasing body of evidence that the stimulatory effects of FSH on granulosa cell growth and differentiation are augmented by IGFs (Adashi *et al.*, 1991; Giudice, 1992; Hammond *et al.*, 1988). The IGFs are low molecular weight, single-chain peptides which share considerable structural and functional homologies with proinsulin (Froesch *et al.*, 1985). IGF-I and IGF-II each cross-react with the insulin receptor and receptors for insulin, and both factors are present on most cells (Rechler and Nissley, 1985), including thecal and granulosa cells in the human ovary (Poretsky *et al.*, 1985; Bergh *et al.*, 1993). Receptors for insulin and IGF-I also share structural and functional similarities, including intracellular signaling involving tyrosine kinases (Sibley *et al.*, 1988). Both types of IGF exert endocrine effects on tissue growth throughout the body, being secreted by the liver under the control of growth hormone (GH). IGFs are also synthesized and act locally in various tissues other than gonads, including muscle and bone (Slack, 1989).

Rat granulosa cells express IGF-I receptors that increase in number following treatment with FSH *in vitro* (Adashi *et al.*, 1991). Stimulation by FSH of steroidogenesis, LH receptors, deposition of proteoglycans, and responsiveness to β-adrenergic agonists is augmented by physiological concentrations of IGF-I or supraphysiological concentrations of insulin in rat granulosa cell culture medium. Such responses to insulin are presumably due to cross-reaction with granulosa cell IGF-I receptors. IGF-I also synergizes with estradiol to promote steroidogenesis in porcine granulosa cells (Veldhuis *et al.*, 1986). The action of IGF-I on FSH-induced cytodifferentiation entails amplification of intracellular cyclic AMP action by a mechanism not yet understood (Adashi *et al.*, 1991).

Human granulosa cells may produce IGF-II rather than IGF-I (Voutilainen and Miller, 1987; Hernandez *et al.*, 1992). IGF-II has been identified in significant amounts in human follicular fluid (Ramasharma *et al.*, 1986) and is secreted into a culture medium by proliferating granulosa cells in culture (Ramasharma and Li, 1987). Stimulation of steroid synthesis in such cells by treatment with FSH, human chorionic gonadotropin (hCG), or dibutyryl cAMP coordinately induces expression of IGF-II mRNA, suggesting an autocrine/paracrine role for IGF-II in stimulating granulosa cell proliferation (Voutilainen and Miller, 1987). Granulosa cells also synthesize IGF-binding proteins that are likely to participate in regulating local responses to IGFs (Suikkari *et al.*, 1989; Ui *et al.*, 1989; Hernandez *et al.*, 1992).

Thecal cells possess IGF-I receptors (Bergh *et al.*, 1993) and also express IGF-I mRNA (Hernandez *et al.*, 1992). Moreover, both basal and LH-stimulated synthesis of androgen in cultured thecal/interstitial cells are enhanced by the presence of insulin or IGF-I (Barbieri *et al.*, 1983, 1986; Cara and Rosenfield, 1988; Hernandez *et al.*, 1988). Under serum-free culture conditions, IGF-I has a striking ability to enhance DNA as well as androgen synthesis in human thecal cell monolayers (Hillier *et al.*, 1991b). Thus IGF-I and/or IGF-II of granulosa cell origin has the potential to exert positive paracrine control over both the growth and endocrine function of the theca interna.

3. Inhibins and activins

Inhibins and activins were first isolated and purified from follicular fluid on the basis of their ability to inhibit (inhibin) (Ling *et al.*, 1985; Miyamoto *et al.*, 1985; Rivier *et al.*, 1985; Robertson et al., 1985) or stimulate (activin) (Vale *et al.*, 1986; Ling *et al.*, 1986) FSH release by pituitary cells *in vitro*. Mature inhibin is a ~32-kDa heterodymeric glycoprotein composed of a common α-subunit and one of two β-subunits, β_A (inhibin-A) and β_B (inhibin-B). Activins are homodimeric forms of inhibin β-subunits: activin-A (β_A,β_A), activin-AB (β_A,β_B). Theoretically, activin-B (β_B,β_B) should also exist but it has not actually been isolated from an ovarian source. The three inhibin/activin subunits are encoded by separate genes (Mason *et al.*, 1985, 1986; Mayo *et al.*, 1986; Forage *et al.*, 1986; Esch *et al.*, 1987), and biologically active recombinant forms of both inhibin and activin have been expressed in mammalian kidney cells.

Intragonadal function(s) for inhibin and activin are implied by the high degree of structural homology (30–40%) between them and other members of a growth/differentiation superfamily that are synthesized from precursors of high molecular weight and whose expression occurs in embryologic, fetal, and adult tissues across a wide range of animal phyla (see Roberts *et al.*, 1988 and Vale *et al.*, 1988 for reviews). Members of this gene family include transforming growth factor-β (TGFβ), Müllerian duct-inhibiting substance (MIS; causes Müllerian duct regression in males), decapentaplegic gene complex (DPP-C; active during insect embryogenesis), and the encoded product of *vg1* (a mesoderm-inducing factor in frog embryos).

Activin is itself implicated as a regulatory factor in several fundamental processes, including erythropoiesis (Eto *et al.*, 1987; Yu *et al.*, 1987), nerve cell survival (Hashimoto *et al.*, 1990), mesoderm induction in amphibian embryos (Van den Eijnden-Van Raaij *et al.*, 1990), and gonadal cell proliferation (Gonzalez-Manchon and Vale, 1989).

There is also direct evidence to suggest intragonadal functions for activin and inhibin (de Jong, 1988; Ying, 1988). Receptors for activin are broadly distributed in immature and adult animal tissues, including the ovary (Vale *et al.*, 1992). Activin and inhibin also exert potent modulatory effects on follicular steroidogenesis *in vitro* (Hsueh *et al.*, 1987), which is discussed further in Section IV,D. Moreover, targeted deletion of the α-inhibin gene by homologous recombination in mouse embryonic stem cells results in sterile progeny that develop stromal gonadal tumors, suggesting a tumor-suppressive function for inhibin in the gonads (Matzuk *et al.*, 1992).

4. Follistatins

Follistatin is a monomeric glycosylated protein that exists in at least three (~32, ~35, and ~39 kDa) forms which are apparently the result of alternative splicing of primary follistatin mRNA transcripts (Shimasaki *et al.*, 1988a,b). Similar to inhibin, follistatin was isolated from follicular fluid on the basis of its ability to inhibit FSH release *in vitro,* albeit with much lower potency (10–30%) (Robertson *et al.*, 1987; Ueno *et al.*, 1987; Inouye *et al.*, 1991). Follistatin is also an activin-binding protein (Inouye *et al.*, 1991; Nakamura *et al.*, 1990) that blocks activin action *in vitro* (Asashima *et al.*, 1991; Shimonaka *et al.*, 1991; Xiao and Findlay, 1991; Nakamura *et al.*, 1992; Xiao *et al.*, 1992). This property qualifies follistatin as a potential intra-ovarian regulator along with activin and inhibin (Ying, 1988).

C. Development-Dependent Expression of Inhibin/Activin and Follistatin Genes

Not only do inhibin, activin, and follistatin modulate gonadotropin action *in vitro* (see Section IV,D), but the inhibin/activin (Woodruff *et al.*, 1988) and follistatin (Nakatani *et al.*, 1991) genes are developmentally regulated *in vivo* (Davis *et al.*, 1986; Woodruff *et al.*, 1987; Shimasaki *et al.*, 1989). Moreover, FSH induces expression of their corresponding mRNAs during granulosa cell differentiation *in vitro,* which is modulated by sex steroids and growth factors (Turner *et al.*, 1989; LaPolt *et al.*, 1990; Michel *et al.*, 1992).

1. Inhibin/activin

In primate (Schwall *et al.*, 1990) and rodent ovaries (Meunier *et al.*, 1988), the expression of inhibin/activin β-subunit mRNAs is relatively abundant in granulosa cells of immature antral follicles compared with the cells in larger, presumptive preovulatory follicles. Conversely, inhibin α-subunit mRNA expression increases during preovulatory development (Schwall *et al.*, 1990) and persists after ovulation both in human (Davis *et al.*, 1986) and in nonhuman primate corpora lutea (Hillier *et al.*, 1989). Such a pattern of inhibin/activin gene expression would be expected to give rise to a relatively high level of ββ homodimeric protein (activin) formation in immature follicles with the heterodimeric αβ congener (inhibin) predominating in preovulatory follicles. It follows that the activin "tone" of a

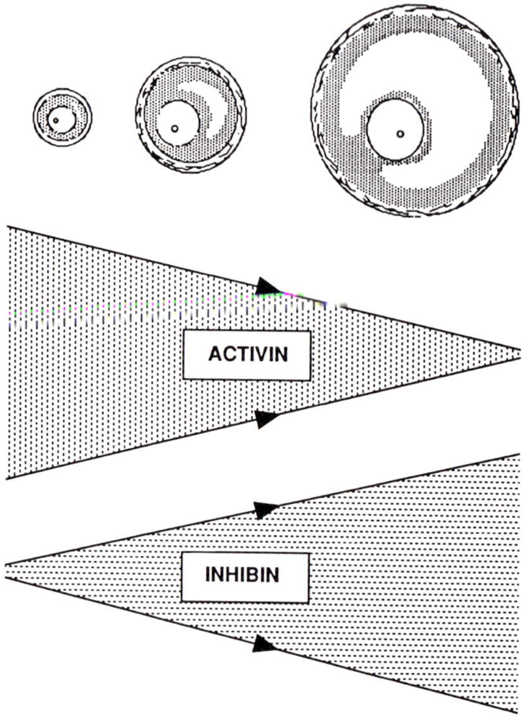

FIGURE 5 Development-related changes in the intrafollicular activin/inhibin tone. Based on the evidence cited that the activin/inhibin β-subunit mRNA is most strongly expressed in immature antral follicles, whereas inhibin α-subunit mRNA and protein are most strongly expressed in healthy Graafian follicles, it is speculated that β,β homodimeric (activin) protein production is initially relatively high but declines relative to α,β heterodimeric (inhibin) protein production during preovulatory follicular development. Follistatin expression also increases in healthy, mature follicles, which would further suppress the activin tone because follistatin acts as an activin-binding protein. Reproduced with permission from Hillier and Miró (1993).

developing follicle must decline as its inhibin tone increases (Figure 5). This development-related process has important implications for paracrine control in the follicle; these are outlined in the next section.

In vitro measurements of immunoactive inhibin have confirmed that human (Hillier *et al.*, 1991a) and nonhuman primate (Hillier *et al.*, 1989) granulosa cells undertake development-related increases in α-inhibin subunit production. Secretion of inhibin protein by rat (Erickson and Hsueh, 1978; Bicsak *et al.*, 1988; Suzuki *et al.*, 1987; Zhang *et al.*, 1988), bovine (Henderson and Franchimont, 1983), human (Tsonis *et al.*, 1987; Hillier *et al.*, 1991a), and nonhuman primate (Hillier *et al.*, 1989) granulosa cells is regulated by gonadotropins and sex steroids *in vitro*. Immunoactive inhibin production by human granulosa cells is induced by FSH *in vitro* and increases during preovulatory development *in vivo*. The onset of LH-responsive inhibin production coincides with the onset of LH-responsive granulosa cell steroid synthesis, consistent with the paracrine role of inhibin proposed later (Hillier *et al.*, 1991a).

The pattern of granulosa cell activin production has not been established at the protein level because there are no suitably specific or sensitive immunoassays. However, assays suitable for this purpose are being developed (Wong *et al.*, 1992).

2. Follistatin

Follistatin mRNA levels in granulosa cells increase with preovulatory follicular development (Shimasaki *et al.*, 1989; Nakatani *et al.*, 1991). Immunoactive follistatin production by rat and bovine granulosa cells is stimulated by FSH, but not LH, *in vitro* (Klein *et al.*, 1991; Saito *et al.*, 1991). The protein is expressed abundantly in the granulosa cells of healthy antral follicles, and the level of expression increases with preovulatory follicular development (Nakatani *et al.*, 1991). Since follistatin may function as an activin-binding protein (see earlier), this pattern of gene expression would be expected to bring about suppression of residual activin action within the preovulatory follicle (Figure 5).

D. Regulatory Actions of Inhibin/Activin and Follistatin

Regulatory roles for activin, inhibin, and follistatin in developing ovarian follicles are also implied by the actions that these proteins have on follicle cells *in vitro*.

1. Inhibin/activin action on thecal cells

Inhibin and activin act directly to modulate androgen synthesis in thecal cells (Hsueh *et al.*, 1987). Treatment of cultured human thecal cells with recombinant activin potently *inhibits* stimulation of androgen (androstenedione, dihydroepiandrosterone, and testosterone) production by LH and IGF-I (Hillier *et al.*, 1991b). Conversely, picomolar amounts of recombinant inhibin markedly *augment* LH/IGF-stimulated androgen production. Inhibin also dose-dependently overrides the inhibitory action of activin on androgen synthesis in rat (Hsueh *et al.*, 1987) and human (Hillier *et al.*, 1991b) thecal cells. Both inhibin and activin selectively modulate thecal C_{19} steroid production, suggesting action at the level of cytochrome P450c17.

2. Inhibin/activin action on granulosa cells

Activin is also a potential autocrine modulator of androgen aromatization in immature granulosa cells. Potent augmentation by activin of FSH-inducible aromatase activity has been convincingly demonstrated both in rat (Hutchison *et al.*, 1987; Miró *et al.*, 1991) and in nonhuman primate (Miró and Hillier, 1992) granulosa cells *in vitro*. Stimulatory effects of activin on FSH receptor levels have also been reported (Nakamura *et al.*, 1992; Xiao *et al.*, 1992). Activin augments FSH-induced progesterone production in immature (i.e., LH-nonresponsive) granulosa cells (Miró *et al.*, 1991; Miró and Hillier, 1992). However, in more mature, LH-responsive cells, activin inhibits progesterone synthesis (Miró *et al.*, 1991; Miró and Hillier, 1992). Activin also inhibits steroidogenesis in human granulosa-lutein cells (Rabinovici *et al.*, 1992; Li *et al.*, 1992). The developmentally regulated mechanism by which the action of activin switches from stimulation to inhibition of steroidogenesis remains to be elucidated.

It is uncertain if inhibin is a physiologically relevant regulator of granulosa cell steroidogenesis, since the protein purified from porcine (Ying *et al.*, 1986) but not bovine (Hutchison *et al.*, 1987) follicular fluid inhibited aromatase activity in rat granulosa cell cultures. Our own studies done on rat (Miró *et al.*, 1991) and nonhuman primate (Miró and Hillier, 1992) granulosa cell cultures showed no effect of recombinant human inhibin-A on basal or gonadotropin-responsive steroid synthesis at concentrations up to 10 ng/ml. At 100 ng/ml, weak (<10%) inhibition by this protein was occasionally observed. The ~25-kDa free inhibin α-subunit purified from bovine follicular fluid was equally inactive (F. Miró, P. Knight, and S. G. Hillier, unpublished observation).

3. Follistatin

Consistent with its role as an activin-binding protein, follistatin inhibits the stimulatory action of activin on FSH-induced differentiation of granulosa cells at all levels examined, including morphology; expression of FSH and LH/hCG receptors; and production of estradiol, progesterone, and inhibin (Xiao *et al.*, 1991, 1992; Nakamura *et al.*, 1992). The effects of follistatin on thecal androgen synthesis have not been reported, but blockade of the inhibitory action of activin is predictable.

V. Paracrine Mechanisms and Preovulatory Follicular Selection

Elucidation of the paracrine mechanisms that give rise to the selection of a preovulatory follicle is essential to a better understanding of the gonadotropic regulation of ovarian estrogen synthesis. When plasma FSH levels rise at the beginning of the follicular phase, multiple follicles with varying potential for FSH-dependent development are present in the ovaries. These follicles therefore require varying degrees of stimulation by FSH to be recruited to further stages of development (Figure 2). The follicle that becomes dominant is the one whose granulosa cells most rapidly acquire an LH-responsive aromatase system (Zeleznik and Hillier, 1984); i.e., the one with the lowest FSH threshold (Brown, 1978). During the midfollicular phase, estradiol secretion by this follicle begins to increase, feeding back through the hypothalamo–pituitary axis to negatively regulate pituitary FSH secretion. This causes a progressive reduction in the circulating FSH level and limits the FSH-dependent development of less mature follicles with relatively high FSH thresholds. In this way, usually a single follicle matures, protected against the fall in circulating FSH by its uniquely high responsiveness to both FSH and LH (Zeleznik and Hillier, 1984).

Local Roles for Activin and Inhibin?

Activin and inhibin could make particularly important contributions to the local control of this process, as is implied by the development-related pattern of inhibin/activin subunit gene expression *in vivo* and by the effects of these proteins on thecal and granulosa cell function *in vitro*. Studies of follicular fluid levels of androgen and estrogen in relation to granulosa cell aromatase activity indicate that the capacity of the theca interna to generate aromatase substrate (androstenedione) increases hand-in-hand with granulosa cell aromatase activity in the follicle that is destined to ovulate (Hillier *et al.*, 1981). It is therefore of considerable interest that immunoreactive inhibin is produced in greatest amounts by granulosa cells from the LH-responsive preovulatory follicle, which also synthesizes the greatest amount of estrogen (Hillier *et al.*, 1991a) (Figure 2).

No data exists on activin production by human granulosa cells. However, in rhesus monkey ovaries, the inhibin-β_B gene encoding activin-B was expressed at particularly high levels by granulosa cells in immature antral follicles and fell to low or undetectable levels in preovulatory follicles (Schwall *et al.*, 1990). Activin, known to potently sensitize granulosa cells to FSH *in vitro* (see earlier), is therefore likely to be produced in greatest amounts during the early stages of follicular development when follicular recruitment occurs (Figure 2).

Thus both direct and indirect lines of evidence exist to support hypothetical functions of activin and inhibin in the follicular paracrine system.

1. Activin and follicular recruitment

During the early follicular phase of the cycle, when recruitment occurs, interfollicular differences in activin production by granlosa cells could contribute to the establishment of variable follicular FSH thresholds (see earlier). The follicle whose granulosa cells express the greatest amounts of activin subunit(s) would be expected to be the most responsive to FSH (Hillier, 1991b). At this stage of development, aromatase is in-

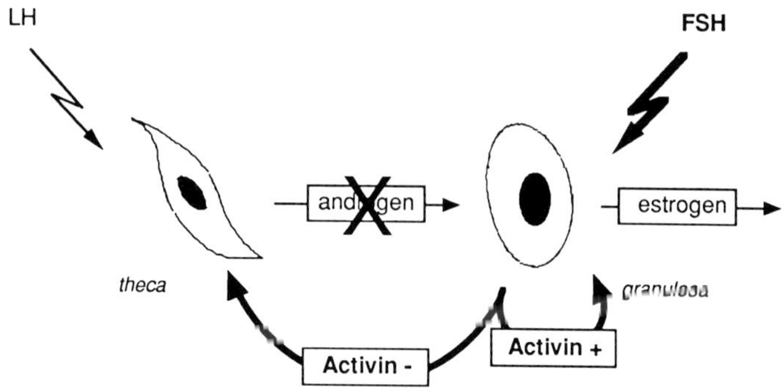

FIGURE 6 Hypothetical regulatory function of activin during follicular recruitment in the early follicular phase of the human menstrual cycle. Activin produced by granulosa cells in immature follicles enhances the inductive action of FSH on aromatase activity while it simultaneously suppresses thecal androgen synthesis at a time when ovarian estrogen synthesis is minimal (Hillier 1991b).

duced/activated by FSH but estrogen secretion is negligible, which is consistent with negative regulation of thecal androgen synthesis by activin (Figure 6).

2. Inhibin and follicular dominance

At midcycle, in the follicle most responsive to FSH, granulosa cell aromatase activity, inhibin production, and LH-receptor expression rise to

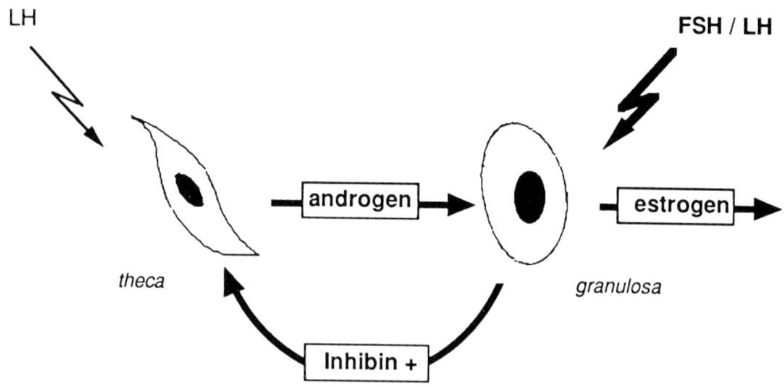

FIGURE 7 Hypothetical regulatory function of inhibin in maintaining follicular dominance during the late follicular phase of the human menstrual cycle. Granulosa cells in the preovulatory follicle respond to stimulation by FSH and LH with increased production of inhibin, paralleling the preovulatory increase in aromatase activity. Inhibin acts locally to promote LH/IGF-stimulated androgen synthesis in the theca interna. As inhibin production increases, possibly in association with reduced production and/or bioavailability of activin, a positive feedback loop is created through which thecal androgen synthesis is amplified to sustain estrogen synthesis in the granulosa cell layer (Hillier, 1991b).

critical levels. Since both aromatase activity and inhibin production are coupled to the LH receptor, the most responsive follicle becomes selected and is thereby destined to ovulate. During the second half of the follicular phase, LH-responsive inhibin production increases only in the dominant follicle, paralleling aromatase activity and estrogen secretion. The inhibin produced enhances LH-stimulated thecal androgen synthesis, creating a paracrine-positive feedback loop by which granulosa cells secure the supply of aromatase substrate required to sustain the preovulatory rise in estrogen secretion (Hillier, 1991b) (Figure 7).

VI. Clinical Uses of FSH and LH

The transition from FSH-dependent to LH-responsive granulosa cell function during preovulatory follicular development also has practical implications for ovulation using FSH and LH (Hillier, 1990; Chappel and Howles, 1991). Once recombinant forms of human FSH and LH are available, it should be possible to administer pure FSH and LH (or isoforms with altered bioactivities and/or metabolic half-lives) separately or combined so that the degree of stimulation given by either hormone can be precisely controlled.

A. FSH Therapy

The mechanism governing selection of the dominant follicle in a spontaneous ovarian cycle can be overridden to achieve multiple follicular development for clinical purposes such as *in vitro* fertilization (IVF) and gamete intrafallopian transfer (GIFT). Until recently, human menopausal gonadotropin (hMG) was the only pharmaceutical-grade human gonadotropin preparation available for this purpose. The hMG preparations currently available usually contain both FSH and LH, either in similar amounts or enriched with respect to FSH. In the future, recombinant FSH is likely to be used instead of urinary FSH to stimulate multiple follicular development. The expected advantages of recombinant FSH over urinary FSH are that it will be easier to produce and, being molecularly pure, it should be safer to administer intravenously. "Designer" FSH preparations with longer or shorter metabolic half-lives might also have specific therapeutic applications.

The usual aim of therapy with hMG is to provide exogenous FSH at a dose exceeding the FSH threshold requirements of multiple immature follicles, so that the number of follicles that enter preovulatory development is increased relative to a spontaneous ovarian cycle. Treatment is then continued during the mid to late follicular phase to override the natural selection process and sustain multiple preovulatory follicular development (Hillier *et al.*, 1985). It is common practice to suppress endogenous gonadotropin secretion by simultaneously administering a luteinizing hormone–releasing hormone (LHRH) agonist during therapy with hMG (Fleming and Coutts, 1986). This prevents the inappropriate secretion of endogenous LH and allows the collection of oocytes to be more optimally timed to an ovulation-inducing injection of hCG.

While many women respond adequately to standard superovulation regimens using urinary gonadotropins, their follicles usually develop asynchronously to yield oocytes of variable quality, some of which do not fertilize or, if they do so, undergo normal postfertilization development. This is probably because sustained high-level stimulation by FSH during the mid to late follicular phase causes immature and intermediately mature follicles to continue to develop, at a time when they would otherwise have undergone spontaneous regression due to withdrawal of stimulation by endogenous FSH. This is when granulosa cells in more mature follicles (those expected to provide the "best" oocytes) become directly responsive to LH.

B. LH as an Adjunct to FSH

The requirement for stimulation by LH is related to the stage of follicular development, *excessive* stimulation by LH serving to arrest the growth and function of all but the dominant follicle. Less mature follicles exposed to inappropriately high levels of LH undergo atresia or become prematurely luteinized, and oocyte development may be compromised (Jacobs, 1991; Chappel and Howles, 1991). Studies of LH action on human granulosa cell proliferation and steroidogenesis *in vitro* hint at the dose dependence of this effect, low-dose stimulation with LH serving to enhance steroidogenesis without inhibiting cell proliferation, but high-dose LH causing enhanced steroidogenesis (Overes *et al.*, 1992) and *inhibition* of cell growth (Yong *et al.*, 1992b) (Figure 3). Thus developing follicles appear to have finite requirements for stimulation by LH beyond which normal development ceases. Whereas each follicle has a threshold beyond which it must be stimulated by FSH to *initiate* preovulatory development (Brown, 1978; Hillier, 1981), it also has a ceiling beneath which LH stimulation must

TABLE 1 The LH "Ceiling" Hypothesis

- Ovarian follicles have development-related requirements for stimulation by LH
- LH, beyond a certain "ceiling" level, suppresses granulosa proliferation, and initiates atresia (immature follicles) or premature luteinization (preovulatory follicles)
- Mature follicles are more resistant (have a higher "ceiling") to LH than immature ones
- During ovulation induction, the LH dose should not exceed the ceiling of the most mature follicle

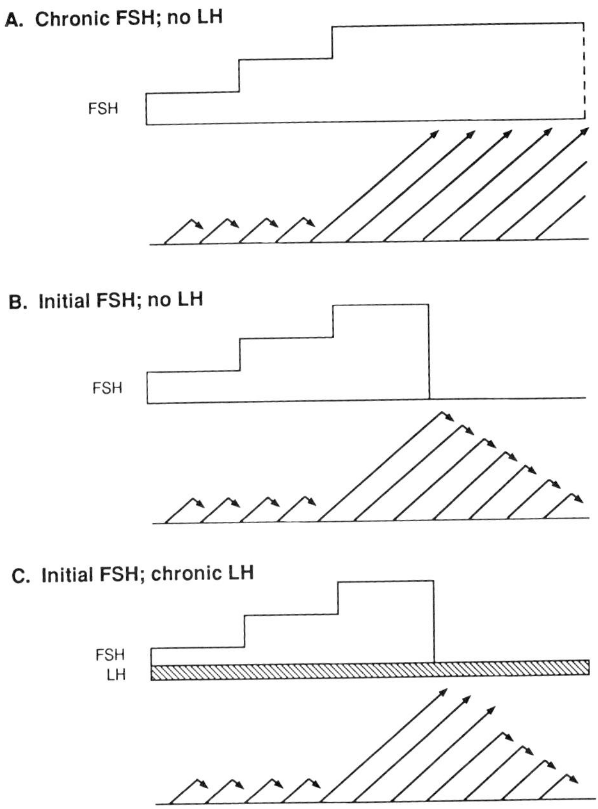

FIGURE 8 Ovulation induction using pure FSH and LH: The theory behind the practice. The abscissas is time (i.e., an arbitrary 14-day follicular phase); ordinates are exogenous gonadotropin doses (open bar, FSH starting at 75 IU/day with a 100% increment every 3 days; hatched bar, LH given throughout at a constant subceiling daily dose). Follicular diameter is indicated by the arrows, starting at about 5 mm and increasing to a maximum of ~20 mm, with the direction of the arrowhead showing continued growth (up) or degeneration (down). (A) When FSH alone is given constantly at a suprathreshold dose, multiple preovulatory follicular development is initiated and persists as long as that dose of FSH is given. (B) If the suprathreshold FSH dose is withdrawn, the responding follicles cease to develop and become atretic. (C) However, chronic administration of a subceiling dose of exogenous LH encourages sustained development of the most mature (i.e., LH-responsive) follicle(s) in the face of FSH withdrawal. Reproduced with permission from Hillier (1993).

remain unless further normal development is to be *terminated* (Hillier, 1993) (Table 1).

While the availability of recombinant FSH is unlikely to alter current strategies for the use of LH-free forms of urinary FSH, the availability of recombinant LH is likely to open up radically new therapeutic options (Figure 8). We have seen that there are good theoretical grounds for using LH as well as FSH to stimulate multiple follicular development for IVF and

GIFT, etc. Ideal treatment with LH would be with a suitably low (i.e., subceiling) dose given at a constant rate throughout a course of treatment with FSH. Administration of FSH would initially be at a high (i.e., suprathreshold) dose, stepping down to a lower dose or withdrawing FSH completely during the late follicular phase once multiple estrogen-secreting follicles had begun to develop. In such a regimen, a restricted number of "leading" follicles capable of synthesizing estrogen in response to tonic stimulation by LH would be expected to complete their development in response to the exogenous LH. Less mature (i.e., nonresponsive to LH) follicles would not develop further, leading to a relatively synchronous group of follicles yielding oocytes best suited for IVF or GIFT. The relative benefit of adjuvant therapy with LH might depend on the level of pituitary suppression achieved by concomitant LHRH analogue therapy, and whether an antagonist or an agonist of LHRH is being used for this purpose.

C. LH Therapy

The availability of pure LH could be particularly beneficial for infertile women with forms of anovulation amenable to gonadotropin therapy. In such patients, the objective of treatment is to stimulate the development of a single ovulatory follicle so that conception might occur *in vivo*. However, a major disadvantage of current therapeutic regimens using hMG is that multiple ovulation frequently occurs. The strategy involving pure LH would be that once a preovulatory follicle had begun to develop in response to a suprathreshold dose of FSH (given in combination with a tonic amount of LH), treatment with FSH could be gradually withdrawn while tonic stimulation with LH was maintained. This would approximate the situation in a natural ovarian cycle through encouraging the terminal maturation of a single LH-responsive preovulatory follicle and, simultaneously, arresting the growth of multiple immature follicles that would otherwise continue to develop in response to FSH. Care would be necessary to avoid using a supraceiling dose of LH in view of the association of above-normal plasma levels of LH with ovulatory dysfunction and/or disordered oocyte development.

VII. Luteinization and Luteolysis

The endocrine function of the corpus luteum (production of progesterone, estradiol, and inhibin) emerges as a continuum of preovulatory follicular development. Luteinization and onset of progesterone secretion is initiated by the ovulation-inducing LH surge. Thereafter tonic stimulation by LH transiently sustains hormone secretion for the duration of the luteal phase [see Zeleznik (1991) for review]. However, LH alone is insufficient to prevent functional and morphological regression of the corpus luteum (i.e., luteolysis). Only when pregnancy occurs and human chorionic gonadotropin is produced by the implanting embryo is the corpus luteum "rescued" and its endocrine function extended. Cellular events underlying these processes are the subjects of this section.

A. Luteinization

The number of steroidogenic cells that populate the corpus luteum is more or less fixed by the time of follicular rupture (Fisch *et al.*, 1989). Recent studies on the control of human granulosa cell replication *in vitro* suggest that LH is capable of acting directly to suppress granulosa cell division, at the same time stimulating production of progesterone and inhibin (Yong *et al.*, 1992a,b) (Figure 3). Both processes appear to involve intracellular actions of cAMP; however, LH also activates inositol lipid hydrolysis/Ca^{2+} signaling in granulosa-luteal cells [see Leung and Steele (1992) for review]. Differential activation of more than one postreceptor pathway is likely to be central to the cellular mechanism by which LH simultaneously terminates preovulatory follicular growth and initiates luteal endocrine function.

Plasma progesterone levels remain low throughout the follicular phase of the cycle but begin to rise shortly before the onset of the LH surge (Yen, 1986). Follicular fluid levels of progesterone and its principal metabolite, 17-hydroxyprogesterone, also increase at this time, heralding transition of the preovulatory follicle into a corpus luteum. When the follicle ovulates 1–2 days later, luteinized granulosa cells acquire a blood supply and begin high rates of progesterone synthesis (see the next section). The luteal progesterone secretion rate becomes maximal in the midluteal phase at over 25 mg (~80 μmol) of progesterone per day (Baird, 1977).

The human corpus luteum also secretes as much if not more estradiol than the preovulatory follicle does during the follicular phase of the cycle. However, the follicular-to-luteal phase transition entails a temporary decline in estrogen secretion. When the midcycle LH surge begins, peripheral and ovarian venous plasma estradiol titers drop precipitously. At the ovarian level, this is reflected in a concomitant decline in the estradiol concentration in the fluid of the ovulatory follicle. Estrogen precursor (androstenedione and testosterone) levels in follicular fluid also fall at this time, reflecting LH-induced suppression of thecal C_{19} steroid synthesis (see Section III,A). However, the aromatase activity of luteinized granulosa cells remains high, facilitating the onset of estrogen secretion by the developing corpus luteum.

Luteal secretion of progesterone, estradiol, and inhibin increases progressively during the first 6 days or so after follicular rupture, reaching peak levels of secretion during the midluteal phase (Figure 1). The functional lifespan of the human corpus luteum is most clearly depicted by the blood level of progesterone since this steroid is not secreted by the ovaries in important amounts until shortly before ovulation. Estradiol and inhibin are also secreted during the follicular phase, being produced by granulosa cells in the developing preovulatory follicle. The synthesis of estradiol and inhibin may therefore be coordinately regulated in the corpus luteum as well as in the preovulatory follicle, as discussed in Section VII,A,2.

1. Progesterone

Luteal steroid secretion rates are controlled mainly by the luteal vasculature, gonadotropic actions of LH and hCG, and paracrine signaling. Progesterone synthesis is only a minor granulosa cell function before onset of the LH surge, even though the LH-responsive enzymes required to sustain luteal progesterone synthesis are induced by FSH during preovulatory development. Thus granulosa cells isolated from preovulatory follicles undertake high rates of progesterone synthesis and respond directly to LH/hCG (i.e., undergo functional luteinization) *in vitro* (Yong *et al.*, 1992b). However, inside the follicle, granulosa cell progesterone synthesis is constrained by restricted access to precursor cholesterol. This is because granulosa-lutein cells require precursor cholesterol derived from plasma in the form of low-density lipoprotein (LDL) to sustain high rates of C_{21} steroid hormone synthesis (Carr *et al.*, 1982). The lamina basalis of the preovulatory follicle allows the passage of plasma proteins across the follicle wall in inverse proportion to molecular size, creating a blood–follicle barrier (Shalgi *et al.*, 1973). Low-density lipoprotein is a large protein molecule ($M_r \sim 1 \times 10^6$) that is effectively excluded from follicular fluid (Hillier and Wickings, 1985). However, when the lamina basalis is breached at the time of ovulation, granulosa-lutein cells become exposed to plasma containing LDL cholesterol and start high rates of progesterone synthesis.

Neovascularization of the granulosa cell layer is stimulated by LH and mediated by locally produced angiogenic factors such as basic fibroblast growth factor (bFGF) (Gospodarowicz and Ferrara, 1989) and vascular endothelial growth factor (VEGF) (Ravindranath *et al.*, 1992b), which are members of the heparin-binding growth factor family (Burgess and Maciag, 1989). This process provides luteal cells with increased access to steroid precursors and permits the rapid drainage of the corpus luteum and dispersal of progesterone. Maximal capillarization is achieved about 1 week after ovulation, coinciding with the peak level of luteal progesterone secretion (Carr *et al.*, 1982).

Paradoxically, as progesterone secretion by the corpus luteum increases in the early luteal phase, its steroidogenic capacity declines (Fisch *et al.*, 1989). At the luteal cell level, this is reflected in reduced steroid synthesis *in vitro* and reduced levels of the mRNAs encoding the enzymes crucial to progesterone synthesis, P450scc and 3β-HSD (Bassett *et al.*, 1991; Ravindranath *et al.*, 1992a). This emphasizes the importance of an extensive and expanding vasculature in facilitating the early luteal phase increase in progesterone secretion, implying that the number of LDL receptors on the surface of luteal cells and their access to LDL-associated precursor cholesterol is of overriding importance in controlling actual progesterone secretion rates (Golos *et al.*, 1987).

2. Estradiol

Granulosa-lutein cells cannot undertake significant estradiol synthesis unless they are supplied with extracellular aromatase substrate, similar to granulosa cells in the preovulatory follicle (Wickings *et al.*, 1986). Theca lutein cells are presumed to provide precursor androgen, which is aromatized by granulosa-lutein cells in an extension of the two-cell type mechanism of estrogen synthesis in the preovulatory follicle (Sasano *et al.*, 1989). High rates of C_{19} steroid synthesis by thecal cells must also depend on appropriate access to blood-borne precursor cholesterol (Carr *et al.*, 1982). As luteal estradiol secretion increases, theca-lutein cells become histologically conspicuous as trabeculae in close proximity to newly formed blood vessels within the granulosa-lutein cell layer. Presumably, therefore, androgen synthesis in theca-lutein cells (hence aromatization in granulosa-lutein cells), like progesterone synthesis in granulosa-lutein cells, is also facilitated by the development of the luteal vasculature. This would explain why the luteal phase pattern of estradiol mirrors progesterone in attaining a maximum in the midluteal phase, even though total luteal aromatase activity is declining at this time (Fisch *et al.*, 1989).

3. Inhibin

By the time that the ovulation-inducing LH surge begins, granulosa cells in the preovulatory follicle are already undertaking inhibin production in response to LH (McLachlan *et al.*, 1990). The parallelism between serum levels of inhibin and estradiol throughout the human menstrual cycle suggests that synthesis of these two factors is closely coordinated. As discussed earlier, there is evidence that granulosa-derived inhibin exerts local positive feedback regulation of androgen synthesis in the theca interna to amplify LH-stimulated estrogen synthesis in the preovulatory follicle. A similar mechanism could operate in the human corpus luteum, since theca-lutein cells also undertake inhibin-responsive androgen synthesis *in vitro* (Hillier et al., 1992).

B. Luteolysis

Both the number of steroidogenic cells that can be recovered from the human corpus luteum and their steroidogenic capacity *in vitro* decline throughout the luteal phase of the menstrual cycle (Fisch *et al.*, 1989; Bassett *et al.*, 1991). However, steroidogenic responsiveness to hCG in-

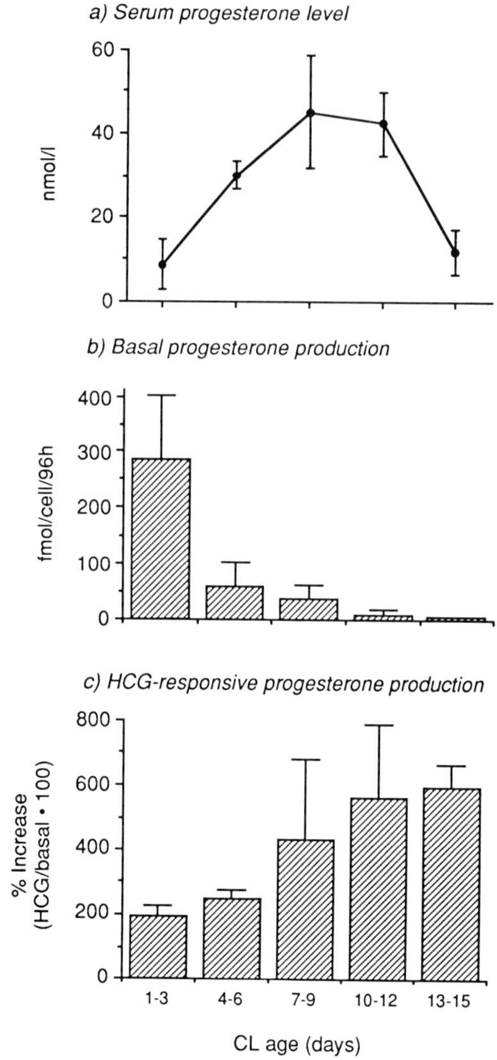

FIGURE 9 Relationship between the circulating progesterone level and basal and hCG-responsive luteal cell progesterone synthesis in the human menstrual cycle. (A) Serum progesterone level at the time of lutectomy. (B) Basal progesterone production by steroidogenic luteal cells *in vitro*. (C) hCG-responsive progesterone production by steroidogenic luteal cells *in vitro*. Corpora lutea (CLs) were enucleated during the luteal phase of spontaneous menstrual cycles (ovulation on Day 0). Steroidogenic luteal cells were isolated by collagenase dispersal and Percoll density-gradient fractionation, and cultured for 96 hours as cell monolayers in the presence and absence of hCG (30 ng/ml). Progesterone accumulation in the culture medium was determined by radioimmunoassay. Basal steroid production is given in relation to cell number; hCG responsiveness is expressed as percentage stimulation due to hCG (hCG/basal × 100). Results are means ± SD. Corpora lutea from three or four patients were studied at each time point. Reproduced with permission from Fisch *et al.* (1989).

creases over the same period (Figure 9). Initially, steroidogenesis is refractory to hCG, but as basal steroidogenic capacity wanes, luteal cells become increasingly responsive to the gonadotropin *in vitro*. This is true both for progesterone synthesis and aromatase activity (Fisch *et al.*, 1989). Interestingly, levels of mRNA encoding the LH/hCG receptor also rise over the same period (Ravindranath *et al.*, 1992c). These results support the theory that the newly formed corpus luteum is transiently autonomous but becomes increasingly dependent upon gonadotropic support as the luteal phase progresses. Maximal responsiveness to HCG is attained during the mid- to late luteal phase when trophoblastic hCG secretion would begin to result in corpus luteum "rescue" in a conception cycle. This agrees with *in vivo* evidence that luteal responsiveness to hCG increases with luteal age, and could explain why luteal progesterone secretion becomes episodic and coincident with plusatile LH release during the second half of the luteal phase in nonconception cycles (Filicori *et al.*, 1984).

Locally produced regulatory factors, including estradiol, prostaglandins, leukotrienes, and cytokines, have been invoked as paracrine triggers of luteolysis [see Behrman *et al* (1991) for review]. However, apoptosis (programmed cell death) is set in train by FSH action on granulosa cells in the follicular phase, well before progesterone secretion by the corpus luteum begins to decline (Zeleznik *et al.*, 1989). This supports the concept that luteolysis is preprogrammed at the cellular level, initiated at around the time of ovulation, and only reversible by the direct action of trophoblastic hCG on the corpus luteum once pregnancy begins (Fisch *et al.*, 1989). In this case, luteolysis can be viewed as a default response to lack of stimulation by hCG. It is unnecessary to invoke an active luteolytic mechanism to explain cyclic initiation of luteal regression. Rather, it is the subcellular mechanism(s) through which hCG impedes this process that may hold the key to understanding how luteal endocrine function is terminated in the human menstrual cycle.

References

Adashi, E. Y., and Resnick, C. E. (1986). Antagonistic interactions of transforming growth factors in the regulation of granulosa cell differentiation. *Endocrinology* 119, 1879–1881.

Adashi, E. Y., Resnick, C. E., Hurwitz, A., Ricciarelli, E., Hernandez, E. R., Roberts, C. T., Leroith, D., and Rosenfeld, R. (1991). Insulin-like growth factors: the ovarian connection. *Hum. Reprod.* 6, 1213–1219.

Amsterdam, A. A., Koch, Y., Lieberman, M. E., and Lindner, H. R. (1975). Distribution of binding sites for human chorionic gonadotropin in the preovulatory follicle of the rat. *J. Biol. Chem.* 67, 894–900.

Amsterdam, A. A., Rotmensch, S., Furman, A., Venter, E. A., and Vlodavsky, I. (1989). Synergistic effect of human chorionic gonadotropin and extracellular matrix of *in vitro* differentiation of human granulosa cells: progesterone production and gap junction formation. *Endocrinology* 124, 1956–1964.

Armstrong, D. T., and Dorrington, J. H. (1979). Estrogen biosynthesis in the ovaries and testes. *In* "Regulatory Mechanisms Affecting Gonadal Hormone Action" (J. A. Thomas and R. L. Singhal, eds.), Vol. 2, pp. 217–258. University Park Press, Baltimore, MD.

Asashima, M., Nakano, H., Uchiyama, H., Sugino, H., Nakamura, T., Eto, Y., Ejima, Davids, D., Plessow, S., Cichocka, I., and Kinoshita, K. (1991). Follistatin inhibits the mesoderm-inducing activity of activin A and the vegetalizing factor from chicken embryo. *Roux's Arch. Dev. Biol.* 200, 4–7.

Baird, D. T. (1977). Synthesis and secretion of steroid hormones by the ovary *in vivo*. *In* "The Ovary" (S. Zuckerman, and B. J. Weir, eds.), 2nd ed., Vol. 3, pp. 305–357. Academic Press, London.

Barbieri, R. L., Makris, A., and Ryan, K. J. (1983). Insulin stimulates androgen accumulation in incubations of human ovarian stroma and theca. *Obstet. Gynecol. Suppl.* 64, 73S–80S.

Barbieri, R. L., Makris, A., Randall, R. W., Daniels, G., Kistner, R. W., and Ryan, K. J. (1986). Insulin stimulates androgen accumulation in incubations of ovarian stroma obtained from women with hyperandrogenism. *J. Clin. Endocrinol. Metab.* 62, 904–910.

Bassett, S. G., Little-Ihrig, L. L., Mason, J. I., and Zeleznik, A. J. (1991). Expression of messenger ribonucleic acids that encode for 3β-hydroxysteroid dehydrogenase and cholesterol side-chain cleavage throughout the luteal phase of the macaque menstrual cycle. *J. Clin. Endocrinol. Metab.* 72, 362–366.

Behrman, H. R., Aten, R. F., and Pepperell, J. R. (1991). Cell-to-cell interactions in luteinization and luteolysis. *In* "Ovarian Endocrinology" (S. G. Hillier, ed.), pp. 190–225. Blackwell, London.

Bendell, J. J., and Dorrington, J. (1988). Rat thecal/intersitial cells secrete a transforming growth factor-β-like factor that promotes growth and differentiation of rat granulosa cells. *Endocrinology* 123, 941–948.

Ben-Ze'ev, A., and Amsterdam, A. (1989). Regulation of heat shock protein synthesis by gonadotropins in cultured granulosa cells. *Endocrinology* 124, 2584–2594.

Bergh, C., Carlsson, B., Olsson, J. H., Selleskog, U., and Hellensjö, T. (1993). Regulation of androgen production in cultured human thecal cells by insulin-like growth factor and insulin. *Fertil. Steril.* 59, 323–331.

Bicsak, T., Cajander, S. B., Vale, W., and Hsueh, A. J. W. (1988). Inhibin: studies of stored and secreted forms by biosynthetic labelling and immunodetection in cultured rat granulosa cells. *Endocrinology* 122, 741–748.

Brown, J. B. (1978). Pituitary control of ovarian function—concepts derived from gonadotropin therapy. *Aust. NZ. J. Obstet. Gynaecol.* 18, 47.

Burgess, W. H., and Maciag, T. (1989). The heparin-binding (fibroblast) growth factor family of proteins. *Annu. Rev. Biochem.* 58, 575–606.

Cara, J. F., and Rosenfield, R. L. (1988). Insulin-like growth factor I and insulin potentiate luteinizing hormone-induced androgen biosynthesis by rat ovarian theca-intersitial cells. *Endocrinology* 123, 733–739.

Carr, B. R., MacDonald, P. C., and Simpson, E. R. (1982). The role of lipoproteins in the regulation of progesterone secretion by the human corpus luteum. *Fertil. Steril.* 38, 303–311.

Chappel, S. C., and Howles, C. (1991). Reevaluation of the roles of luteinizing hormone and follicle-stimulating hormone in the ovulatory process. *Hum. Reprod.* 9, 1206–1212.

Daniel, S. A. J., and Armstrong, D. T. (1986). Androgens in the ovarian microenvironment. *Semin. Reprod. Endocrinol.* 4, 89–100.

Davis, J. S., Tedesco, T. A., West, L. A., Maroulis, G. B., and Weakland, L. L. (1989). Effects of human chorionic gonadotropin, prostaglandin F-2-alpha and protein kinase C activators on the cyclic AMP and inositol phosphate 2nd messenger systems in cultured human granulosa-luteal cells. *Mol. Cell. Endocrinol.*, 65, 187–94.

Davis, S. R., Dench, F., Nikolaidis, I., Clements, J. A., Forage, R. G., Krozowski, Z., and

Burger, H. G. (1986). Inhibin A-subunit gene expression in the ovaries of immature female rats is stimulated by pregnant mare serum gonadotropin. *Biochem. Biophys. Res. Commun.* 138, 1191–1195.

Dehennin, L., Blacker, C., Reifsteck, A., and Scholler, R. (1984). Estrogen 2-, 4-, 6- or 16-hydroxylation by human follicles shown by gas chromatography associated with stable isotope dilution. *J. Steroid Biochem.* 20, 465–471.

Derynck, R. (1986). Transforming growth factor-α: structure and biological activities. *J. Cell. Biochem.* 32, 293–304.

Dorrington, J., Chuma, A. V., and Bendell, J. J. (1988). Transforming growth factor β and follicle-stimulating hormone promote rat granulosa cell proliferation. *Endocrinology* 123, 353–359.

Van den Eijnden-Van Raaij, A. J. M., van Zoelant, E. J. J., Van Nimmen, K., Koster, C. H., Snoek, G. T., Durston, A. P., and Huylebroe, P. (1990). Activin-like factor from a *Xenopus laevis* cell line responsible for mesoderm induction. *Nature (London)* 245, 732–734.

Erickson, G. F., and Hsueh, A. J. W. (1978). Secretion of inhibin by rat granulosa cells *in vitro*. *Endocrinology* 103, 1960–1963.

Erickson, G. F., Magoffin, D. A., Dyer, C. A., and Hofeditz, C. (1985). The ovarian androgen producing cells: a review of structure/function relationships. *Endocr. Rev.* 6, 371–399.

Esch, F. S., Shimasaki, S., Cooksey, K., Mercado, M., Mason, A. J., Ying, S. Y., Ueno, N., and Ling, N. (1987). Complementary deoxyribonucleic acid (cDNA) cloning and DNA sequence analysis of rat ovarian inhibins. *Mol. Endocrinol.* 1, 388–396.

Eto, Y., Tsuji, Y., Takezawa, M., Takano, S., Yokogawa, Y., and Shibai, H. (1987). Purification and characterization of Erythroid Differentiation Factor (EDF) isolated from human leukemia cell line THP-1. *Biochem. Biophys. Res. Commun.* 142, 1095–1103.

Feng, P., Catt, K., and Knecht, M. (1986). Transforming growth factor β regulates the inhibitory actions of epidermal growth factor during granulosa cell differentiation. *J. Biol. Chem.* 261, 14167–14170.

Feng, P., Knecht, M., and Catt, K. (1987). Hormonal control of epidermal growth factor receptors by gonadotropins during granulosa cell differentiation. *Endocrinology* 120, 1121–1126.

Filicori, M., Butler, J. P., and Crowley, W. F. (1984). Neuroendocrine regulation of the corpus luteum in the human. *J. Clin. Invest.* 73, 1638–1647.

Fisch, B., Margara, R. A., Winston, R. M. L., and Hillier, S. G. (1989). Cellular basis of luteal steroidogenesis in the human ovary. *J. Endocrinol.* 122, 303–311.

Fleming, R., and Coutts, J. R. T. (1986). Induction of multiple follicular growth in normally menstruating women with endogenous gonadotropin suppression. *Fertil. Steril.* 45, 226–232.

Forage, R. G., Ring, J. W., Brown, R. W., McInerey, B. V., Cobon, G. S., Gregson, R. P., Robertson, D. M., Morgan, F. J., Hearn, M. T. W., Findlay, J. K., Wettenhall, R. E. H., Burger, H. G., and de Kretser, D. M. (1986). Cloning and sequence analysis of cDNA species coding for the two subunits of inhibin from bovine follicular fluid. *Proc. Natl. Acad. Sci. USA* 83, 3091–3095.

Franchimont, P., Hazee-Hagelstein, M. T., Charlet-Renard, Ch., and Jasper, J. M. (1986). Effect of mouse epidermal growth factor on DNA and protein synthesis, progesterone and inhibin production by bovine granulosa cell cultures. *Acta Endocrinol. (Copenh.)* 111, 122–127.

Froesch, E. R., Schmid, Chr., Schwander, J., and Zapf, J. (1985). Actions of insulin-like growth factors. *Annu. Rev. Physiol.* 47, 443–467.

Gill, G. N., Bertics, P. J., and Santon, J. (1987). Epidermal growth factor and its receptor. *Mol. Cell. Endocrinol.* 51, 169–186.

Giudice, L. C. (1992). Insulin-like growth factors and ovarian follicular development. *Endocr. Rev.* 13, 641–669.

Goldenberg, R. L., Reiter, E. O., and Ross, G. T. (1973). Follicle response to exogenous gonadotropins: an estrogen-mediated phenomenon. *Fertil. Steril.* 24, 121–125.

Golos, T. G., Strauss, J. F., and Millier, W. L. (1987). Regulation of lipoprotein receptor and cytochrome P-450scc mRNA levels in human granulosa cells. *J. Steroid Biochem.* 27, 767–773.

Gonzalez-Manchon, C., and Vale, W. (1989). Activin-A, inhibin and transforming growth factor-β modulate growth of two gonadal cell lines. *Endocrinology* 125, 1666–1672.

Gospodarowicz, D., and Balecki, H. (1979). Fibroblast and epidermal growth factors are mitogenic agents for cultured granulosa cells of rodent, porcine, and human origin. *Endocrinology* 104, 757–764.

Gospodarowicz, D., and Ferrara, N. (1989). Fibroblast growth factor and the control of pituitary and gonad development and function. *J. Steroid Biochem.* 32, 183–191.

Gougeon, A., and Lefévre, B. (1983). Evolution of the largest healthy and atretic follicles during the human menstrual cycle. *J. Reprod. Fertil.* 69, 497–502.

Green, A. R. (1989). Peptide regulatory factors: multifunctional mediators of cellular growth and differentiation. *Lancet* i, 705–707.

Gwynne, J. T., and Strauss, J. F. (1982). The role of lipoproteins in steroidogenesis and cholesterol metabolism in steroidogenic glands. *Endocr. Rev.* 3, 299–329.

Hammond, J. M., Hersey, R. M., Walega, M. A., and Weisz, J. (1986). Catecholestrogen production by porcine ovarian cells. *Endocrinology* 118, 2292–2299.

Hammond, J. M., Hsu, C.-J., Mondschein, J. S., and Canning, S. H. (1988). Paracrine and autocrine functions of growth factors in the ovarian follicle. *J. Anim. Sci.* 66, 21–31.

Harlow, C. R., Shaw, H. J., Hillier, S. G., and Hodges, J. K. (1988). Factors influencing FSH-induced steroidogenesis in marmoset granulosa cells: effects of androgens and stages of follicular development. *Endocrinology* 122, 2780–2787.

Hashimoto, M., Kondo, S., Sakurai, T., Eto, Y., Shibai, H., and Maramutsu, M. (1990). Activin/EDF as an inhibitor of neural differentiation. *Biochem. Biophys. Res. Commun.* 126, 220–226.

Henderson, K. M., and Franchimont, P. (1983). Inhibin production by bovine ovarian tissues *in vitro* and its regulation by androgens. *J. Reprod. Fertil.* 67, 291–298.

Hernandez, E. R., Hurwitz, A., Vera, A., Pellicer, A., Adashi, E. Y., LeRoith, D., and Roberts, C. R. Jr. (1992). Expression of the genes encoding insulin-like growth factors and their receptors in the human ovary. *J. Clin. Endocrinol. Metab.* 74, 419–425.

Hickey, G. J., Chen, S., Besman, M. J., Shively, J. E., Hall, P. F., Gaddy-Kurten, D., and Richards, J. S. (1988). Hormonal regulation, tissue distribution, and content of aromatase cytochrome P450 messenger ribonucleic acid and enzyme in rat ovarian follicles and corpora lutea: relationship to estradiol biosynthesis. *Endocrinology* 122, 1426.

Hild-Petito, S., Stouffer, R. L., and Brenner, R. M. (1988). Immunocytochemical localization of estradiol and progesterone receptors in the monkey ovary throughout the menstrual cycle. *Endocrinology* 123, 2896–2905.

Hillier, S. G. (1981). Regulation of follicular estrogen biosynthesis: a survey of current concepts. *J. Endocrinol.* 89, 3P–18P.

Hillier, S. G. (1985). Sex steroid metabolism and follicular development in the ovaries. *In* "Oxford Reviews of Reproductive Biology" (J. R. Clarke, ed.), Vol. 7, pp. 168–222. Clarendon Press, Oxford.

Hillier, S. G. (1990). Ovarian manipulation with pure gonadotropins. *J. Endocrinol.* 127, 1–4.

Hillier, S. G. (1991a). Cellular basis of follicular endocrine function. *In* "Ovarian Endocrinology" (S. G. Hillier, ed.), pp. 73–106. Blackwell, London.

Hillier, S. G. (1991b). Regulatory functions for inhibin and activin in human ovaries. *J. Endocrinol.* 131, 171–175.

Hillier, S. G. (1993). Ovarian stimulation with recombinant gonadotropins: LH as an adjunct to FSH. *In* "The New Frontier in Ovulation Induction" (H. S. Jacobs, ed.), pp. 39–47. Parthenon Publishing, Carnforth, UK.

Hillier, S. G., and Miró, F. (1993). Inhibin, activin and follistatin: potential roles in ovarian physiology. *Ann. N.Y. Acad. Sci.* 687, 29–38.

Hillier, S. G., and Wickings, E. J. (1985). Cellular aspects of corpus luteum function. *In* "The Luteal Phase" (S. L. Jeffcoate, ed.), pp. 1–23. Wiley, London.

Hillier, S. G., Zeleznik, A. J., Knazek, R. A., and Ross, G. T. (1980). Hormonal regulation of preovulatory follicle maturation in the rat. *J. Reprod. Fertil.* 60, 219–229.

Hillier, S. G., Reichert, L. E. R. Jr., and van Hall, E. V. (1981). Control of preovulatory follicular estrogen biosynthesis in the human ovary. *J. Clin. Endocrinol. Metab.* 52, 847–856.

Hillier, S. G., Afnan, A. M. M., Margara, R. A., and Winston, R. M. (1985). Superovulation strategy before IVF. *Clinics Obstet. Gynaecol.*, 12, 687–723.

Hillier, S. G., Wickings, E. J., Saunders, P. T. K., Shimasaki, S., Reichert Jr., L. E., and McNeilly, A. S. (1989). Hormonal control of inhibin production by primate granulosa cells. *J. Endocrinol.* 123, 65–73.

Hillier, S. G., Wickings, E. J., Illingworth, P. I., Yong, E. L., Reichert, L. E. Jr., Baird, D. T., and McNeilly, A. S. (1991a). Control of immunoreactive inhibin production by human granulosa cells. *Clin. Endocrinol. (Oxford).* 35, 71–78.

Hillier, S. G., Yong, E. L., Illingworth, P. I., Baird, D. T., Schwall, R. H., and Mason, A. J. (1991b). Effect of recombinant activin on androgen synthesis in cultured human thecal cells. *J. Clin. Endocrinol. Metab.* 7, 1206–1211.

Hillier, S. G., Yong, E. L., Illingworth, P. I., Baird, D. T., Schwall, R. H., and Mason, A. J. (1991c). Effect of recombinant inhibin on androgen synthesis in cultured human thecal cells. *Mol. Cell. Endocrinol.* 75, R1–R6; 79, 177.

Hillier, S. G., Yong, E. L., Baird, D. T., and Fisch, B. (1992). Stertoid synthesis in the corpus luteum. *In* "Local Regulation of Ovarian Function" (N.-O. Sjöberg, L. Hamberger, P. O. Janson, Ch. Owman, and H. J. T. Coelingh Bennink, eds.), pp. 287–295. Parthenon Publishing, Carnforth, Lancs., UK.

Hofeditz, C., Magoffin, D. A., and Erickson, G. F. (1988). Evidence for protein kinase C regulation of ovarian theca-interstitial cell androgen biosynthesis. *Biol. Reprod.* 39, 873–881.

Horie, K., Takakura, K., Fujiwara, H., Suginami, H., Liao, S., and Mori, T. (1992). Immunohistochemical localization of androgen receptor in the human ovary throughout the menstrual cycle in relation to estrogen and progesterone receptor expression. *Hum. Reprod.* 7, 184–190.

Hsueh, A. J. W., Adashi, E. Y., Jones, P. B. C., and Welsh, T. J. Jr. (1984). Hormonal regulation of the differentiation of cultured granulosa cells. *Endocr. Rev.* 5, 76–126.

Hsueh, A. J. W., Dahl, K. D., Vaughan, J., Tucker, E., Rivier, J., Bardin, C. W., and Vale, W. (1987). Heterodimers and homodimers of inhibin subunits have different paracrine action in the modulation of luteinizing hormone-stimulated androgen biosynthesis. *Proc. Natl. Acad. Sci. USA.* 84, 5082–5086.

Hudson, K. E., and Hillier, S. G. (1985). Catechol estradiol control of FSH-stimulated granulosa cell steroidogenesis. *J. Endocrinol.* 106, R1–R3.

Hutchison, L. A., Findlay, J. K., de Vos, F. L., and Robertson, D. M. (1987). Effects of bovine inhibin, transforming growth factor-β and bovine activin-A on granulosa cell differentiation. *Biochem. Biophys. Res. Commun.* 146, 1405–1412.

Inouye, S., Guo, Y., DePaolo, L., Shimonaka, M., Ling, N., and Shimasaki, S. (1991). Recombinant expression of human follistatin with 315 and 288 amino acids: chemical and biological comparison with native porcine follistatin. *Endocrinology* 129, 815–822.

Jacobs, H. S. (1991). The LH hypothesis. *In* "Polycystic Ovaries: A Disorder or a Symptom" (R. W. Shaw, ed.), Vol. 3, pp. 91–98. Advances in Reproductive Endocrinology. Parthenon Publishing, Carnforth, Lancs., UK.

Jones, P. B. C., Welsh, T. H. Jr., and Hsueh, A. J. W. (1982). Regulation of ovarian progestin production by epidermal growth factor in cultured granulosa cells. *J. Biol. Chem.* 257, 11268–11273.

de Jong, F. H. (1988). Inhibin. *Physiol. Rev.* 68, 555–607.

Kim, I.-C., and Schomberg, D. W. (1989). The production of transforming growth factor-β by rat granulosa cell cultures. *Endocrinology* 124, 1345–1351.

Klein, R., Robertson, D. M., Shukovski, L., Findlay, J. K., and de Kretser, D. M. (1991). The radioimmunoassay of follicle-stimulating hormone (FSH)-suppressing protein (FSP): Stimulation of bovine granulosa cell FSP production by FSH. *Endocrinology* 128, 1048–1056.

Knecht, M. (1988). Plasminogen activator is associated with the extracellular matrix of ovarian granulosa cells. *Mol. Cell. Endocrinol.* 56, 1–9.

Knecht, M., and Catt, K. J. (1983). Modulation of cAMP-mediated differentiation in ovarian granulosa cells by epidermal growth factor and platelet-derived growth factor. *J. Biol. Chem.* 258, 2789–2794.

Knecht, M., Feng, P., and Catt, K. J. (1986). Transforming growth factor-beta regulates the expression of luteinizing hormone receptors in ovarian granulosa cells. *Biochem. Biophys. Res. Commun.* 139, 800–807.

Knecht, M., Feng, P., and Catt, K. (1987). Bifunctional role of transforming growth factor-β during granulosa cell development. *Endocrinology* 120, 1243–1249.

Kudlow, J. E., Kobrin, M. S., Purchio, A. F., Twardzik, D. R., Hernandez, E. R., Asa, S. L., and Adashi, E. Y. (1987). Ovarian transforming growth factor-α gene expression: immunohistochemical localization to the theca-interstitial cells. *Endocrinology* 121, 1577–1579.

LaPolt, P. S., Piquette, G., Soto, D., Sincich, C., and Hsueh, A. J. W. (1990). Regulation of inhibin subunit messenger ribonucleic acid levels by gonadotropins, growth factors, and gonadotropin-releasing hormone in cultured rat granulosa cells. *Endocrinology* 127, 823–831.

Le Nestour, E., Marraoui, J., Lahlou, N., Roger, M., de Ziegler, D., and Bouchard, P. (1993). Role of estradiol in the rise in follicle-stimulating hormone levels during the luteal-follicular transition. *J. Clin. Endocrinol. Metab.* 77, 439–442.

Leung, P. C. K., and Armstrong, D. T. (1980). Interactions of steroids and gonadotropins in the control of steroidogenesis in the ovarian follicle. *Annu. Rev. Physiol.* 42, 71–82.

Leung, P. C. K., and Steele, G. L. (1992). Intracellular signaling in the gonads. *Endocr. Rev.* 13, 476–498.

Li, W., Yuen, B. H., and Leung, P. C. K. (1992). Inhibition of progestin accumulation by activin-A in human granulosa cells. *J. Clin. Endocrinol. Metab.* 75, 285–289.

Ling, N., Ying, S.-Y., Ueno, N., Esch, F., Denoroy, L., and Guillemin, R. (1985). Isolation and partial characterization of a Mr 32,000 protein with inhibin activity from porcine follicular fluid. *Proc. Natl. Acad. Sci. USA* 82, 7217–7221.

Ling, N., Ying, S. Y., Ueno, N., Shimasaki, S., Esch, F., Hotta, M., and Guillemin, R. (1986). Pituitary FSH is released by a heterodimer of the β-subunits from the two forms of inhibin. *Nature (London)* 321, 779–782.

Lischinsky, A., Khalil, M. W., Hobkirk, R., and Armstrong, D. T. (1983). Formation of androgen conjugates by porcine granulosa cells. *J. Steroid Biochem.* 19, 1435–1440.

McLachlan, R. I., Cohen, N. L., Dahl, K. D., Bremner, W. J., and Soules, M. R. (1990). Serum inhibin levels during the periovulatory interval in normal women: relationships with sex steroid and gonadotrophin levels. *Clin. Endocrinol. (Oxford)* 32, 39–48.

McNatty, K. P. (1981). Hormonal correlates of follicular development in the human ovary. *Aust. J. Biol. Sci.* 34, 249–268.

McNatty, K. P., Reinhold, V. N., DeGrazia, C., Osathanondh, R., and Ryan, K. (1979). Metabolism of androstenedione by human ovarian tissue *in vitro* with particular reference to reductase and aromatase activity. *Steroids* 34, 429–443.

Magoffin, D. A., Grancedo, B., and Erickson, G. F. (1989). Transforming growth factor −β promotes differentiation of ovarian thecal-interstitial cells but inhibits androgen production. *Endocrinology* 125, 1951–1958.

Mason, A. J., Hayflick, J. S., Ling, N., Esch, F., Ueno, N., Ying, S.-Y., Guillemin, R., *et al.*

(1985). Complementary DNA sequences of ovarian follicular fluid inhibin show precursor structure and homology with transforming growth factor-β. *Nature* 318, 659–63.

Mason, A. J., Hayflick, J. S., Ling, N., Esch, F., Ueno, N., Ying, S. Y., Guillemin, R., Niall, H., and Seeburg, P. H. (1985). Complementary DNA sequences of ovarian follicular fluid inhibin show precursor structure and homology with transforming growth factor-β. *Nature (London)* 318, 659–663.

Mason, A. J., Niall, H. D., and Seeburg, P. H. (1986). Structure of two human ovarian inhibins. *Biochem. Biophys. Res. Commun.* 135, 957–964.

Massagué, J. (1983). Epidermal growth factor-like transforming growth factor. II. Interaction with epidermal growth factor receptors in human placenta membranes and A431 cells. *J. Biol. Chem.* 258, 13614–13620.

Massagué, J. (1987). The TGF β family of growth and differentiation factors. *Cell* 49, 437–438.

Matzuk, M. M., Finegold, M. J., Su, J.-G., Hsueh, A. J. W., and Bradley, A. (1992). α-Inhibin is a tumour-suppressor gene with gonadal specificity in mice. *Nature (London)* 360, 313–319.

Mayo, K. E., Cerelli, G. M., Spiess, J., Rivier, J., Rosenfeld, M. G., Evans, R. M., and Vale, W. (1986). Inhibin A-subunit cDNAs from porcine ovary and human placenta. *Proc. Natl. Acad. Sci. USA* 83, 5849–5853.

Meunier, H., S. B. Cajander, H. J. Roberts, C. Rivier, P. E. Sawchenko, A. J. W. Hsueh, and W. Vale. (1988). Rapid changes in the expression of inhibin α-, βA-, and βB-subunits in ovarian cell types during the rat estrous cycle. *Mol. Endocrinol.* 2, 1352–1363.

Michel, U., McMaster, J. W., and Findlay, J. K. (1992). Regulation of steady-state follistatin mRNA levels in rat granulosa cells *in vitro*. *J. Mol. Endocrinol.* 9, 147–156.

Miró, F., and Hillier, S. G. (1992). Relative effects of activin and inhibin on steroid hormone synthesis in primate granulosa cells. *J. Clin. Endocrinol. Metab.* 75, 1556–1561.

Miró, F., Smyth, C. D., and Hillier, S. G. (1991). Development-related effects of recombinant activin on steroid synthesis in rat granulosa cells. *Endocrinology* 129, 3388–3394.

Miyamoto, K., Hasegawa, Y., Fukuda, M., Nomura, M., Igarashi, M., Kangawa, K., and Matsu, H. (1985). Isolation of porcine follicular fluid inhibin of 32K daltons. *Biochem. Biophys. Res. Commun.* 129, 396–403.

Mondschein, J. S., Canning, S. F., Miller, D. Q., and Hammond, J. (1989). Insulin-like growth factors (IGFs) as autocrine/paracrine regulators of granulosa cell differentiation and growth: studies with a neutralizing monoclonal antibody to IGF-1. *Biol. Reprod.* 40, 79–85.

Nakamura, T., Takio, K., Eto, Y., Shibai, H., Titani, K., and Sugino, H. (1990). Activin-binding protein from rat ovary is follistatin. *Science* 247, 836–838.

Nakamura, T., Hasegawa, Y., Sugino, K., Kogawa, K., Titani, K., and Sugino, H. (1992). Follistatin inhibits activin-induced differentiation of rat follicular granulosa cells *in vitro*. *Biochim. Biophys. Acta* 1135, 103–109.

Nakatani, A., Shimasaki, S., DePaolo, L. V., Erickson, G. F., and Ling, N. (1991). Cyclic changes in follistatin messenger ribonucleic acid and its protein in the rat ovary during the estrous cycle. *Endocrinology* 129, 603–611.

Ny, T., Bjersing, L., Hsueh, A. J. W., and Loskutoff, D. J. (1985). Cultured granulosa cells produce two plasminogen activators and an antiactivator, each regulated differently by gonadotropins. *Endocrinology* 116, 1666–1668.

O'Connell, M. L., Canipari, R., and Strickland, S. (1987). Hormonal regulation of tissue plasminogen activator secretion and mRNA levels in rat granulosa cells. *J. Biol. Chem.* 262, 2339–2444.

Oliver, J. E., Aitman, T. J., Powell, J. F., Wilson, C. A., and Clayton, R. N. (1989). Insulin-like growth factor I gene expression in the rat ovary is confined to the granulosa cells of developing follicles. *Endocrinology* 224, 2671–2679.

Overes, H. W. T. M., de Leeuw, R., and Kloosterboer, H. J. (1992). Regulation of aromatase activity in FSH-primed rat granulosa cells *in vitro* by follicle-stimulating hormone and various amounts of human chorionic gonadotropin. *Hum. Reprod.* 7, 191–196.

Poretsky, L., Grigorescu, F., Seibel, M., Moses, A. C., and Flier, J. S. (1985). Distribution and characterization of insulin and insulin-like growth factor receptors in normal human ovary. *J. Clin. Endocrinol. Metab.* 61, 728–734.

Ramasharma, K., Caberera, C. M., and Li, C. H. (1986). Identification of insulin-like growth factor II in human seminal and follicular fluids. *Biochem. Biophys. Res. Commun.* 140, 536–42.

Ramasharma, K., and Li, C. H. (1987). Human pituitary and placental hormones control human insulin-like growth factors secretion in human granulosa cells. *Proc. Natl. Acad. Sci. USA* 84, 2643.

Rabinovici, J., Spencer, S., Doldi, N., Goldsmith, P. C., Schwall, R., and Jaffe, R. B. (1992). Activin-A as an intraovarian modulator: actions localization, and regulation of the intact dimer in human ovarian granulosa cells. *J. Clin. Invest.* 89, 1528–1536.

Ravindranath, N., Little-Ihrig, L. L., Fairchild-Benyo, D., and Zeleznik, A. J. (1992a). Role of luteinizing hormone in the expression of cholesterol side-chain cleavage cytochrome P450 and 3β-hydroxysteroid dehydrogenase, Δ^{4-5} isomerase messenger ribonucleic acids in the primate corpus luteum. *Endocrinology* 131, 2065–2070.

Ravindranath, N., Little-Ihrig, L. L., Phillips, H. S., Ferrara, N., and Zeleznik, A. J. (1992b). Vascular endothelial growth factor messenger ribonucleic acid expression in the primate ovary. *Endocrinology* 131, 254–260.

Ravindranath, N., Little-Ihrig, L. L., and Zeleznik, A. J. (1992c). Characterization of the levels of messenger ribonucleic acid that encode for luteinizing hormone receptor during the luteal phase of the primate menstrual cycle. *J. Clin. Endocrinol. Metab.* 74, 779–785.

Rechler, M. M., and Nissley, S. P. (1985). The nature and regulation of the receptors for insulin-like growth factors. *Annu. Rev. Physiol.* 47, 425–442.

Richards, J. S., Jahnsen, T., Hedin, L., Lifka, J., Ratoosh, S. L., Durica, J. M., and Goldring, N. B. (1987). Ovarian follicular development: from physiology to molecular biology. *Recent Prog. Horm. Res.* 43, 231–270.

Rivier, J., Spiess, J., McClintock, R., Vaughan, J., and Vale, W. (1985). Purification and partial characterization of inhibin from porcine follicular fluid. *Biochem. Biophys. Res. Commun.* 133, 120–127.

Roberts, A. B., Flanders, K. C., Kondaiah, P., Thomson, N. L., Van Obberghen-Schilling, E., Wakefield, L., Rossi, P., de Crombrugghe, B., Heine, U., and Sporn, M. B. (1988). Transforming growth factor β: biochemistry and roles in embryogenesis, tissue repair and remodeling, and carcinogenesis. *Recent Prog. Horm. Res.* 44, 157–193.

Robertson, D. M., Foulds, L. M., Leversha, L., Morgan, F. J., Hearn, M. T. W., Burger, H. G., Wettenhall, R. E. H., and de Kretser, D. M. (1985). Isolation of inhibin from bovine follicular fluid. *Biochem. Biophys. Res. Commun.* 126, 220–226.

Robertson, D. M., Klein, R., de Vos, F. L., McLachlan, R. I., Wettenhall, R. E. H., Hearn, M. T. W., Burger, H., and de Kretser, D. M. (1987). The isolation of polypeptides with FSH suppressing activity from bovine follicular fluid which are structurally different to inhibin. *Biochem. Biophys. Res. Commun.* 149, 744–749.

Saito, S., Nakamura, T., Tatani, K., and Sugino, H. (1991). Production of activin-binding protein by rat granulosa cells *in vitro*. *Biochem. Biophys. Res. Commun.* 176, 413–422.

Sasano, H., Okamoto, M., Mason, J. I., Simpson, E. R., Mendelson, C. R., Sasano, N., and Silverberg, S. G. (1989). Immunolocalization of aromatase, 17α-hydroxylase and side-chain cleavage cytochromes P-450 in the human ovary. *J. Reprod. Fertil.* 85, 163–169.

Schwall, R. H., Mason, A. J., Wildox, J. N., Bassett, S. G., and Zeleznik, A. J. (1990). Localization of inhibin/activin subunit mRNA's within the primate ovary. *Mol. Endocrinol.* 4, 75–79.

Segaloff, D. L., Wang, H., and Richards, J. S. (1990). Hormonal regulation of luteinizing hormone/chorionic gonadotropin receptor mRNA in rat ovarian cells during follicular development and luteinization. *Mol. Endocrinol.* 4, 1856–1865.

Shalgi, R., Kraicer, P., Rimon, A., Pinto, M., and Soferman, N. (1973). Proteins of human follicular fluid: the blood–follicle barrier. *Fertil. Steril.* 24, 429–434.

Shimasaki, S., Koga, M., Esch, F., Cooksey, K., Mercado, M., Koba, A., Ueno, N., Ying, S.-Y., and Ling, N. (1988a). Primary structure of the human follistatin precursor and its genomic organization. *Proc. Natl. Acad. Sci. USA* 85, 4218–4222.

Shimasaki, S., Koga, M., Esch, F., Mercado, M., Cooksey, K., Koba, A., and Ling, N. (1988b). Porcine follistatin gene structure supports two forms of mature follistatin produced by alternative splicing. *Biochem. Biophys. Res. Commun.* 152, 717–723.

Shimasaki, S., Koga, M., Buscalia, M. L., Simmons, D. L., Bicsak, T., and Ling, N. (1989). Follistatin gene expression in the ovary and extragonadal tissues. *Mol. Endocrinol.* 3, 651–659.

Shimonaka, M., Inouye, S., Shimasaki, S., and Ling, N. (1991). Follistatin binds to both activin and inhibin through the common beta-subunit. *Endocrinology* 128, 3313–3315.

Sibley, D. R., Benovic, J. L., Caron, M., and Lefkowitz, R. L. (1988). Phosphorylation of cell surface receptors: a mechanism for regulating signal transduction pathways. *Endocr. Rev.* 9, 38–56.

Skinner, M. K., McKeracher, H. L., and Dorrington, J. H. (1985). Fibronectin as a marker of granulosa cell cytodifferentiation. *Endocrinology* 117, 886–892.

Skinner, M. K., Keski-Oja, J., Osteen, K. G., and Moses, H. L. (1987). Ovarian theca cells produce transforming growth factor-β which regulates granulosa cell growth. *Endocrinology* 121, 786–792.

Slack, J. M. W. (1989). Peptide regulatory factors in embryonic development. *Lancet* i, 1312–1315.

Steinkampf, M. P., Mendelson, C. R., and Simpson, E. R. (1987). Regulation by follicle-stimulating hormone of the synthesis of aromatase cytochrome P-450 in human granulosa cells. *Mol. Endocrinol.* 1, 465–471.

Steinkampf, M. P., Mendelson, C. R., and Simpson, E. R. (1988). Effects of epidermal growth factor and insulin-like growth factor I on the levels of mRNA encoding aromatase P-450 of human ovarian granulosa cells. *Mol. Cell. Endocrinol.* 59, 93–99.

Suzuki, T., Miyamoto, K., Hasegawa, Y., Abe, U., Ui, M., Ibuki, Y., and Igarashi, M. (1987). Regulation of inhibin production by rat granulosa cells. *Mol. Cell. Endocrinol.* 54, 185–195.

Suikkari, A. M., Jalkanen, J., Koistinen, R., Bützow, R., Ritvos, O., Ranta, T., and Seppäla, M. (1989). Human granulosa cells synthesize low molecular weight insulin-like growth factor binding protein. *Endocrinology* 124, 1088–1090.

Tamura, T., Kitawaki, J., Yamamoto, T., Osawa, Y., Kominami, S., Takemori, S., and Okada, H. (1992). Immunohistochemical localization of 17α-hydroxylase/C17-20 lyase and aromatase cytochrome P-450 in the human ovary during the menstrual cycle. *J. Endocrinol.* 135, 589–595.

Tsonis, C. G., Hillier, S. G., and Baird, D. T. (1987). Production of inhibin bioactivity by human granulosa–lutein cells: stimulation by LH and testosterone *in vitro*. *J. Endocrinol.* 112, R11–14.

Turner, I. M., Saunders, P. T. K., Shimasaki, S., and Hillier, S. G. (1989). Regulation of inhibin subunit gene expression by FSH and estradiol in cultured rat granulosa cells. *Endocrinology* 125, 2790–2792.

Ueno, N., Ling, N., Ying, S.-Y., Esch, F., Shimasaki, S., and Guillemin, R. (1987). Isolation and partial characterization of follistatin: a single-chain M_r 35,000 monomeric protein that inhibits the release of follicle-stimulating hormone. *Proc. Natl. Acad. Sci. USA* 84, 8282–8286.

Ui, M., Shimonaka, M., Shimasaki, S., and Ling, N. (1989). An insulin-like growth factor binding protein in ovarian follicular fluid blocks follicle-stimulating hormone-stimulated steroid production by ovarian granulosa cells. *Endocrinology* 125, 912–916.

Vale, W., Rivier, J., Vaughan, J., McClintock, R., Corrigan, A., Woo, W., Karr, D., and Spiess, J. (1986). Purification and characterization of an FSH releasing protein from porcine ovarian follicular fluid. *Nature (London)* 321, 776–779.

Vale, W., Rivier, C., Hsueh, A., Campen, C., Meunier, H., Bicsak, T., Vaughan, J., Corrigan, A., Bardin, W., Sawchenko, P., Petraglia, F., Yu, J., Plotsky, P., Spiess, J., and Rivier, J. (1988). Chemical and biological characterization of the inhibin family of protein hormones. *Recent Prog. Horm. Res.* 44, 1–30.

Vale, W., Struthers, S., Gaddy-Kurten, D., Smith, B., Kintner, C., and Mathews, L. (1992). Biological and molecular characterization of activins and their receptors. *In* "Proc. IX Ovarian Workshop," Chapel Hill, NC Serono Symposia USA, Norwell Mass. (in press).

Veldhuis, J. D., Rogers, R. J., and Fulanetto, R. W. (1986). Synergistic actions of estradiol and somatomedin-C on swine ovarian (granulosa) cells. *Endocrinology* 119, 530–538.

Voutilainen, R., and Miller, W. L. (1987). Coordinate tropic hormone regulation of mRNAs for insulin-like growth factor II and the cholesterol side-chain cleavage enzyme, P450scc, in human steroidogenic tissues. *Proc. Natl. Acad. Sci. USA* 84, 1590–1594.

Waterfield, M. D. (1989). Epidermal growth factor and related molecules. *Lancet* i, 1243–21246.

Wheeler, M. B., and Veldhuis, J. (1989). Facilitative actuons of the protein kinase-C effector system on hormonally stimulated adenosine 3′,5′-monophosphate production by swine luteal cells. *Endocrinology* 125, 2414–2420.

Whitelaw, P. F., Smyth, C. D., Howles, C. M., and Hillier, S. G. (1992). Cell-specific expression of aromatase and LH receptor mRNAs in rat ovary. *J. Mol. Endocrinol.* 9, 309–313.

Wickings, E. J., Hillier, S. G., and Reichert, L. E., Jr. (1986). Gonadotropic control of steroidogenesis in human granulosa-lutein cells. *J. Reprod. Fertil.* 76, 677–684.

Wong, W. L., Lofgren, J., Garg, S., Bald, L., Woodruff, T., and Fendly, B. (1992). Specific immunoassays for measurement of activin A and activin B in biological fluids. *In* "Proc. 74th Annual Meet. Endocr. Soc." San Antonio, Texas. [Abstract 200]

Woodruff, T., Meunier, H., Jones, P. B., Hsueh, A. J. W., and Mayo, K. E. (1987). Rat inhibin: molecular cloning of α- and β-subunit complementary deoxyribonucleic acids and expression in the ovary. *Mol. Endocrinol.* 1, 561–569.

Woodruff, T. K., D'Agostino, J. B., Schwartz, N. B., and Mayo, K. E. (1988). Dynamic changes in inhibin messenger RNAs in rat ovarian follicles during the reproductive cycle. *Science* 239, 1296–1299.

Xiao, S., and Findlay, J. K. (1991). Interactions between activin and follicle-stimulating hormone-suppressing protein and their mechanisms of action on cultured rat granulosa cells. *Mol. Cell. Endocrinol.* 79, 99–107.

Xiao, S., Robertson, D. M., and Findlay, J. K. (1992). Effects of activin and FSH-suppressing protein-follistatin on FSH receptors and differentiation of cultured rat granulosa cells. *Endocrinology* 131, 2365–2370.

Yen, S. S. C. (1986). The human menstrual cycle. *In* "Reproductive Endocrinology" (S. S. C. Yen and R. B. Jaffe, eds.), pp. 200–236. Saunders, Philadelphia.

Ying, S.-Y. (1988). Inhibins, activins, and follistatins: gonadal proteins modulating the secretion of follicle-stimulating hormone. *Endocr. Rev.* 9, 267–293.

Ying, S.-Y., Becker, A., Ueno, N., and Guillemin, R. (1986). Inhibin and beta type transforming growth factor (TGFβ) have opposite modulating effects on the follicle stimulating hormone (FSH)-induced aromatase activity of cultured rat granulosa cells. *Biochem. Biophys. Res. Commun.* 136, 969–975.

Yong, E. L., Baird, D. T., and Hillier, S. G. (1992a). Mediation of gonadotropin-stimulated growth and differentiation of human granulosa cells by adenosine-3′,5′-monophosphate: one molecule, two messages. *Clin. Endocrinol. (Oxford)* 37, 51–58.

Yong, E. L., Baird, D. T., Yates, R., Reichert, L. E., Jr., and Hillier, S. G. (1992b). Hormonal regulation of the growth and steroidogenic function of human granulosa cells. *J. Clin. Endocrinol. Metab.* 74, 842–849.

Yu, J., Shao, L., Lemas, V., Yu, A. L., Vaughan, J., Rivier, J., and Vale, W. (1987). Importance of FSH-releasing protein and inhibin in erythroid differentiation. *Nature (London)* 330, 765–767.

Zeleznik, A. J. (1991). Control of luteal endocrine function. *In* "Ovarian Endocrinology" (S. G. Hillier, ed.), pp. 167–189. Blackwell, London.

Zeleznik, A. J., and Hillier, S. G. (1984). The role of gonadotropins in the selection of the preovulatory follicle. *Clin. Obstet. Gynecol.* 27, 927–940.

Zeleznik, A. J., and Kubik, C. J. (1986). Ovarian responses in macaques to pulsatile infusion of follicle-stimulating hormone (FSH) and luteinizing hormone (LH): increased sensitivity of the maturing follicle to FSH. *Endocrinology* 119, 2025–2032.

Zeleznik, A. J., Schuler, H. M., and Reichert, L. E., Jr. (1981). Gonadotropin-binding sites in the rhesus monkey ovary: role of the vasculature in the selective distribution of human chorionic gonadotropin to the preovulatory follicle. *Endocrinology* 109, 356–362.

Zeleznik, A. J., Ihrig, L. L., and Bassett, S. G. (1989). Developmental expression of Ca^{++}/Mg^{++} sensitive endonuclease activity in rat granulosa and luteal cells. *Endocrinology* 125, 2218–2220.

Zhang, Z., Lee, V. W. K., Carson, R. S., and Burger, H. G. (1988). Selective control of rat granulosa cell inhibin production by FSH and LH *in vitro*. *Mol. Cell. Endocrinol.* 56, 35–40.

Zoller, L. C., and Weisz, J. (1978). Identification of cytochrome P-450, and its distribution in the membrana granulosa of the preovulatory follicle using quantitative cytochemistry. *Endocrinology* 103, 310–313.

2

Control Systems of Theca-Interstitial Cells

DENIS A. MAGOFFIN & GREGORY F. ERICKSON

I. Introduction

The periodic ovulation of fertilizable oocytes by the mammalian ovary is the culmination of the process of follicle development. The ovarian follicle, in addition to the oocyte, contains two principal endocrine cell types, the theca and the granulosa. During follicular growth and development, the presumptive theca and granulosa cells (GC) proliferate and differentiate in a tightly controlled sequence of events which results in the timely acquisition of new functional capabilities and the loss of others. Among these functions is the ability to secrete steroid hormones and regulatory peptides. These changes require the orderly expression of patterns of genes which are characteristic of theca and granulosa cells. Certain genes are unique to one cell type while other genes are expressed in both cell types. In this chapter the control systems that regulate gene expression and steroidogenesis in theca cells, and the hormones that activate these control mechanisms are described and discussed.

The process of thecogenesis begins soon after primordial follicles have been recruited into the pool of growing follicles and continues throughout follicle development, ovulation, and luteinization (Figure 1). The theca cells first become recognizable in preantral follicles after the oocyte has reached full size and there are multiple layers of proliferating granulosa cells (Peters, 1969). It has been suggested, but not proven, that the theca-interstitial cells (TIC) are derived from a nonsteroidogenic precursor cell (Weakly, 1966; Hoage and Cameron, 1976). As the follicle develops through the

Molecular Biology of the Female Reproductive System

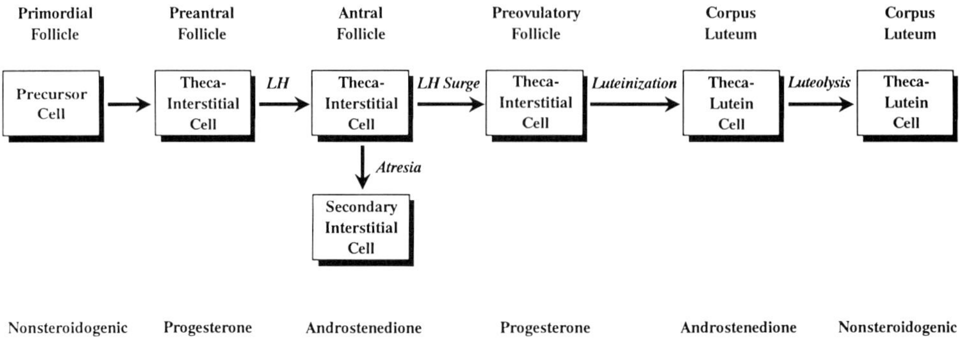

FIGURE 1 Steroidogenesis during thecogenesis. Theca-interstitial cells are thought to differentiate from fibroblast-like precursor cells to first become active progesterone-producing and then androgen-producing cells as the dominant follicle grows to preovulatory size. After the ovulatory surge of LH, the theca luteinize. In follicles that are not selected for ovulation, even though the granulosa compartment degenerates, the theca-interstitial cells retain their differentiated state and become secondary interstitial cells.

preantral stages, the TIC express genes coding for luteinizing hormone (LH) receptors and the cholesterol side chain cleavage (P450scc), and 3β-hydroxysteroid dehydrogenase (3β-HSD) enzymes, which confer LH responsiveness and the capacity to produce progesterone on the TIC. Physiologically low concentrations of LH, which are present in the follicular phase of the cycle, stimulate the early antral follicles to progress through the antral stages of development and become preovulatory follicles (Richards *et al.*, 1980; Bogovich *et al.*, 1981). During this time the TIC express the 17α-hydroxylase/C17–20 lyase (P45017α) gene and change from progesterone-producing cells to cells which produce primarily androstenedione (Bogovich and Richards, 1982). The androstenedione is aromatized to estrogen by the granulosa cells and leads to the preovulatory LH surge. In follicles which undergo atresia, the TIC do not die but remain in the interstitial compartment of the ovary (Kingsbury, 1939; Guraya and Greenwald, 1968; Deanesly, 1972) and are known as secondary interstitial cells (SIC); they retain LH responsiveness and the capacity to secrete androstenedione (Rice and Savard, 1966; McNatty *et al.* 1979). When the preovulatory surge of LH occurs, the TIC lose the capacity to express P45017α (Lieberman *et al.*, 1975; Hillenso *et al.*, 1976; Hamburger *et al.*, 1978; Ahren *et al.*, 1979) and revert to progesterone-producing cells as the follicle collapses and luteinization begins. The TIC contribute to the formation of the corpus luteum and become theca-lutein cells which in women once again express P45017α and produce androstenedione. Finally, when luteolysis occurs, the theca-lutein cells lose the capacity to synthesize steroid hormones.

II. Regulators

A. The Endocrine System

Two principal regulatory systems have been shown to control the process of thecogenesis. The classic endocrine system is the fundamental mechanism controlling thecogenesis. It has been proposed that an autocrine/paracrine system within the ovary modulates the effects of the endocrine system in both a positive and a negative manner. In this section we discuss the evidence that supports a role for the endocrine regulators LH, prolactin (PRL), and estradiol (E_2) in thecogenesis.

1. LH

Evidence from studies with dispersed whole rat ovarian cells cultured in defined medium (Erickson *et al.*, 1985) and more recent studies with isolated TIC (Magoffin, 1991) support the conclusion that LH is the primary regulator of thecogenesis.

There is a large body of evidence supporting the existence of LH/human chorionic gonadotropin (hCG) receptors on the TIC. Autoradiographic studies with [^{125}I]hCG demonstrated that binding sites are present in TIC and SIC (Rajaneimi and Vanha-Perttula, 1972; Magoffin and Erickson, 1982a, 1988a). Quantitative binding studies with [^{125}I]hCG reveal that TIC contain a single class of high-affinity receptors ($K_d = 1.0 \pm 0.08 \times 10^{-10}$ M) and that each cell has approximately 16,500 binding sites in the plasma membrane (Magoffin and Erickson, 1982a). When undifferentiated TIC are treated with hCG *in vitro*, the LH receptors are initially coupled to cyclic adenosine monophosphate (cAMP) production but not to androgen biosynthesis (Magoffin and Erickson, 1982a; Magoffin, 1989). Only after approximately 20 hours of continuous LH treatment do the TIC differentiate into active androgen-producing cells (Magoffin, 1989; Magoffin and Weitsman, 1993a). It is clear that stimulation with LH alone in defined medium induces thecogenesis by stimulating the expression in TIC of each of the steroidogenic enzymes necessary for androgen biosynthesis, namely P450scc, 3β-HSD, and P45017α (Magoffin and Weitsman, 1993a,b,c).

The process of thecogenesis is central to the "two-cell, two-gonadotropin" concept of ovarian estrogen biosynthesis (Falck, 1959; Bjersing, 1967). Luteinizing hormone stimulates the theca cells to differentiate and secrete androgens, which diffuse across the basal lamina, where they are aromatized to estrogens by granulosa cells. Follicle-stimulating hormone (FSH) is needed to stimulate differentiation and induce the aromatase enzyme in the granulosa cells.

In addition to being a potent stimulator of thecogenesis, LH stimulation is associated with an increase in TIC proliferation. As the follicle grows, there is a large increase in the number of TIC produced by the mitosis of theca stem cells (Lane and Davis, 1939). A stimulatory role for LH in TIC proliferation is suggested by the observation that long-term treatment with hCG causes the theca interna to hypertrophy (Selye and Collip, 1933).

2. Prolactin

There is convincing evidence indicating the existence of prolactin receptors on TIC. Autoradiographic studies using [^{125}I]prolactin have suggested that prolactin-binding sites are present on TIC and SIC (Midgley, 1973; Oxberry and Greenwald, 1982). In addition, [^{125}I]prolactin specifically binds to isolated interstitial tissue (Richards and Midgley, 1976). Quantitative binding studies demonstrated that TIC contain a single class of high-affinity prolactin receptors ($K_d = 2.42 \times 10^{-10}$ M), with approximately 4700 binding sites per cell (Magoffin and Erickson, 1982b).

The presence of prolactin receptors on TIC strongly suggests that prolactin could play a role in regulating thecogenesis. When TIC are treated with prolactin *in vitro*, there is a rapid and irreversible inhibition of LH-stimulated steroid biosynthesis. In conditions where the circulating concentrations of prolactin are high, such as during lactation and hyperprolactinemia, folliculogenesis is inhibited and estrogen biosynthesis is suppressed (McNeilly *et al.*, 1982). Thus it appears that inhibition of thecal androgen production may play a physiological role in the ovarian response to prolactin.

3. Estrogen

The evidence that E_2 may play a role in regulating thecogenesis begins with the observation that [^{3}H] E_2 is concentrated in the nuclei of TIC and SIC (Stumpf, 1969). Subsequently, *in vivo* experiments demonstrated that E_2 treatment suppressed the ability of LH to stimulate ovarian androgen biosynthesis (Leung *et al.*, 1978; Leung and Armstrong, 1979; Magoffin and Erickson, 1981) but not progesterone production, suggesting that high concentrations of E_2, on the order of those found in dominant follicles, caused a selective inhibition of the P45017α enzyme. To overcome the question of indirect effects, ovarian cells were treated with E_2 in culture (Magoffin and Erickson, 1982c). The results of these experiments demonstrated a direct effect of E_2 in the ovary; it rapidly caused a selective inhibition of P45017α activity with no changes in the LH receptor, LH-stimulated cAMP production, or progesterone production. The observation that the inhibition is rapid ($t_{1/2}$ = 10 min) indicates that E_2 may initially act as a competitive inhibitor, but the long-lasting nature of the inhibition after washing E_2 from the culture medium suggests that it may also affect gene expression.

B. The Autocrine/Paracrine Systems

It is clear that the hormones LH, PRL, and E_2 are major endocrine regulators of thecogenesis in the rat ovary (Erickson *et al.*, 1985). However, research done in the past few years has demonstrated that the biological responses of the endocrine system can be either amplified or attenuated by intrinsically produced proteins that have growth factor activity (Mondschein and Schomberg, 1981). Recent evidence strongly suggests that intrinsic growth factors may mediate the biological actions of the endocrine system (Ui *et al.*, 1989; Bicsak *et al.*, 1990). This is the principle underlying our current concept of autocrine/paracrine control systems in the ovary.

Much of what we know about ovarian growth factors has come from studies in the rat. All five growth factor families are expressed intrinsically in developing rat follicles and the growth factors cause the TIC, GC, and oocyte to change their differentiated state. In the rat ovary, the intrinsic growth factor (IGF) systems include the IGF-I family (Adashi *et al.*, 1985), the transforming growth factor-β (TGFβ) family (Knecht *et al.*, 1989; Ying, 1988), the TGFα/epidermal growth factor (EGF) family (May and Schomberg, 1989), the fibroblast growth factor family (Gospodarowicz *et al.*, 1987), and the cytokine family (Adashi, 1990). In this section we discuss the evidence for the autocrine/paracrine systems in the rat ovary and consider the question of what role TICs play in the systems. This evidence is summarized in Table 1.

1. IGF-I Family

The IGF system, complete with ligands, receptors, binding proteins, and biological responses, has been demonstrated in the rat ovary (Table 1). The genes encoding IGF-I (Murphy *et al.*, 1987; Oliver *et al.*, 1989; Hernandez *et al.*, 1989; Carlsson *et al.*, 1989) and IGF-II (Hernandez *et al.*, 1990b, 1991) have been identified in mature and immature rat ovaries, being localized to the GC (Oliver *et al.*, 1989; Zhou *et al.*, 1991) and TIC, respectively (Hernandez *et al.*, 1990a, 1991). The expression of IGF-II in TIC is inhibited by E_2 (Hernandez *et al.*, 1990b). Therefore, one could speculate that intrinsic IGF-II production by TIC in dominant and atretic follicles might be low and high, respectively. The physiological role of TIC-derived IGF-II is unknown. The expression of IGF-I is confined to the GC of dominant follicles (Oliver *et al.*, 1989; Zhou *et al.*, 1991), thus implying a possible causal relationship between high rates of IGF-I synthesis in the GC and follicle selection in the rat ovary. This theory is supported by the evidence that IGF-I levels are increased by the hormone E_2 (Hernandez *et al.*, 1989), and decreased by hypophysectomy (Zhou *et al.*, 1991).

The genes encoding the type I IGF receptor and the type II IGF/mannose-6-phosphate receptor (Table 1) have been identified in rat ovaries, being localized to both the TIC and GC (Zhou *et al.*, 1991; Hernandez *et al.*, 1990b, 1991). Further, the proteins that bind to these hormones have also been identified, indicating that the mRNAs are translated into functional receptors (Adashi *et al.*, 1985; Cara and Rosenfield, 1988). As with the ligands, receptor gene expression is under hormone control; e.g., the type I IGF receptor is increased markedly by FSH and E_2, but the type II IGF

TABLE 1 Endocrine and Autocrine/Paracrine Control Systems in Rat Ovaries

Control systems	Signal transduction mechanism	Cellular source	Effects		
			Theca	Granulosa	Oocyte
Endocrine					
LH	cAMP/PKA	Pituitary	+	+/−	+
PRL	Unknown	Pituitary	−	+/−	Unknown
Estrogen	Intracellular receptor	Granulosa	−	+	Unknown
Autocrine/paracrine					
IGF Family					
IGF-I	Tyrosine kinase	Granulosa	+	+	Unknown
IGF-II	Unknown	Theca	Unknown	Unknown	Unknown
IGFBPs	NA	GC and TIC	Unknown	−	Unknown
TGFβ family					
TGFβ	Serine/threonine kinase	Theca	+/−	+	+
Activin	Serine/threonine kinase	Granulosa	−	+/−	Unknown
Inhibin	Unknown	Granulosa	+	+/−	Unknown
MIS	Unknown	Granulosa	Unknown	Unknown	−
EGF/TGFα family			−	+/−	+
EGF	Tyrosine kinase	Unknown	+/−	+	Unknown
TGFα	Tyrosine kinase	Theca			
FGF family		Stroma	−	+/−	+
bFGF	Tyrosine kinase	Unknown	(−)	(−)	
aFGF	Tyrosine kinase				
Cytokine family					
IL-1β	Unknown	Theca	−	−/+	Unknown
TNFα	Unknown	Unknown	−	−	Unknown

receptor is much less affected (Hernandez *et al.*, 1990b, 1991). These observations support the idea that the genes encoding IGF-I and its receptor are upregulated coordinately during selection, and implicate E_2 and FSH in the process.

The IGF-I receptor is a heterodimer which, when activated, expresses intrinsic protein tyrosine kinase activity. Dramatic evidence that IGF-I and the gonadotropins act synergistically to control follicle cytodifferentiation has been reported. In general, IGF-I by itself has little or no effect on the differentiated function of either the TIC or GC, but when IGF-I is added together with LH (Magoffin and Erickson, 1988b; Magoffin *et al.*, 1990; Cara and Rosenfield, 1988; Hernandez *et al.*, 1988) and FSH (Adashi *et al.*, 1985), the activities of the TIC and GC, respectively, are stimulated dramatically. In the case of GC, IGF-I and FSH act synergistically to control the levels of cAMP (Adashi *et al.*, 1986a) steroidogenesis (Adashi *et al.*, 1985; Davoren *et al.*, 1986), LH receptor (Adashi *et al.*, 1985; Davoren *et al.*, 1986), proteoglycan (Adashi *et al.*, 1986b), inhibin (Zhiwen *et al.*, 1987), and DNA synthesis (Adashi *et al.*, 1985; Bley *et al.*, 1992). Similar studies with TIC indicate that IGF-I and LH act synergistically to control the level of androgen expression. The details of these studies are discussed in Section III, B. Little is known about the intraovarian effects of TIC-derived IGF-II.

Evidence for an IGF-I system in the rat corpus luteum also exists. First, the messenger RNA for IGF-I is present in some CL (Oliver *et al.*, 1989; Zhou *et al.*, 1991). Second, the CL in ovaries of lactating rats strongly expresses the mRNA encoding the IGF-I receptor (Zhou *et al.*, 1991). Third, IGF-I and the IGF-I receptor mRNAs are abundantly expressed in rat CL throughout pregnancy, and E_2 can modulate the levels of both messages (Parmer *et al.*, 1991). Fourth, IGF-I amplifies the stimulatory effects of E_2 and hCG on progesterone production by isolated rat lutein cells (Nelson *et al.*, 1992). Thus, a strong argument can be made for the existence of an endocrine-regulated IGF-I system in the rat CL which may be involved in the process of luteogenesis during the estrous cycle and pregnancy.

An important concept generated in the past few years is that the regulation of the activity of the IGFs can involve specific binding proteins termed the IGFBPs. Under normal physiological conditions, IGF-I and IGF-II in body fluids are bound to a family of homologous proteins that can either stimulate or inhibit the activity of the IGFs (Shimasaki and Ling, 1991). Six different rat IGFBP cDNAs have been cloned and their complete primary structure deduced from the clones: IGFBP-1, containing 247 amino acids, M_r = 26,801 Da; IGFBP-2, containing 270 amino acids, M_r = 29,564 Da; IGFBP-3, containing 265 amino acids, M_r = 28,856 Da; IGFBP-4, containing 233 amino acids, M_r = 25,681 Da; IGFBP-5, containing 252 amino acids, M_r 28,428 Da; and IGFBP-6, containing 201 amino acids, M_r = 21,461 Da. Each IGFBP has a different amino acid sequence and presumably a different set of physiological functions.

Compelling evidence demonstrating the effects of IGFBPs on ovarian function has come from the studies of Nicholas Ling and Shunichi Shimasaki. They discovered that follicular fluid IGFBP-2 and -3 have antigonadotropic effects on FSH stimulation of GC cytodifferentiation (Ui *et al.*, 1989; Shimasaki *et al.*, 1990). Significantly, the mechanism of inhibition involved the specific binding of these IGFBPs to the intrinsic IGF-I produced by the GC (Bicsak *et al.*, 1990). These results have led to the novel theory that the biological responses of FSH on the GC depend directly on the local production of IGF-I and, when present, the IGFBPs will bind IGF-I and negate its tropic activity. The potential importance of the TIC and SIC in this theory has been nicely illustrated by *in situ* hybridization studies (Nakatani *et al.*, 1991). These experiments showed that all rat TIC and SIC strongly express the gene encoding IGFBP-2 (Figure 2). As indicated in other studies, cultured rat TIC-SIC appear capable of expressing IGFBP-3,

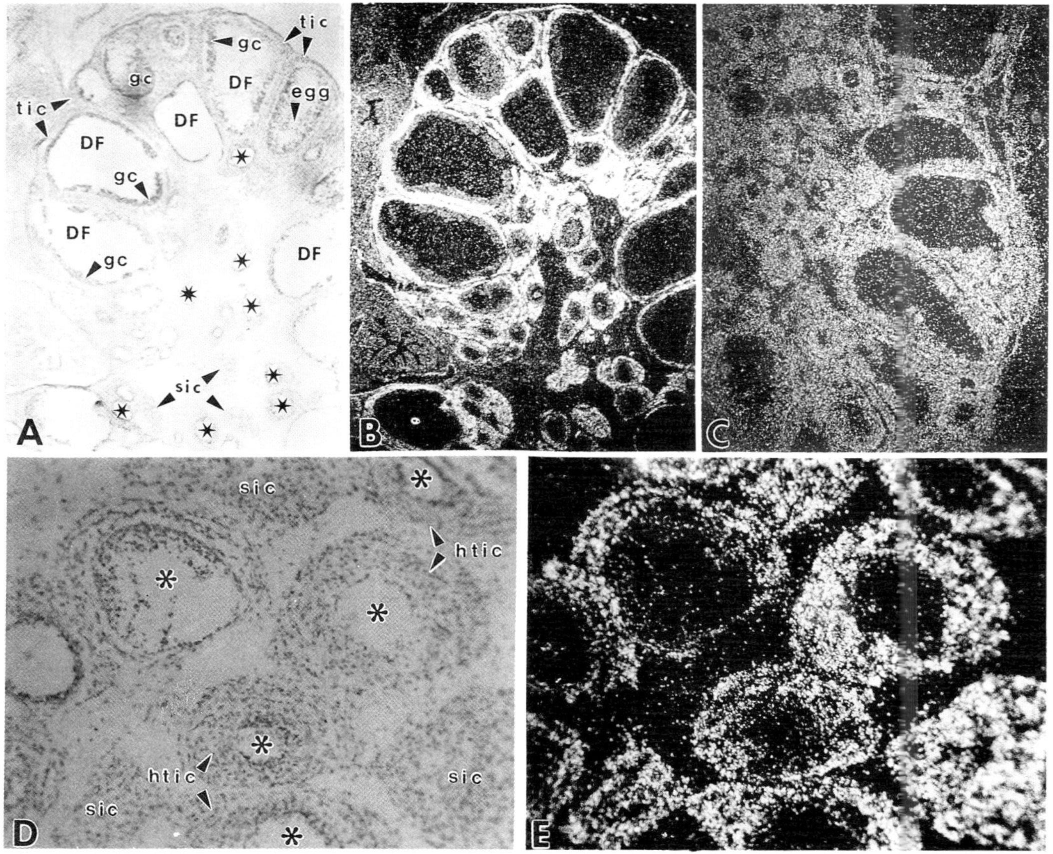

A
tic
gc
gc
DF
DF
egg
tic
DF
gc
DF
gc
DF
sic
B
C
D
sic
htic
htic
sic
sic
E

and the levels were increased by E_2 (Ricciarelli *et al.*, 1992). Considering that IGFBP-2 and -3 are potent inhibitors of FSH action, they are likely to have interesting and important consequences for the ability of granulosa cells to respond to stimulation by FSH. The dramatic results demonstrating the synthesis of these proteins in the interstitial tissue provides the basis for hypotheses involving novel roles for TIC/SIC-derived IGFBPs in regulating folliculogenesis, particularly atresia.

2. TGFβ Family

Included in the TGFβ family are TGFβ itself, activin, inhibin, and Müllerian inhibiting substance (MIS). They are all dimeric molecules composed of disulfide-bonded homologous polypeptide chains denoted A and B. Only the dimer is biologically active. Typically, all three combinations, AA, AB, and BB chains, are present in cells. Signal transduction of TGFβ and activin is through receptors with protein serine/threonine kinase activity.

There is ample evidence (Table 1) that an intrinsic TGFβ system exists in the rat Graafian follicle (Adashi *et al.*, 1989a). The mRNAs and proteins for TGFβ1 and TGFβ2 are expressed in the follicle wall, being localized predominantly to the TIC (Mulheron and Schomberg, 1990; Hernandez *et al.*, 1990a; Mulheron *et al.*, 1991). Further, cultured TIC synthesize and secrete TGFβ (Bendell and Dorington, 1988; Skinner *et al.*, 1987a). TGFβ derived from TIC can interact with receptors on follicle cells and upregulate their activity. That is, TGFβ acts together with FSH to stimulate GC mitosis (Bendell and Dorrington, 1988; Skinner *et al.*, 1987a; Dorrington *et al.*, 1988), LH receptor (Dodson and Schomberg, 1987; Knecht *et al.*, 1987), EGF/TGFα receptor expression (Feng *et al.*, 1988), and P450arom and P450scc expression (Adashi *et al.*, 1989a; Hernandez, 1990a). Finally, TGFβ induces meiotic maturation in the oocyte (Feng *et al.*, 1986) and causes TIC to differentiate as progesterone-producing cells (Magoffin *et al.*, 1989). Thus, the concept is emerging that TGFβ produced by the TIC is a potent positive regulatory molecule that affects the differentiated state of all follicle tissues. Although the physiological significance of these findings is not clear, it is interesting to speculate that the TGFβ produced by the TIC could play an important role in regulating folliculogenesis in the rat. The cellular mechanisms of this response are discussed in Section III,C.

Studies have shown (Table 1) that a complete inhibin, activin, and follistatin (FS) system is present in developing rat follicles (DePaolo *et al.*,

---

FIGURE 2 *In situ* hybridization analysis of IGFBP-2 mRNA in the rat ovary. Ovary sections are from an immature (28-day-old) rat which was injected once with 10 IU PMSG and killed 48 hours later. DF, dominant follicles; gc, granulosa cells; tic, theca-interstitial cells; sic, secondary interstitial cells; *, atretic preantral follicle; htic, hypertrophied theca-interstitial cells. (A) Bright-field photomicrograph. (B) Dark-field photomicrograph of same tissue hybridized with antisense cRNA probe. (D and E) Higher magnification of A. Reproduced with permission from Nakatani *et al.* (1991). Copyright © by The Endocrine Society.

1991). During folliculogenesis, the granulosa cells express the inhibin α, inhibin βA, and inhibin βB subunits (Meunier *et al.*, 1988). It is important to note that the pattern of expression of these subunits changes dramatically over the estrous cycle and that FSH and LH are involved in their regulation (Meunier *et al.*, 1989). The inhibin α- and βA-subunits are also expressed transiently in the TIC and SIC during the luteal phase of the estrous cycle. This suggests that the TIC/SIC tissue could produce free inhibin α, inhibin αβA, or activin A. The significance of these subunits in the TIC/SIC is unclear.

In vitro evidence indicates that rat follicle cells are also targets for activin, inhibin, and MIS (Table 1). In cultured TIC, activin suppresses the LH-stimulated androgen production (Hsueh *et al.*, 1987). In cultured rat GC, activin alone is without effect on steroidogenesis, but activin interacts with FSH to amplify the production of E_2 and progesterone (Xiao *et al.*, 1990; Xiao and Findlay, 1991; Miro *et al.*, 1991). The biological function of these proteins in the ovary is confusing and seems conflicting. Woodruff *et al.* (1990) concluded from their study that activin and inhibin promote atresia and selection respectively *in vivo*. At the opposite extreme, Findlay *et al.* (1990) and Doi *et al.* (1992) conclude that activin is a positive regulatory molecule in folliculogenesis. Further work is needed before one can decide whether activin is a cause of selection and/or atresia. Because FS is an activin-binding protein, it is implicated in possible activin/inhibin interactions (Findlay *et al.*, 1990; Nakamura *et al.*, 1990; DePaolo *et al.*, 1991). The cellular sites of FS gene expression are the GC and CL (Nakatani *et al.*, 1991). Follistatin alone has no observable effects on GC steroidogenesis; however, FS amplifies the effects of FSH on progesterone, but attenuates the influence of FSH on E_2 synthesis (Xiao *et al.*, 1990; Xiao and Findlay, 1991; Saito *et al.*, 1991). Follistatin was shown to block the GC responses to exogenous activin (Xiao and Findlay, 1991). By virtue of its ability to bind activin and inhibin, FS could have a regulatory role in controlling the biological activity of inhibin and activin in the TLC.

Finally, rat GC of healthy preantral and antral follicles have been shown to produce MIS by *in situ* hybridization and immunocytochemistry techniques (Ueno *et al.*, 1989a,b; Hirobe *et al.*, 1992). Because MIS is found in dividing GC in the periantral and cumulus regions, MIS has been implicated in the local control of GC mitosis. Little is known concerning the possible function of MIS in folliculogenesis and there is no information concerning the possible regulatory effects of MIS on thecogenesis.

3. TGFα/EGF Family

EGF and TGFα are homologous proteins that bind to the same receptor. The EGF/TGFα receptor is a monomeric protein with intrinsic protein tyrosine kinase. It is generally assumed that the two ligands are functionally equivalent, but evidence is emerging that suggests this assumption may not be valid. There is substantial evidence that a TGFα/EGF system, complete with ligand receptor and biological response, operates in the rat follicle (Table 1), and that the TIC play a central role in the system (Mondschein and Schomberg, 1981; Chabot *et al.*, 1986; Gangrade *et al.*, 1991). First, it has been demonstrated by Northern blot analysis that ovaries of immature estrogen-primed rats contain the mRNA encoding TGFα (Kudlow *et al.*, 1987). Other types of studies point to the TIC as the cellular site of TGFα production (Skinner *et al.*, 1987b). It is interesting that TGFα/EGF inhibits the effects of LH on TIC differentiation *in vitro* (Erickson and Case, 1983; Weitsman and Magoffin, 1993); TGFα/EGF could then serve as a negative autocrine regulator of the TIC tissue. The potential importance of TIC-derived TGFα in controlling GC growth and development is clear. That is, TGFα/EGF can stimulate GC DNA synthesis (Knecht and Catt, 1982) and 20α-dihydroprogesterone production (Knecht and Catt, 1982; Adashi *et al.*, 1985), and inhibit FSH-induced cAMP, progesterone, LH, and FSH receptor induction (Adashi *et al.*, 1985; Tilly *et al.*, 1992). Finally, TGFα/EGF can induce meiotic maturation (Dekel and Sherizly, 1985). Thus, the concept is emerging that TIC synthesize and secrete TGFα and that TGFα has great impact on oogenesis, thecogenesis, and cytodifferentiation of granulosa.

4. FGF Family

The FGF family includes proteins encoded by at least seven distinct genes, including basic FGF (bFGF) and acidic FGF (aFGF). Both of these FGFs stimulate mitosis in a variety of cell types, including GC. The FGFs mediate their biological responses by binding to cell surface receptors with tyrosine-specific protein kinase activity. These growth factors also exhibit a high affinity to heparin sulfate proteoglycans; however, the significance of this binding in the ovary has yet to be established.

Several lines of evidence indicate that a bFGF system exists in the rat ovary (Table 1). First, by using a reverse transcription polymerase chain reaction, the message encoding bFGF has been demonstrated in whole ovaries of immature rats primed with pregnant mare's serum gonadaptropin (PMSG) but not in the GC of the preovulatory follicle (Koos and Olson, 1989). This result implies that the bFGF gene is expressed in the stromal compartment of the ovary, perhaps in the interstitial tissue. The inability to detect the bFGF message by Northern analysis suggests that bFGF mRNA levels are low, perhaps being localized to subtypes of cells (Shimasaki *et al.*, 1988). In culture, bFGF alone has essentially no effect on either basal or LH-stimulated androgen production by TIC; however, bFGF is a potent inhibitor of androgen synthesis induced by IGF-I plus LH (Hurwitz *et al.*, 1990). The importance of the direct inhibitory effects of bFGF on IGF-I action in TIC is unknown. Direct evidence for pleiotropic effects of bFGF on the GC has also been obtained. Shikone and colleagues (1992) demonstrated that GC contain bFGF receptors and FSH increases their number. *In vitro,* bFGF is a potent inhibitor of FSH-induced E_2, LH receptor, and inhibin synthesis (Baird, *et al.*, 1986; Adashi *et al.*, 1988; LaPolt *et al.*, 1990a) but stimulates the production of progesterone and tissue plasminogen activator, and induces meiotic maturation in follicle-enclosed oocytes (LaPolt *et al.*, 1990b). Thus, it seems that ovarian bFGF, perhaps of TIC origin, could be an important regulatory factor in TIC, GC, and oocyte development.

5. Cytokine Family

Cytokines are a diverse family of regulatory proteins that have a marked effect on the proliferation and differentiation of different types of tissue, most notably during angiogenesis. The major cytokines include the interleukins (IL), tumor necrosis factors (TNF), colony-stimulating factors (CSF), and the interferons. In the area of cytokine research, a new line of investigation has emerged with evidence for the expression of IL-1 and TNF in rat ovaries (Table 1).

There are two types of IL-1: IL-1α, containing 159 amino acids, and IL-1β, containing 153 amino acids. Even though these peptides exhibit only 26% homology, they appear to exert the same biological responses through the same receptor. The mRNA for IL-1β was not detected in immature control or PMSG-primed ovaries; however, the message was found 6 hours after hCG was injected and, interestingly, it was localized to the theca interna (Hurwitz *et al.*, 1991b). These results suggest that the ovulatory stimulus induces IL-1β gene expression in theca tissue of dominant follicles. The cellular localization of the IL-1β mRNA is not known, but could be TIC and/or resident white blood cells. In these experiments, IL-1, but not LH/hCG, elicited increases in IL-1β mRNA levels (Hurwitz *et al.*, 1991b). Thus, theca IL-1β gene activity appears to be induced by an autocrine control mechanism that does not involve LH. At the level of the follicle, IL-1β is a potent inhibitor of TIC and GC functions *in vitro.* Experiments with cultured TIC demonstrated that, although IL-1β showed no effect alone, it elicited a dramatic dose-dependent inhibition of the synergistic effects of LH and IGF-I on androgen production (Hurwitz *et al.*, 1991a). Collectively, these results suggest that the theca interna of dominant follicles becomes committed to producing IL-1 during the ovulatory sequence, and that when IL-1 is secreted, it acts in an autocrine manner to markedly inhibit LH and IGF-I-stimulated steroidogenesis.

The ability of IL-1β to influence the activities of cultured GCs has also been demonstrated. Recombinant IL-1β inhibits FSH-dependent induction of LH receptor, progesterone, and P450arom (Gottschall, *et al.*, 1987; Kasson and Gorospe, 1989), but stimulates the production of 20α-dihydroprogesterone and prostaglandin biosynthesis (Kokia *et al.*, 1992). The nature of the IL-1β signal transduction mechanism in any target cell is unclear.

The mRNA for TNFα has been demonstrated within the ovaries of adult cycling rats. Although the precise cell type which expresses TNFα has yet to be established, it is not related to the white blood cells or CL (Sancho-Tello *et al.*, 1992). Adding nanomolar amounts of TNFα to TIC *in vitro* almost completely inhibits the LH-dependent induction of androgen synthesis (Roby and Terranova, 1988, 1990; Andreani *et al.*, 1991). TNFα also profoundly affects the GC, inhibiting the induction of P450arom and P450scc by FSH (Emoto and Baird, 1988; Darbon *et al.*, 1989; Adashi *et al.*, 1989b), and stimulating prostaglandin production (Kokia *et al.*, 1992). Thus, intrinsic ovarian TNFα has a variety of inhibitory effects on follicle cells, including acting as an antigonadotropin for TIC. The nature of the signal transduction pathway that mediates biological responses by TNFα is unknown in any cell type.

There is little doubt that the cytokine system is present in the rat ovary. We predict that more cytokine regulatory elements will be identified in the future. The challenge is to identify the physiological significance of the cytokine system in regulating ovarian homeostasis.

III. Mechanisms: Gene Expression and Steroidogenesis

A. Cyclic AMP/Protein Kinase A Pathway

There is substantial evidence that the cAMP/protein kinase A (PKA) pathway is one of the most important regulatory mechanisms in TIC. The LH/hCG receptor present in the plasma membrane of TIC is a member of the G-protein-coupled family of cell surface receptors (McFarland *et al.*, 1989). Evidence from a wide variety of systems has shown that these receptors are positively coupled to adenyl cyclase by a stimulatory guanosine triphosphate (GTP)-binding protein, G_s (Spiegel, 1987; Birnbaumer *et al.*, 1991). Studies with dispersed whole ovarian cells (Magoffin and Erickson, 1982a) and with isolated TIC (Magoffin and Erickson, 1988a; Magoffin, 1989) demonstrated that hormone binding to the LH/hCG receptor stimulates an increase in cAMP production by the TIC. The intracellular receptor for cAMP in mammalian cells is PKA. Northern blot analysis has shown that the genes for both the type I and type II PKA isoenzymes are expressed in TIC (Hedin *et al.*, 1987). Thus, the principal elements of the cAMP/PKA pathway are present in the ovarian TIC.

There is convincing evidence that the actions of LH on TIC steroidogenesis are mediated through the cAMP/PKA pathway. In addition to the observation that LH stimulates the production of cAMP, the stimulatory effects of LH on steroidogenesis can be mimicked by treating TIC with cAMP analogs *in vitro* (Magoffin and Erickson, 1982a; Bogovich *et al.*, 1986; Richards *et al.*, 1986; Magoffin, 1989). By using certain combinations of cAMP analogs which can selectively activate either the type I or type II PKA, it was shown that both isoenzymes of PKA are involved in regulating steroidogenesis and expression of the P45017α and P450scc enzyme genes in TIC (Magoffin, 1989). Thus, the evidence strongly indicates that the stimulation of TIC differentiation and steroidogenesis by LH is mediated through the cAMP/PKA signal transduction pathway.

Activation of cAMP/PKA signal transduction pathway in undifferentiated TIC stimulates thecogenesis, which is characterized by a marked increase in the cells' capacity to synthesize androgens (Table 2). Most of the information regarding the sequential events which occur in thecogenesis are derived from studies of isolated TIC cultured *in vitro* under defined conditions. Initially there is a brief period of about 2 hours during which there is no significant effect on TIC steroidogenesis (Magoffin, 1989; Magoffin and Weitsman, 1993b). The molecular events that occur during this time are unknown, but somehow the TIC begin to acquire the capacity to produce progesterone. During the next approximately 18 hours the TIC become active progesterone-producing cells (Magoffin, 1989; Magoffin and Weitsman, 1993b) but do not produce androgens (Magoffin, 1989; Magoffin and Weitsman, 1993a). After approximately 20 hours of stimulation, the TIC rapidly increase their capacity to produce androgens while progesterone levels decline (Magoffin, 1989; Magoffin and Weitsman, 1993a,b), presumably because progesterone is metabolized to androgens. Androgen production can be maintained above basal levels for at least 30 days if the cAMP/PKA pathway is continuously activated (Erickson *et al.*, 1985), but if stimulation is interrupted, the rate of steroid hormone production rapidly declines to basal levels.

Recent studies in isolated TIC have demonstrated that activation of the cAMP/PKA pathway stimulates the expression of the steroidogenic enzyme genes required for androgen biosynthesis (Table 2). Stimulation with LH causes an increase in the levels of the mRNAs for the P450scc, 3β-HSD, and P45017α enzymes (Magoffin and Weitsman, 1993a,b,c). Both P450scc and P45017α mRNA levels are stimulated in the range of 15- to 20-fold whereas 3β-HSD mRNA increases only 3-fold. The reason that 3β-HSD mRNA increases relatively little compared with the P450 enzyme mRNAs is unknown but may reflect an increased stability of the message for 3β-HSD in TIC. In the cases of P450scc and P45017α, activation of the cAMP/PKA pathway with LH or direct activation of type I or type II PKA stimulates an increase in protein content in the TIC (Magoffin, 1989; Magoffin *et al.*, 1989, 1990). The increases in steroidogenic enzyme mRNAs occur approximately 2 hours prior to increases in steroidogenesis, indicating that translation of the mRNAs and processing and targeting of the newly synthesized proteins to the appropriate intracellular organelles take place without significant delays during the 2-hour lag period (Magoffin and Weitsman, 1993a,b,c).

TABLE 2 Effects of Endocrine and Autocrine/Paracrine Signal Transduction Mechanisms on TIC Steroidogenesis and Gene Expression

Signal transduction mechanism	Steroid production		Protein expression		Gene expression		
	Androgen	Progestin	$P450_{scc}$	$P450_{17\alpha}$	$P450_{scc}$	3β-HSD	$P450_{17\alpha}$
cAMP/PKA	+[a]	+	+	+	+	+	+
Tyrosine kinase							
IGF-I (−/+ LH)	0/+	0/+	+/+	0/+	+/+	+/+	0/+
EGF/TGF-α (−/+ LH)	0/−	0/+	Unknown	Unknown	+/+	0/+	0/−
Serine kinase (−/+ LH)	0/−	0/+	0/+	0/0	0/0	0/0	0/0
Protein kinase C (−/+ LH)	0/+	Unknown	Unknown	Unknown	Unknown	Unknown	Unknown

[a] +, stimulation; −, inhibition; 0, no effect.

Taken together, the evidence strongly suggests that activation of the cAMP/PKA pathway is sufficient to stimulate thecogenesis by increasing the transcription and translation of the genes for the enzymes required for androgen biosynthesis, P450scc, 3β-HSD, and P45017α. Once the TIC have expressed the steroidogenic enzymes, activation of the cAMP/PKA pathway also stimulates the rate-limiting step in steroid hormone biosynthesis, namely transfer of cholesterol across the mitochondrial membranes. As shown in the following sections, the cAMP/PKA regulatory pathway is unique by virtue of its ability to stimulate the expression of each of the steroidogenic enzyme genes and catalytic activity as well.

B. Tyrosine Kinase Pathway

Members of two families of the autocrine/paracrine regulators bind to cell surface receptors with intrinsic protein tyrosine kinase activity in the intracellular domain—namely insulin and IGF-I (Rechler and Nissley, 1986), and TGFα and EGF (Carpenter and Zendegui, 1986). The fact that IGF-I increases LH-stimulated androgen production by TIC whereas TGFα inhibits androgen production (Figure 3) raises the intriguing question: How do two regulators, which seemingly activate the same intracellular signaling pathway, cause opposite effects on androgen biosynthesis? Examination of the evidence demonstrates that these two regulators exert similar effects on certain biological responses in TIC and opposite effects on others.

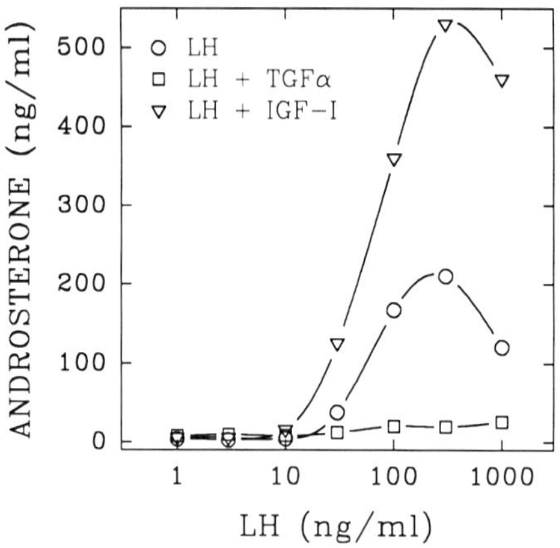

FIGURE 3 Differential regulation of LH-stimulated androgen biosynthesis in TIC by IGF-I and TGFα. Isolated TICs were cultured in the presence of increasing concentrations of LH with and without saturating concentrations of IGF-I (30 ng/ml) or TGFα (100 ng/ml) for 2 days. IGF-I augmented whereas TGFα inhibited LH-stimulated androgen production.

When the tyrosine kinase pathway is activated by IGF-I or EGF/TGFα *in vitro*, it does not affect steroidogenesis (Hernandez *et al.*, 1988; Cara and Rosenfield, 1988; Magoffin and Erickson, 1988b; Magoffin *et al.*, 1990; Erickson and Case, 1983). In the presence of activators of the cAMP/PKA pathway such as LH, IGF-I causes a synergistic stimulation of androgen biosynthesis (Hernandez *et al.*, 1988; Cara and Rosenfield, 1988; Magoffin and Erickson, 1988b; Magoffin *et al.*, 1990). This effect appears to be a general stimulation of steroidogenesis (Table 2) since both androgen and progesterone production are increased (Magoffin *et al.*, 1990). In contrast, activation of the tyrosine kinase pathway by EGF/TGFα in the presence of activators of cAMP/PKA causes a marked inhibition of androgen production (Erickson and Case, 1983) but merely decreases the sensitivity of the TIC to LH stimulation of progesterone production (Weitsman and Magoffin, unpublished observations). It is unclear how these differential effects are mediated, but it is reasonable to speculate that the substrate proteins of the receptor tyrosine kinases may be different or that there may be differences in tyrosine phosphatase activity.

Examination of the effects of the tyrosine kinase pathway on steroidogenic gene expression provides some interesting insight into differences in the biological responses to the IGF-I and TGFα receptor tyrosine kinases. Activation of the tyrosine kinase pathway by either IGF-I or TGFα alone causes increases in P450scc and 3β-HSD mRNA (Magoffin and Weitsman, 1993b,c; Weitsman and Magoffin, unpublished observation). The P450scc mRNA is translated into functional protein, but no increase in catalytic activity occurs because the enzyme is compartmentalized in the inner mitochondrial membrane. The tyrosine kinase pathway does not increase transfer of cholesterol substrate across the mitochondrial membranes (Magoffin *et al.*, 1990). It is not known whether the tyrosine kinase pathway alone stimulates the translation of 3β-HSD. There are no significant alterations to P45017α mRNA when the tyrosine kinase pathway is activated alone by either ligand (Magoffin and Weitsman, 1993a; Weitsman and Magoffin, unpublished observation). When there is a concomitant activation of the cAMP/PKA and tyrosine kinase pathways, P450scc mRNA and protein are elevated above the levels stimulated by either pathway alone (Magoffin *et al.*, 1990; Magoffin and Weitsman, 1993b). The expression of 3β-HSD mRNA is increased at submaximal concentrations but at saturating concentrations of LH and IGF-I, 3β-HSD mRNA levels are equivalent to the levels stimulated by either pathway alone (Magoffin and Weitsman, 1993a). As with steroidogenesis, a significant difference is observed with respect to P45017α expression (Table 2). IGF-I augments LH-stimulated P45017α mRNA and protein (Magoffin *et al.*, 1990; Magoffin and Weitsman, 1993a), whereas TGFα blocks the LH stimulation of P45017α mRNA (Weitsman and Magoffin, unpublished observation). Thus, it appears that the expression of P450scc and of 3β-HSD in TIC coordinately regulated in a similar manner by the tyrosine kinase pathway, whether activated by IGF-I or TGFα. P45017α expression is regulated in a markedly different manner, being stimulated by IGF-I and blocked by TGFα.

In summary, the tyrosine kinase pathway is capable of stimulating expression of the genes which enable undifferentiated TIC to become progestin-producing cells. It does not regulate the transfer of cholesterol across the mitochondrial membranes and consequently has no direct effect on the rate of steroid hormone biosynthesis. It is interesting to note that the expression of P450scc and 3β-HSD could be stimulated in interstitial cells which do not yet contain LH receptors. These actions may be physiologically important during the first stages of thecogenesis in preantral follicles when the TIC differentiate into progesterone-producing cells (Figure 1). Finally, tyrosine kinase-mediated signals may be important in fine tuning the differentiated state of the TIC. IGF-I could send a paracrine signal to the TIC that increased androgen is needed for aromatization, and TGFα could send an autocrine signal which could help prevent excessive androgen production in normal ovaries.

C. Serine/Threonine Kinase Pathway

The serine/threonine kinase signaling pathway activated by TGFβ or activin is clearly different from the cAMP/PKA pathway, even though PKA is a serine/threonine kinase. Unlike the stimulatory actions of the cAMP/PKA pathway and similar to the tyrosine kinase pathway, this serine/threonine kinase pathway has no effect on steroidogenesis (Table 2) when activated alone (Magoffin *et al.*, 1989; Hernandez *et al.*, 1990a). When the serine/threonine kinase pathway is activated concomitantly with the cAMP/PKA pathway, the stimulatory actions of the cAMP/PKA pathway on androgen biosynthesis are blocked (Magoffin *et al.*, 1989; Hernandez *et al.*, 1990a) but progesterone production is increased, probably as a result of the decreased metabolism of progesterone to androgen (Magoffin *et al.*, 1989). Thus, the evidence supports the conclusion that the TGFβ serine/threonine kinase mediates a specific inhibition of the P45017α enzyme activity (Hernandez *et al.*, 1990a).

When the effects of the TGFβ serine/threonine kinase pathway on steroidogenic enzyme gene expression in TIC are examined, a distinct pattern of responses is revealed. At a concentration of TGFβ which causes more than 80% inhibition of androgen production, there is no significant alteration of LH-stimulated P450scc or P45017α mRNA expression (Fournet and Magoffin, unpublished observation). There is also no significant alteration in the amount of P45017α protein in the TIC but there is an increase in LH-stimulated P450scc protein (Magoffin *et al.*, 1989). The available evidence supports the conclusion that the TGFβ serine/threonine kinase pathway exerts only minimal regulation of TIC gene expression, but is a potent and specific inhibitor of P45017α activity. The physiological role of TGFβ remains to be proven, but activation of the TGFβ serine/threonine kinase pathway could provide an interesting mechanism by which the TIC of a dominant follicle might suppress androgen biosynthesis of adjacent follicles and, by limiting the availability of androgen substrate in those adjacent follicles, inhibit E_2 production and cause them to become atretic. This pathway may also mediate an autocrine feedback loop which would limit excessive androgen production without suppressing the expression of the steroidogenic enzymes in the TIC.

D. Protein Kinase C Pathway

In vitro studies with phorbol esters indicated that protein kinase C might be important in regulating TIC androgen biosynthesis (Hofeditz *et al.*, 1988). Long-term treatment with phorbol ester caused a marked inhibition of LH-stimulated androgen production and a general suppression of steroidogenesis. In many cell types, long-term treatment with phorbol ester depletes protein kinase C in the cells, suggesting that in TIC, protein kinase C may be important for maintaining LH responsiveness. This concept is supported by the discovery that angiotensin II, which activates protein kinase C, sensitizes TIC to LH stimulation (Magoffin, 1991). Interestingly, angiotensin II does not alter basal or maximally stimulated androgen production, but at intermediate concentrations of LH increases the sensitivity of the TIC to LH stimulation (Table 2). The mechanism of this sensitization is unclear; however, there is evidence that activation of protein kinase C can decrease the inhibitory activity of the Gi protein on adenylate cyclase (Bell and Brunton, 1986), thereby releasing the inhibitory tone on the cyclase. Another intriguing observation is that protein kinase C can phosphorylate the catalytic subunit of adenylate cyclase and perhaps increase its catalytic activity (Yoshimasa *et al.*, 1987). Either one or both of these mechanisms could explain the effects of angiotensin II on TIC androgen biosynthesis.

The physiological role of the protein kinase C pathway in TIC is unknown. There is ample evidence that a complete renin–angiotensin system exists within the ovary and that the TIC contain angiotensin II receptors (Fernandez *et al.*, 1985; Culler *et al.*, 1986; Glorioso *et al.*, 1986; Husain *et al.*, 1987; Speth and Husain, 1988). The importance of this system is suggested by the discovery that *in vivo* administration of an angiotensin antagonist blocks ovulation in the rat (Pellicer *et al.*, 1988). Further experiments are necessary before we will be able to understand how this potential autocrine mechanism functions during follicle development.

IV. Conclusion

Because TIC play a central role in regulating ovarian homeostasis, an understanding of the mechanisms that control their differentiated functions is an important goal of research on the ovary. There is now substantial evidence that all of the major receptor–signal transduction systems, the G-protein-linked receptors coupled to protein kinase A and C pathways, catalytic tyrosine kinase and serine/threonine kinase receptors, and the nuclear steroid receptor family, function in the TIC in response to various endocrine and autocrine/paracrine ligands. It is important to note that these receptor signaling systems do not act independently, but interact with each other to regulate the biological responses of the TIC. How these systems interact in regulating TIC growth and differentiation, and how these interactions lead to the mechanisms that control selection and atresia in the pool of developing follicles need further research.

Acknowledgment

This research was supported by NICHD Research Grants HD-28154 (D.M.), HD-28953 (D.M.), HD-24585 (G.E.), and NIH PHS HD-29008 (G.E.).

References

Adashi, E. Y. (1990). The potential relevance of cytokines to ovarian physiology: the emerging role of resident ovarian cells of the white blood cell series. *Endocr. Rev.* 11, 454–464.

Adashi, E. Y., Resnick, C. E., D'Ercole, A. J., Svoboda, M. E., and Van Wyk, J. J. (1985). Insulin-like growth factors as intraovarian regulators of granulosa cell growth and function. *Endocr. Rev.* 6, 400–420.

Adashi, E. Y., Resnick, C. E., Svoboda, M. E., and Van Wyk, J. J. (1986a). Synergistic interactions of somatomedin-C with adenosine 3′, 5′ cyclic monophosphate dependent granulosa cell agonists. *Biol. Reprod.* 34, 81–88.

Adashi, E. Y., Resnick, C. E., Svoboda, M. E., Van Wyk, J. J., Hascall, V. C., and Yanagishita, M. (1986b). Independent and synergistic actions of somatomedin-C in the stimulation of proteoglycan biosynthesis by cultured rat granulosa cells. *Endocrinology* 118, 456–458.

Adashi, E. Y., Resnick, C. E., Croft, C. S., May, J. V., and Gospodarowicz, D. (1988). Basic fibroblast growth factor as a regulator of ovarian granulosa cell differentiation: a novel non-mitogenic role. *Mol. Cell. Endocrinol.* 55, 7–14.

Adashi, E. Y., Resnick, C. E., Hernandez, E. R., May, J. V., Purchio, A. F., and Twardzik, D. R. (1989a). Ovarian transforming growth factor-β (TGFβ): cellular sites and mechanisms of action. *Mol. Cell. Endocrinol.* 61, 247–256.

Adashi, E. Y., Resnick, C. E., Croft, C. S., and Payne, D. W. (1989b). Tumor necrosis factor α inhibits gonadotropin hormonal action in nontransformed ovarian granulosa cells. *J. Biol. Chem.* 264, 11591–11597.

Ahren, K., Hamburger, L., Hillensjo, T., Nilsson, L., and Nordenstrom, K. (1979). Control of steroidogenesis in the preovulatory rat follicle. *J. Steroid Biochem.* 11, 791–798.

Andreani, C. L., Payne, D. W., Parkman, J. N., Resnick, C. E., Hurwitz, A., and Adashi, E. Y. (1991). Cytokine mediated regulation of ovarian function: TNFα inhibits gonadotropin supported ovarian androgen biosynthesis. *J. Biol. Chem.* 266, 6761–6766.

Baird, A., Esch, F., Mormede, P., Ueno, N., Ling, N., Bohlen, P., Ying, S-Y., Wehrenberg, W.

B., and Guillemin, R. (1986). Molecular characterization of fibroblast growth factor: distribution and biological activities in various tissues. *Recent Prog. Horm. Res.* 42, 143–205.

Bell, J. D., and Brunton, L. L. (1986). Enhancement of adenylate cyclase activity in S49 lymphoma cells by phorbol esters. *J. Biol. Chem.* 261, 12036–12041.

Bendell, J. S., and Dorrington, J. (1988) Rat thecal/interstitial cells secrete transforming growth factor-β which can regulate granulosa cell growth. *Endocrinology* 121, 788–792.

Bicsak, T. A., Shimonaka, M., Malkowski, M., and Ling, N. (1990). Insulin-like growth factor-binding protein (IGF-BP) inhibition of granulosa cell function: effect on cyclic adenosine 3′, 5′-monophosphate deoxyribonucleic acid synthesis and comparison with the effect of an IGF-I antibody. *Endocrinology* 126, 2184–2189.

Birnbaumer, L., Perez-Reyes, E., Bertrand, P., Gudermann, T., Wei, X.-Y, Kim, H., Castellano, A., and Codina, J. (1991). Molecular diversity and function of G proteins and calcium channels. *Biol. Reprod.* 44, 207–224.

Bjersing, L. (1967). On the morphology and endocrine function of granulosa cells of ovarian follicles in corpora lutea. *Acta Endocrinol. (Suppl.) (Copenhagen)* 125, 5–23.

Bley, M. A., Simón, J. C., Estévez, A. G., de Asina, L. J., Barañao, J. L. (1992). Effect of follicle-stimulating hormone on insulin-like growth factor-I-stimulated rat granulosa deoxyribonucleic acid synthesis. *Endocrinology* 131, 1223–1229.

Bogovich, K., and Richards, J. S. (1982). Androgen biosynthesis in developing ovarian follicles: evidence that luteinizing hormone regulates theca 17α-hydroxylase and C17-20 lyase activity. *Endocrinology* 111, 1201–08.

Bogovich, K., Richards, J. S., and Reichert, L. E. (1981). Obligatory role of luteinizing hormone (LH) in the initiation of preovulatory follicular growth in the pregnant rat: specific effects of human chorionic gonadotropin and follicle-stimulating hormone on LH receptors and steroidogenesis in theca, granulosa, and luteal cells. *Endocrinology* 109, 860–867.

Bogovich, K., Scales, L. M., Higginbottom, E., Ewing, L. L., and Richards, J. S. (1986). Short term androgen production by rat ovarian follicles and long term steroidogenesis by thecal explants in culture. *Endocrinology* 118, 1379–1386.

Cara, J. F., and Rosenfield, R. L. (1988). Insulin-like growth factor I and insulin potentiate luteinizing hormone-induced androgen synthesis by rat ovarian theca-interstitial cells. *Endocrinology* 123, 733–739.

Carlsson, B., Carlsson, L., and Billig, H. (1989) Estrous cycle-dependent co-variation of insulin-like growth factor I (IGF-I) messenger ribonucleic acid and protein in the rat ovary. *Mol. Cell. Endocrinol.* 64, 271–275.

Carpenter, G., and Zendegui, J. G. (1986). Epidermal growth factor, its receptor, and related proteins. *Exp. Cell Res.* 164, 1–10.

Chabot, J-G., Arnaud, R. S., Walker, P., and Pelletier, G. (1986). Distribution of epidermal growth factor receptor in the rat ovary. *Mol. Cell. Endocrinol.* 44, 99–108

Culler, M. D., Tarlatzis, B. C., Lightman, A., Fernandez, L. A., DeCherney, A. H., Negro-Vilar, A., and Naftolin, F. (1986). Angiotensin II-like immunoreactivity in human ovarian follicular fluid. *J. Clin Endocrinol. Metab.* 62, 613–615.

Darbon, J. M., Oury, F., Laredo, J., and Bayard, F. (1989). Tumor necrosis factor α inhibits FSH induced differentiation in cultured granulosa cells. *Biochem. Biophys. Res. Commun.* 163, 1038–1043.

Davoren, J. B., Kasson, B. G., Li, C. H., and Hsueh, A. J. W. (1986). Specific insulin-like growth factor (IGF) I and II binding sites on rat granulosa cells: relation to IGF action. *Endocrinology* 119, 2155–2162.

Deanesly, R. (1972). Origins and development of interstitial tissue in ovaries of rabbit and guinea pig. *J. Anat.* 113, 251–260.

Dekel, N., and Sherizly, I. (1985). Epidermal growth factor induces maturation of rat follicle-enclosed oocytes. *Endocrinology* 116, 406–409.

DePaolo, L. V., Bicsak, T. A., Erickson, G. F., Shimasaki, S., and Ling, N. (1991). Follistatin and activin: a potential intrinsic regulatory system within diverse tissues. *Proc. Soc. Exp. Biol. Med.* 198, 500–512.

Dodson, W. C., and Schomberg, D. W. (1987). The effect of transforming growth factor-β on follicle stimulating hormone induced differentiation of cultured rat granulosa cells. *Endocrinology* 120, 512–516.

Doi, M., Igarashi, M., Hasegawa, Y., Eto, Y., Shibai, H., Muira, T., and Ibuki, Y. (1992). *In vivo* action of activin-A on pituitary-gonadal system. *Endocrinology* 130, 139–144.

Dorrington, J., Chuma, A. V., and Bendell, J. J. (1988). Transforming growth factor-β and FSH promote rat granulosa cell proliferation. *Endocrinology* 123, 353–359.

Emoto, N., and Baird, A. (1988). The effect of TNFα/cachectin on FSH induced aromatase activity in cultured rat granulosa cells. *Biochem. Biophys. Res. Commun.* 153, 792–798.

Erickson, G. F., and Case, E. (1983). Epidermal growth factor antagonizes ovarian theca-interstitial cytodifferentiation. *Mol. Cell. Endocrinol.* 31, 71–76.

Erickson, G. F., Magoffin, D. A., Dyer, C. A., and Hofeditz, C. (1985). The ovarian androgen producing cells: a review of structure/function relationships. *Endocr. Rev.* 6, 371–399.

Falck, B. (1959). Site of production of oestrogen in rat ovary as studied by micro-transplants. *Acta Physiol. Scand. (Suppl. 163)* 47, 1–101.

Feng, P., Catt, K. J., and Knecht, M. (1986). Transforming growth factor-β regulates the inhibitory action of epidermal growth factor during granulosa cell differentiation. *J. Biol. Chem.* 261, 14167–14170.

Feng, P., Catt, K. J., and Knecht, M. (1988). TGFβ stimulates meiotic maturation of the rat oocyte. *Endocrinology* 122, 181–186.

Fernandez, L. A., Tarlatzis, B. C., Rzasa, P. J., Caride, V. J., Laufer, N., Negro-Vilar, A. F., DeCherney, A. H., and Naftolin, F. (1985). Renin-like activity in ovarian follicular fluid *Fertil. Steril.* 44, 219–223.

Findlay, J. K., Xiao, S., and Shukovski, L. (1990). The role of inhibin-related peptides as intragonadal regulators. *Reprod. Fertil. Dev.* 2, 205–218.

Gangrade, B. K., Davis, J. S., and May, J. V. (1991). A novel mechanism for the induction of aromatase in ovarian cell *in vitro:* role of TGFα-induced protein tyrosine kinase. *Endocrinology* 129, 2790–2792.

Glorioso, N., Atlas, S. A., Laragh, J. H., Jewelewicz, R., and Sealey, J. E. (1986). Prorenin in high concentrations in human ovarian follicular fluid. *Science* 233, 1422–1424.

Gospodarowicz, D., Ferrara, N., Schweigerer, L., and Neufeld, G. (1987). Structural characterization and biological functions of fibroblast growth factor. *Endocr. Rev.* 8, 95–114.

Gottschall, P. E., Uchara, A., Hoffman, S. T., and Arimura, A. (1987). Interleukin-1 inhibits FSH differentiation of rat granulosa cells *in vitro*. *Biochem. Biophys. Res. Commun.* 149, 502–509.

Guraya, S. S., and Greenwald, G. S. (1968). A comparative histochemical study of interstitial tissue and follicular artresia in the mammalian ovary. *Anat. Rec.* 149, 411–417.

Hamburger, L., Hillensjo, T., and Ahren, K. (1978). Steroidogenesis in isolated cells of preovulatory rat follicles. *Endocrinology* 103, 771–777.

Hedin, L., McKnight, G. S., Lifka, J., Durica, J. M., and Richards, J. S. (1987). Tissue distribution and hormonal regulation of messenger ribonucleic acid for regulatory and catalytic subunits of adenosine 3′, 5′-monophosphate-dependent protein kinases during ovarian follicular development and luteinization in the rat. *Endocrinology* 120, 1928–1935.

Hernandez, E. R., Resnick, C. E., Svoboda, M. E., Van Wyk, J. J., Payne, D. W., and Adashi, E. Y. (1988). Somatomedin-C/insulin-like growth factor I as an enhancer of androgen biosynthesis by cultured rat ovarian cells. *Endocrinology* 122, 1603–1612.

Hernandez, E. R., Roberts, C. T., LeRoith, D., and Adashi, E. Y. (1989). Rat ovarian insulin-like growth factor (IGF-I) gene expression is granulosa cell-selective: 5′-untranslated mRNA variant representation and hormonal regulation. *Endocrinology* 125, 572–574.

Hernandez, E. R., Hurwitz, A., Payne, D. W., Dharmarajan, A. M., Purchio, A. F., and Adashi, E. Y. (1990a). TGFβ1 inhibits ovarian androgen production: gene expression, cellular localization, mechanism and sites of action. *Endocrinology* 127, 2804–2811.

Hernandez, E. R., Roberts, G., Hurwitz, A., LeRoith, D., and Adashi, E. Y. (1990b). Rat ovarian insulin-like growth factor II gene expression is theca interstitial cell-exclusive: hormonal regulation and receptor distribution. *Endocrinology* 127, 3249–3251.

Hernandez, E. R., Hurwitz, A., Botero, L., Ricciarelli, E., Werner, H., Roberts, L. T., LeRoith, D., and Adashi, E. Y. (1991). Insulin-like growth factor receptor gene expression in the rat ovary: divergent regulation of distinct receptor species. *Mol. Endocrinol.* 5, 1799–1805.

Hillensjo, T., Bauminger, S., and Ahren, K. (1976). Effect of luteinizing hormone on the pattern of steroid production by preovulatory follicles of pregnant mares serum gonadotropin-injected immature rats. *Endocrinology* 99, 996–1002.

Hirobe, S., He, W-W., Lee, M. M., and Donahoe, P. K. (1992). Mullerian inhibiting substance messenger ribonucleic acid expression in granulosa cells and sertoli cells coincide with their mitotic activity. *Endocrinology* 131, 854–862.

Hoage, T. R., and Cameron, I. L. (1976). Folliculogenesis in the ovary of the mature mouse: an autoradiographic study. *Anat. Rec.* 184, 699–709.

Hofeditz, C., Magoffin, D. A., Erickson, G. F. (1988). Evidence for protein kinase C regulation of ovarian theca-interstitial cell androgen biosynthesis. *Biol. Reprod.* 39, 873–881.

Hsueh, A. J. W., Dahl, K. D., Vaughan, J., Tucker, E., Rivier, J., Bardin, C. W., and Vale, W. (1987). Heterodimers and homodimers of inhibin subunits have different paracrine action in the modulation of luteinizing hormone-stimulated androgen biosynthesis. *Proc. Natl. Acad. Sci. USA* 84, 5082–5086.

Hurwitz, A., Hernandez, E. R., Resnick, C. E., Packman, J. N., Payne, D. W., and Adashi, E. Y. (1990). Basic fibroblast growth factor inhibits gonadotropin supported ovarian androgen biosynthesis: mechanism(s) and site(s) of action. *Endocrinology* 126, 3089–3095.

Hurwitz, A., Payne, D. W., Parkman, J. N., Andreani, C. L., Resnick, C. E., Hernandez, E. R., and Adashi, E. Y. (1991a). Cytokine-mediated regulation of ovarian function: IL-1 inhibits gonadotropin-induced androgen biosynthesis. *Endocrinology* 129, 1250–1256.

Hurwitz, A., Ricciarelli, E., Botero, L., Rohan, R. M., Hernandez, E. R., and Adashi, E. Y. (1991b). Endocrine and autocrine mediated regulation of rat ovarian (theca-interstitial) interleukin-1β gene expression: gonadotropin-dependent preovulatory acquisition. *Endocrinology* 129, 3427–3429.

Husain, A., Bumpus, F. M., De Silva, P., and Speth, R. (1987). Localization of angiotensin II receptors in ovarian follicles and the identification of angiotensin II in rat ovaries. *Proc. Natl. Acad. Sci. USA* 84, 2489–2493.

Kasson, B., and Gorospe, W. (1989). Effects of interleukins 1, 2, 3 on FSH induced differentiation of rat granulosa cells. *Mol. Cell. Endocrinol.* 62, 103–111.

Kingsbury, B. F. (1939). Atresia and the interstitial cells of the ovary. *Am. J. Anat.* 65, 309–336.

Knecht, M., and Catt, K. J. (1982). Modulation of cAMP mediated differentiation in ovarian granulosa cells by EGF and PDGF. *J. Biol. Chem.* 258, 2789–2799.

Knecht, M., Feng, P., and Catt, K. (1987). Bifunctional role of transforming growth factor-β during granulosa cell development. *Endocrinology* 120, 1243–1249.

Knecht, M., Feng, P., and Catt, K. J. (1989). Transforming growth factor-β: autocrine, paracrine, and endocrine effects in ovarian cells. *Sem. Reprod. Endocrinol.* 7, 12–20.

Kokia, E., Hurwitz, A., Ricciarelli, E., Tedeschi, C., Resnick, C. E., Mitchell, M. D., and Adashi, E. Y. (1992). Interleukin-1 stimulates ovarian prostaglandin biosynthesis: evidence for heterologous contact-dependent cell-cell interaction. *Endocrinology* 130, 3095–3097.

Kudlow, J. E., Korbin, M. S., Purchio, A. F., Twardzik, D. R., Hernandez, E. R., Asa, S. L., and Adashi, E. Y. (1987). Ovarian transforming growth factor-α gene expression. Immunohistochemical localization to the theca interstitial cells. *Endocrinology* 121, 1577–1579.

Koos, R. D., and Olson, C. E. (1989). Expression of basic fibroblast growth factor in the rat ovary. Detection of mRNA using reverse transcription-polymerase chain reaction amplification. *Mol. Endocrinol.* 3, 2041–2048.

Lane, C. E., and Davis, F. R. (1939). The ovary of the adult rat. I. Changes in growth of the follicle and in volume and mitotic activity of the granulosa and theca during the estrous cycle. *Anat. Rec.* 73, 429–442.

LaPolt, P. S., Piquette, G. N., Soto, D., Sinchich, C., and Hsueh, A. J. W. (1990a). Regulation of inhibin subunit messenger RNA levels by gonadotropins, growth factors, and gonadotropin-releasing hormone in cultured granulosa cells. *Endocrinology* 127, 823–831.

LaPolt, P. S., Yamato, M., Veljkovic, M., Sinchich, C., Tsafriri, A., and Hsueh, A. J. W. (1990b). Basic fibroblast growth factor induction of granulosa cell tissue-type plasminogen activator expression and oocyte maturation: potential role as a paracrine ovarian hormone. *Endocrinology* 127, 2357–2363.

Leung, P. C. K., and Armstrong, D. T. (1979). Estrogen treatment of immature rats inhibits ovarian androgen production *in vitro*. *Endocrinology* 104, 1411–1417.

Leung, P. C. K., Goff, A. K., Kennedy, T. G., and Armstrong, D. T. (1978). An intraovarian inhibitory action of estrogen on androgen production *in vivo*. *Biol. Reprod.* 19, 641–647.

Lieberman, M. E., Barnea, A., Bauminger, S., Tsafriri, A., Collins, W. P., and Lindner, H. R. (1975). LH effect on the pattern of steroidogenesis in cultured Graafian follicles of the rat: dependence on macromolecular synthesis. *Endocrinology* 96, 1533–1542.

Magoffin, D. A. (1989). Evidence that luteinizing hormone-stimulated differentiation of purified ovarian thecal-interstitial cells is mediated by both type I and type II adenosine 3′, 5′-monophosphate-dependent protein kinases. *Endocrinology* 125, 1464–1473.

Magoffin, D. A. (1991). Luteinizing hormone stimulation of ovarian thecal-interstitital cell differentiation: sensitizing effect of angiotensin II. *In* "Signaling Mechanisms and Gene Expression in the Ovary" (G. Gibori, ed.), pp. 417–422. Serono Symposia, Norwell, MA.

Magoffin, D. A. (1992). Regulation of differentiated functions in ovarian theca cells. *Sem. Reprod. Endocrinol.* 9, 321–331.

Magoffin, D A., and Erickson, G. F. (1981). Mechanism by which 17β-estradiol inhibits ovarian androgen production in the rat. *Endocrinology* 108, 962–969.

Magoffin, D. A., and Erickson, G. F. (1982a). Primary culture of differentiating ovarian androgen-producing cells in defined medium. *J. Biol. Chem.* 257, 4507–4513.

Magoffin, D. A., and Erickson, G. F. (1982b). Prolactin inhibition of luteinizing hormone-stimulated androgen synthesis in ovarian interstitial cells cultured in defined medium: mechanism of action. *Endocrinology* 111, 2001–2007.

Magoffin, D. A., and Erickson, G. F. (1982c). Direct inhibitory effect of estrogen on LH-stimulated androgen synthesis by ovarian cells cultured in defined medium. *Mol. Cell. Endocrinol.* 28, 81–89.

Magoffin, D. A., and Erickson, G. F. (1988a). Purification of ovarian theca-interstitial cells by density gradient centrifugation. *Endocrinology* 122, 2345–2347.

Magoffin, D. A., and Erickson, G. F. (1988b). An improved method for primary culture of ovarian androgen-producing cells in serum-free medium: effect of lipoproteins, insulin, and insulin-like growth factor I. *In Vitro Cell. Dev. Biol.* 24, 862–870.

Magoffin, D. A., and Weitsman, S. R. (1993a). Differentiation of ovarian theca-interstitial cells in vitro: regulation of 17α-hydroxylase messenger ribonucleic acid expression by luteinizing hormone and insulin-like growth factor-I. *Endocrinology* 132, 1945–1951.

Magoffin, D. A., and Weitsman, S. R. (1993b). Insulin-like growth factor-I stimulates the expression of 3β-hydroxysteroid dehydrogenase messenger ribonucleic acid in ovarian theca-interstitial cells. *Biol. Reprod.* 48, 1166–1173.

Magoffin, D. A., and Weitsman, S. R. (1993c). Effect of insulin-like growth factor-I on cholesterol side chain cleavage messenger ribonucleic acid expression in ovarian theca-intestitial cells stimulated to differentiate in vitro. *Mol. Cell. Endocrinol.* (in press)

Magoffin, D. A., Gancedo, B., and Erickson, G. F. (1989). Transforming growth factor-β promotes differentiation of ovarian theca-interstitial cells but inhibits androgen production. *Endocrinology* 125, 1951–1958.

Magoffin, D. A., Kurtz, K. M., and Erickson, G. F. (1990). Insulin-like growth factor-I selectively stimulates cholesterol side-chain cleavage expression in ovarian theca-interstitial cells. *Mol. Endocrinol.* 4, 489–496.

May, J. V., and Schomberg, D. W. (1989). The potential relevance of epidermal growth factor (EGF) and transforming growth factor-alpha (TGFα) to ovarian physiology. *Sem. Reprod. Endocrinol.* 7, 1–11.

McFarland, K. C., Sprengel, R., Phillips, H. S., Kohler, M., Rosemblit, N., Nikolics, K., Segaloff, D. L., and Seeburg, P. H. (1989). Lutropin-choriogonadotropin receptor: an unusual member of the G protein-coupled receptor family. *Science* 245, 494–499.

McNatty, K. P., Makris, A., DeGrazia, G., Osathanondh, R., and Ryan, K. J. (1979). The production of progesterone, androgens, and estrogens by granulosa cells, thecal tissue, and stromal tissue from human ovaries *in vitro*. *J. Clin. Endocrinol. Metab.* 49, 687–699.

McNeilly, A. S., Glasier, A., Jonassen, J., and Howie, P. W. (1982). Evidence for direct inhibition of ovarian function by prolactin. *J. Reprod. Fertil.* 65, 559–569.

Meunier, H., Cajander, S. B., Roberts, V. J., Rivier, C., Sawchenko, P. E., Hsueh, A. J. W., and Vale, W. (1988). Rapid changes in the expression of inhibin α-, βA-, and βB- subunits in ovarian cell types during the rat estrous cycle. *Mol. Endocrinol.* 2, 1352–1363.

Meunier, H., Roberts, V. J., Sawchenko, P. E., Cajander, S. B., Hsueh, A. J. W., and Vale, W. (1989). Periovulatory changes in the expression of inhibin α- βA-, and βB- subunits in hormonally induced immature female rats. *Mol. Endocrinol.* 3, 2062–2069.

Midgley, A. R. (1973). Autoradiographic analysis of gonadotropin binding to rat ovarian tissue sections. *Adv. Exp. Med. Biol.* 36, 365–378.

Miro, F., Smyth, C. D., and Hillier, S. G. (1991). Development-related effects of recombinant activin on steroid synthesis in rat granulosa cells. *Endocrinology* 129, 3388–3394.

Mondschein, J. S., and Schomberg, D. W. (1981). Growth factors modulate receptor induction in granulosa cell cultures. *Science* 211, 1179–1180.

Mulheron, G. W., and Schomberg, D. W. (1990). Rat granulosa cells express transforming growth factor β type 2 mRNA which is regulatable by FSH in vitro. *Endocrinology* 126, 1777–1779.

Mulheron, G. W., Danielpow, D., and Schomberg, D. W. (1991). Rat thecal/interstitial cells express transforming growth factor-β type I and 2, but only type 2 is regulated by gonadotropin in vitro. *Endocrinology* 129, 368–374.

Murphy, L. J., Bell, G. I., and Friesen, H. G. (1987). Tissue distribution of insulin-like growth factor I and II messenger ribonucleic acid in the adult rat. *Endocrinology* 120, 1279–1282.

Nakamura, T., Takio, K., Eto, Y., Shibai, H., Titani, K., and Sugino, H. (1990). Activin-binding protein from rat ovary is follistatin. *Science* 247, 836-838.

Nakatani, A., Shimasaki, S., Erickson, G. F., and Ling, N. (1991). Tissue specific expression of four insulin-like growth factor binding proteins (1, 2, 3 and 4) in the rat ovary. *Endocrinology* 129, 1521–1529.

Nelson, S. E., McLean, M. P., Jayatilak, P. G., and Gibori, G. (1992). Isolation, characterization, and culture of cell subpopulations forming the pregnant rat corpus luteum. *Endocrinology* 130, 954–966.

Oliver, J. E., Aitman, T. J., Powell, J. F., Wilson, C. A., and Clayton, R. N. (1989). Insulin-like growth factor I gene expression in the rat ovary is confined to the granulosa cells of developing follicles. *Endocrinology* 124, 2671–2679.

Oxberry, V. A., and Greenwald, G. S. (1982). An autoradiographic study of the binding of ^{125}I-labeled follicle-stimulating hormone, human chorionic gonadotropin, and prolactin to the hamster ovary throughout the estrous cycle. *Biol. Reprod.* 27, 505–516.

Parmer, T. G., Roberts, C. T., LeRoith, D., Adashi, E. Y., Khan, I., Solan, N., Nelson, S., Zilberstein, M., and Gibori, G. (1991). Expression, action and steroidal regulation of insulin-like growth factor-I (IGF-I) and IGF-I receptor in the rat corpus luteum: their differential role in the two cell populations forming the corpus luteum. *Endocrinology* 129, 2924–2932.

Pellicer, A., Palumbo, A., DeCherney, A. H., and Naftolin, F. (1988). Blockage of ovulation by an angiotensin antagonist. *Science* 240, 1660–1661.

Peng, X.-R., Hsueh, A. J. W., LaPolt, P. S., Bjersing, L., and Ny, T. (1991). Localization of luteinizing hormone receptor messenger ribonucleic acid expression in ovarian cell types during follicle development and ovulation. *Endocrinology* 129, 3200–3207.

Peters, H. (1969). The development of the mouse ovary from birth to maturity. *Acta Endocrinol.* 62, 98–116.

Rajaneimi, H., and Vanha-Perttula, T. (1972). Specific receptor for LH in the ovary: evidence by autoradiography and tissue fractionation. *Endocrinology* 90, 1–9.

Rechler, M. M., and Nissley, S. P. (1986). Insulin-like growth factor (IGF)/somatomedin receptor subtypes: structure, function, and relationships to insulin receptors and IGF carrier proteins. *Horm. Res.* 24, 152–159.

Ricciarelli, E., Hernandez, E. R., Tedeschi, C., Botero, L. F., Kokia, E., Rohan, R. M., Rosenfeld, R. G., Albiston, A. L., Herington, A. C., and Adashi, E. Y. (1992). Rat ovarian insulin-like growth factor binding protein-3: a growth hormone dependent theca-interstitial cell-derived antigonadotropin. *Endocrinology* 130, 3092–3094.

Rice, B. F., and Savard, K. (1966). Steroid hormone formation in the human ovary. IV. Ovarian stromal compartment; formation of radioactive steroids from acetate-1-^{14}C and action of gonadotropins. *J. Clin. Endocrinol. Metab.* 26, 593–609.

Richards, J. S., and Midgley, A. R. (1976). Protein hormonal action: a key to understanding ovarian follicular and luteal cell development. *Biol. Reprod.* 14, 82–94.

Richards, J. S., Jongssen, J. A., and Kersey, K. A. (1980). Evidence that changes in tonic luteinizing hormone secretion determine the growth of preovulatory follicles in the rat. *Endocrinology* 107, 641–648.

Richards, J. S., Hedin, L., and Caston, L. (1986). Differentiation of rat ovarian thecal cells: Evidence for functional luteinization. *Endocrinology* 118, 1660–1668.

Roby, K. F., and Terranova, P. F. (1988). Tumor necrosis factor alpha alters follicular steroidogenesis *in vitro*. *Endocrinology* 123, 2952–2954.

Roby, K. F., and Terranova, P. F. (1990). Effects of tumor necrosis factor-α *in vitro* on steroidogenesis of healthy and atretic follicles of the rat: theca as a target. *Endocrinology* 126, 2711–2718.

Saito, S., Nakamura, T., Titani, K., and Sugino, H. (1991). Production of activin-binding protein by rat granulosa cells *in vitro*. *Biochem. Biophys. Res. Commun.* 176, 413–422.

Sancho-Tello, M., Perez-Roger, I., Imakawa, K., Tilzer, L., and Terranova, P. F. (1992). Expression of TNFα in the rat ovary. *Endocrinology* 130, 1359–1364.

Selye, H., and Collip, J. B. (1933). Production of exclusively thecal luteinization and continuous oestrus with anterior pituitary like hormone. *Proc. Soc. Exp. Biol. Med.* 30, 647–649.

Shikone, T., Yamoto, M., and Nakano, R. (1992). Follicle stimulating hormone induces functional receptors for basic fibroblast growth factor in rat granulosa cells. *Endocrinology* 131, 1063–1068.

Shimasaki, S., and Ling, N. (1991). Identification and molecular characterization of insulin-like growth factor binding proteins (IGFBP-1, -2, -3, 4, -5, and -6). *Growth Factor Res.* 3, 243–246.

Shimasaki, S., Emoto, N., Mercado, M., Shibata, F., Cooksey, K., Baird, A., and Ling, N.

(1988). Complimentary DNA cloning and sequencing of rat ovarian basic fibroblast growth factor and tissue distribution study of its mRNA *Biochem. Biophys. Res. Commun.* 157, 256–263.

Shimasaki, S., Shimonaka, M., Ui, M., Inouye, S., Shibata, F., and Ling, N. (1990). Structural characterization of a follicle stimulating hormone action inhibitor in porcine ovarian follicular fluid. *J. Biol. Chem.* 265, 2198–2202.

Skinner, M. K., Keski-Oja, J., Osteen, K. G., and Moses, H. C. (1987a). Ovarian theca cells produce transforming growth factor β which can regulate granulosa cell growth. *Endocrinology* 121, 788–792.

Skinner, M. K., Lobb, D., and Dorrington, J. H. (1987b). Ovarian thecal interstitial cells produce an epidermal growth factor-like substance. *Endocrinology* 121, 1892–1899.

Speth, R. C., and Husain, A. (1988). Distribution of angiotensin-converting enzyme and angiotensin II-receptor binding sites in the rat ovary. *Biol. Reprod.* 38, 695–702.

Spiegel, A. M. (1987). Signal transduction by guanine nucleotide binding proteins. *Mol. Cell. Endocrinol.* 49, 1–16.

Stumpf, W. E. (1969). Nuclear concentration of [^{3}H]-estradiol in target tissues: dry-mount autoradiography of vagina, oviduct, ovary, testis, mammary tumor, liver and adrenal. *Endocrinology* 85, 31–37.

Tilly, J. L., LaPolt, P. S., and Hsueh, A. J. W. (1992). Hormonal regulation of FSH receptor mRNA levels in cultured rat granulosa cells. *Endocrinology* 130, 1296–1302.

Ueno, S., Takahashi, M., Manganaro, T. F., Ragin, R. C., and Donahoe, P. K. (1989a). Cellular localization of Mullerian inhibiting substance in the developing rat ovary. *Endocrinology* 124, 1000–1006.

Ueno, S., Kuroda, T., MacLaughlin, D. T., Ragin, R. C., Manganaro, T. F., and Donahoe, P. K. (1989b). Mullerian inhibiting substance in the adult rat ovary during various stages of the estrous cycle. *Endocrinology* 125, 1060–1066.

Ui, M., Shimonaka, M., Shimasaki, S., and Ling, N. (1989). An insulin-like growth factor-binding protein in ovarian follicular fluid blocks follicle stimulating hormone-stimulated steroid production by ovarian granulosa cells. *Endocrinology* 125, 912–916.

Weakly, B. S. (1966). Electron microscopy of the oocyte and granulosa cells in the developing ovarian follicles of the golden hamster. *J. Anat.* 100, 503–534.

Weitsman, S. R., and Magoffin, D. A. (1993). Transforming growth factor-α inhibits luteinizing hormone-stimulated androgen production by blocking 17α-hydroxylase/$C_{17\text{-}20}$ lyase activity in ovarian theca-interstitial cells. *Endocr. J.* 1, 109–115.

Woodruff, T. K., Lyon, R. J., Hansen, S. E., Rice, G. C., and Mather, J. P. (1990). Inhibin and activin regulate ovarian folliculogenesis. *Endocrinology* 127, 3196–3205.

Xiao, S., and Findlay, J. K. (1991). Interactions between activin and follicle-stimulating hormone-suppressing protein and their mechanisms of action on cultured rat granulosa cels. *Mol. Cell. Endocrinol.* 79, 99–107.

Xiao, S., Findlay, J. K., and Robertson, D. M. (1990). The effect of bovine activin and follicle-stimulating hormone (FSH) suppressing protein/follistatin on FSH-induced differentiation of rat granulosa cells *in vitro*. *Mol. Cell. Endocrinol.* 69, 1–8.

Ying, S.-Y. (1988). Inhibins, activins and follistatins: gonadal proteins modulating the secretion of FSH. *Endocr. Rev.* 9, 267–293.

Yoshimasa, T., Sibley, D. R., Bouvier, M., Lefkowitz, R. J., and Caron, M. G. (1987). Cross-talk between cellular signaling pathways suggested by phorbol-ester-induced adenylate cyclase phosphorylation. *Nature (London)* 327, 67–70.

Zhiwen, Z., Carson, R. A., Herington, A. C., Lee, V. W. K., and Burger, H. G. (1987). Follicle stimulating hormone and somatomedin-C stimulate inhibin production by rat granulosa cells in vitro. *Endocrinology* 120, 1633–1638.

Zhou, J., Chin, E., and Bondy, C. (1991). Cellular pattern of insulin-like growth factor-I (IGF-I) and IGF-I receptor gene expression in the developing and mature ovarian follicle. *Endocrinology* 129, 3281-3288.

3

Growth Factor-Mediated Cell–Cell Interactions in the Ovary

MICHAEL K. SKINNER & JEFF A. PARROTT

I. Introduction

The local production and integrated actions of various growth factors are required for the growth and development of all tissues. Growth factors mediate critical cell–cell interactions that control cell proliferation and organ development and the number of specific growth factors identified has increased dramatically. Growth factors often exist in families composed of unique gene products that have similarities in structure and function, but often differ in the way gene expression is controlled. The existence of multiple members in a growth factor family allows for unique sites of expression and more precise developmental and hormonal regulation of growth factor production. Table 1 contains a partial list of several of the major types of growth factors, including nomenclature and major functions attributed to them.

Ovarian development also requires growth factor-mediated cell–cell interactions as a general mechanism for controlling cellular proliferation. Most of the information available on growth factors and the ovary deals with follicular development and adult ovarian function. These are the primary stages of development discussed here. However, it is likely that many of the same cell–cell interactions and growth factors are also important for other stages of ovarian development (e.g., embryonic or prepubertal stages). Since research has been focused primarily on identifying the sites of production and actions of specific growth factors, this chapter emphasizes the growth factors potentially involved in ovarian cell–cell interac-

Molecular Biology of the Female Reproductive System

TABLE 1 Properties and Nomenclature of Several Common Growth Factors

Growth Factor		Approx. size (kDA)	Examples of physiological action
Insulin-like growth factor-I	IGF-I	7.5	Skeletal growth
Insulin-like growth factor-II	IGF-II	7.5	Fetal development
Epidermal growth factor	EGF	6	Tissue growth
Transforming growth factor α	TGFα	5	Tissue growth
Transforming growth factor β	TGFβ	25/dimer	Growth inhibition/tissue repair
Fibroblast growth factor	FGF	17	Angiogenesis/tissue growth
Vascular endothelial growth factor	VEGF	25–50/dimer	Angiogenesis/tissue growth
Nerve growth factor	NGF	13	Neuronal development
Interleukin-1	IL-1	17	Immune response/inflammation
Platelet-derived growth factor	PDGF	30/dimer	Tissue growth
Stem cell factor (c-kit ligand)	SCF	30	Tissue growth/fetal development

tions. The specific functions of these growth factors *in vivo* in large part remain to be elucidated.

II. Growth Factors in the Ovary

Ovarian physiology requires rapid and continuous regulation of the growth associated with folliculogenesis. Growth factor-mediated interactions among theca cells, granulosa cells, and the oocyte are needed to maintain ovarian function and oogenesis. Granulosa cells provide the cytoarchitectural support for the developing oocyte and also help to form the follicle and antrum. Theca cells surround and form the exterior wall of the follicle. The interactions between theca cells and granulosa cells provide an example of a mesenchymal (stromal)–epithelial cell interaction. The effects of this cellular interaction on oocyte development and the influence of the endocrine system on this cellular interaction are essential to female reproductive biology. The theca cells and granulosa cells of the preantral and small antral follicles must undergo extensive proliferation and functional differentiation prior to ovulation (Hsueh *et al.*, 1984). In most large animals, follicle size increases from millimeters to centimeters. The primary expansion in cell number takes place in both the granulosa and theca cell populations. In addition to the cell proliferation required during follicle development, follicles at various stages of development become atretic and cell growth is arrested. Therefore, the regulation of cell proliferation in the follicle requires both stimulatory and inhibitory growth factors.

TABLE 2 Growth Factors in the Ovary

Growth factor	Proposed site of synthesis	Proposed site of action	Proposed function[a]
IGF-1	Granulosa	Granulosa	+Growth/+differentiation
		Theca	+Growth/+differentiation
FGF	Granulosa	Granulosa	+Growth
		Endothelium	Angiogenesis
TGFβ	Theca	Granulosa	−Growth/+differentiation
	Granulosa	Theca	−Growth/+differentiation
TGFα	Theca	Granulosa	+Growth/−differentiation
		Theca	+growth
VEGF	Granulosa	Endothelium	Angiogenesis/+growth
NGF	Ovary	Neurons	Innervation
SCF	Granulosa	Oocyte	Oocyte maturation

[a]A plus sign denotes an increase and a minus sign indicates a decrease.

Follicle-stimulating hormone (FSH) and estrogen have been shown to stimulate proliferation of follicle cells *in vivo* (Goldenberg *et al.*, 1972; Louvet and Vaitukaitis, 1976; Richards, 1979). These hormones, however, have negligible effects on cell growth *in vitro*. This implies that *in vivo* hormone actions are most likely indirectly mediated by the local production of growth factors. The mechanisms and specific growth factors involved in the control of ovarian cell proliferation remain to be fully elucidated (Carson *et al.*, 1989). However, several major growth factors have been identified as being produced and/or acting in the ovary. These are summarized in Table 2.

A. Insulin-like Growth Factor

The insulin-like growth factors (IGF) derive their name from their structural similarity to insulin (Froesch *et al.*, 1985). IGF-I (previously termed somatomedin C) is considered essential for cellular replication and is a progression factor for cell growth and DNA synthesis. Production and secretion of IGF-I by the liver accounts for the high levels of IGF-I in serum and interstitial fluid (Daughaday and Rotwein, 1989). IGF-II is another member of this family and may act as a growth factor during fetal development.

IGF-I is produced by granulosa cells under the control of growth hormone (Davoren and Hsueh 1986), FSH, and estradiol (Hammond et al., 1985; Hsu and Hammond, 1987). The gene expression of IGF-1 has also been localized in granulosa cells (Hernandez *et al.*, 1989; Oliver *et al.*, 1989) but not theca cells. The receptors for IGF-1 have been localized to granulosa cells (Baranao and Hammond, 1984; Davoren *et al.*, 1986; Adashi *et al.*,

1988b) and are affected by the actions of FSH (Adashi *et al.*, 1986, 1988c,d). The regulation of IGF receptor gene expression has also been examined (Hernandez *et al.*, 1991, 1992). IGF-1 stimulates granulosa cell oxytocin production (Schams *et al.*, 1988), the P450 side chain cleavage enzyme (Veldhuis *et al.*, 1986), the aromatase gene expression (Steinkampf *et al.*, 1988), lipoprotein metabolism (Veldhuis *et al.*, 1987), adenylate cyclase activity (Adashi *et al.*, 1986), plasminogen activator production (Tilly and Johnson, 1990), and LH receptor induction (Adashi *et al.*, 1985b).

In addition to affecting cellular function, IGF stimulates the proliferation of bovine (Savion *et al.*, 1981) and porcine (Baranao and Hammond, 1984) granulosa cells, but not rat granulosa cells *in vitro* (Adashi *et al.*, 1984). Although theca cells do not appear to produce IGF-1 (Oliver *et al.*, 1989), IGF-II gene expression has been localized to theca cells and not granulosa cells (Hernandez *et al.*, 1990a). Theca cells also contain IGF receptors and respond to IGF through an alteration in steroidogenesis (Hernandez *et al.*, 1988) and LH receptor binding (Cara *et al.*, 1990). Potential interactions between granulosa cells and theca cells through the local production and action of IGF have been suggested (Adashi *et al.*, 1985a; Geisthovel *et al.*, 1990). The localization of IGF-1 expression to granulosa and not theca cells implies a potential IGF-1-mediated paracrine interaction between granulosa and theca cells. IGF-1 can also play a role as an autocrine factor for granulosa cells. A physiological parameter to consider, however, is the high circulatory levels of liver-derived IGF-1 (>100 ng/ml) available to both cell types. This is an additional source of IGF-1 that needs to be considered in understanding the importance of IGF-mediated cell–cell interactions.

IGF binding proteins (IGFBP) are produced by ovarian cell types and are present in the follicle. These binding proteins can reduce the effective concentration and modulate the actions of IGF. Several forms of IGFBP have been identified and are present in the ovary. Both granulosa cells and theca cells produce IGFBP(2) (Samares *et al.*, 1992; Ricciarelli *et al.*, 1991; Nakatani *et al.*, 1991), IGFBP(3) (Ricciarelli *et al.*, 1992; Samares *et al.*, 1992; Mondschein *et al.*, 1990), and IGFBP(4), which appears to be species-specific for cellular localization. IGFBP(4) and IGFBP(5) appear to be expressed primarily by granulosa cells (Erickson *et al.*, 1992a,b; Nakatani *et al.*, 1991). All the forms are present in follicular fluid at various stages of development. Although the specific function(s) of these IGFBPs remains to be elucidated, it has been postulated that they may inhibit or control the actions of IGF (Ui *et al.*, 1989).

B. Transforming Growth Factor-α/Epidermal Growth Factor

Transforming growth factor-α (TGFα) is one of the structurally related peptides belonging to the epidermal growth factor (EGF) family (Derynck, 1988; Carpenter and Cohen, 1990). Because they have a similar protein structure, these factors act at the same receptor to stimulate cell growth (Carpenter, 1987). TGFα is synthesized as a transmembrane precursor, which may activate EGF receptors on neighboring cells or be proteolytically cleaved to release mature peptide. TGFα appears to be produced by non-transformed cells, and may play an important role as a growth regulator in normal tissues.

Although EGF was not found to be produced in the ovary, an EGF-like substance was found in theca cells (Skinner *et al.*, 1987b) and was identified as TGFα (Skinner and Coffey, 1988; Kudlow *et al.*, 1987). Granulosa cells do not express TGFα (Skinner and Coffey, 1988; Lobb *et al.*, 1989) but have been shown to contain the EGF receptor (Mondschein and Schomberg, 1981; Chabot *et al.*, 1986; Feng *et al.*, 1986). EGF generally is inhibitory for adenylate cyclase (Dodson and Schomberg, 1987), LH receptor activity (Mondschein and Schomberg, 1981; Knecht and Catt, 1983a; May *et al.*, 1987), and FSH-induced aromatase activity (Hsueh *et al.*, 1981; May *et al.*, 1982). TGFα has inhibitory effects on granulosa cells (Adashi *et al.*, 1987) that are similar to those of EGF. The effects of TGFα, however, can vary among species (Gangrade *et al.*, 1991). Theca cells also contain the EGF receptor (Skinner and Coffey, 1988), and EGF/TGF-α influences theca cell steroidogenesis (Erickson and Case, 1983). One of the initial observations on growth regulation in the ovary was the ability of EGF to stimulate granulosa cell proliferation (Gospodarowicz *et al.*, 1977). The growth of theca cells can also be stimulated by TGFα/EGF (Skinner and Coffey, 1988). The actions of TGFα/EGF and the potential presence of EGF in the ovaries of various species can vary and remain to be fully elucidated.

Circulatory levels of EGF/TGFα are negligible, therefore, the factors must be produced locally within a specific organ. The ability of the theca cell to produce TGFα that can stimulate the growth of both granulosa and theca cells implies that TGFα may have an important role in promoting cell proliferation during follicle development. TGFα has been localized in developing ovarian follicles (Lobb *et al.*, 1989; Chegini and Williams, 1992). An interesting observation is that this appears to be a mesenchymal/stromal-controlled growth process (Skinner, 1990). Therefore, TGFα is postulated to mediate a paracrine interaction between theca and granulosa cells and an autocrine interaction between theca cells. The ability of hormones to influence TGFα production remains to be elucidated and may provide a mechanism through which hormones can regulate ovarian follicle cell growth.

C. Transforming Growth Factor-β

Transforming growth factor-β (TGFβ) is a multifunctional regulatory molecule that can stimulate or inhibit aspects of cellular growth and differentiation (Roberts and Sporn, 1988). TGFβ acts as a growth inhibitor by inhibiting the actions of growth factors such as EGF/TGFα. TGFβ can also promote cellular differentiation, extracellular matrix production, and chemotaxis. Different subtypes of TGFβ are produced as latent secreted precursors. Most cell types contain receptors for this ubiquitous factor.

Skinner et al. (1987a) and Gangrade and May (1990) demonstrated that ovarian theca cells express and produce TGFβ *in vivo,* immunocytochemical localization of TGFβ is primarily confined to the theca cell layer (Thompson et al., 1989). Several recent studies have confirmed the immunocytochemical localization of TGFβ isoforms in follicle cells (Chegini and Williams, 1992; Chegini and Flanders, 1992); this may vary with follicle development and hormone treatment (Roy *et al.*, 1992). Although freshly isolated bovine granulosa cells do not appear to express TGFβ (Skinner *et al.*, 1987a), cultured rat granulosa cells produce TGFβ which can be suppressed by FSH (Kim and Schomberg, 1989; Mulheron and Schomberg, 1990). Therefore, theca cells appear to be a predominant source of ovarian TGFβ, but granulosa cells also have the capacity to express TGFβ. The specific types of TGFβ expressed and their hormonal and developmental regulation remain to be fully elucidated (Mulheron *et al.*, 1991, 1992).

Local production of TGFβ allows it to act on various ovarian cell types. TGFβ stimulates a number of granulosa cell functions, including FSH-induced LH receptors (Knecht *et al.*, 1986; Dodson and Schomberg, 1987), EGF actions (Feng *et al.*, 1986), FSH-induced aromatase activity (Ying *et al.*, 1986; Hutchinson *et al.*, 1987), IGF-1 production (Mondschein *et al.*, 1988), and inhibin production (Zhiwen *et al.*, 1988). TGFβ can also influence theca cell function and steroidogenesis (Magoffin *et al.*, 1989; Caubo *et al.*, 1989; Hernandez *et al.*, 1990b), and oocyte maturation (Feng *et al.*, 1988; Tsafriri *et al.*, 1989). However, it is not known if TGFβ acts directly or indirectly on oocytes. In addition to effects on cellular differentiation, TGFβ can also influence ovarian cell growth. TGFβ has been shown to inhibit TGFα/EGF-induced bovine and porcine granulosa cell growth (Skinner *et al.*, 1987a; Mondschein *et al.*, 1988). Conflicting data were found with rat granulosa cells (Dorrington *et al.*, 1988). TGFβ can also inhibit TGFα/EGF-induced theca cell growth (Roberts and Skinner, 1991). Observations suggest that TGFβ may have an important role as a growth inhibitor in the ovary. The ability of TGFβ to inhibit cell growth allows for a more differentiated state of the cell that is reflected in the generally stimulatory effects of TGFβ on cell function. Therefore, the influence of TGFβ on cell function may be indirectly mediated through the inhibition of cellular proliferation. Growth inhibition may be important in preventing premature cell growth of the preantral follicle, arresting cell growth during atresia, and controlling cell growth during follicle cell expansion.

The local production and action of TGFβ within the developing ovarian follicle implies that TGFβ is an important paracrine and autocrine factor for ovarian cell–cell interactions. The hormonal regulation of TGFβ production

(Bendell and Dorrington, 1991) may also have a role in the endocrine regulation of ovary growth. The physiological significance of TGFβ in the ovarian follicle remains to be elucidated.

D. Fibroblast Growth Factor

Fibroblast growth factor (FGF) can influence aspects of both cellular growth and differentiation (Gospodarowicz *et al.*, 1987). Aside from growth stimulation, recent studies indicate that FGF may play a critical role in angiogenesis and tissue repair. FGF has many cellular targets and is important in many organ systems, including the ovary (Gospodarowicz and Ferrara, 1989).

Basic FGF is produced by granulosa cells in the developing embryonic gonad (Gonzalez *et al.*, 1990) and in the adult ovary (Neufeld *et al.*, 1987; Koos and Olson, 1989; Guthridge *et al.*, 1992). The angiogenic factor in the ovary and corpus luteum has been identified as FGF (Gospodarowicz *et al.*, 1985). FGF can act on granulosa cells to alter the steroidogenic capacity of the cell (Baird and Hsueh, 1986; Adashi *et al.*, 1988a), gonadotropin receptors (Mondschein and Schomberg, 1981), and plasminogen activator expression (LaPolt *et al.*, 1990; Tilly and Johnson, 1990). In addition to these effects on cell function, FGF can stimulate granulosa cell proliferation (Gospodarowicz *et al.*, 1977; Gospodarowicz and Bialecki, 1979), and may indirectly cause many of the effects observed on cell function. An additional role for FGF production in the ovary is to act as an angiogenic factor and promote vascularization of the developing follicle and corpus luteum.

E. Platelet-Derived Growth Factor and Vascular Endothelial Growth Factor

Platelet-derived growth factor (PDGF) is a common growth factor that allows cells to become competent to enter the growth cycle. PDGF acts on granulosa cells to enhance FSH-induced progesterone secretion, adenylate cyclase activity (Knecht and Catt, 1983b), plasminogen activator production (Tilly and Johnson, 1990), and LH receptor induction (Knecht and Catt, 1983b; Mondschein and Schomberg, 1984). The effects of PDGF on ovarian cell growth remain to be elucidated, but action as a potential competence factor for cell proliferation is a plausible activity. The local production of PDGF in the ovary remains to be examined. A factor that is structurally related to PDGF has been identified as vascular endothelial growth factor (VEGF) (Leung *et al.*, 1989; Conn *et al.*, 1990). This growth factor is expressed in the ovary, particularly in luteal tissue, and is postulated to have a role in angiogenesis of the follicle (Phillips *et al.*, 1990; Ravindranath *et al.*, 1992).

F. Nerve Growth Factor

Nerve growth factor (NGF) is another mitogen that may mediate intercellular interactions involving growth (Yanker and Shooter, 1982). NGF is important for the development and maintenance of sympathetic neurons in the peripheral nervous system and cholinergic neurons in the central nervous system. Its expression typically correlates with the amount of sympathetic innervation.

NGF is also expressed in the ovary (Lara *et al.*, 1990a) and is affected by ovarian innervation (Lara *et al.*, 1990a,b). NGF antibodies inhibit ovarian sympathetic innervation (Lara *et al.*, 1990c). The low-affinity NGF receptor is expressed in the ovary and is regulated upon ovulation (Dissen et al., 1991). NGF production in the ovary is therefore likely to have actions on ovarian function through sympathetic innervation.

G. Additional Growth Factors

Several additional types of growth factors act and/or are produced by ovarian cells. One such factor is stem cell factor (SCF)/c-kit ligand. Zsebo *et al.* (1990a) characterized stem cell factor and found that it influences stem cell growth and development. SCF acts at the c-kit tyrosine kinase receptor (Zsebo *et al.*, 1990b) and therefore is also referred to as the c-kit ligand. The c-kit tyrosine kinase receptor is expressed in the ovary by oocytes at various stages of development (Manova *et al.*, 1990; Horie *et al.*, 1991). The c-kit receptor expression appears to decline with the onset of meiotic maturation, suggesting a role for SCF in meiotic arrest (Horie *et al.*, 1991). Expression of SCF by follicular cells suggests a role for SCF to mediate cell–cell interaction with the oocyte. It is anticipated that many additional growth factors will be identified upon further investigation. For example, keratinocyte growth factor and hepatocyte growth factor are present in the ovary and appear to mediate theca-granulosa all interactions (unpublished observation).

III. Summary

It is apparent that a large number of growth factors are produced and act in the ovary. Most of the research to date has focused on how specific growth factors affect differentiated functions of gonadal cell types. Factors that promote cell growth generally have suppressive effects on differentiation and attenuate hormone responsiveness. Factors that inhibit growth generally enhance differentiation and increase hormone responsiveness. When considering the function and physiology of locally produced growth factors, a distinction needs to be made between growth and differentiation. A factor that promotes cell proliferation and the cell cycle will indirectly reduce the differentiated state of the cell, while the inverse is true of a factor that arrests cell proliferation and inhibits the cell cycle. Although specific growth factors may have a role in regulating differentiated functions, the possibility exists that many of the actions observed may be the indirect result of effects on cell growth. Therefore, the physiological importance of growth factor regulation of differentiated function remains to be elucidated. The control of cell growth, however, is a major function of specific growth factors. The integrated actions of various factors such as TGFα and TGFβ could provide an efficient mechanism for regulating the cell proliferation required in gonadal development. Further investigation of the developmental regulation of the expression, production, and action of individual growth factors will provide insight into the potential physiological roles for the various growth factors. Evidence obtained to date implies that growth factors will be critical regulatory agents involved in ovarian cell–cell communication.

The endocrine regulation of ovarian cell growth has been well documented *in vivo*. The actions of gonadotropins and reproductive steroids, however, are distinct from the pharmacology of most growth factors. In addition, these hormones often have negligible effects on cell proliferation *in vitro*. The possibility that the actions of reproductive hormones on gonadal cell growth are indirectly mediated through alterations in the expression of locally produced growth factors needs to be seriously considered. Current work suggests that hormones may regulate growth factor production. Further investigation of the hormonal regulation of the production and actions of growth factors will help elucidate the mechanisms involved in the endocrine regulation of gonadal development.

References

Adashi, E. Y., Resnick, C. E., Svoboda, M. E., and Van Wyk, J. J. (1984). A novel role for somatomedin-C in the cytodifferentiation of ovarian granulosa cells. *Endocrinology* 155, 1227–1229.

Adashi, E. Y., Resnick, C. E., D'Ercole, A. J., Svoboda, M. E., and Van Wyk, J. J. (1985a). Insulin-like growth factors as intraovarian regulators of granulosa cell growth and function. *Endocr. Rev.* 6, 400–420.

Adashi, E. Y., Resnick, C. E., Svoboda, M. E., and Van Wyk, J. J. (1985b). Somatomedin-C enhances induction of luteinizing hormone receptors by follicle-stimulating hormone in cultured rat granulosa cells. *Endocrinology* 116, 2369–2375.

Adashi, E. Y., Resnick, C. E., Svoboda, M. E., and Van Wyk, J. J. (1986). Somatomedin-C as an amplifier of follicle-stimulating hormone action: enhanced accumulation of adenosine 3′,5′-monophosphate. *Endocrinology* 118, 149–155.

Adashi, E. Y., Resnick, C. E., and Twardzik, D. R. (1987). Transforming growth factor-α attenuates the acquisition of aromatase activity by cultured rat granulosa cells. *J. Cell. Biochem.* 33, 1–13.

Adashi, E. Y., Resnick, C. E., Croft, C. S., May, J. V., and Gospodarowicz, D. (1988a). Basic fibroblast growth factor as a regulator of ovarian granulosa cell differentiation: a novel non-mitogenic role. *Mol. Cell. Endocrinol.* 55, 7–14.

Adashi, E. Y., Resnick, C. E., Hernandez, E. R., Svoboda, M. E., and Van Wyk, J. J. (1988b). Characterization and regulation of a specific cell membrane receptor for somatomedin-C/insulin-like growth factor I in cultured rat granulosa cells. *Endocrinology* 122, 194–201.

Adashi, E. Y., Resnick, C. E., Hernandez, E. R., Svoboda, M. E., and Van Wyk, J. J. (1988c). In vivo regulation of granulosa cell somatomedin-C/insulin-like growth factor 1 receptors. *Endocrinology* 122, 1383–1389.

Adashi, E. Y., Resnick, C. E., Hernandez, E. R., May, J. V., Knecht, M., Svoboda, M. E., and Van Wyk, J. J. (1988d). Insulin-like growth factor-1 as an amplifier of follicle-stimulating hormone action: studies on mechanism(s) and site(s) of action in cultured rat granulosa cells. *Endocrinology* 122, 1583–1591.

Baird, A., and Hsueh, A. J. W. (1986). Fibroblast growth factor as an intraovarian hormone: differential regulation of steroidogenesis by an angiogenic factor. *Regul. Pept.* 16, 243–250.

Baranao, J. L., and Hammond, J. M. (1984). Comparative effects of insulin and insulin-like growth factors on DNA synthesis and differentiation of porcine granulosa cells. *Biochem. Biophys. Res. Commun.* 124, 484–490.

Bendell, J. J., and Dorrington, J. (1991). Estradiol-17 beta stimulates DNA syntheses in rat granulosa cells: action mediated by transforming growth factor-beta. *Endocrinology* 128, 2663–2665.

Cara, J. F., Fan, J., Azzarello, J., and Rosenfield, R. L. (1990). Insulin-like-growth factor-I enhances luteinizing hormone binding to rat ovarian theca-interstitial cells. *J. Clin. Invest.* 86, 560–565.

Carpenter, G. (1987). Receptors for epidermal growth factor and other polypeptide mitogens. *Annu. Rev. Biochem.* 56, 881–914.

Carpenter, G., and Cohen, S. (1990). Epidermal growth factor. *J. Biol. Chem.* 265, 7709–7712.

Carson, R. S., Zhang, Z., Hutchinson, L. A., Herington, A. C., and Findlay, J. K. (1989). Growth factors in ovarian function. *J. Reprod. Fertil.* 85, 735–746.

Caubo, B., DeVinna, R. S., and Tonetta, S. A. (1989). Regulation of steroidogenesis in cultured porcine theca cells. *Endocrinology* 125, 321–326.

Chabot, J. G., St.-Arnaud, R., Walker, P., and Pelletier, G. (1986). Distribution of epidermal growth factor receptors in the rat ovary. *Mol. Cell. Endocrinol.* 44, 99–108.

Chegini, N., and Flanders, K. C. (1992). Presence of transforming growth factor-beta and their selective cellular localization in human ovarian tissue of various reproductive stages. *Endocrinology* 130, 1707–1715.

Chegini, N., and Williams, R. S. (1992). Immunocytochemical localization of transforming growth factors (TGFs) TGF-alpha and TGF-beta in human ovarian tissues. *J. Clin. Endocrinol. Metab.* 74, 973–980.

Conn, G., Bayne, M. L., Soderman, D. D., Kwok, P. W., Sullivan, K. A., Palisi, T. M., Hope, D. A., and Thomas, K. S. (1990). Amino acid and cDNA sequences of a vascular endothelial cell mitogen that is homologous to platelet-derived growth factor. *Proc. Natl. Acad. Sci. USA* 87, 2628–2632.

Daughaday, W. H., and Rotwein, P. (1989). Insulin-like growth factors I and II: peptide, messenger RNA and gene structure, serum, and tissue concentrations. *Endocr. Rev.* 10, 68–91.

Davoren, J. B., and Hsueh, A. J. W. (1986). Growth hormone increases ovarian levels of immunoreactive somatomedin C/insulin-like growth factor 1 in vivo. *Endocrinology* 119, 2155–2162.

Davoren, J. B., Kasson, B. G., Li, C. H., and Hsueh, A. J. W. (1986). Specific insulin-like growth factor (IGF) I- and II-binding sites on rat granulosa cells: relation to IGF action. *Endocrinology* 119, 2155–2162.

Derynck, R. (1988). Transforming growth factor alpha. *Cell* 54, 593–595.

Dissen, G. A., Hill, D. F., Costa, M. E., Ma, Y. J., and Ojeda, S. R. (1991). Nerve growth factor receptors in the peripubertal rat ovary. *Mol. Endocrinol.* 5, 1642–1650.

Dodson, W. C., and Schomberg, D. W. (1987). The effect of transforming growth factor-beta on follicle-stimulating hormone-induced differentiation of cultured rat granulosa. *Endocrinology* 120, 512–516.

Dorrington, J. H., Chuma, A. V., and Bendall, J. J. (1988). Transforming growth factor-β and follicle-stimulating hormone promote rat granulosa cell proliferation. *Endocrinology* 123, 352–359.

Erickson, G. F., and Case, E. (1983). Epidermal growth factor antagonizes ovarian theca-interstitial cytodifferentiation. *Mol. Cell. Endocrinol.* 31, 71–76.

Erickson, G. F., Nakatani, A., Ling, N., and Shimasaki, S. (1992a). Localization of insulin-like growth factor-binding protein-5 messenger ribonucleic acid in rat ovaries during the estrous cycle. *Endocrinology* 130, 1867–1878.

Erickson, G. F., Nakatani, A., Ling, N., and Shimasaki, S. (1992b). Cyclic changes in insulin-like growth factor-binding protein-4 messenger ribonucleic acid in the rat ovary. *Endocrinology* 130, 625–636.

Feng, P., Catt, K. J., and Knecht, M. (1986). Transforming growth factor beta regulates the inhibitory actions of epidermal growth factor during granulosa cell differentiation. *J. Biol. Chem.* 261, 14167–14170.

Feng, P., Catt, K. J., and Knecht, M. (1988). Transforming growth factor-β stimulated meiotic maturation of the rat oocyte. *Endocrinology* 122, 181–188.

Froesch, E. R., Schmid, C., Schwander, J., and Zapf, J. (1985). Actions of insulin-like growth factors. *Annu. Rev. Physiol.* 47, 443–467.

Gangrade, B. K., and May, J. V. (1990). The production of transforming growth factor-beta in the porcine ovary and its secretion *in vitro*. *Endocrinology* 127, 2372–2380.

Gangrade, B. K., Davis, J. S., and May, J. V. (1991). A novel mechanism for the induction of aromatase in ovarian cells *in vitro*: role of transforming growth factor alpha-induced protein tyrosine kinase. *Endocrinology* 129, 2790–2792.

Geisthovel, F., Moretti-Rojas, I., Rojas, F. J., and Asch, R. H. (1990). Insulin-like growth factors and thecal-granulosa-cell function. *Hum. Reprod.* 5, 785–799.

Goldenberg, R. L., Vaitukaitis, J. L., and Ross, G. T. (1972). Estrogen and follicle-stimulating hormone interactions on follicle growth in rats. *Endocrinology* 90, 1492–1497.

Gonzales, A. M., Buscaglia, M., Ong, M., and Baird, A. (1990). Distribution of basic fibroblast growth factor in the 18-day rat fetus: localization in the basement membranes of diverse tissues. *J. Cell. Biol.* 110, 753–765.

Gospodarowicz, D., III, and Birdwell, C. R. (1977). Effects of fibroblast and epidermal growth factors on ovarian cell proliferation *in vitro*. 1. Characterization of the response of granulosa cells to FGF and EGF. *Endocrinology* 100, 1108–1120.

Gospodarowicz, D., and Bialecki, H. (1979). Fibroblast and epidermal growth factors are mitogenic agents for cultured granulosa cells of rodent, porcine, and human origin. *Endocrinology* 104, 757–764.

Gospodarowicz, D., and Ferrara, N. (1989). Fibroblast growth factor and the control of pituitary and gonad development and function. *J. Steroid Biochem.* 32, 183–191.

Gospodarowicz, D., Cheng, J., Lui, G.-M., and Bohlen, P. (1985). Corpus luteum angiogenic factor is related to fibroblast growth factor. *Endocrinology* 117, 201–213.

Gospodarowicz, D., Ferrara, N., Schweigerer, L., and Neufeld, G. (1987). Structural characterization and biological functions of fibroblast growth factor. *Endocr. Rev.* 8, 95–114.

Guthridge, M., Schmitt, J., Bertolini, J., Cowling, J., Runting, A., Katsahambas, S., Drummond, A. E., and Hearn, M. T. (1992). Studies on basic fibroblast growth factor (FGF-beta) gene expression in the rat and pig ovary using in situ hybridization and quantitative reverse transcriptase-polymerase chain reaction techniques. *Exs. Angiogenesis* 61, 219–229.

Hammond, J. M., Baranoa, J. L. S., Skaleris, D., Knight, A. B., Ronanus, J. A., and Rechler, M. M. (1985). Production of insulin like growth factors by ovarian granulosa cells. *Endocrinology* 117, 2553–2555.

Hernandez, E. R., Resnick, C. E., Svoboda, M. E., Van Wyk, J. J., Payne, D. W., and Adashi, E. Y. (1988). Somatomedin-C/insulin-like growth factor I as an enhancer of androgen biosynthesis by cultured rat ovarian cells. *Endocrinology* 122, 1603–1612.

Hernandez, E. R., Roberts, C. T. Jr., LeRoith, D., and Adashi, E. Y. (1989). Rat ovarian insulin-like growth factor I (IGF-I) gene expression is granulosa cell-selective: 5′-untranslated mRNA variant representation and hormonal regulation. *Endocrinology* 125, 572–574.

Hernandez, E. R., Roberts, C. T. Jr., Hurwitz, A., LeRoith, D., and Adashi, E. Y. (1990a). Rat ovarian insulin-like growth factor II gene expression is theca-interstitial cell-exclusive: Hormonal regulation and receptor distribution. *Endocrinology* 127, 3249–3251.

Hernandez, E. R., Hurwitz, A., Payne, D. W., Dharmarajan, A. M., Purchio, A. F., and Adashi, E. Y. (1990b). Transforming growth factor-β1 inhibits ovarian androgen production: gene expression, cellular localization, mechanism(s), and site(s) of action. *Endocrinology* 127, 2804–2811.

Hernandez, E. R., Hurwitz, A., Botero, L., Ricciarelli, E., Werner, H., Roberts, C. T. Jr., LeRoith, D., and Adashi, E. Y. (1991). Insulin-like growth factor receptor gene expression in the rat ovary: divergent regulation of distinct receptor species. *Mol. Endocrinol.* 5, 1799–1805.

Hernandez, E. R., Hurwitz, A., Vera, A., Pellicer, A., Adashi, E. Y., LeRoith, D., and Roberts, C. R. Jr. (1992). Expression of the genes encoding the insulin-like growth factors and their receptors in the human ovary. *J. Clin. Endocrinol. Metab.* 74, 419–425.

Horie, K., Takakura, K., Taii, S., Narimoto, K., Noda, Y., Nishikawa, S., Nakayama, H., Fujita, J., and Mori, T. (1991). The expression of c-kit protein during oogenesis and early embryonic development. *Biol. Reprod.* 45, 547–552.

Hsu, C., and Hammond, J. M. (1987). Gonadotropins and estradiol stimulate immunoreactive insulin-like growth factor-I production by porcine cells *in vitro*. *Endocrinology* 120, 198–207.

Hsueh, A. J. W., Welsh, T. H., and Jones, P. B. C. (1981). Inhibition of ovarian and testicular steroidogenesis by epidermal growth factor. *Endocrinology* 108, 2002–2004.

Hsueh, A. J. W., Adashi, E. Y., Jones, P. B. C., and Welsh, T. H. Jr. (1984). Hormonal regulation of the differentiation of cultured ovarian granulosa cells. *Endocr. Rev.* 5, 76–127.

Hutchinson, L. A., Findlay, J. K., de Vos, F. L., and Robertson, D. M. (1987). Effects of bovine inhibin, transforming growth factor-beta and bovine Activin-A on granulosa cell differentiation. *Biochem. Biophys. Res. Commun.* 146, 1405–1412.

Kim, I.-C., and Schomberg, D. W. (1989). The production of transforming growth factor-β activity by the rat granulosa cells. *Endocrinology* 124, 1345–1351.

Knecht, M., and Catt, K. (1983a). Epidermal growth factor and gonadotropin-releasing hormone inhibit cyclic AMP-dependent luteinizing hormone receptor formation in ovarian granulosa cells. *J. Cell Biol.* 21, 209–217.

Knecht, M., and Catt, K. J. (1983b). Modulation of cAMP-mediated differentiation in ovarian granulosa cells by epidermal growth factor and platelet-derived growth factor. *J. Biol. Chem.* 258, 2789–2794.

Knecht, M., Feng, P., and Catt, K. J. (1986). Transforming growth factor-beta regulates the expression of luteinizing hormone receptors in ovarian granulosa cells. *Biochem. Biophys. Res. Commun.* 139, 800–807.

Koos, R. D., and Olson, C. E. (1989). Expression of basic fibroblast growth factor in the rat ovary: detection of mRNA using reverse transcription-polymerase chain reaction amplification. *Mol. Endocrinol.* 3, 2041–2048.

Kudlow, J. E., Kobrin, M. S., Purchio, A. F., Twardzik, D. R., Hernandez, E. R., Asa, S. L., and Adashi, E. Y. (1987). Ovarian transforming growth factor-a gene expression: immunohistochemical localization to the theca-interstitial cells. *Endocrinology* 121, 1577–1579.

LaPolt, P. S., Yamoto, M., Veljkovic, M., Sincich, C., Tor, N. Y., Tsafriri, A., and Hsueh,

A. J. W. (1990). Basic fibroblast growth factor induction of granulosa cell tissue-type plasminogen activator expression and oocyte maturation: potential role as a paracrine ovarian hormone. *Endocrinology* 127, 2357–2363.

Lara, H. E., Hill, D. F., Katz, K. H., and Ojeda, S. R. (1990a). The gene encoding nerve growth factor is expressed in the immature rat ovary: effect of denervation and hormonal treatment. *Endocrinology* 126, 357–363.

Lara, H. E., McDonald, J. K., Ahmed, C. E., and Ojeda, S. R. (1990b). Guanethidine-mediated destruction of ovarian sympathetic nerves disrupts ovarian development and function in rats. *Endocrinology* 127, 2199–2209.

Lara, H. E., McDonald, J. K., and Ojeda, S. R. (1990c). Involvement of nerve growth factor in female sexual development. *Endocrinology* 126, 364–375.

Leung, D. W., Cachianes, G., Kuang, W. J., Goeddel, D. V., and Ferrara, N. (1989). Vascular endothelial growth factor is a secreted angiogenic mitogen. *Science* 246, 1306–1309.

Lobb, D. K., Kobrin, M. S., Kudlow, J. E., and Dorrington, J. H. (1989). Transforming growth factor-alpha in the adult bovine ovary: identification in growing ovarian follicles. *Biol. Reprod.* 40, 1087–1093.

Louvet, J.-P., and Vaitukaitis, J. L. (1976). Induction of follicle-stimulating hormone (FSH) receptors in rat ovaries by estrogen priming. *Endocrinology* 99, 758–764.

Magoffin, D. A., Gancedo, B., and Erickson, G. F. (1989). Transforming growth factor-β promotes differentiation of ovarian thecal-interstitial cells but inhibits androgen production. *Endocrinology* 125, 1951–1958.

Manova, K., Nocka, K., Besmer, P., and Bachvarova, R. F. (1990). Gonadal expression of c-kit encoded at the W locus of the mouse. *Development* 110, 1057–1069.

May, J. V., Gilliam, F. R., Rein, M. S., Mondschein, J. S., and Schomberg, D. W. (1982). Growth and differentiated functions of porcine and rat granulosa cells following cryopreservation. *Biol. Reprod.* 27, 641–651.

May, J. V., Buck, P. A., and Schomberg, D. W. (1987). Epidermal growth factor enhances [125]iodo-follicle-stimulating hormone binding by cultured porcine granulosa cells. *Endocrinology* 120, 2413–2420.

Mondschein, J. S., and Schomberg, D. W. (1981). Growth factors modulate gonadotropin receptor induction in granulosa cell cultures. *Science* 211, 1179–1180.

Mondschein, J. S., and Schomberg, D. W. (1984). Effects of partially and highly purified platelet-derived growth factor preparations in luteinizing hormone receptor induction in granulosa cell cultures. *Biol. Reprod.* 30, 603–608.

Mondschein, J. S., Canning, S. F., and Hammond, J. M. (1988). Effects of transforming growth factor-β on the production of immunoreactive insulin-like growth factor I and progesterone and on [^{3}H]thymidine incorporation in porcine granulosa cell cultures. *Endocrinology* 123, 1970–1976.

Mondshein, J. S., Smith, S. A., and Hammond, J. M. (1990). Production of insulin-like growth factor binding proteins (IGFBPs) by porcine granulosa cells: identification of IGFBP-2 and -3 and regulation by hormones and growth factors. *Endocrinology* 127, 2298–2306.

Mulheron, G. W., and Schomberg, D. W. (1990). Rat granulosa cells express transforming growth factor-β type 2 messenger ribonucleic acid which is regulatable by follicle-stimulating hormone *in vitro*. *Endocrinology* 126: 1777–1779.

Mulheron, G. W., Danielpour, D., and Schomberg, D. W. (1991). Rat thecal/interstitial cells express transforming growth factor-β type 1 and 2, but only type 2 is regulated by gonadotropin *in vitro*. *Endocrinology* 129, 368–374.

Mulheron, G. W., Bossert, N. L., Lapp, J. A., Walmer, D. K., and Schomberg, D. W. (1992). Human granulosa-luteal and cumulus cells express transforming growth factors-beta type 1 and 2 mRNA. *J. Clin. Endocrinol. Metab.* 74, 458–460.

Nagy, F. (1972). Cell division kinetics and DNA synthesis in the immature Sertoli cells of the rat testis. *J. Reprod. Fertil.* 28, 389–395.

Nakatani, A., Shimasaki, S., Erickson, G. F., and Ling, N. (1991). Tissue-specific expression of four insulin-like growth factor-binding proteins (1,2,3, and 4) in the rat ovary. *Endocrinology* 129, 1521–1529.

Neufeld, G., Ferrara, N., Mitchell, R., Schweigerer, L., and Gospodarowicz, D. (1987). Granulosa cells produce basic fibroblast growth factor. *Endocrinology* 121, 597–603.

Oliver, J. E., Aitman, T. J., Powell, J. R., Wilson, C. A., and Clayton, R. N. (1989). Insulin-like growth factor I gene expression in the rat ovary is confined to the granulosa cells of developing follicles. *Endocrinology* 124, 2671–2679.

Phillips, H. S., Hains, J., Leung, D. W., and Ferrara, N. (1990). Vascular endothelial growth factor is expressed in rat corpus luteum. *Endocrinology* 127, 965–967.

Ravindranath, N., Little-Ihrig, L., Phillips, H. S., Ferrara, N., and Zeleznik, A. J. (1992). Vascular endothelial growth factor messenger ribonucleic acid expression in the primate ovary. *Endocrinology* 131, 254–260.

Ricciarelli, E., Hernandez, E. R., Hurwitz, A., Kokia, E., Rosenfeld, R. G., Schwander, J., and Adashi, E. Y. (1991). The ovarian expression of the antigonadotropic insulin-like growth factor binding protein-2 is theca-interstitial cell-selective: evidence for hormonal regulation. *Endocrinology* 129, 2266–2268.

Ricciarelli, E., Hernandez, E. R., Tedeschi, C., Botero, L. F., Kokia, E., Rohan, R. M., Rosenfeld, R. G., Albiston, A. L., Herington, A. C., and Adashi, E. Y. (1992). Rat ovarian insulin-like growth factor binding protein-3: a growth hormone-dependent theca-interstitial cell-derived antigonadotropin. *Endocrinology* 130, 3092–3094.

Richards, J. S. (1979). Hormonal control of ovarian follicular development: a 1978 perspective. *Rec. Prog. Horm. Res.* 35, 343–350.

Roberts, A. B., and Sporn, M. B. (1988). Transforming growth factor beta. *Adv. Cancer Res.* 51, 107–145.

Roberts, A. J., and Skinner, M. K. (1991). Transforming growth factors-a and -β differentially regulate growth and steroidogenesis of bovine thecal cells during antral follicle development. *Endocrinology* 129, 2041–2048.

Roy, S. K., Ogren, C., Roy, C., and Lu, B. (1992). Cell-type-specific localization of transforming growth factor-beta 2 and transforming growth factor-beta 1 in the hamster ovary: differential regulation by follicle-stimulating hormone and luteinizing hormone. *Biol. Reprod.* 46, 595–606.

Samares, S. E., Hagen, D. R., Shimasaki, S., Ling, N., and Hammond, J. M. (1992). Expression of insulin-like growth factor-binding protein-2 and -3 messenger ribonucleic acid in the porcine ovary: localization and physiological changes. *Endocrinology* 130, 2739–2744.

Savion, N., Lui, G., Laherty, R., and Gospodarowicz, D. (1981). Factors controlling proliferation and progesterone production by bovine granulosa cells in serum-free medium. *Endocrinology* 109, 409–420.

Schams, D., Koll, R., and Li, C. H. (1988). Insulin-like growth factor-I stimulates oxytocin and progesterone production by bovine granulosa cells in culture. *J. Endocrinol.* 116, 97–100.

Skinner, M. K. (1990). Mesenchymal (stromal)-epithelial cell interactions in the testis and ovary which regulate gonadal function. *Reprod. Fertil. Dev.* 2, 237–243.

Skinner, M. K., and Coffey, R. J. (1988). Regulation of ovarian cell growth through the local production of transforming growth factor-a by theca cells. *Endocrinology* 123, 2632–2638.

Skinner, M. K., Keski, Oja, J., Osteen, K. G., and Moses, H. L. (1987a). Ovarian thecal cells produce transforming growth factor-β which can regulate granulosa cell growth. *Endocrinology* 121, 786–792.

Skinner, M. K., Lobb, D., and Dorrington, J. H. (1987b). Ovarian thecal/interstitial cells produce an epidermal growth factor-like substance. *Endocrinology* 121, 1892–1899.

Smith, E. P., Svoboda, M. E., VanWyk, J. J., Kierszenbaum, A. L., and Tres, L. T. (1987).

Partial characterization of a somatomedin-like peptide from the medium of cultured rat Sertoli cells. *Endocrinology* 120, 186–193.

Steinkampf, M. P., Mendelson, C. R., and Simpson, E. R. (1988). Effects of epidermal growth factor and insulin-like growth factor I on the levels of mRNA encoding aromatase cytochrome P-450 of human ovarian granulosa cells. *Mol. Cell. Endocrinol.* 59, 93–99.

Thompson, N. L., Flanders, K. C., Smith, J. M., Ellingsworth, L. R., Roberts, A. B., and Sporn, M. B. (1989). Expression of transforming growth factor-β1 in specific cells and tissues of adult and neonatal mice. *J. Cell Biol.* 108, 661–669.

Tilly, J. L., and Johnson, A. L. (1990). Effect of several growth factors on plasminogen activator activity in granulosa and theca cells of the domestic hen. *Poultry Sci.* 69, 292–299.

Tsafriri, A., Vale, W., and Hsueh, A. J. W. (1989). Effects of transforming growth factors and inhibin-related proteins on rat preovulatory Graafian follicles in *vitro*. *Endocrinology* 125, 1857–1862.

Ui, M., Shimonaka, M., Shimasaki, S., and Ling, N. (1989). An insulin-like growth factor-binding protein in ovarian follicular fluid blocks follicle-stimulating hormone-stimulated steroid production by ovarian granulosa cells. *Endocrinology* 125, 912–916.

Veldhuis, J. D., Rodgers, R. J., Dee, A., and Simpson, E. R. (1986). The insulin-like growth factor, somatomedin C, induces the synthesis of cholesterol side-chain cleavage cytochrome P-450 and adrenodoxin in ovarian cells. *J. Biol. Chem.* 261, 2499–2502.

Veldhuis, J. D., Nestler, J. E., and Strauss, J. F. (1987). The insulin-like growth factor, somatomedin-C, modulates low density lipoprotein metabolism by swine granulosa cells. *Endocrinology* 121, 340–346.

Yanker, B. A., and Shooter, E. M. (1982). The biology and mechanism of action of nerve growth factor. *Annu. Rev. Biochem.* 51, 845–868.

Ying, S. T., Becker, A., Ling, N., Ueno, N., and Guillemin, R. (1986). Inhibin and beta type transforming growth factor (TGFβ) have opposite modulating effects on the follicle stimulating hormone (FSH)-induced aromatase activity of cultured rat granulosa cells. *Biochem. Biophys. Res. Commun.* 136, 969–975.

Zhiwen, Z., Findlay, J. K., Carson, R. S., Herington, A. C., and Burger, H. G. (1988). Transforming growth factor β enhances basal and FSH-stimulated inhibin production by rat granulosa cells in vitro. *Mol. Cell. Endocrinol.* 58, 161–166.

Zsebo, K. M., Wypych, J., McNiece, I. K., Lu, H. S., Smith, K. A., Karkare, S. B., Sachdev, R. K., Yuschenkoff, V. N., Birkett, N. C., Williams, L. R., Satyagal, V. N., Tung, W., Bosselman, R. A., Mendiaz, E. A., and Langley, K. E. (1990a). Identification, purification, and biological characterization of hematopoietic stem cell factor from buffalo rat liver conditioned medium. *Cell* 6, 195–201.

Zsebo, K. M., Williams, D. A., Geissler, E. N., Broudy, V. C., Martin, F. H., Atkins, H. L., Hsu, R., Birkett, N. C., Okino, K. H., Murdock, D. C., Jacobsen, F. W., Langley, K. E., Smith, K. A., Takeishi, T., Cattanach, B. M., Galli, S. J., and Snuggs, S. V. (1990b). Stem cell factor is encoded at the SI locus of the mouse and is the ligand for the c-kit tyrosine kinase receptor. *Cell* 63, 213–224.

4

Potential Relevance of Cytokines to Ovarian Physiology

ELI Y. ADASHI, EHUD KOKIA, & ARYE HURWITZ

I. Introduction

While most attention has been directed thus far at the somatic cellular components of the ovary, the potential role(s) and relative importance of the resident ovarian white blood cell and its cytokine product(s) have received relatively limited attention. Efforts are under way to reconcile traditional ovarian physiology with observations relevant to intraovarian components of the white blood cell series.

II. The Resident White Blood Cell

Unlike some gonadal compartments (e.g., the testicular seminiferous tubule), the ovary does not constitute an immunologically privileged site. Thus, resident ovarian (i.e., extravascular) mononuclear phagocytes (macrophages), lymphocytes, and polymorphonuclear granulocytes can be observed at various stages of the ovarian life cycle. For example, macrophages, but not other representatives of the white blood cell series, are now known to constitute a major cellular component of the interstitial (i.e., interfollicular) ovarian compartment (Hume *et al.*, 1984). In part, these macrophages are present within the ovarian stroma near perifollicular capillaries (Katabuchi *et al.*, 1989). Unfortunately, little is known at this time about this apparently permanent (i.e., noncyclic) presence. Nevertheless, it is tempting to speculate that interstitial macrophages could exert paracrine effects at the level of the adjacent somatic ovarian cells with which they have been observed to establish physical contact (Hume *et al.*, 1984).

With the exception of macrophages, few if any other white blood cells have been observed in the early phases of follicular development. However, this situation is precipitously altered as preovulatory events (or atresia) prompt massive ovarian infiltration by several representatives of the white blood cell series (Parr, 1974). Paving the way are mast cells (Parr, 1974; Jones *et al.*, 1980; Shinohara *et al.* 1987; Nakamura *et al.*, 1987; Schmidt *et al.*, 1988; Krishna *et al.*, 1986), which, relative to other cells, increase progressively during the latter portion of the follicular phase until they degranulate (and release histamine) in response to the proestrous LH surge (Krishna and Terranova, 1985). The resultant follicular hyperemia (Szego and Gitin, 1964; Wurtman, 1964; Krishna *et al.*, 1986; Murdoch *et al.*, 1983), coupled with chemotactic signaling (Seow *et al.*, 1988; Halterman and Murdoch, 1986), appears to not only play a critical role in subsequent luteal function (Cavender and Murdoch, 1988) but also heralds an orderly sequence of periovulatory events reminiscent of an acute inflammatory response (Espey, 1980).

Eosinophils and T lymphocytes, which migrate into the corpus luteum, are the next to arrive (Murdoch *et al.*, 1988). Interestingly, ovine corpora lutea have been shown to secrete a specific chemoattractant for eosinophils (Halterman and Murdoch, 1986; Murdoch, 1987), the precise nature of which has not been identified. However, increased numbers of eosinophils have been observed in corpora lutea of sheep treated with a luteolytic dose of prostaglandin $F_{2\alpha}$ ($PGF_2\alpha$) (Murdoch, 1987). Conceivably then, the chemoattractant could be $PGF_{2\alpha}$ inducible (Espey, 1980). Interestingly, infiltration and subsequent degranulation of these cells has been reported to occur prior to either functional or structural luteal regression (Murdoch, 1987). It has therefore been suggested that cytotoxins released by eosinophils could injure luteal cells, cause physical changes in membranes, and/or activate degradation of mRNA, thereby resulting in luteal demise (Murdoch *et al.*, 1988). Moreover, unmasking of cellular antigens by removal of sialic acid residues and/or exposure by a change in the physical state of the cellular membrane could lead to an antibody-mediated reaction involving complement, cellular destruction, and phagocytosis (Murdoch *et al.*, 1988).

Activated T cells are known to produce lymphokines that attract and activate monocytes/macrophages. Accordingly, T lymphocytes are followed in short sequence by phagocytic monocytes with ultrastructural, histochemical, and functional features indistinguishable from those observed in other body sites (Kirsch *et al.*, 1981). These attributes include intense staining for nonspecific esterase activity, phagocytosis of IgG-coated red blood cells through an Fc receptor, ingestion of latex beads, and trypsin resistance (as assessed by adherence to the surface of culture dishes). Furthermore, ovarian macrophages have been shown to specifically bind a monoclonal antibody against a plasma membrane protein of mononuclear phagocytes. Macrophages within the pericapillary spaces represent, in all likelihood, the dark stellate "K" cells that have been seen scattered among the granulosa cells in light micrographs (Gillim *et al.*, 1969). In this connection, it appears worth noting that cocultures of ovarian macrophages and luteal cells were seen to make discrete cell–cell contacts (Kirsch *et al.*, 1981). This *in vitro* phenomenon may in fact be analogous to the activity observed *in vivo* in which macrophages send out processes that contact several adjacent luteal cells, raising the possibility of cell–cell signaling (Adams and Hertig, 1969).

Yet other, phagocytically active macrophages, highly represented in older (regressing) but not young corpora lutea, appear to be involved in heterophagy of structurally damaged luteal cells that results from the disintegration of the corpus luteum (Bulmer, 1964). As the corpus luteum matures, these macrophages, are characterized by the presence of many electron-dense lysosomal granules and phagocytotic granules of variable configuration and structure throughout their cytoplasm (Lobel *et al.*, 1961). The invasion of T lymphocytes and macrophages into the corpus luteum is delayed by pregnancy and thus is not strictly a function of age of the corpus luteum. Indeed, only a few macrophages and T lymphocytes were observed in rabbit corpora lutea of Day 19 pregnancy, 6- to 8-fold higher numbers having been noted in corpora lutea of Day 19 pseudopregnancy. By the day of parturition, however, macrophages do begin to infiltrate the corpora lutea of pregnancy (Bagavandoss *et al.*, 1988).

Although macrophages have been observed as a prominent cell type within the corpus luteum of several species (Gillim *et al.*, 1969; Adams and Hertig, 1969; Bulmer, 1964; Lobel *et al.*, 1961; Bagavandoss *et al.*, 1988; VanLennep and Madden, 1965; Leavitt *et al.*, 1973; Stacy *et al.*, 1976; Paavola, 1977a,b, 1979a,b), little attention has been given to the properties and role of these cells (Adashi, 1989). The only role previously considered has been heterophagy of damaged luteal cells. That resident macrophages may function as potential *in situ* regulators of ovarian function has been suggested, at least in part, by finding numerous macrophages in regressing corpora luteum while only a few were observed in their young counterparts. Moreover, regressing (but not young) corpora lutea readily produced tumor necrosis factor α (TNFα) (Bagavandoss *et al.*, 1987, 1988) in response to *in vitro* stimulation with lipopolysaccharide (LPS). The latter antigen is the cell-wall component of gram-negative bacteria and is a potent stimulus for TNFα production by macrophages. Although the mechanism(s) by which macrophages influence ovarian function (rather than structure) remain uncertain, it is generally assumed that local paracrine secretion of regulatory molecule(s) (i.e., cytokines) may be at play.

The potential relevance of macrophages becomes immediately evident when consideration is given to their ability to elaborate several growth factors previously implicated as putative intraovarian regulators (Takemura and Werb, 1984; Rappolee *et al.*, 1988). Indeed, noncytokine secretory products of the macrophage such as basic fibroblast growth factor, transforming growth factor (TGFα and TGFβ1), have been shown to exert profound modulatory effects on the growth and functional development of the ovarian granulosa luteal cell. More recently, however, increasing attention has been paid to the potential role of macrophage-generated cytokines. The potential relevance of these cytokines to ovarian physiology is discussed in the following section.

III. The Interleukin-1 Example

Interleukin-1 (IL-1), a polypeptide cytokine (previously referred to as lymphocyte-activating factor) predominantly produced and secreted by activated macrophages, has been shown to possess a wide range of biological functions as well as playing a role as an immune mediator (Dinarello, 1988). Although the relevance of IL-1 to ovarian physiology remains uncertain, an increasing body of evidence supports such a possibility. First, measurable amounts of IL-1-like activity have been documented in both porcine (Takakura *et al.*, 1989) and human (Kanh *et al.*, 1988) follicular fluid. Second, *in vitro* studies at the level of the murine and porcine ovary revealed IL-1 to possess antigonadotropic (Gottschall *et al.*, 1987, 1989a,b; Kasson and Gorospe, 1989; Fukuoka *et al.*, 1988, 1989; Yasuda *et al.*, 1990) or steroidogenic (Nakamura *et al.*, 1990) properties contingent upon the experimental circumstances under study. Accordingly, it is tempting to speculate that locally derived IL-1, possibly originating from somatic ovarian cells or resident ovarian macrophages, may be the centerpiece of an intraovarian regulatory loop. Since IL-1 is an established mediator of inflammation (Dinarello, 1988) and since ovulation may constitute an inflammatory-like reaction (Espey, 1980), it is possible that IL-1 may play an intermediary role in the preovulatory developmental cascade and the terminal ovulatory process. Such speculation is supported by the recognition that IL-1 has been shown in multiple (nonovarian) tissues to promote several ovulation-associated phenomena such as prostaglandin biosynthesis, plasminogen activator production, glycosaminoglycan generation, collagenase activation, and vascular permeability enhancement (Dinarello, 1988).

To evaluate this hypothesis, we set out to evaluate rat ovarian IL-β gene expression, to determine its cellular localization, and to study its modulation by key endocrine and autocrine regulatory signals (Hurwitz *et al.*, 1991b). A solution hybridization/RNase protection assay was used in which rat ovarian total RNA (20 μg) was hybridized with a ^{32}P-labeled 272 base rat IL-1β antisense riboprobe. To assess rat ovarian IL-1β gene expression under *in vivo* circumstances, an established experimental model capable of simulating naturally occurring follicular maturation, ovulation, and corpus luteum formation was used. Specifically, a single subcutaneous injection of pregnant mare serum gonadotropin (PMSG) (15 IU/rat) was followed (48 hours) later by an ovulatory dose (15IU) of human chorionic gonadotropin (hCG). A faint protected fragment 222 bases long that corresponded to the IL-1B message was detectable in whole ovarian material prior to gonadotropic stimulation (Figure 1). Treatment with PMSG for 48 hours resulted in a modest, albeit measurable increase in the densitometrically quantified steady-state levels of the ovarian IL-1β message. Most striking, however, were the increases in the relative abundance of ovarian IL-1β transcripts following a 6 hour exposure to hCG, producing a 4- to 5-fold increase ($P < 0.05$) over the untreated state approximately 6 hours prior to projected follicular rupture. Subsequent evaluation of ovarian IL-1β transcripts, 24 and 48 hours following hCG administration, revealed significant ($P < 0.05$) decreases (relative to the 6-hour peak), to a level comparable to that seen at the conclusion of 48 hours of treatment with PMSG. Cellular localization studies revealed that the gonadotropin-dependent IL-1β mRNA excluded theca-interstitial cells.

To assess rat ovarian IL-1β gene expression under *in vitro* circumstances, we attempted to determine whether IL-1 itself may influence the relative level of its own message. Treatment of whole ovarian dispersates with rhIL-1β (10 ng/ml) for 4 and 24 hours resulted in a marked ($P < 0.05$) time-dependent increase (up to 12-fold) in the relative abundance of IL-1β transcripts when compared with untreated controls.

Taken together, these observations establish the rat ovarian theca-interstitial cell as a site of IL-1β gene expression, the preovulatory acquisition of which is gonadotropin dependent. In addition, our findings document the ability of IL-1β to exert a positive upregulatory effect on its own expression, an autocrine action potentially concerned with self-amplification. As such, this temporal (potentially self-amplifying) sequence of events provides strong indirect support for the proposal that intraovarian IL-1β may play an intermediary role in the preovulatory developmental cascade.

To delineate the scope of the human intraovarian IL-1 system, we explored the possibility that the genes encoding IL-1, its receptor, and its receptor antagonist are expressed at the level of the human ovary (Hurwitz *et al.*, 1992). Solution hybridization/RNase protection assays were used in which total RNA (20 μg) was hybridized with the corresponding ^{32}P-labeled riboprobes. No detectable IL-1 signal was evident in whole ovarian material from Days 4 or 12 of an unstimulated menstrual cycle (Figure 2). However, as in the epidermal carcinoma cell line A431 which was used as a positive control, the expected protected fragments corresponding to IL-1α (670 bp) and IL-1β (175 bp) transcripts (IL-1β $\gg$ IL-1α) were detected in

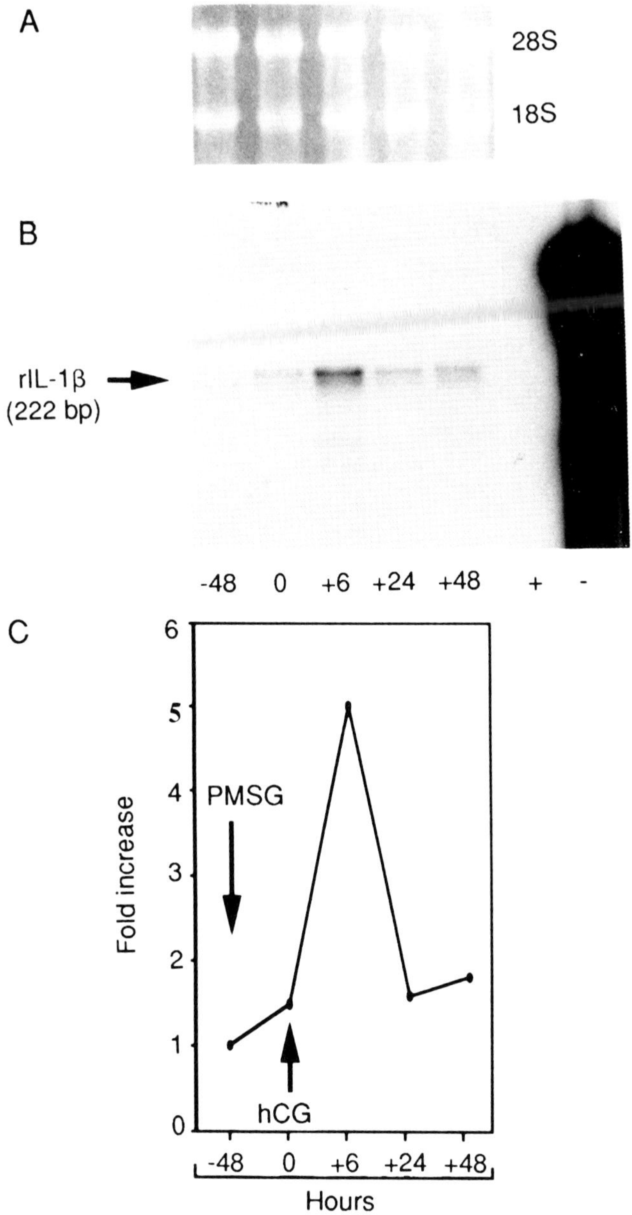

FIGURE 1 Rat ovarian Il-1β gene expression. A solution hybridization/RNase protection assay was performed using a rat IL-1β riboprobe. Total RNA (20 μg) was extracted from ovaries before (−48 hours) and after (0) PMSG (15 IU/rat) treatment as well as 6, 24, and 48 hours after hCG (15 IU/rat) was administered (0). The duration of autoradiographic exposure was 7 days. The plus and minus symbols designate riboprobe lanes treated with or without RNase, respectively. Reproduced with permission from Hurwitz *et al.* (1991b). Copyright © 1991 by Williams & Wilkins.

FIGURE 2 Human ovarian IL-1α and IL-1β gene expression. A solution hybridization/RNase protection assay was performed using the IL-1α and IL-1β riboprobes. Total RNA (20 μg) was extracted from preovulatory peripheral monocytes. M, pool of three patients; T, theca cells; G, granulosa cells; or K, A431 keratinocytes. Autoradiographic exposure was 7 days. The plus and minus signs designate riboprobe lanes treated with or without RNase, respectively. The data reflect a representative experiment, qualitatively comparable results having been obtained in two additional experiments using distinct specimens. Reproduced with permission from Hurwitz *et al.* (1992).

preovulatory follicular aspirates secured in the course of a gonadotropin-stimulated cycle.

Given that preovulatory follicular aspirates may constitute a complex mix of somatic ovarian and circulating cellular elements, we undertook to further identify the cell type responsible for IL-1 gene expression. In this connection, preovulatory peripheral monocytes obtained at the time of oocyte retrieval proved negative, thereby effectively eliminating the possibility that contaminating peripheral monocytes contribute to IL-1 transcripts detected in preovulatory follicular aspirates. Moreover, IL-1β transcripts were detected in cultured forskolin (25 μ*M*)-treated (macrophage-poor) granulosa (but not theca-interstitial) cells, suggesting (but not conclusively proving) that the granulosa cell may be a site of IL-1β gene expression. That this indeed may be the case was further documented by observing IL-1β transcripts in macrophage-depleted follicular aspirates prepared by magnetically driven immune sorting and validated by flow cytometry analysis. A single protected fragment (projected to be 477 bases long) corresponding to type I IL-1 receptor transcripts was detected in whole ovaries from Days 4 and 12 of an unstimulated menstrual cycle, in preovulatory follicular aspirates, and in term placenta used as a positive control. Type I IL-1 receptor transcripts were also detected in cultured granulosa and theca-interstitial cell preparations, but not in preovulatory peripheral monocytic cells obtained at the time of oocyte retrieval.

Treatment of cultured granulosa or theca-interstitial cells with forskolin (25 μM) resulted in a 2- to 3-fold increase in the steady-state levels of type I IL-1 receptor transcripts. A single protected fragment (147 bases long) corresponding to IL-1 receptor antagonist transcripts was detected in whole ovarian material from Day 4 of an unstimulated menstrual cycle as well as in macrophage-free preovulatory follicular aspirates. No detectable signal was noted in granulosa or theca-interstitial cells cultured in the absence or presence of forskolin.

Taken together, these findings reveal the existence of a complete, highly compartmentalized, hormonally dependent intraovarian IL-1 system replete with ligands, receptor, and receptor antagonist. The apparent midcycle induction of ovarian IL-1 gene expression and the reported ability of IL-1 to promote a host of ovulation-associated phenomena (in multiple nonovarian tissues) gives rise to the speculation that locally derived IL-1 may be the centerpiece of an intraovarian regulatory loop concerned with the genesis and maintenance of the preovulatory cascade of follicular events.

We also examined the possibility that the theca-intestitial (androgen-producing) cell may also be a site of IL-1 reception and action (Hurwitz *et al.*, 1992). The basal accumulation of androsterone, the major androgenic steroid synthesized by whole ovarian dispersates from immature rats, in the presence of insulin (1 μg/ml), increased 8- to 9-fold after treatment with hCG (1 ng/ml). AlthoughIL-1α or IL-1β (10 ng/ml) by themselves were without effect on basal androsterone accumulation, both cytokines (IL-1β $<$ IL-1α) inhibited hCG hormonal action (in the presence of insulin) in a dose-dependent manner, the maximal inhibitory effect being 75%.

Similar results were obtained when highly purified theca-interstitial cells derived from the same animal model were used, suggesting that IL-1-attenuated androgen biosynthesis is due, at least in part, to IL-1 acting directly at the level of the theca-interstitial cell. The IL-1 effect proved relatively specific since all other known interleukins (IL-1, IL-3, IL-4, IL-5, and IL-6) were without effect. Moreover, IL-1β action was effectively immunoneutralized when concurrently applied with anti-IL-1β (but not nonimmune) sera. Significantly, the antigonadotropic action of IL-1 could not be accounted for by a decrease in the viable cell mass. Tracer studies with radiolabeled steroid substrates suggested that IL-1-attenuated accumulation of ovarian androsterone is due, if only in part, to inhibition of

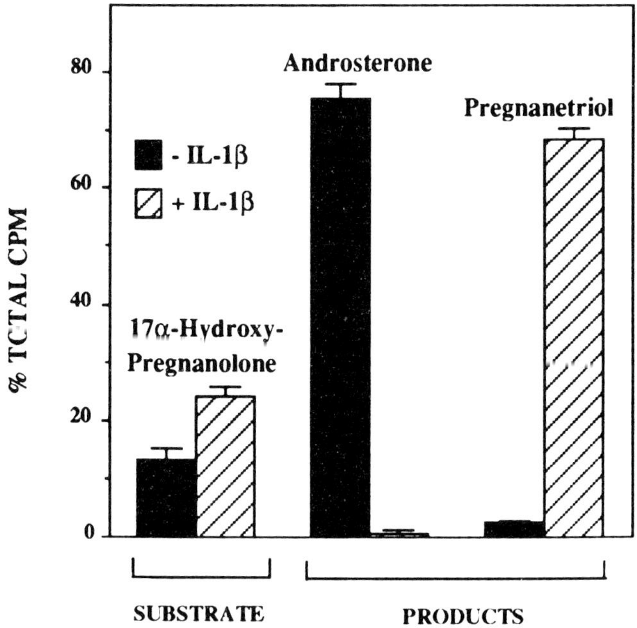

FIGURE 3 Effect of IL-1 on the steroidogenic pathway of theca-interstitial cells: Cellular labeling with a 17:20 lyase substrate, [^{3}H]17α-hydroxypregnanolone. Highly purified theca-interstitial cells were initially cultured for 72 hours in the presence of hCG (1 ng/ml), with or without IL-1β (10 ng/ml). Cells were then washed, the hormones replaced, and [^{3}H]17α-hydroxypregnanolone (10^5cpm/culture) added. After 24 hours, media were collected, extracted, fractionated by HPLC, and the radiolabeled steroid peaks identified by coelution with known standards. Each bar represents the means ± SE of four measurements. Reproduced with permission from Hurwitz *et al.* (1991a). Copyright © 1991 by Williams & Wilkins.

transformations catalyzed by (theca-interstitial) 17α-hydroxylase/17:20 lyase, stimulation of theca-interstitial (or granulosa) 20α-hydroxysteroid dehydrogenase-mediated conversions, or both (Figure 3).

Taken together, these findings indicate that relatively low concentrations of IL-1, possibly originating from somatic ovarian cells or resident ovarian macrophages, are capable of inhibiting gonadotropin-supported androgen production. As such, these and previous observations suggest that intraovarian IL-1 may play a dual regulatory role in the developing ovarian follicle by targeting both the granulosa and theca-interstitial cells as its sites of action.

IV. The Tumor Necrosis Factor-α Example

The potential ovarian relevance of another macrophage product, tumor necrosis factor, has also been explored (Bagavandoss *et al.*, 1987, 1988;

Emoto and Baird, 1988; Roby and Terranova, 1988; Adashi *et al.*, 1989, 1990; Sancho-Tello and Terranova, 1991; Veldhuis *et al.*, 1991; Zolti *et al.*, 1990; Darbon *et al.*, 1989). TNFα, a 157-amino acid polypeptide, was originally named for its oncolytic activity as displayed in the serum of BCG-immunized, endotoxin-challenged mice (Pennica *et al.*, 1984; Old, 1985; Aggarwal and Kohr, 1985; Beutler and Cerami, 1986, 1987). Indeed, TNFα proved capable of inducing tumor necrosis *in vivo* and of exerting nonspecies-specific cytolytic or cytostatic effects on a broad range of transformed cell lines *in vitro*. Although TNFα was initially thought to be tumor-selective, it has become clear that certain nontumor cells possess TNFα receptors and that TNFα may be a regulatory monokine with pleiotropic noncytotoxic activities in addition to its antitumor properties. Most importantly, TNFα has been shown to be involved in the differentiation of a variety of cell types.

At the level of the ovary, TNFα is capable of attenuating the differentiation of cultured granulosa cells from immature rats through virtual neutralization of the action of follicle-stimulating hormone (FSH) at a site(s) proximal but not distal to cyclic adenosine monophosphate (cAMP) generation (Emoto and Baird, 1988). In other studies, TNFα has been found to effect complex dose-dependent alterations in the elaboration of progesterone and androstenedione, but not estrogen, by explanted preovulatory follicles of murine origin (Roby and Terranova, 1988; Sancho-Tello and Terranova, 1991).

TNFα may also regulate the function of the androgen-producing theca-interstitial cell (Andreani *et al.*, 1991). The basal accumulation of androsterone (the major androgenic steroid) synthesized by whole ovarian dispersates from immature rats remained unchanged following treatment with TNFα (30 ng/ml) alone. In contrast, concurrent treatment with increasing concentrations of TNFα (0.03–30 ng/ml) yielded dose-dependent inhibition of the human chorionic gonadotropin (1 ng/ml)-stimulated accumulation of androsterone (Figure 4). This reversible and immunoneutralizable effect of TNFα was characterized by a minimal effective dose of 0.1 ng/ml, a median inhibitory dose of 0.9 ng/ml, a maximal inhibitory effect of 90%, and a minimal time requirement of ≤48 hours. Comparable results were obtained when using highly purified theca-interstitial cells, thereby indicating that TNFα is capable of exerting a direct inhibitory effect at the level of the ovarian androgen-producing cell. TNFα action was not accounted for by alterations in the plated viable cell mass. Instead, treatment with TNFα significantly inhibited the hCG-supported accumulation of cAMP, the putative second messenger of gonadotropin hormonal action. TNFα action at sites distal to cAMP generation was associated with stimulation of 20α-hydroxysteroid dehydrogenase activity, inhibition of 17α-hydroxylase/17:20 lyase activity, or both.

These findings indicate that TNFα, acting at relatively low concentra-

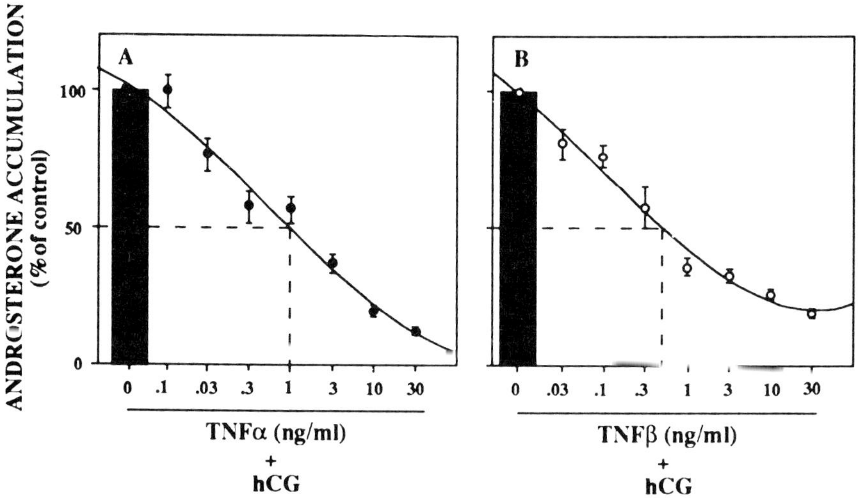

FIGURE 4 Effect of treatment with TNFα and TNFβ on hCG-stimulated androsterone accumulation by cultured whole ovarian dispersates. Whole ovarian dispersates (2.5×10^4 viable cells/culture) obtained from immature intact rats were cultured for 96 hours under serum-free conditions in the presence of hCG (1 ng/ml), with varying concentrations (0.03–30 ng/ml) of TNFα (A) and TNFβ (B). Each data point represents the means ± SE of three separate experiments (10–12 assays total at each point). Reproduced with permission from Andreani *et al.* (1991).

tions, is capable of inhibiting gonadotropin-supported ovarian androgen biosynthesis by selectively modulating the activity of relevant key steoridogenic enzymes. As such, these observations suggest that the theca-interstitial cell is a site of TNFα reception and action, and that TNFα, possibly originating in ovarian macrophages, may help regulate ovarian androgen production, an effect due in part to inhibition of the activity of the key steroidogenic enzymes, 17α-hydroxylase/17:20 lyase.

The role of TNFα was also examined in cultured granulosa cells from immature rats by monitoring their differentiation at the level of progestin biosynthesis (Adashi *et al.*, 1989, 1990). Whereas treatment with FSH (100 ng/ml) resulted in a substantial increase in the accumulation of (radioimmunoassayable) progesterone, concurrent exposure to TNFα (10 ng/ml) proved inhibitory, with the TNFα effect being characterized by an apparent median inhibitory dose of 1.1 ± 0.3 ng/ml and a minimal time requirement of ≤48 hours. Qualitatively similar results were obtained using "luteinized" (i.e., hCG-treated/FSH-primed granulosa cells) cells.

Quantification of the more distal metabolite 20α-hydroxy-4-pregnen-3-one (20α-DHP) in HPLC—fractionated media by enzyme-based (nonradiometric) assays revealed that TNFα also decreased the overall (FSH-

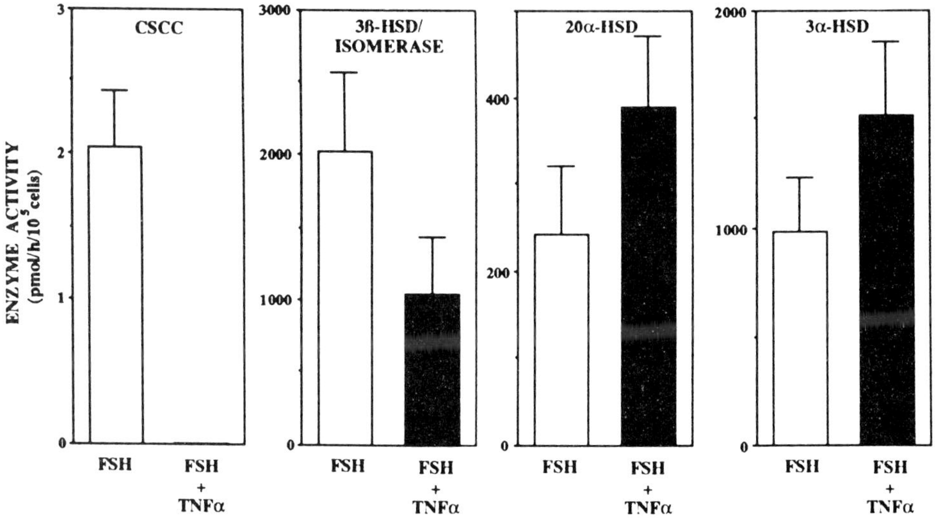

FIGURE 5 TNFα-attenuated progesterone accumulation. Effect of TNFα on the activities of relevant steroidogenic enzymes. Granulosa cells from immature rats were cultured under serum-free conditions for 72 hours in the absence or presence of FSH (100 ng/ml), with or without TNFα (10 ng/ml). The progesterone content of the media was determined by radioimmunoassay. Results represent the means ± SE of three separate determinations. Reproduced with permission from Adashi *et al.* (1990).

driven) accumulation of 20α-DHP (70%). Cellular radiolabeling with [^{3}H]25-hydroxycholesterol (or [^{3}H]pregnenolone) revealed that TNFα produces an apparent diminution in the (FSH-supported) accumulation rates of pregnenolone, progesterone, and 20α-DHP. Concurrent evaluation of relevant key steroidogenic enzymes revealed TNFα to be a potent inhibitor of the FSH-stimulated activities of cholesterol side chain cleavage and of 3β-hydroxysteroid dehydrogenase (HSD)/isomerase (>90 and 42% inhibition, respectively). The activity of the steroidogenic enzymes 20α-HSD and 3α-HSD was not significantly altered (Figure 5).

Taken together, these findings suggest that the ability of TNFα to diminish the gonadotropin-supported accumulation of progesterone by granulosa-luteal cells is largely due to attenuation of key biosynthetic steps that lead to progesterone production. These findings show that TNFα, possibly of intraovarian (e.g., macrophage or granulosa cell) origin, may be the centerpiece of a regulatory loop designed to attenuate the hormonal action of gonadotropin. Acting at or adjacent to its site of synthesis, TNFα may thus partake in the modulation of the ovarian progestin economy, possibly in connection with the demise of the corpus luteum.

If TNFα plays a role in ovarian physiology, its *in vivo* origin(s) must be determined. In principle, two general possibilities are worth considering.

First, TNFα may be locally derived from (activated) resident ovarian macrophages, as has been shown for regressing (but not young) corpora lutea (Bagavandoss *et al.*, 1987, 1988). Although basal TNFα activity was undetectable in corpora lutea of both pregnancy and pseudopregnancy, TNFα activity was markedly stimulated in the presence of LPS. However, the detection of TNFα activity in some luteal tissue on Day 5 and the scarcity of macrophages at this stage raise the possibility that cells other than macrophages may also produce TNFα in the corpus luteum. Indeed, TNFα may be of granulosa cell origin, as suggested by immunohistochemical studies in which antral or atretic granulosa cells have been implicated as possible sites of TNFα gene expression (Roby and Terranova, 1989). Given such strong association between TNFα elaboration and follicular and luteal decline, it is tempting to speculate that TNFα may play a role in the still-enigmatic processes of atresia and/or luteolysis. In this capacity, TNFα of intraovarian origin may exert its effect(s) at or adjacent to its site of synthesis, interacting with specific granulosa-luteal cell surface receptors to modulate gonadotropin hormonal action. Undoubtedly, future studies of the regulation of the TNFα receptor and the elucidation of the *in vivo* source of its ligand will shed new light on the relevance of this system to the process of follicular development and/or demise.

V. Conclusion

While the rudimentary nature of current observations is clearly apparent, there is every reason to believe that continued investigations will provide new and meaningful insights relevant to understanding the complex interactions among the various cellular components of the ovary. Now that the necessary tools are available, additional efforts in this area are to be anticipated. If progress to date is any indication, the odds are that the next decade will reveal that the resident ovarian white blood cell and its cytokine messengers play major roles throughout the ovarian life cycle.

References

Adams, E. C., and Hertig, A. T. (1969). Studies on the human corpus luteum: observations on the ultrastructure of development and regression of the luteal cells during the menstrual cycle. *J. Cell Biol.* 41, 696–715.

Adashi, E. Y. (1989). Cytokine-mediated regulation of ovarian function: encounters of a third kind. *Endocrinology* 124, 2043–2045.

Adashi, E. Y., Resnick, C. E., Croft, C. S., and Payne, D. W. (1989). Tumor necrosis factor α inhibits gonadotropin hormonal action in non-transformed ovarian granulosa cells. *J. Biol. Chem.* 264, 11591–11597.

Adashi, E. Y., Resnick, C. E., Packman, J. N., Hurwitz, A., and Payne, D. W. (1990). Cytokine-mediated regulation of ovarian function: tumor necrosis factor α inhibits gonadotropin-supported progesterone accumulation by differentiating and luteinized murine granulosa cells. *Am. J. Obstet. Gynecol.* 162, 889–899.

Aggarwal, B. B., and Kohr, W. J. (1985). Human tumor necrosis factor, *In* "Methods in Enzymology" (G. Di Sabato, J. J. Langone, and H. Van Vunakis, eds.), Vol. 116, pp. 448–456. Academic Press, San Diego, CA.

Andreani, C. L., Resnick, C. E., Packman, J. N., Payne, D. W., Hurwitz, A., and Adashi, E. Y. (1991). Cytokine-mediated regulation of ovarian function: tumor necrosis factor α inhibits gonadotropin-supported ovarian androgen biosynthesis. *J. Biol. Chem.* 266, 6761–6766.

Bagavandoss, P., Kunkel, S. L., Wiggins, R. C., and Keyes, P. L. (1987). Tumor necrosis factor-α (TNF-α) production and localization of macrophages and T lymphocytes in the rabbit corpus luteum. *Endocrinology* 122, 1185–1187.

Bagavandoss, P., Wiggins, R. C., Kunkel, S. L., and Keyes, P. L. (1988). Localization of macrophages and T lymphocytes and induction of tumor necrosis factor-α (TNF-α) in corpora lutea of pregnant and pseudo-pregnant rabbits. *Biol. Reprod.* 38(Suppl.1) 56.

Beutler, B., and Cerami, A. (1986). Cachectin and tumor necrosis factor as two sides of the same biological coin. *Nature* 320, 584–588.

Beutler, B., and Cerami, A. (1987). Cachectin: more than a tumor necrosis factor. *N. Engl. J. Med.* 316, 379–385.

Bulmer, D. (1964). The histochemical distribution of certain ovarian enzymes. *J. Anat.* 98, 27–36.

Cavender, J. L., and Murdoch, W. J. (1988). Morphological studies of the microcirculatory system of periovulatory ovine follicles. *Biol. Reprod.* 39, 389–397.

Darbon, J. M., Oury, F., Laredo, J., and Bayard, F. (1989). Tumor necrosis factor-α inhibits follicle-stimulating hormone-induced differentiation in cultured rat granulosa cells. *Biochem. Biophys. Res. Commun.* 163, 1038–1045.

Dinarello, C. A. (1988). Biology of interleukin-1. *FASEB J.* 2, 108–115.

Emoto, N., and Baird, A. (1988). The effect of tumor necrosis factor/cachectin on follicle-stimulating hormone-induced aromatase-activity in cultured rat granulosa cells. *Biochem. Biophys. Res. Commun.* 153, 792–798.

Espey, L. (1980). Ovulation as an inflammatory reaction—a hypothesis. *Biol. Reprod.* 22, 73–106.

Fukuoka, M., Mori, T., Taii, S., and Yasuda, K. (1988). Interleukin-1 inhibits luteinization of porcine granulosa cells in culture. *Endocrinology* 122, 367–369.

Fukuoka, M., Taii, S., Yasuda, K., Takakura, K., and Mori, T. (1989). Inhibitory effects of interleukin-1 on luteinizing hormone-stimulated adenosine 3′,5′-monophosphate accumulation by cultured porcine granulosa cells. *Endocrinology* 125, 136–143.

Gillim, S. W., Christensen, A. K., and McLennan, C. E. (1969). Fine structure of the human menstrual corpus luteum at its stage of maximum secretory activity. *Am. J. Anat.* 126, 409–427.

Gottschall, P. E., Uehara, A., Hoffman, S. T., and Arimura, A. (1987). Interleukin-1 inhibits FSH-induced differentiation in rat granulosa cells *in vitro*. *Biochem. Biophys. Res. Commun.* 149, 502–509.

Gottschall, P. E., Katsuura, G., and Arimura, A. (1989a). Interleukin-1 beta is more potent than interleukin-1 alpha in suppressing follicle-stimulating hormone-induced differentiation of ovarian granulosa cells. *Biochem. Biophys. Res. Commun.* 163, 764–770.

Gottschall, P. E., Katsuura, G., and Arimura, A. (1989b). Interleukin-1 suppresses follicle-stimulating hormone-induced estradiol secretion from cultured ovarian granulosa cells. *J. Reprod. Immunol.* 15, 281–290.

Halterman, S. D., and Murdoch, W. J. (1986). Ovarian function in ewes treated with antihistamines. *Endocrinology* 119, 2417–2421.

Hume, D. A., Halpin, D., Charlton, H., and Gordon, S. (1984). The mononuclear phagocyte system of the mouse defined by immunohistochemical localization of antigen F4/80: macrophages of endocrine organs. *Proc. Natl. Acad. Sci. USA* 81, 4174–4147.

Hurwitz, A., Payne, D. W., Packman, J. N., Andreani, C. L., Resnick, C. E., Hernandez, E. R., and Adashi, E. Y. (1991a). Cytokine-mediated regulation of ovarian function: interleukin-1 inhibits gonadotropin-induced androgen biosynthesis. *Endocrinology* 129, 1250–1256.

Hurwitz, A., Ricciarelli, E., Botero, L., Rohan, R. M., Hernandez, E. R., and Adashi, E. Y. (1991b). Endocrine- and autocrine-mediated regulation of rat ovarian (theca-interstitital) interleukin-1β gene expression: gonadotropin-dependent preovulatory acquisition. *Endocrinology* 129, 3427–3429.

Hurwitz, A., Loukides, J., Ricciarelli, E., Botero, L., Katz, E., McAllister, J. M., Garcia, J. E., Rohan, R. M., Adashi, E. Y., and Hernandez, E. R. (1992). The human intraovarian interleukin-1 (IL-1) system: highly-compartmentalized and hormonally-dependent regulation of the genes encoding IL-1, its receptor, and its receptor antagonist. *J. Clin. Invest.* 89, 1746–1754.

Jones, R. E., Duvall, D., and Guillette, L. J. (1980). Rat ovarian mast cells: distribution and cyclic changes. *Anat. Rec.* 197, 489–493.

Kasson, B. G., and Gorospe, W. C. (1989). Effects of interleukins 1, 2, and 3 on follicle stimulating hormone induced differentiation of rat granulosa cells. *Mol. Cell. Endocrinol.* 62, 103–111.

Katabuchi, H., Fukumatsu, Y., and Okamura, H. (1989). Immunohistochemical and morphological observations of macrophages in the human ovary. *In* "Growth Factors and the Ovary" (A. N. Hirshfield, ed.), 7th Ovarian Workshop on Paracrine Communication in the Ovary: Ontogenesis and Growth Factors. Plenum, New York.

Khan, S. A., Schmid, K., Hallin, P., Paul, R. D., Geyter, C. D., and Nieschlag, E. (1988). Human testis cytosol and ovarian follicular fluid contains high amounts of interleukin-1-like factor(s). *Mol. Cell Endocrinol.* 58, 221–230.

Kirsch, T. M., Friedman, A. C., Vogel, R. L., and Flickinger, G. L. (1981). Macrophages in corpora lutea of mice: characterization and effects on steroid secretion. *Biol. Reprod.* 25, 629–638.

Krishna, A., and Terranova, P. F. (1985). Alterations in mast cell degranulation and ovarian histamine in the proestrous hamster. *Biol. Reprod.* 32, 1211–1217.

Krishna, A., Terranova, P. F., Matteri, R. L., and Papkoff, H. (1986). Histamine and increased ovarian blood flow mediate LH-induced superovulation in the cyclic hamster. *J. Reprod. Fertil.* 76, 23–29.

Krishna, A., Beesley, K., and Terranova, P. F. (1989). Histamine, mast cells and ovarian function. *J. Endocrinol.* 120, 363–371.

Leavitt, W., Basom, C., Bagwell, J., and Blaha, G. (1973). Structure and function of the hamster corpus luteum during the estrous cycle. *Am. J. Anat.* 136, 235–249.

Lobel, B. L., Rosenbaum, R. M., and Deane, H. W. (1961). Enzymic correlates of physiological regression of follicles and corpora lutea in ovaries of normal rats. *Endocrionology* 68, 232–247.

Murdoch, W. J. (1987). Treatment of sheep with prostaglandin F_{2a} enhances production of a luteal chemoattractant for eosinophils. *Am. J. Reprod. Immunol. Microbiol.* 15, 52–56.

Murdoch, W. J., Nix, K. J., and Dunn, T. G. (1983). Dynamics of ovarian blood supply to periovulatory follicles of the ewe. *Biol. Reprod.* 28, 1001–1006.

Murdoch, W. J., Steadman, L. E., and Belden, E. L. (1988). Immunoregulation of luteolysis. *Med. Hypotheses* 27, 197–199.

Nakamura, Y., Smith, M., Krishna, A., and Terranova, P. F. (1987). Increased number of mast cells in the dominant follicle of the cow: relationships among luteal, stromal, and hilar regions. *Biol. Reprod.* 37, 546–549.

Nakamura, Y., Kato, H., and Terranova, P. F. (1990). Interleukin-α increases thecal progesterone production of preovulatory follicles in cyclic hamsters. *Biol. Reprod.* 43, 169–173.

Old, L. (1985). Tumor necrosis factor (TNF). *Science* 230, 630–632.

Paavola, L. G. (1977a). The role of macrophages in corpus luteum regression. *Anat. Rec.* 187, 673.

Paavola, L. G. (1977b). The corpus luteum of the guinea pig: fine structure at the time of maximum progesterone secretion and during regression. *Am. J. Anat.* 150, 565–603.

Paavola, L. G. (1979a). The corpus luteum of the guinea pig. *Am. J. Anat.* 154, 337–364.

Paavola, L. G. (1979b). Cellular mechanisms involved in luteolysis. *Adv. Exp. Med. Biol.* 1127, 527–533.

Parr, E. L. (1974). Histological examination of the rat ovarian follicle wall prior to ovulation. *Biol. Reprod.* 11, 483–503.

Pennica, D., Nedwin, G., Hayflick, J., Seeburg, P., Derynck, R., Palladino, M., Kohr, W., Aggarwal, B., and Goeddel, D. (1984). Human tumour necrosis factor: precursor structure, expression and homology to lymphotoxin. *Nature* 312, 724–729.

Rappolee, D. A., Mark, D., Banda, M. J., and Werb, Z. (1988). Wound macrophages express TFG-a and other growth factors in vivo: analysis by mRNA phenotyping. *Science* 241, 708–712.

Roby, K. F., and Terranova, P. F. (1988). Tumor necrosis factor alpha alters follicular steroidogenesis *in vitro*. *Endocrinology* 123, 2952–2954.

Roby, K. F., and Terranova, P. F. (1989). Localization of tumor necrosis factor in the rat and bovine ovary using immunocytochemistry and cell blot: evidence for granulosal production. *In* "Growth Factors and the Ovary" (A. N. Hirshfield, ed.), pp. 273–278. 7th Ovarian Workshop on Paracrine Communication in the Ovary: Ontogenesis and Growth Factors. Plenum, New York.

Sancho-Tello, M., and Terranova, P. F. (1991). Involvement of protein kinase C in regulating tumor necrosis factor α-stimulated progesterone production in rat preovulatory follicles in vitro. *Endocrinology* 128, 1223–1228.

Schmidt, G., Owman, C. H., and Sjoberg, N. O. (1988). Cellular localization of ovarian histamine, its cyclic variations and histaminergic effects on ovulation in the rat ovary perfused in vitro. *J. Reprod. Fertil.* 82, 409–417.

Seow, W. K., Thong, Y. H., Waters, M. J., Walters, M., and Cummins, J. M. (1988). Isolation of a chemotactic protein for neutrophils from human ovarian follicular fluid. *Int. Arch. Allegy. Appl. Immunol.* 86, 331–336.

Shinohara, H., Nakatani, T., Morisawa, S., Matsuda, T., and Naruse, Y. (1987). Mast cells in the ovarian bursa of the golden hamster. *Biol. Reprod.* 36, 445-450.

Stacy, B., Gemmell, R., and Thorburn, G. (1976). Morphology of the corpus luteum in the sheep during regression induced by prostaglandin F_{2a} *Biol. Reprod.* 14, 280–291.

Szego, C. M., and Gitin, E. S. (1964). Ovarian histamine depletion during acute hyperaemic response to luteinizing hormone. *Nature* 201, 682–684.

Takakura, K., Taii, S., Fukuoka, M., Yasuda, K., Tayaga, Y., Yodoi, J., and Mori, T. (1989). IL-2 Receptor/p55 (Tac)-inducing activity in porcine follicular fluid. *Endocrinology* 125, 618–623.

Takemura, R., and Werb, Z. (1984). Secretory products of macrophages and their physiological functions. *Am. J. Physiol.* 246, C1–C9.

VanLennep, E., and Madden, L. (1965). Electron microscopic observations on the involution of the human corpus luteum of menstruation. *Z. Zellforsch.* 66, 365-380.

Veldhuis, J. D., Garmey, J. C., Urban, R. J., Demers, L. M., and Aggarwal, B. B. (1991). Ovarian actions of tumor necrosis factor-α (TNFα): pleiotropic effects of TNFα on differentiated functions of untransformed swine granulosa cells. *Endocrinology* 129, 641–648.

Wurtman, R. J. (1964). An effect of luteinizing hormone on the fractional perfusion of the rat ovary. *Endocrinology* 75, 927–933.

Yasuda, K., Fukuoka, M., Taii, S., Takakura, K., and Mori, T. (1990). Inhibitory effects of interleukin-1 on follicle stimulating hormone induction of aromatase activity, progesterone secretion, and functional luteinizing hormone receptors in cultures of porcine granulosa cells. *Biol. Reprod.* 43, 905–912.

Zolti, M., Meirom, R., Shemesh, M., Wollach, D., Mashiach, S., Shore, L, and Rafael, Z. B. (1990). Granulosa cells as a source and target organ for tumor necrosis factor-α. *FEBS Lett.* 261, 253–255.

5

Role of Insulin-like Growth Factors (IGF) and IGF-Binding Proteins in Folliculogenesis

GREGORY F. ERICKSON, AKIRA NAKATANI,
X.-J. LIU, SHUNICHI SHIMASAKI,
& NICHOLAS LING

I. Introduction

A major goal of research on the ovary is understanding how a few follicles are selected to grow and develop into dominant follicles which secrete their eggs into the oviducts to be fertilized (Erickson, 1993). The process begins when a cohort of primordial follicles is recruited to initiate growth. Successive recruitments give rise to a large number of developing follicles—primary, secondary, tertiary and Graafian—that are present in the ovaries. Not all follicles become dominant, and those that are not selected die by atresia (Ingram, 1962; Erickson *et al.*, 1985). For example, only a very small proportion (as few as 0.1%) of the total follicles survive atresia and give rise to dominant follicles (Aria, 1920). Recognition of the low survival rate has led to the conclusion that folliculogenesis is a highly selective process.

The most important question in research on the ovary concerns the mechanisms by which a cohort follicle is either selected to become dominant or destined for programmed cell death by atresia. The concept of growth factors, particularly the insulin-like growth factors (IGFs), is prominent in current attempts to explain the nature and control of selection and atresia. This chapter discusses the basic findings on the IGFs and their binding proteins (IGFBPs) as they relate to folliculogenesis in the rat. First, however, we briefly discuss the nature of the control of folliculogenesis.

II. The Phenomenon of Folliculogenesis

Once a primordial follicle has entered into the pool of growing follicles, it is either selected to become a dominant preovulatory follicle or it degenerates by a process called atresia. At the level of endocrine action, the process of selection and atresia is controlled by gonadotropins.

A. Selection

A dominant follicle evolves according to a precise developmental program that involves a defined sequence and temporal pattern of gene expression. This is the principle underlying selection. The gonadotropic hormones, follicle-stimulating hormone (FSH) and luteinizing hormone (LH), are obligatory regulatory molecules for the expression of the genes in the granulosa (GC) and theca-interstitial (TIC) cells that lead to selection (Erickson *et al.*, 1985; Richards *et al.*, 1987). This is called the "two gonadotropin, two cell concept" (Erickson, 1978) (Figure 1). One key function of FSH and LH is to coordinate the expression of the P450scc, P45017α, and P450arom genes, which ensures the proper timing of progesterone (P), androgen (A), and estrogen (E) synthesis (Richards and Hedin, 1988; Waterman and Simpson, 1989). In the rat, there is evidence that E acts as a gonadotropic hormone by interacting synergistically with FSH to establish maximum levels of GC differentiation (Richards *et al.*, 1987; Richards and Hedin, 1988) and proliferation (Rao *et al.*, 1978); as a result atresia is prevented (Hirshfield, 1989; Hurwitz and Adashi, 1992). Equally important, this synergy leads to the induction of LH and prolactin receptors in the GC (Richards *et al.*, 1987; Richards and Hedin, 1988; Erickson, 1983). Because these receptors are needed to trigger ovulation and luteinization (Smith *et al.*, 1975), their expression has an important function in the selection process. Thus, the gonadotropic hormones, FSH, E, LH, and prolactin (PRL), play a central role in the mechanisms controlling the temporal pattern of gene activation required to generate follicle dominance.

B. Atresia

In order for it to become dominant, a follicle must complete all steps involved in selection. Interruption of the sequence at any step will result in atresia (Erickson, 1993; Erickson *et al.*, 1985). A fundamental problem in ovarian physiology is understanding the mechanisms by which negative or atretogenic agents function to cause a follicle to undergo atresia. Despite the importance of atresia to ovarian homeostasis, very little knowledge is available, and most of what is known has been gleaned from studies in rat models (reviewed in Erickson *et al.*, 1985).

Studies in immature, hypophysectomized, estrogen-primed rats have

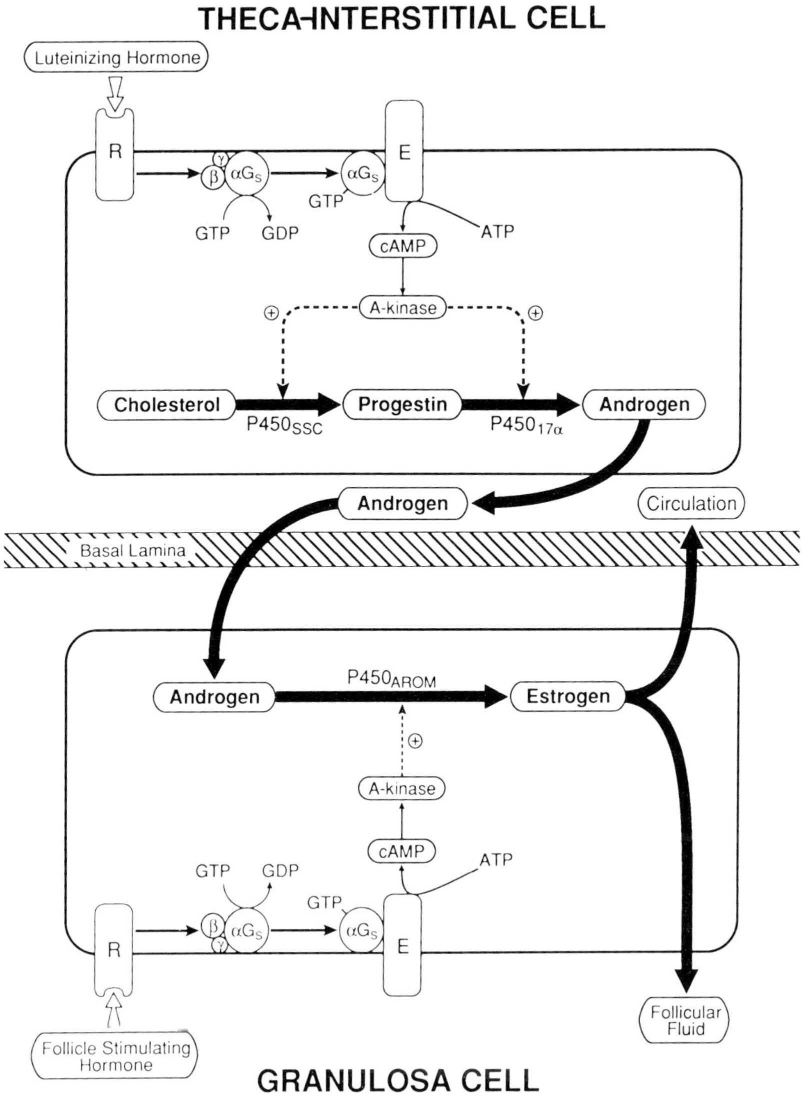

FIGURE 1 Diagram illustrating the two-cell, two-gonadotropin concept of follicle estrogen biosynthesis. Reproduced with permission from Erickson (1993). Copyright © by McGraw-Hill.

shown that preantral follicles can be induced to undergo atresia by (1) exposure to a low dose of LH human chorionic gonadotropin (hCG) (Louvet *et al.*, 1975a); (2) treatment with relatively high levels of testosterone dihydrotestosterone (T/DHT) (Louvet *et al.*, 1975b); (3) withdrawal of estrogen (Hillier and Ross, 1979; Schwall and Erickson, 1981); or

(4) exposure to gonadotropin-releasing hormone (GnRH) (Banka and Erickson, 1985). These results are compatible with the following hypothesis (reviewed in Erickson *et al.*, 1985): Theca-interstitial cells surrounding preantral follicles respond to a low dose of LH by synthesizing and secreting T/DHT; the T/DHT diffuses into the preantral GC where it binds and activates the androgen receptor; the androgen receptor complex acts to inhibit expression of the E receptor gene in the GC. Since this leads to a loss of E stimulation, the GC stop dividing and differentiation ceases. A corollary of this hypothesis is that each of the atretogenic hormones induces atresia through a common mechanism that directly causes to E withdrawal. If so, then rat preantral atresia would be analogous to apoptosis (programmed cell death) in the rat ventral prostate induced by androgen withdrawal (Biehl and Miesteld, 1991).

Experimental studies have also yielded data that implicate LH in the mechanism of atresia in Graafian (antral) follicles (reviewed in Erickson *et al.*, 1985). The classical theory of antral atresia is as follows: LH acts on surrounding TIC to cause a selective decrease in the level of P45017α activity and this is followed by a significant decrease in androgen levels; when androgen production falls, there is a concomitant loss of E production. A similar sequence is most likely the explanation for antral atresia induction, when the activity of the gonadotropic hormones, FSH/LH, is compromised, e.g., after hypophysectomy (Talbert *et al.*, 1951), with specific antisera (Schwartz, 1974), or treatment with inhibin (Vale *et al.*, 1988). Again, the argument could be made that antral atresia might also be caused by E withdrawal, possibly in a way that is functionally analogous to the phenomenon occurring in preantral atresia.

Thus, the available evidence demonstrates a possible relationship between E withdrawal and the atretogenic signals which cause programmed atresia in both preantral and antral follicles of the rat. In addition to the classic atretogenic agents such as low-dose LH, T/DHT, and E withdrawal, other regulatory hormones, namely angiotensin II (Daud *et al.*, 1988), and prostaglandin $F_{2\alpha}$ ($PGF_{2\alpha}$) (Pang and Behrman, 1981), have been implicated in the atretic process.

III. The Concept of Autocrine/Paracrine Modulators in Folliculogenesis

More recently, growth factors which are themselves the products of gonadotropin action appear to have a role in folliculogenesis and act locally on follicle cells to modulate the timing and direction of their differentiation. Evidence is emerging that the intrinsic growth factors mediate hormone-dependent follicle differentiation by either amplifying or attenuating the biological responses of the selectogenic and atretogenic regulators. This is the principle underlying the autocrine/paracrine concept of folligulogenesis and luteogenesis. Understanding the nature of this paracrine/autocrine control process and its physiological significance are fundamental objectives of ovarian research today.

The search for ovarian growth factors has been very productive, particularly in the rat. All major growth factor families are intrinsically expressed in rat follicles during normal development, and growth factors act on the follicle itself to modulate the timing and direction of its developmental process. These include the IGF system (Adashi *et al.*, 1985d), the transforming growth factor-β (TGFβ) system (Knecht *et al.*, 1989), the TGFα/epidermal growth factor (EGF) system (May and Chomberg, 1989), the fibroblast growth factor (FGF) system (Gospodarowicz, *et al.*, 1987), and the cytokine system (Adashi, 1990). We focus our attention on recent advances in the IGF system.

IV. The Ovarian IGF System

There is increasing evidence that an IGF system, complete with ligand, receptor, binding proteins and biological response, functions in regulating ovarian homeostasis in the rat. What is the nature of the follicle IGF system? What are the IGFBPs and to what extent do they regulate the developmental program of folliculogenesis?

A. The Rat Follicle

In the rat ovary, normal follicle development involves a tissue-specific and temporal pattern of expression of the IGFs. The messenger ribonucleic acids (mRNAs) for both IGF-I (Murphy *et al.*, 1987; Oliver *et al.*, 1989; Hernandez *et al.*, 1989; Carlsson *et al.*, 1989; Zhou *et al.*, 1991) and IGF-II (Hernandez *et al.*, 1990) have been identified in mature and immature rat ovaries. Detailed studies of the cellular sites of IGF gene expression have revealed that the IGF-I and IGF-II mRNAs are localized to the GC (Oliver *et al.*, 1989; Hernandez *et al.*, 1989; Carlsson *et al.*, 1989; Zhou *et al.*, 1991) and TIC (Hernandez *et al.*, 1990) respectively. Following recruitment, the GC begin to express IGF-I (Oliver *et al.*, 1989; Hernandez *et al.*, 1989; Carlsson *et al.*, 1989; Zhou *et al.*, 1991), and a high level of IGF-I activity is present in the GC of all healthy follicles, regardless of their developmental state. Conversely, if a cohort follicle undergoes atresia, the expression of IGF-I activity ceases (Oliver *et al.*, 1989; Hernandez *et al.*, 1989; Carlsson *et al.*, 1989; Zhou *et al.*, 1991). Such evidence supports the conclusion that selection requires the continual expression of IGF-I in the GC. At present, very little is known about the hormonal control of IGF-I expression in the rat follicle; however, the evidence that GC in hypophysectomized rats express high levels of IGF-I mRNA strongly suggests the presence of a primary intrinsic control mechanism (Zhou *et al.*, 1991). Other experiments suggest that the gonadotropic hormones, FSH/LH (Carlsson *et al.*, 1989; Zhou *et al.*, 1991) and E/diethylstilbestrol (DES) (Hernandez *et al.*, 1990), stimulate IGF-I expression. There is also evidence that growth hormone (GH) can stimulate the production of intrinsic IGF-I in rat ovaries (Davoren and Hsueh, 1986).

Not only are the ligands produced, but the receptors for IGF-I and IGF-II are expressed in normal developing rat follicles. Studies with adult rat ovaries have demonstrated that GC, TIC, and oocytes strongly express IGF-I receptor mRNA (Zhou *et al.*, 1991; Hernandez *et al.*, 1990, 1991). A similar pattern of expression was observed for the IGF-II receptor (Hernandez *et al.*, 1991). Using radiolabeled IGFs, specific high-affinity type I and type II receptors have been identified in GC (Davoren *et al.*, 1986; Adashi *et al.*, 1988a,b). This result demonstrates that the mRNAs are translated into receptor protein. As with the ligands, there is evidence that the expression of IGF receptors is under hormone control, e.g., the biosynthesis of the IGF receptors is decreased by hypophysectomy, but up-regulated by treatment with gonadotropic hormones, FSH and LH (Hernandez *et al.*, 1990, 1991; Davoren *et al.*, 1986; Adashi *et al.*, 1988a, 1988b). Collectively, the evidence strongly suggests that intrinsic follicular IGF activity and the endocrine system are functionally related.

Considerable effort has been devoted to identifying the biological effects of IGF-I on the rat follicle. In serum-free tissue culture model systems, exogenous IGF-I has proven to be highly active in stimulating both GC and TIC differentiation. Regarding the GC, IGF-I alone elicits dramatic dose-dependent increases in proteoglycan (Adashi *et al.*, 1986a) and inhibin (Zhiwen *et al.*, 1987) biosynthesis, but otherwise IGF-I by itself appears to have little or no effect on the differentiated functions of the GC. Significantly, IGF-I and FSH act synergistically to establish maximal production of cAMP (Adashi, *et al.*, 1986b,c), E, P, 20α dihydroprogesterone (20α DHP) (Davoren *et al.*, 1986; Adashi *et al.*, 1984, 1985a,b), LH receptor (Davoren *et al.*, 1986; Adashi *et al.*, 1985c), proteoglycan (Adashi *et al.*, 1986a), inhibin (Zhiwen *et al.*, 1987), and DNA synthesis (Adashi *et al.*, 1984) in the GC. In addition, GH-bound receptor in rat GC amplifies production of FSH-stimulated estrogen and progesterone (Hutchinson *et al.*, 1988). In regard to TIC, IGF-I has been shown to act synergistically with LH to enhance androgen biosynthesis (Magoffin and Erickson, 1988; Magoffin *et al.*, 1990). Thus, from a physiological standpoint, an intrinsic IGF system could be involved in selection and atresia.

B. The Rat Corpus Luteum

Several lines of evidence indicate that an IGF-I system, also complete with ligand, receptor, and biological action, operates in the rat corpus luteum (CL). First, *in situ* hybridization experiments show that IGF-I mRNA is present in some CL (Oliver *et al.*, 1989; Zhou *et al.*, 1991). Second, *in situ* hybridization experiments (Zhou *et al.*, 1991) indicate that CL in ovaries of lactating rats strongly express the mRNA encoding the IGF-I receptor. In this connection, Parmer and associates (1991), using the solution hybridization/RNAase protection assay, found that IGF-I and the IGF-I receptor mRNAs are abundantly expressed in rat CL throughout pregnancy and that E modulates the levels of both messages. A third line of evidence comes from *in vitro* experiments showing that IGF-I significantly amplifies the stimulatory effects of E and hCG on P production by isolated rat lutein cells (Nelson *et al.*, 1992). Thus, a strong argument can be made for the existence of an endocrine-regulated IGF-I system in the rat CL which may be involved in luteogenesis during the cycle and pregnancy.

These and other observations support the concept that an intrinsic IGF-I system in the rat ovary functions in specific endocrine cell types (GC, TIC, SIC, CL) to control the mechanisms of ovarian homeostasis.

V. The Ovarian IGF-Binding Proteins

Under normal physiological conditions, the IGFs in body fluids are bound to a family of homologous proteins termed the IGFBPs (Shimasaki and Ling, 1991). Thus, to understand the relationship between the IGFs and normal ovarian function, one needs to consider the IGFBPs. Six different IGFBP complementary DNAs (cDNAs) have been cloned and their complete primary structure deduced from the clones (Shimasaki and Ling, 1991). Each IGFBP has a different amino acid sequence and presumably a different set of physiological functions. How were the IGFBPs isolated? What are their structures? What are their effects on follicular granulosa cells?

A. Isolation

Although IGFBPs were detected in ovarian follicular fluid prior to 1988 (Seppälä *et al.*, 1984; Hammond *et al.*, 1985), their isolation and physiological significance were not realized until it was serendipitously discovered that they inhibit steroidogenesis in rat GC in culture (Ui *et al.*, 1989; Ling *et al.*, 1990). The discovery originated from the search for the endogenous FSH-action inhibitors in porcine follicular fluid, using the well-defined rat GC culture assay to monitor the inhibitory effect of the active substances on FSH-stimulated E production. Using a purification scheme that involved precipitation of the high-molecular-weight proteins in porcine follicular fluid with ammonium sulfate, followed by dialysis of the supernatant in 30% (vol/vol) acetic acid, gel filtration chromatography of the dialyzed material in Sephacryl S-200 superfine under acidic conditions, and several steps of reversed-phase high-performance liquid chromatography, a polypeptide was isolated which exhibited a potent inhibition of FSH-stimulated E production in rat GC (Ui *et al.*, 1989).

Amino-terminal amino acid sequence analysis of the purified inhibitor, however, revealed that its sequence was very similar to the amino acid sequence of a recently characterized 53-kDA growth hormone-dependent IGFBP isolated from human serum (Wood *et al.*, 1988) and later designated as IGFBP-3. The possibility that the FSH-action inhibitor purified from follicular fluid could be an IGFBP was strengthened later by molecular cloning of the inhibitor from a porcine ovarian cDNA library. The deduced amino acid sequence of the isolated cDNA confirmed that the inhibitor was indeed the porcine homologue of the human IGFBP-3 (Shimasaki *et al.*, 1990a).

At the time when IGFBP-3 was identified as the FSH-action inhibitor in porcine follicular fluid, two other IGFBPs designated IGFBP-1 and -2 had already been fully characterized (Shimasaki *et al.*, 1990a). Moreover, IG-FBP-1 had been detected in human follicular fluid by radio immunoassay (RIA) (Seppälä *et al.*, 1984). However, it was not certain whether IGFBP-1 and -2 were also present in porcine follicular fluid or additional unknown IGFBPs could be found in the ovary. As a first step to fully comprehend the physiological roles played by the IGFBPs in the ovary, it was necessary to isolate and identify all IGFBPs present in porcine follicular fluid.

By incorporating an affinity chromatography step into the procedure used to purify the FSH-action inhibitor (Ui *et al.*, 1989), IGFBP-2 and -3 plus three novel IGFBPs were isolated from porcine follicular fluid (Shimasaki *et al.*, 1991b). The three novel IGFBPs were later designated as IGFBP-4, -5, and -6 (reviewed by Shimasaki and Ling, 1991). In contrast to human follicular fluid, IGBP-1 was not found in the porcine fluid. Fractionation of adult rat serum also yielded the same number of IGFBPs (BP-2, -3, -4, -5, and -6) but not IGFBP-1 (Shimasaki *et al.*, 1991a). Subsequent cloning of the IGFBP-4 (Shamasaki *et al.*, 1990b), -5 (Shimasaki *et al.*, 1991a) and -6 (Shimasaki *et al.*, 1991b) from rat and human tissue cDNA libraries yielded their complete primary structures.

B. Structure

Figure 2 shows the most favorable alignment of the deduced amino acid sequences from the cDNAs of the novel IGFBP-4, -5, and -6, together with the previously characterized IGFBP-1, -2, and -3 in rat and human species. None of the IGFBPs bears any amino acid sequence homology to the type I or type II IGF receptors. The homologous amino acid sequence regions among these six IGFBPs are located at the amino-terminal and carboxyl-teminal thirds of the molecules, while the middle portion of the molecules is the most divergent. There are 18 cysteines in rat and human

```
Rat IGFBP-1          APQPWHCAPCTAERLELCPPVP-AS-CPEISRPAGCGCCPTCALPLGAACGVATAACAQGLSCRALPGEPRPLHALTRGQGAC
Rat IGFBP-2           EVLFRCPPCTPERLAACGPPPDAP-CAELVREPGCGCCSVCARQEGEACGVYIPRCAQTLRCYPNPGSELPLKALVTGAGTC
Rat IGFBP-3 GAGAVGAGPVVRCEPCDARALAQCAPPTAPACTELVREPGCGCCLTCALREGDACGVYTERCGTGLRCQPRPAEQYPLKALLNGRGFC
Rat IGFBP-4            DEAIHCPPCSEEKLARCRPPVG---CEELVREPGCGCCATCALGLGMPCGVYTPRCGSGMRCYPPRGVEKPLRTLMHGQGVC
Rat IGFBP-5          LGSFVHCEPCDEKALSMC-PPSP-LGC-ELVKEPGCGCCMTCALAEGQSCGVYTERCAQGLRCLPRQDEEKPLHALLHGRGVC
Rat IGFBP-6            ALAGCPGC---(GPGVQEEDAGSPAD)----GCAETGGCFRREGQPCGVYIPKCAPGLQCQPRENEETPLRALLIGQGRC

(VLEPPPPATSSLSGSQHEEAKAAVASEDELAESPEMTEEQLLDSFHLMAPSREDQPILWNAISTYSSMRAREITDLKKWKE)
(EKRRVGATPQQVADSEDDHSEGGLVENHVDGTMNMLGGSSAGRKPPKSGMKELAVFREKVNEQHRQMGKGAKHLSLEEPKKLRPPPART)
(ANASAASNLSAYLPSQPSPGNTTESEEDHNAGSVESQVVPSTHRVTDSKFHPLHSKMEVIIKGQARDSQRYKVDYESQSTDTQNFSSESKRETEYG)
(TELSEIEAIQESLQTSDKDESEHPNNSFNPCSAHDHRCLQKHMAKVRDRSKMKVVGTPREEPRPVPQG)
(LNEKSYGEQTKIERDSREHEEPTTSEMAEETYSPKVFRPKHTRISELKAEAVKKDRRKKLTQSKFVGGAENTAHPRVIPAPEMRQESDQG)
(QRARGPSEETTKESKPHGGASRPRDRDRCKNPRTSAAPIRPSPVQDGEMG)

PCQRELYKVLERLAAAQQKA--GD-E-IYKFYLPNCNKNGFYHSKQCETSLDGEAGLCWCVYPWSGKKIPGSLETRGDPNCHQYFNVQN
PCQQELDQVLERISTMRLPDDRGPLEHLYSLHIPNCDKHGLYNLKQCKMSLNGQRGECWCVNPNTGKPIQGAPTIRGDPECHLFYNEQQENDGVHAQRVQ
PCRREMEDTLNHLKFLNVLSPRGV-------HIPNCDKKGFYKKKQCRPSKGRKRGFCWCVDKYGQPLPGYDTKGKDDVHCLSVQSQ
SCQSELHRALERLAASQ--S-RTH-EDLFIIPIPNCDRNGNFHPKQCHPALDGQRGKCWCVDRKTGVKLPGGLEPKGELDCHQLADSIQE
PCRRHMEASLQEFKASPRMVPRAV-------YLPNCDRKGFYKRKQCKPSRGRKRGICWCVDKY-GMKLPGMEYVDGDFQCHAFDSSNVE
PCRRHLDSVLQQLQTE-VF--RGGANGL---YVPNCDLRGFYRKQQCRSSQGNRRGPCWCVDPM-GQPLPVSPDGQGSSQCSARSSG

Human IGFBP-1          APWQCAPCSAEKLALCPPVSAS---------------CSE----VTRSAGCGCCPMCALPLGAACGVATARCARGLSCRALPGEQQPLHALTRGQGAC
Human IGFBP-2        EVLFRCPPCTPERLAACGPPPVAPPAAVAAVAGGARMPCAE----LVREPGCGCCSVCARLEGEACGVYTPRCGQGLRCYPHPGSELPLQALVMGEGTC
Human IGFBP-3 GASSGGLGPVVRCEPCDARALAQCAPPPAV---------------CAE----LVREPGCGCCLTCALSEGQPCGIYTERCGSGLRCQPSPDEARPLQALLDGRGLC
Human IGFBP-4        DEAIHCPPCSEEKLARCRPPVG---------------CEE----LVREPGCGCCATCALGLGMPCGVYTPRCGSGLRCYPPRGVEKPLHTLMHGQGVC
Human IGFBP-5      LGSFVHCEPCDEKALSMC-PPSPLG--------------C-E----LVKEPGCGCCMTCALAEGQSCGVYTERCAQGLRCLPRQDEEKPLHALLHGRGVC
Human IGFBP-6         ALARCPGCGQGVQAGC-PGG---------------CVEEEDGGSPAEGCAEAEGCLRREGQECGVYTPNCAPGLQCHPPKDDEAPLRALLLGRGRC

(VQESDASAPHAAEAGSPESPESTEITEEELLDNFHLMAPSEEDHSILWDAISTYDGSKALHVTNIKKWKE)
(EKRRDAEYGASPEQVADNGDDHSEGGLVENHVDSTMNMLGGGGSAGRKPLKSGMKELAVFREKVTEQHRQMGKGGKHHLGLEEPKKLRPPPART)
(VNASAVSRLRAYLLPAPPAPGNASESEEDRSAGSVESPSVSSTHRVSDPKFHPLHSKIIIIKKGHAKDSQRYKVDYESQSTDTQNFSSESKRETEYG)
(MELAEIEAIQESLQPSDKDEGDHPNNSFSPCSAHDRRCLQKHFAKIRDRSTSGGKMKVNGAPREDARPVPQG)
(LNEKSYREQVKIERDSREHEEPTTSEMAEETYSPKIFRPKHTRISELKAEAVKKDRRKKLTQSKFVGGAENTAHPRIISAPEMRQESEQG)
(LPARAPAVAEENPKESKPQAGTARPQDVNRRDQQRNPGTSTTPSQPNSAGVQDTEMG)

PCRIELYRVVESLAKAQETS--GE-E-ISKFYLPNCNKNGFYHSRQCETSMDGEAGLCWCVYPWNGKRIPGSPEIRGDPNCQMYFNVQN
PCQQELDQVLERISTMRLPDERGPLEHLYSLHIPNCDKHGLYNLKQCKMSLNGQRGECWCVNPNTGKLIQGAPTIRGDPECHLFYNEQQEARGVHTQRMQ
PCRREMEDTLNHLKFLNVLSPRGV-------HIPNCDKKGFYKKKQCRPSKGRKRGFCWCVDKYGQPLPGYTTKGKEDVHCYSMQSK
SCQSELHRALERLAASQ--S-RTH-EDLYIIPIPNCDRNGNFHPKQCHPALDGQRGKCWCVDRKTGVKLPGGLEPKGELDCHQLADSFRE
PCRRHMEASLQELKASPRMVPRAV----Y---LPNCDRKGFYKRKQCKPSRGRKRGICWCVDKY-GMKLPGMEYVDGDFQCHTFDSSNVE
PCRRHLDSVLQQLQTEVY---RG-AQTLY---VPNCDHRGFYRKRQCRSSQGQRRGPCWCVDRM-GKSLPGSPDGNGSSSCPTGSSG
```

FIGURE 2 Alignment of the amino acid sequences of six rat and human IGFBPs. Amino acids are shown in one-letter code and gaps are inserted to allow maximal homology alignment. Cysteines are shown by shaded boxes, including the two additional cysteines present in the midregion of IGFBP-4. Potential Asn-linked glycosylation sites are underlined.

IGFBP-1, -2, -3, and -5, and their locations in the molecules are totally conserved. Among those 18 cysteines, 12 are located at the amino-terminal of the molecule, spanning one-third of the total amino acid sequence, and the remaining 6 lie in the carboxyl-terminal region, spanning the last third of the protein sequence. IGFBP-4 contains 2 extra cysteines at the mid-region of the molecule in addition to the 18 conserved cysteines.

In contrast, rat and human IGFBP-6 contain only 14 and 16 cysteines, respectively. Eight of the 14 cysteines in the rat protein are located at the amino-terminal region of the molecule, whereas the human homologue contains 10 cysteines in the same region. The carboxyl-terminal region of both rat and human IGFBP-6 contains the remaining 6 conserved cysteines. The absence of 4 and 2 cysteines in the rat and human IGFBP-6 amino acid sequences, respectively, resulted in the loss of the invariant Gly–Cys–Gly–Cys–Cys sequence present in all of the other five IGFBPs. Human and rat IGFBP-3 contain three and four potential Asn-linked glycosylation sites, respectively, and human and rat IGFBP-4 possess one potential Asn-linked glycosylation site in the midregion of the moleclue. Only human IGFBP-6 has a potential Asn-linked glycosylation site at the extreme carboxyl-terminal of the molecule, but this site does not bear a carbohydrate chain. instead, IGFBP-6 is O-linked glycosylated (Bach *et al.*, 1992).

C. Effects on Granulosa Cells

The effects of IGFBPs on GC function have been examined using serum-free cultures of rat GC obtained from immature, DES-treated rats (Bicsak *et al.*, 1990). Both porcine IGFBP-2 and -3 dose-dependently inhibit FSH-stimulated E and P production, with EC_{50} values of 4.1–7.6 n*M* for IGFBP-3 and 12.6–12.9 n*M* for IGFBP-2. Adding rat IGFBP-4 and -5 to the GC culture also dose-dependently inhibits the FSH-stimulated production of the two gonadal steroids with EC_{50} values of 8.0–12.5 n*M* for IGFBP-4 and 18–20 n*M* for IGFBP-5 (Liu *et al.*, 1992). Experiments using cells that were primed with FSH to induce functional LH receptors showed that the inhibitory action of IGFBP-3 was specific to FSH but not LH (Bicsak *et al.*, 1990). Porcine IGFBP-3 and -2 were also capable of inhibiting both forskolin- and cholera toxin-stimulated steroidogensis, confirming that the binding proteins were not competing with FSH for binding to its receptor (Bicsak *et al.*, 1990). In addition, neither of the IGFBPs had any effect on either basal or FSH-stimulated cyclic adenosine monophosphate (cAMP) production, while exogenously added IGF-I stimulated cAMP production (Bicsak *et al.*, 1990). However, both IGFBP-2 and -3 inhibited FSH-stimulated [^{3}H]thymidine uptake by the granulosa cells, while IGF-I had no effect on this parameter (Bicsak *et al.*, 1990). Hence the biological effects of the IGFBPs in GC function are mainly inhibitory.

VI. Expression of IGFBPs in Rat Ovaries

A. Northern Analysis

Evidence that adult rat ovaries contain the mRNAs encoding IGFBPs has come from our recent studies using Northern blot analysis and *in situ* hybridization techniques (Nakatani *et al.*, 1991; Erickson *et al.*, 1992a,b). As shown in Figure 3, extracts prepared from ovaries of adult cycling rats contain detectable messenger RNA transcripts for five of the six IGFBPs. Northern blotting analysis of the rat ovaries did not exhibit detectable IGBP-1 mRNA. This result suggests that expression of the IGFBP-1 gene is low or absent in adult ovary tissue; however, the mRNAs for the other five IGFBPs were clearly identified in the rat ovary. Electrophoresis of poly(A^+) RNA showed intense single bands when hybridized with the IGFBP-2, IGFBP-4 and IGFBP-6 antisense cRNA probes at 2.0, 2.6, and 1.3 kb, respectively (Figure 3). A distinct, but less intense single mRNA band migrating at 2.6 kb was identified when the ovary extract was hybridized with the IGFBP-3 antisense cRNA probe. Two mRNA transcripts, 6.0 kb and a less intense 1.6 kb species, were found when hybridized with the IGFBP-5 cRNA probe (Figure 3). Because the differences in their sizes lies within the

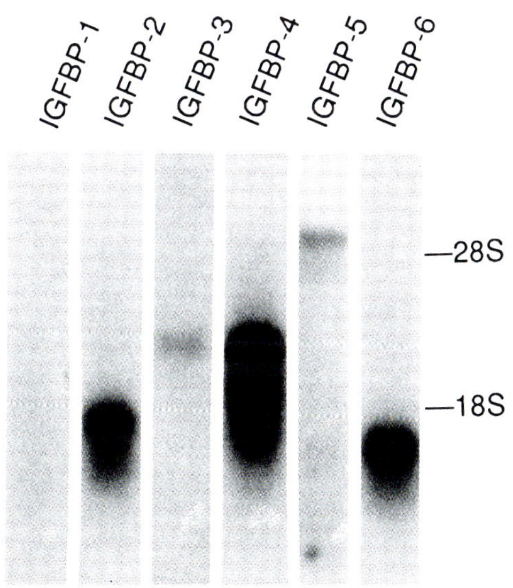

FIGURE 3 Northern blot analysis of the six IGFBP mRNAs in adult rat ovaries. Poly(A^+) RNA (1 μg per lane) was electrophoresed, blotted onto a nylon membrane filter, and then probed for the IGFBP mRNAs. The 18S and 28S indicate the positions of ribosomal RNA.

3′-noncoding region (Kiefer *et al.*, 1991), these two IGFBP-5 mRNAs are most likely due to polymorphism. Further studies are necessary to determine whether there are regulatory mechanisms which alter the ratio of these two IGFBP-5 mRNA species during ovary function, and their potential significance.

Based on these data, it is clear that the messenger RNAs for IGFBP-2, -3, -4, -5, and -6 are expressed in adult rat ovary tissue. These results are compatible with a hypothesis that normal activity of the rat ovary involves the expression of five IGFBP genes.

B. *In Situ* Hybridization

A key question concerns what cell types express the five IGFBP mRNAs. The cellular sites of synthesis of the family of IGFBPs in the adult rat ovary *in situ* hybridization techniques have been identified (Nakatani *et al.*, 1991; Erickson *et al.*, 1992a,b).

1. IGFBP-1

Using Northern analysis and *in situ* hybridization (Nakatani *et al.*, 1991), we found no evidence for IGFBP-1 expression in the adult rat ovary (Figure 3 and Table 1). Based on this finding, it is reasonable to conclude that locally produced IGFBP-1 may not play an important role in the mechanisms of ovarian homeostasis in the adult cycling rat. However, it does not rule out the possibility that locally produced IGFBP-1 may function in embryonic, neonatal, and/or juvenile rat ovaries. It should be noted that the IGFBP-1 gene is expressed in human granulosa and luteal cells (Koistinen *et al.* 1990). Therefore there appear to be major differences among species in the kinds of IGFBPs that the ovaries synthesize. This conclusion may extend to other tissues as well. The significance of this concept remains to be determined.

2. IGFBP-2

IGFBP-2 possesses antigonadotropin activity, and therefore may be involved in the control of ovarian homeostasis (Bicsak *et al.*, 1990). The gene encoding IGFBP-2 is expressed strongly in the rat ovary (Nakatani *et al.*, 1991; Erickson *et al.*, unpublished observations), being present in three major cell types, namely, the theca-interstitial cells, secondary interstitial cells (SIC), and surface epithelium; the message is not present in oocytes, granulosa cells (healthy or atretic), theca externa, corpora lutea, or stromal tissue (Table 1, Figure 4).

An important feature of IGFBP-2 gene expression in interstitial cells is that it changes from inactivity to activity (and vice versa) during thecogenesis (Erickson *et al.*, unpublished observations). Thecogenesis occurs in several steps (Erickson *et al.*, 1985). In the first step, a set of pre-

TABLE 1 Cellular Localization of mRNAs for the Family IGFBPs in Adult Rat Ovary Tissue

Tissue	IGFBP-1	IGFBP-2	IGFBP-3	IGFBP-4	IGFBP-5	IGFBP-6
Granulosa cells						
Healthy						
Primordial	−[a]	−	−	−	−	−
Primary	−	−	−	−	−	−
Secondary	−	−	−	−	−	−
Tertiary	−	−	−	−	−	−
Dominant	−	−	−	−	−	−
Preovulatory	−	−	−	−	−	−
Atretic						
Preantral	−	−	−	+	+	−
Antral	−	−	−	+	+	−
Theca-interstitial cells						
Secondary	−	+	−	−	−	−
Tertiary	−	+	−	−	−	−
Dominant	−	+	−	−	−	−
Preovulatory	−	+	−	+/−	−	−
Atretic	−	+	−	−	−	−
Secondary interstitial cells	−	+	−	−	+/−	−
Corpora lutea	−	−	+/−	+/−	+/−	−
Oocyte	−	−	−	−	−	−
Theca externa	−	−	−	−	−	+
Stroma	−	−	−	+/−	−	+
Smooth muscle	−	−	−	−	−	+
Surface epithelium	−	+	−	−	+	−

[a] −, mRNA was undetectable; +, mRNA was detectable and the hybridization signal was strong; +/−, there was heterogeneity in mRNA expression, i.e., some of the histological units contained the mRNA, others did not. A bold + indicates a strong, consistent hybridization signal. A lightface + indicates inconsistent hybridization signal.

sumptive or undifferentiated theca cells surround the primary and secondary preantral follicles. When this happens, IGFBP-2 mRNA is not expressed, implying that the gene is turned off.

The second step takes place in the late secondary or early tertiary follicle (300–400 μm in diameter) when the presumptive theca transform into differentiated TIC. When this happens, they become positive for IGFBP-2, implying that there may be a stimulus turning on IGFBP-2 gene expression at this step in thecogenesis. Whether a Graafian (antral) follicle develops to the preovulatory stage or dies by atresia, the TIC continuously express IGFBP-2 mRNA. This observation argues that IGFBP-2 expression is a property (molecular marker) of the differentiated rat TIC. Physiologically, LH is the primary regulator of thecogenesis (Erickson *et al.*,

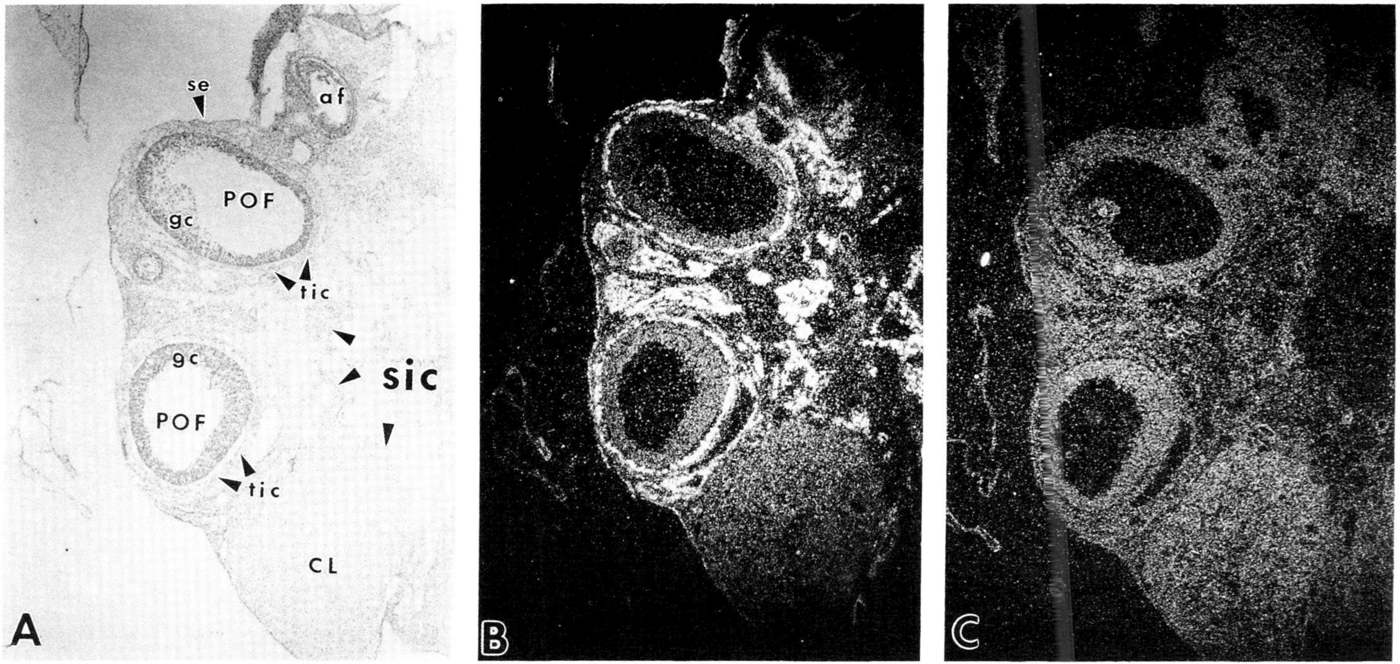

FIGURE 4 Localization of IGFBP-2 mRNA in the adult rat ovary at proestrus 2,000 hours by *in situ* hybridization. POF, preovulatory follicle; gc, granulosa cells; tic, theca-interstitial cells; sic, secondary interstitial cells; af, atretic follicle; se, surface epithelium; CL, corpus luteum. (A) Bright-field photomicrograph. (B) Dark-field photomicrograph hybridized with an antisense cRNA probe. (C) Control: dark-field photomicrograph hybridized with a sense cRNA probe. Reproduced with permission from Nakatani *et al.* (1991). Copyright © 1991 The Endocrine Society.

1985). Perhaps LH is involved in the induction and maintenance of steady-state expression levels of IGFBP-2 mRNA in the TIC.

The third step occurs after ovulation when the TIC differentiate into theca-lutein cells (TLC). When this happens, message levels decrease and eventually no IGFBP-2 mRNA is detected in the TLC. It appears that IGFBP-2 gene expression is inhibited (switched off) at the last step in thecogenesis. Since IGFBP-2 can block gonadotropin-dependent E and P synthesis (Bicsak *et al.*, 1990), understanding IGFBP-2 gene induction and repression in the interstitial cells may have considerable physiological importance.

During the estrous cycle there appear to be striking changes in the level of IGFBP-2 mRNA in the TIC and SIC. The mechanisms by which IGFBP-2 transcriptional control is exerted are unknown; however, the mRNA levels follow fairly closely the cyclic changes in E_2 concentrations, e.g., IGFBP-2 mRNA levels in the interstitial cells appear to increase when E levels increase to peak values at proestrus and decline to their lowest levels at estrus 1000 hours when E secretion reaches baseline (Erickson *et al.*, unpublished observations). In the study by Ricciarelli *et al.* (1991), E was shown to act directly on rat TIC/SIC to stimulate IGFBP-2 mRNA synthesis. Thus, the hypothesis that IGFBP-2 expression in these cells is controlled by E may have validity.

The discovery that dominant follicles are encapsulated by a thick layer of TIC which are actively expressing IGFBP-2 mRNA (Figure 4) is paradoxical in view of the evidence that IGFBP-2 is a potent inhibitor of action of the gonadotropic hormone FSH (Bicsak *et al.*, 1990). Hence the question: How can selection occur? It is possible that the IGFBP-2 mRNA is not translated in dominant follicles. If true, then some physiological mechanism operates at the translational level to inhibit IGFBP-2 protein synthesis in those follicles destined for ovulation. Another possibility is that conditions exist in the dominant follicle that prevent the expression of IGFBP-2 bioactivity; that is, a mechanism that could block or destroy IGFBP-2 activity. In this regard, data indicate that FSH induces the synthesis of a protease in GC which cleaves and inactivates the IGFBPs (Liu *et al.*, 1992). Therefore, an FSH-dependent protease may play an important role in preventing IGFBP-2 from functioning in the microenvironment of dominant follicles.

It is clear that rat surface epithelial cells constitutively express IGFBP-2 mRNA. The importance of IGFBP-2 expression in these cells is unknown. However, one could speculate that the IGFs function to control surface epithelium growth and degeneration during the cycle, and that IGFBP-2 might have a role as well.

3. IGFBP-3

In the adult rat ovary (Nakatani *et al.*, 1991), only corpora lutea express IGFBP-3 mRNA (Table 1, Figure 5). The finding that only certain CL contain

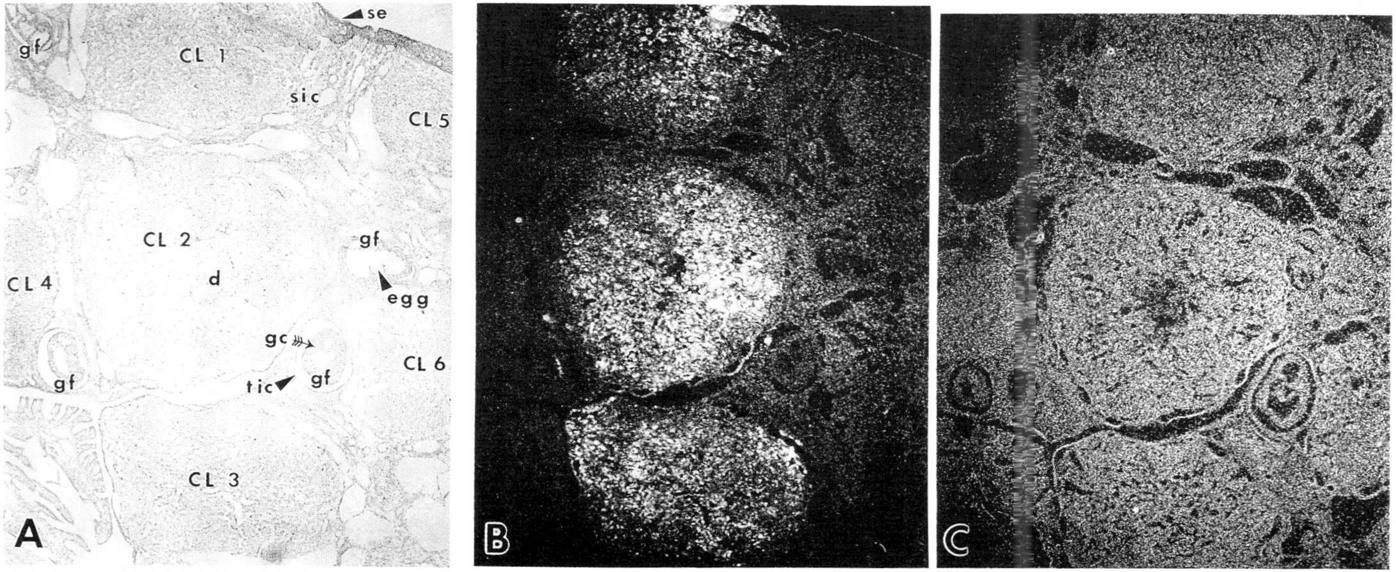

FIGURE 5 Localization of IGFBP-3 mRNA in the rat ovary at proestrus 1000 hours. CL 1–6, corpora lutea 1 to 6; gf, Graafian follicle; sic, secondary interstitial cells; tic, theca-interstitial cells; gc, granulosa cells; se, surface epithelium. (A) Bright-field photomicrograph. (B) Dark-field photomicrograph hybridized with an antisense cRNA probe. (C) Control: dark-field photomicrograph hybridized with a sense cRNA probe. Reproduced with permission from Nakatani *et al.* (1991).

the message argues that luteogenesis involves variable IGFBP-3 gene activity.

In a study of IGFBP-3 expression over the estrous cycle, (Erickson *et al.*, 1993), we discovered that IGFBP-3 gene activity is restricted to CL undergoing involution. In the early stages of luteinization on the morning of estrus—the time that the TIC and invading capillary cells undergo extensive hyperplasia and when sizable quantities of P are secreted by the new CL—there is no detectable IGFBP-3 mRNA. This lack of IGFBP-3 expression in CL during luteinization fits well with the evidence that IGFBP-3 can block IGF-I dependent increases in P synthesis by granulosa-lutein cells (Ui *et al.*, 1989; Bicsak *et al.*, 1990). Interestingly, the expression of luteolysis is preceded by a burst of IGFBP-3 mRNA expression in the new CL; that is, a sharp increase in IGFBP-3 mRNA occurs in the lutein cells directly before the young CL show decreased P synthesis and intense necrosis. The activation of the IGFBP-3 gene at the beginning of luteolysis may imply a causal relation by mechanisms involving a blockade of IGF-I action. Experiments are in progress to test this hypothesis.

A novel finding in this study (Erickson *et al.*, 1993) is that endothelial cells of the CL are very active in expressing IGFBP-3 mRNA. The mRNA is first detected on diestrus II in a subpopulation of endothelial cells near the central cavity and then occurs in almost all endothelial cells on proestrus. The evidence indicates that IGFBP-3 gene expression is activated in endothelial cells lining the sinusoidal vessels close to the onset of luteolysis, and that the process continues during CL destruction. IGFBP-3 mRNA is identified in CL endothelial cells about 24 hours before it is detected in the granulosa and theca-lutein cells. The significance of this tissue-specific and temporal pattern of IGFBP-3 gene expression in the rat ovary is unknown, but we might predict that the regulated production of IGFBP-3 by the blood tissue is involved in luteolysis. If true, then IGFBP-3 might function as an autocrine/paracrine luteolysin by binding and inactivating the tropic effects of IGF-I required for lutein cells to be optimally active.

4. IGFBP-4

An important point to emerge from the *in situ* hybridization studies is that IGFBP-4 mRNA is found primarily in granulosa cells of atretic antral follicles and some CL. In contrast, there is no detectable IGFBP-4 mRNA in GC of healthy follicles, or in SIC, oocytes, theca externa, or surface epithelium (Table 1, Figure 6). A relatively high level of IGFBP-4 mRNA can be detected in some stromal cells, but only on the morning of estrus (Erickson *et al.*, 1992a). The significance of the expression of IGFBP-4 in the stromal compartment on the morning of estrus is unknown; however, it is noteworthy that this is the time when dominant follicles are selected. Perhaps the two events are related in some way.

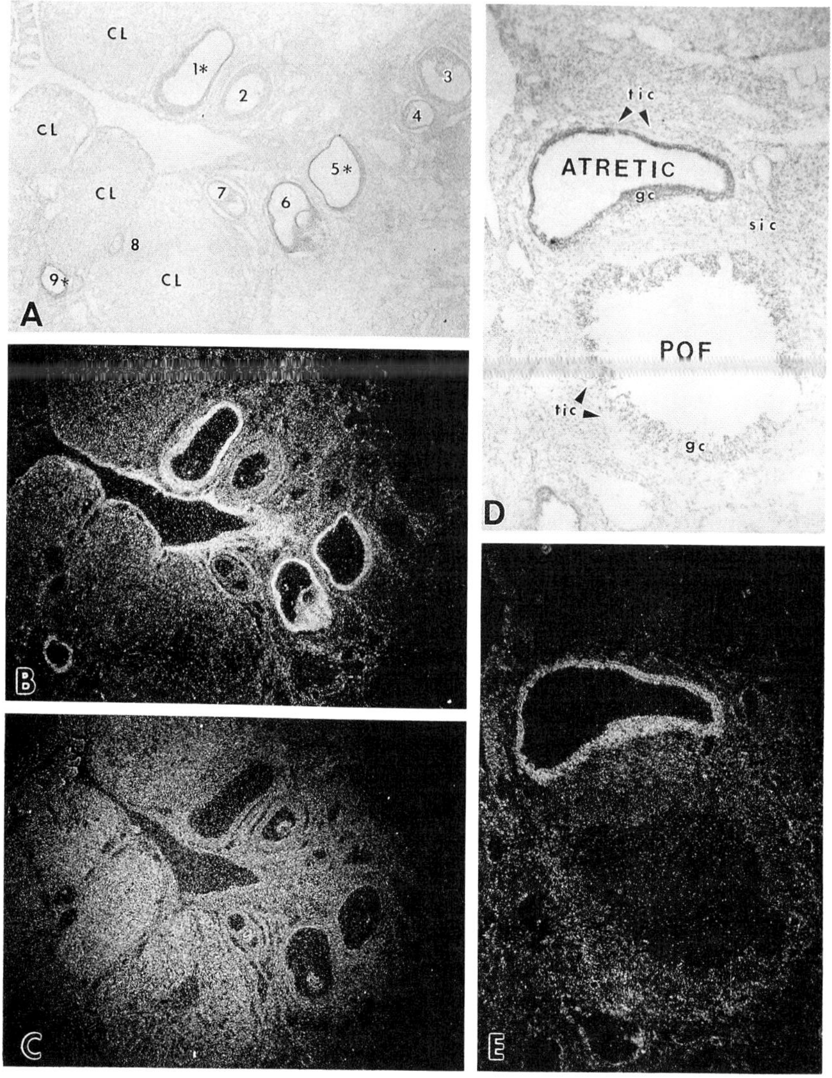

FIGURE 6 Localization of IGFBP-4 mRNA in the rat ovary at estrus 200 hours (A, B, and C) and at proestrus 2,000 hours (D and E). 1–9, developing follicles (asterisk indicates atresia); CL, corpora lutea; gc, granulosa cells; tic, theca-interstitial cells; sic, secondary interstitial cells. (A and D) Bright-field photomicrographs. (B and E) Dark-field photomicrographs hybridized with an anti-sense cRNA probe. (C) Control: dark-field photomicrograph of same section shown in A and B, hybridized with a sense cRNA probe. Reproduced with permission from Erickson *et al.* (1992). Copyright © 1992 The Endocrine Society.

The ability of granulosa cells in a cohort of atretic follicles to express the IGFBP-4 gene is interesting because it suggests a possible connection between IGFBP-4 and atresia. It has been shown by Oliver and his colleagues (1989) that follicle selection is accompanied by the continued synthesis of IGF-I in rat GC, and there is dramatic evidence demonstrating the ability of IGF-I to regulate the growth and differentiation of the GC (Adashi *et al.*, 1985d) and TIC (Magoffin and Erickson, 1988). Furthermore, there is direct evidence that the IGFBPs, including IGFBP-4, can block the positive stimulatory effects of FSH on GC differentiation by binding and inactivating the intrinsically produced IGF-I (Bicsak *et al.*, 1990; Liu *et al.*, 1992). Collectively, these results have provided the foundation for the theory that IGF-I, produced by the GC in response to FSH stimulation, is an obligatory autocrine/paracrine regulatory element in the selection process. Because IGFBP-4 is produced by atretic GC, it is possible that it may act at the local level to promote atresia by sequestering IGF-I.

At the opposite extreme, granulosa cells of dominant follicles do not appear to express IGFBP-4 mRNA (Table 1, Figure 6). This finding argues that selection might involve the inhibition of IGFBP-4 gene activity. Indeed, a characteristic feature of the dominant follicle in the rat is the absence of any of the IGFBPs (IGFBP-1, -2, -3, -4, -5, or -6) in the GC. Thus, once a cohort follicle is selected to become dominant, signals may act at the level of all six IGFBP genes to inhibit transcription. How might this occur? There is evidence from *in vitro* studies (Adashi *et al.*, 1991; Lui *et al.*, 1992) that high doses of FSH inhibit IGFBP-4 (and -5) production by cultured GC. Also, there is evidence that dominant follicles in humans contain high concentrations of FSH (McNatty *et al.*, 1975). These results are consistent with the hypothesis that one function of high levels of FSH in the microenvironment is to inhibit the expression of the IGFBP genes, thereby allowing selection to proceed. Figure 7 is a diagrammatic representation of this model.

5. IGFBP-5

IGFBP-5 mRNA is expressed at high levels in adult rat ovaries throughout the estrous cycle (Erickson *et al.*, 1992b), and interestingly, the GC of antral and preantral atretic follicles are actively engaged in IGFBP-5 expression (Table 1, Figure 8). The message is also localized to the SIC, CL, and surface epithelium, but none was found in healthy GC, oocytes, TIC, theca externa, stroma, or smooth muscle cells (Table 1, Figure 8). The relevance of IGFBP-5 gene activity in the SIC, CL, and surface epithelium is unknown, but clearly it could be involved in modulating IGF-I and/or IGF-II activity in these cell types. Because the levels of IGFBP-5 change during the cycle, we predict that there is hormonal regulation of IGFBP-5 gene activity in these important cell types.

With respect to folliculogenesis, an important observation is that the

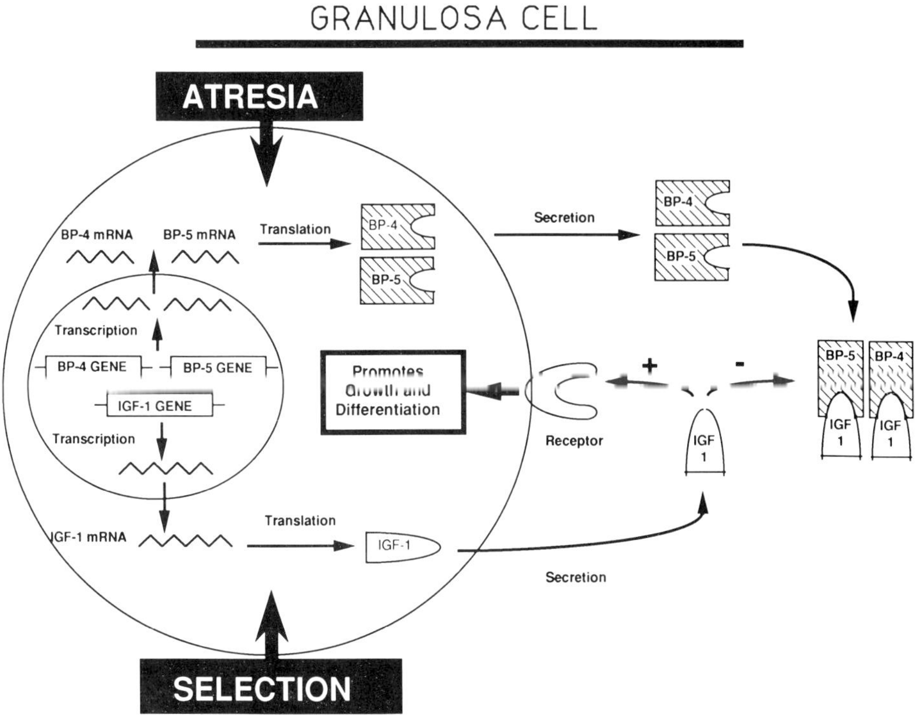

FIGURE 7 **Diagram of a theoretical model of the sequence of IGF-1 and IGFBP-4 and -5 expression in granulosa cells during selection and atresia in the rat ovary.**

IGFBP-5 gene transcription is activated in atretic preantral follicles. This observation is important because it suggests that the process of IGFBP-5 induction might be functionally involved in the physiological mechanism of preantral artresia. In the cycle, this event is most marked on estrous morning, a time when most (≈80%) of the preantral follicles are atretic (Oakberg, 1979). The mechanism of induction of IGFBP-5 in preantral follicles is unknown. Liu *et al.* (1992) have demonstrated that FSH is involved in the expression of this IGFBP. Oliver and colleagues (1989) reported that rat preantral follicles ≥130 μm synthesize IGF-I. Collectively, these data support a model in which IGFBP-5 could promote preantral atresia by binding and negating the tropic activities of locally produced IGF-I (Figure 7).

Finally, it should be noted that IGFBP-5 mRNA is found in GC of atretic antral follicles, but it is relatively rare and restricted to only certain atretic antral follicles around the morning of estrus (Erickson *et al.*, 1992b). We stated earlier that virtually all atretic antral follicles strongly express IGFBP-4 mRNA. Therefore, in the adult cycling rat, IGFBP-4 and IGFBP-5 are

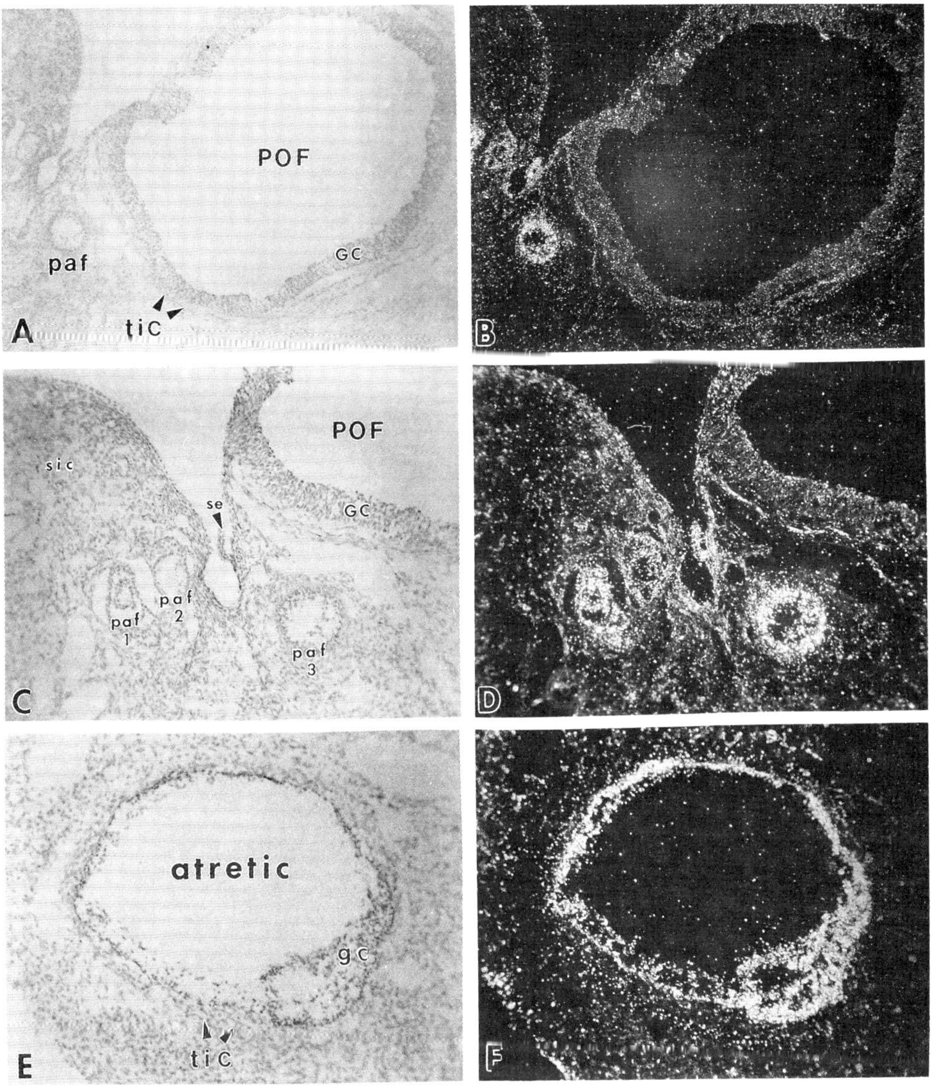

FIGURE 8 Localization of IGFBP-5 mRNA in the rat ovary at proestrus 2000 hours (A–D) and estrus 200 hours (E and F). POF, preovulatory follicle; gc, granulosa cells; tic, theca-interstitial cells; paf, preantral follicle; sic, secondary interstitial cells; se, surface epithelium. (A, C, and E) Bright-field photomicrographs. (B, D, and F) Dark-field photomicrographs hybridized with an antisense cRNA probe. Reproduced with permission from Erickson *et al.* (1992). Copyright © 1992 The Endocrine Society.

primarily expressed in atretic antral and preantral follicles, respectively. These observations argue that IGFBP-4 and -5 gene activity can be regulated differently in folliculogenesis and thus may have different physiological activities. Nevertheless, the findings support the novel concept that IGFBP-5 may be an autocrine/paracrine regulator of ovarian physiology, particularly in relation to preantral follicle atresia.

6. IGFBP-6

Preliminary analyses of IGFBP-6 expression in the adult rat ovary indicate that the message is selectively expressed in the smooth muscle cells of large blood vessels and in the theca externa of developing Graafian follicles (Table 1). The hypothesis emerging from this evidence is that growth and differentiation of the ovarian smooth muscle cells may be governed by the activity of IGFBP-6 protein, possibly by mechanisms involving IGF-I and/or II. However, there is as yet no evidence to support this hypothesis.

VII. Summary and Conclusions

Historically, ovarian homeostasis has been thought to be primarily under endocrine control. However, research done in many laboratories in the past few years has demonstrated that the hormonal signals can be locally modulated, either amplified or attenuated, by growth factors. Hence, our current theory of ovarian homeostasis, which encompasses the entire span from recruitment to luteolysis, is that the events are under the control of endocrine and autocrine/paracrine regulatory elements. The potential role of the intrinsic IGF system in normal folliculogenesis and luteogenesis has been clearly demonstrated. The data presented in the preceding sections of this chapter support the concept that IGF-I has an important regulatory role in selection and luteinization in the rat ovary. In the past few years, our group has made several novel discoveries that provided important new information on the ovarian IGF concept. These discoveries include (1) the identification of a novel FSH inhibitor in follicular fluid which proved to be IGFBP-3; (2) the fact that the IGFBP functions by sequestering intrinsically produced IGF-I required for GC differentiation; this exciting result strongly suggests that intrinsic IGF-I is an obligatory mediator of FSH-dependent rat GC differentiation; and (3) normally occurring atresia involves the tissue-specific expression of two IGFBPs (IGFBP-4 and -5) in the granulosa cells, and normal luteolysis involves the tissue-specific expression of IGFBP-3. Conversely, no IGFBP appear in GC of selected follicles or the functioning CL. Collectively, these novel findings have led us to propose that the induction of IGFBP gene activity could perhaps negatively regulate the tropic IGF-I activity required to prevent dominant follicles and functioning CL from undergoing a programmed death process. The next important problem will be to understand the regulation of inducible and time-specific expression of ovarian IGFBPs (-2, -3, -4, -5, and -6) and to show how the regulations are integrated into the overall physiological programs of ovarian homeostasis.

Acknowledgment

This work was supported by NICHD Program Project Grants HD-09690, NICHD Contract NO1-HD-0-2902, and NICHD Research Grant HD-24585.

References

Adashi, E. Y. (1990). The potential relevance of cytokines to ovarian physiology: the emerging role of resident ovarian cells of the white blood cell series. *Endocr. Rev.* 11, 454–464.

Adashi, E. Y., Resnick, C. E., Svoboda, M. E., and Van Wyk, J. J. (1984). A novel role for somatomedin-C in the cytodifferentiation of the ovarian granulosa cell. *Endocrinology* 115, 1227–1229.

Adashi, E. Y., Resnick, C. E., Brodie, A. M. H., Svoboda, M. E., and Van Wyk, J. J. (1985a). Somatomedin-C mediated potentiation of follicle stimulating hormone-induced aromatase activity of cultured rat granulosa cells. *Endocrinology* 117, 2313–2320.

Adashi, E. Y., Resnick, C. E., Svoboda, M. E., and Van Wyk, J. J. (1985b). Somatomedin-C synergizes with follicle stimulating hormone in the acquisition of progestin biosynthetic capacity by cultured rat granulosa cells. *Endocrinology* 116, 2135–2142.

Adashi, E. Y., Resnick, C. E., Svoboda, M. E., and Van Wyk, J. J. (1985c). Somatomedin-C enhances induction of luteinizing hormone receptors by follicle stimulating hormone in cultured rat granulosa cells. *Endocrinology* 116, 2369–2375.

Adashi, E. Y., Resnick, C. E., D'Ercole, A. J., Svoboda, M. E., and Van Wyk, J. J. (1985d). Insulin-like growth factors as intraovarian regulators of granulosa cell growth and function. *Endocr. Rev* 6, 400–420.

Adashi, E. Y., Resnick, C. E., Svoboda, M. E., Van Wyk, J. J., Hascall, V. C., and Yanagishita, M. (1986a). Independent and synergistic actions of somatomedin-C in the stimulation of proteoglycan biosynthesis by cultured rat granulosa cells. *Endocrinology* 118, 456–458.

Adashi, E. Y., Resnick, C. E., Svoboda, M. E., and Van Wyk, J. J. (1986b). Somatomedin-C as an amplifier of follicle-stimulating hormone action: enhanced accumulation of adenosine 3'5'-monophosphate. *Endocrinology* 118, 149–155.

Adashi, E. Y., Resnick, C. E., Svoboda, M. E., and Van Wyk, J. J. (1986c). Synergistic interactions of somatomedin-C with adenosine 3',5'-cyclic-monophosphate-dependent granulosa cell agonists. *Biol. Reprod.* 34, 81–88.

Adashi, E. Y., Resnick, C. E., Hernandez, E. R., Svoboda, M. E., and Van Wyk, J. J. (1988a). Characterization and regulation of a specific cell membrane receptor for somatomedin-C/insulin-like growth factor I in cultured rat granulosa cells. *Endocrinology* 122, 194–201.

Adashi, E. Y., Resnick, C. E., Hernandez, E. R., Svoboda, M. E., and Van Wyk, J. J. (1988b). *In vivo* regulation of granulosa cell somatomedin C/insulin-like growth factor I receptors. *Endocrinology* 122, 1383–1389.

Adashi, E. Y., Resnick, C. E., Hurwitz, A., Ricciarelli, E., Hernandez, and E. R., Rosenfeld, R. G. (1991). Ovarian granulosa cell-derived insulin-like growth factor binding proteins: modulatory role of follicle stimulating hormone. *Endocrinology* 128, 754–760.

Aria, H. (1920). On the postnatal development of the ovary (albino rat) with special reference to the number of ova. *Am. J. Anat.* 27, 405–411.

Bach, L. A., Thotakura, N. R., and Rechler, M. W. (1992). Human insulin-like growth factor binding protein-6 is O-glycosylated. *Biochem. Biophys. Res. Commun.* 186, 301-307.

Banka, C. L., and Erickson, G. F. (1985). Gonadotropin releasing hormone induces meiotic maturation in subpopulations of atretic preantral follicles. *Endocrinology* 117, 1500–1507.

Bicsak, T. A., Shimonaka, M., Malkowski, M., and Ling, N. (1990). Insulin-like growth factor-binding protein (IGF-BP) inhibition of granulosa cell function: effect on cyclic adenosine 3′,5′-monophosphate, deoxyribonucleic acid synthesis and comparison with the effect of an IGF-I antibody. *Endocrinology* 126, 2184-2189.

Biehl, M. M., and Miesteld, R. L. (1991). Isolation and characterization of transcripts induced by androgen withdrawal and apoptotic cell death in the rat ventral prostate. *Mol. Endocrinl.* 5, 138 1388.

Carlsson, B., Carlsson, L., and Billig, H. (1989). Estrous cycle-dependent co-variation of insulin-like growth factor I (IGF-I) messenger ribonucleic acid and protein in the rat ovary. *Mol. Cell. Endocrinl.* 64, 271–275.

Daud, A. I., Bumpus, F. M., and Husain, A. (1988). Evidence for selective expression of angiotensin II receptors in atretic follicles in the rat ovary: an autoradiographic study. *Endocrinology* 122, 2727–2734.

Davoren, J. B., Kasson, B. G., Li, C. H., and Hsueh, A. J. W. (1986). Specific insulin-like growth factor (IGF) I and II binding sites on rat granulosa cells: relation to IGF action. *Endocrinology* 119, 2155–2162.

Davoren, J. B., Kasson, B. G., Li, C. H., and Hsueh, A. J. W. (1986). Specific insulin-like growth factor (IGF) I and II binding sites on rat granulosa cells; relation to IGF action. *Endocrinology* 119, 2155–2162.

Erickson, G. F. (1978). Normal ovarian function. *Clin. Obstet. Gynecol.* 21, 31–52.

Erickson, G. F. (1983). Primary cultures of ovarian cells in serum-free medium as models of hormone-dependent differentiation. *Mol. Cell. Endocrinl.* 29, 21–49.

Erickson, G. F. (1994). The ovary: Basic principles and concepts. *In* "Endocrinology and Metabolism" (P. Felig, J. D. Baxter, A. E. Broadus, and L. A. Frohman, eds.), 3rd Ed. McGraw-Hill, New York (in press).

Erickson, G. F., Magoffin, D. A., Dyer, C. A., and Hofeditz, C. (1985). The ovarian androgen producing cells: a review of structure/function relationships. *Endocr. Rev.* 6, 371–399.

Erickson, G. F., Nakatani, A., Ling, N., and Shimasaki, S. (1992a). Cyclic changes in insulin-like growth factor binding protein-4 (IGFBP-4) mRNA in the rat ovary. *Endocrinology* 130, 625–636.

Erickson, G. F., Nakatani, A., Ling, N., and Shimasaki, S. (1992b). Localization of insulin-like growth factor binding protein-5 (IGFBP-5) mRNA in rat ovaries during the estrous cycle. *Endocrinology* 130, 1867–1878.

Erickson, G. F., Nakatani, A., Ling, N., and Shimasaki, S. (1993). Insulin-like growth factor binding protein-3 gene expression is restricted to involuting corpora lutea in rat ovaries. *Endocrinology* 133, 1147–1157.

Gospodarowicz, D., Ferrara, N., Schweigerer, L., and Neufeld, G. (1987). Structural characterization and biological functions of fibroblast growth factor. *Endocr. Rev.* 8, 95–114.

Hammond, J. M., Baranao, J. L. S., Skaleris, D., Knight, A. B., Romanus, J. A., Rechler, M. M. (1985). Production of insulin-like growth factors by ovarian granulosa cells. *Endocrinology* 117, 2553–2555.

Hernandez, E. R., Roberts, C. T., Jr., LeRoith, D., and Adashi, E. Y. (1989). Rat ovarian insulin-like growth factor I (IGF-I) gene expression is granulosa cell-selective: 5′-untranslated mRNA variant representation and hormonal regulation. *Endocrinology* 125, 572–575.

Hernandez, E. R., Roberts, C. T., Jr., Hurwitz, A., LeRoith, D., and Adashi, E. Y. (1990). Rat ovarian insulin-like growth factor II gene expression is theca-interstitial cell-exclusive: hormonal regulation and receptor distribution. *Endocrinology* 127, 3249–3251.

Hernandez, E. R., Hurwitz, A., Botero, L., Ricciarelli, E., Werner, H., Roberts, C. T., Jr., LeRoith, D., and Adashi, E. Y. (1991). Insulin-like growth factor receptor gene expression in the rat ovary: divergent regulation of distinct receptor species. *Mol. Endocrinol.* 5, 1799–1805.

Hillier, S. H., and Ross, G. T. (1979). Effects of exogenous testosterone on ovarian weight, follicular morphology and intraovarian progesterone concentration in estrogen-primed hypophysectomized immature female rats. *Biol. Reprod.* 20, 261-268.

Hirshfield, A. N. (1989). Rescue of atretic follicles *in vitro* and *in vivo*. *Biol. Reprod.* 40, 181–190.

Hurwitz, A., and Adashi, E. Y. (1992). Ovarian follicular atresia is an apoptotic process: a paradigm for programmed cell death in endocrine tissues. *Mol. Cell. Endocrinol.* 84, 19–23.

Hutchinson, L. A., Findlay, J. K., and Herrington, A. C. (1988). Growth hormone and insulin-like growth factor-I accelerate PMSG-induced differentiation of granulosa cells. *Mol. Cell. Endocrinol.* 55, 61-69.

Ingram, D. L. (1962). Atresia. *In* "The Ovary" (S. Zuckerman, A. M. Mandly, and P. Eckstein, eds.), Vol. 1, pp. 247–268. Academic Press, New York.

Kiefer, M. C., Ioh, R. S., Bauer, D. M., and Zapf, J. (1991). Molecular cloning of a new human insulin-like growth factor binding protein. *Biochem. Biophys. Res. Commun.* 176, 219–225.

Knecht, M,. Feng, P., and Catt, K. J. (1989). Transforming growth factor-β: autocrine, paracrine, and endocrine effects in ovarian cells. *Sem. Reprod. Endocrinol.* 7, 12–20.

Koistenen, R., Suikkari, A.-M., Tiitinen, A., Kontula, K., and Seppälä, M. (1990). Human granulosa cells contain insulin-like growth factor-binding protein (IGFBP-1) mRNA. *Clin. Endocrinol.* 32, 635–640.

Ling, N., Ui, M., Shimonaka, M., Shimasaki, S., and Bicsak, T. (1990). An IGF-binding protein in ovarian follicular fluid blocks FSH-stimulated steroid production by ovarian granulosa cells. *In* "Neuroendocrine Regulation of Reproduction" (S. S. C. Yen and W. Vale, eds.), pp. 207–215. Serono Symposia Publishers, New York.)

Liu, X.-J., Malkowski, M., Guo, Y., Erickson, G. F., Shimasaki, S., and Ling, N. (1992). Development of specific antibodies to rat insulin-like growth factor binding proteins (IGFBP-2 to 6): analysis of IGFBP production by rat granulosa cells. *Endocrinology* 132, 1176–1183.

Louvet, J. P., Harman, S. M., and Ross, G. T. (1975a). Effects of hCG, LH and human FSH on ovarian weights in estrogen-primed hypophysectomized immature female rats. *Endocrinology* 96, 1179–1186.

Louvet, J. P., Harman, S. M., Schreiber, J. K., and Ross, G. T. (1975b). Evidence for a role of androgens in follicular maturation. *Endocrinology* 97, 366–372.

Magoffin, D. A., and Erickson, G. F. (1988). An improved method for primary culture of ovarian androgen-producing cells in serum-free medium: effect of lipoproteins, insulin, and insulin-like growth factor I. *In Vitro Cell. Dev. Biol.* 24, 862-870.

Magoffin, D. A., Kurtz, K. M., and Erickson, G. F. (1990). Insulin-like growth factor-1 selectively stimulates cholesterol side-chain cleavage expression in ovarian theca-interstitial cells. *Mol. Endocrinol.* 4, 489–496.

May, J. V., and Schomberg, D. W. (1989). The potential relevance of epidermal growth factor (EGF) and transforming growth factor-alpha (TGFα) to ovarian physiology. *Sem. Reprod. Endocrinol.* 7, 1–11.

McNatty, K. P., Hunter, W. M., McNeilly, A. S., and Sawers, R. S. (1975). Changes in the concentration of pituitary and steroid hormones in the follicular fluid of human Graafian follicles throughout the menstrual cycle. *J. Endocrinol.* 64, 551–571.

Murphy, L. J., Bell, G. I., and Friesen, H. G. (1987). Tissue distribution of insulin-like growth factor I and II messenger ribonucleic acid in the adult rat. *Endocrinology* 120, 1279–1282.

Nakatani, A., Shimasaki, S., Erickson, G. F., and Ling, N. (1991). Tissue specific expression of four insulin-like growth factor binding proteins (1, 2, 3, and 4) in the rat ovary. *Endocrinology* 129, 1521-1529.

Nelson, S. E., McLean, M. P., Jayatilak, P. G., and Gibori, G. (1992). Isolation, characterization, and culture of cell subpopulations forming the pregnant rat corpus luteum. *Endocrinology* 130, 954–966.

Oakberg, E. F. (1979). Follicular growth and atresia in the mouse. *In Vitro* 15, 41–49.

Oliver, J. E., Aitman, T. J., Powell, J. F., Wilson, C. A., and Clayton, R. N. (1989). Insulin-like growth factor I gene expression in the rat ovary is confined to the granulosa cells of developing follicles. *Endocrinology* 124, 2671–2679.

Pang, C. Y., and Behrman, H. R. (1981). Acute effects of $PGF_{2\alpha}$ on ovarian and luteal blood flow, luteal gonadotropin uptake *in vivo* and gonadotropin binding *in vitro*. *Endocrinology* 108, 2239–2244.

Parmer, T. G., Roberts, C. T., LeRoith, D., Adashi, E. Y., Khan, I., Solan, N., Nelson, S., Zilberstein, M., and Gibori, G. (1991). Expression, action and steroidal regulation of insulin-like growth factor-I (IGF-I) and IGF-I receptor in the rat corpus luteum: their differential role in the two cell populations forming the corpus luteum. *Endocrinology* 129, 2924–2932.

Rao, M. C., Midgley, A. R., and Richards, J. S. (1978). Hormonal regulation of ovarian cellular proliferation. *Cell* 14, 71–78.

Ricciarelli, E., Hernandez, E. R., Hurwitz, A., Kokia, E., Rosenfeld, R. G., Schwander, J., and Adashi, E. Y. (1991). The ovarian expression of the antigonadotropic insulin-like growth factor binding protein-2 is theca interstitial cell-selective: evidence for hormonal regulation. *Endocrinology* 129, 2266–2268.

Richards, J. S., and Hedin, L. (1988). Molecular aspects of hormone action in ovarian follicular development, ovulation, and luteinization. *Annu. Rev. Physiol.* 50, 411–463.

Richards, J. S., Jahnsen, T., and Hedin, L. (1987). Ovarian follicular development: from physiology to molecular biology. *Recent Prog. Horm. Res.* 43, 231-270.

Schwall, R., and Erickson, G. F. (1981). Functional and morphological changes in rat theca cells during atresia. *In* "Dynamics of Ovarian Function" (N. B. Schwartz and M. Hunzicker-Dunn, eds.), pp. 29–34. Raven Press, New York.

Schwartz, N. B. (1974). The role of FSH and LH and their antibodies on follicle growth and ovulation. *Biol. Reprod.* 10, 236–252.

Seppälä, M., Wahlström, T., Koskimies, A. I., Tenhunen, A., Rutanen, E. M., Koistinen, R., Huhtaniemi, I., Bohn, H., and Stenman, U.-H. (1984). Human preovulatory follicular fluid, luteinized cells of hyperstimulated preovulatory follicles and corpus luteum contain placental protein 12. *J. Clin. Endocrinol. Metab.* 58, 505–510.

Shimasaki, S., and Ling, N. (1991). Identification and molecular characterization of insulin-like growth factor binding proteins (IGFBP-1,-2,-3,-4,-5 and -6). *Prog. Growth Factor Res.* 3, 243–266.

Shimasaki, S., Shimonaka, M., Ui, M., Inouye, S., Shibata, F., and Ling, N. (1990a). Structural characterization of a follicle stimulating hormone action inhibitor in porcine ovarian follicular fluid: its identification as the insulin-like growth factor binding protein. *J. Biol. Chem.* 265, 2198–2202.

Shimasaki, S., Uchiyama, F., Shimonaka, M., and Ling, N. (1990b). Molecular cloning of the cDNAs encoding a novel insulin-like growth factor-binding protein from rat and human. *Mol. Endocrinol.* 4, 1451–1458.

Shimasaki, S., Shimonaka, M., Zhang, H.-P., and Ling, N. (1991a). Identification of five different insulin-like growth factor binding proteins (IGFBPs) from adult rat serum and molecular cloning of a novel IGFBP-5 in rat and human. *J. Biol. Chem.* 266, 10,646–10,653.

Shimasaki, S., Gao, L., Shimonaka, M., and Ling, N. (1991b). Isolation and molecular cloning of insulin-like growth factor-binding protein-6. *Mol. Endocrinol.* 4, 938–948.

Smith, M. S., Freeman, M. E., and Neill, J. D. (1975). The control of progesterone secretion during the estrous cycle and early pseudopregnancy in the rat: prolactin, gonadotropin and steroid levels associated with rescue of the corpus luteum of pseudopregnancy. *Endocrinology* 96, 219–226.

Talbert, G. B., Meyer, R. K., and McShan, W. H. (1951). Effect of hypophysectomy at the beginning of proestrus on maturing follicles in the ovary of the rat. *Endocrinology* 49, 687–694.

Ui, M., Shimonaka, M., Shimasaki, S., and Ling, N. (1989). An insulin-like growth factor-binding protein in ovarian follicular fluid blocks follicle stimulating hormone-stimulated steroid production by ovarian granulosa cells. *Endocrinology* 125, 912–916.

Vale, W., Rivier, C., Hsueh, A. J. W., Campen, C., Meunier, H., Bicsak, T., Vaughan, J., Corrigan, A., Bardin, W., and Sawchenko, P. (1988). Chemical and biological characterization of the inhibin family of proteins. *Recent Prog. Horm. Res.* 44, 1–34.

Waterman, M. R., and Simpson, E. R. (1989). Regulation of steroid hydroxylase gene expression is multifactorial in nature. *Recent Prog. Horm. Res.* 45, 533–566.

Wood, W. I., Cachianes, G., Henzel, W. J., Winslow, G. A., Spencer, S. A., Hellmiss, R., Martin, J. L., and Baxter, R. C. (1988). Cloning and expression of the growth hormone-dependent insulin-like growth factor-binding protein. *Mol. Endocrinol.* 2, 1176–1185.

Zhiwen, Z., Carson, R. S., Herington, A. C., Lee, V. W. K., and Burger, H. G. (1987). Follicle stimulating hormone and somatomedin-C stimulate inhibin production by rat granulosa cells *in vitro*. *Endocrinology* 120, 1633–1638.

Zhou, J., Chin, E., and Bondy, C. (1991). Cellular pattern of insulin-like growth factor-I (IGF-I) and IGF-I receptor gene expression in the developing and mature ovarian follicle. *Endocrinology* 129, 3281–3288.

6

Expression of Genes Encoding Steroidogenic Enzymes in the Ovary

MARGARET M. HINSHELWOOD, MICHELLE DEMETER-ARLOTTO, GARY D. MEANS, & EVAN R. SIMPSON

I. Introduction

Steroid secretion throughout the ovarian cycle occurs in a highly coordinated and episodic fashion that is determined, in part, by the pattern of gonadotropin secretion. The ovarian cycle can be divided into two parts on a temporal and functional basis: the follicular phase, which is characterized by selection and growth of the ovulatory follicle(s), and the luteal phase of the cycle, in which the corpus luteum (CL) is dominant. Throughout the follicular phase, the steroidogenic pathway is geared toward the production of estradiol-17β by the dominant follicle(s). After the ovulatory surge of gonadotropins, luteinization takes place. The primary steroidal product of the corpus luteum is progesterone, although in some species, namely humans and higher primates, estradiol-17β is also produced by the luteal cells. Figure 1 shows the steroidogenic pathways expressed in the ovary. The amount of a specific steroid synthesized in any tissue depends on two factors—the quantity of substrate available, cholesterol in the case of progesterone or C_{19} steroid in the case of estrogen, and the activities of the particular steroidogenic enzymes involved (Strauss *et al.*, 1981). Therefore, in order to understand the episodic changes in steroid production that occur throughout the ovarian cycle, one must gain insight into the changes in the levels of steroidogenic enzymes throughout the cycle and the mechanisms responsible for regulating these changes.

Molecular Biology of the Female Reproductive System

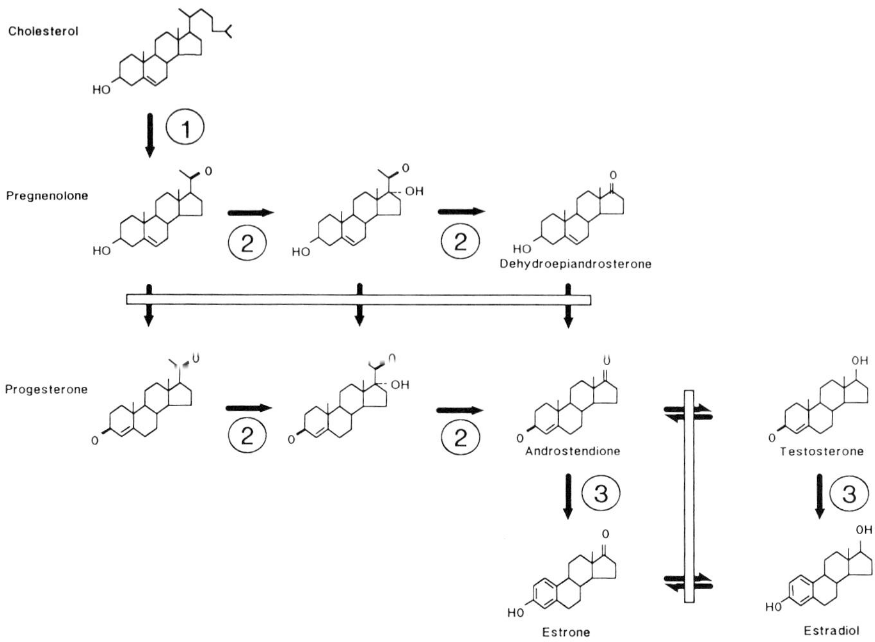

FIGURE 1 Schematic representation of ovarian steroid biosynthesis. The cytochrome P450 enzymes that are involved in steroid biosynthesis are indicated by numbers. Each of the hydroxylation reactions catalyzed by P450 enzymes is characterized by the subsequent removal of either six carbons (P450scc), two carbons (P45017α), or one carbon (P450arom). 3β-HSD catalyzes the oxidation of the hydroxyl at C3 and the subsequent isomerization of Δ^5 steroids to their Δ^4 products. The reversible conversion of androstenedione to testosterone or estrone to estradiol is a consequence of the 17β-hydroxysteroid dehydrogenase/oxidoreductase (17β-HSD) family of enzymes. 1, P450scc; 2, P45017α; 3, P450arom. Horizontal bar is 3β-HSD, vertical bar is 17β-HSD.

II. Role of P450 Cytochromes in Ovarian Steroidogenesis

The successive steps in estrogen biosynthesis in the ovarian follicle occur as a result of the progressive loss of carbon atoms from the precursor cholesterol, which is metabolized in a series of hydroxylation reactions. These are catalyzed by three members of a superfamily of genes known collectively as cytochrome P450 (Nebert *et al.*, 1991). They are cholesterol side chain cleavage cytochrome P450 (P450scc), 17α-hydroxylase cytochrome P450 (P45017α), and aromatase cytochrome P450 (P450arom). A number of complementary deoxyribonucleic acid (cDNA) clones specific for various members of the cytochrome P450 superfamily, which is currently divided into 12 families, have been identified (Estabrook, 1989).

Members within each family exhibit greater than 36% amino acid sequence identity, whereas sequence identity across families is less than 36%. The heme binding domain of all P450 cytochromes, however, is highly conserved. Cytochrome P450scc, along with its electron donors, adrenodoxin and adrenodoxin reductase, which are localized to the mitochondria, catalyzes the conversion of cholesterol (a C_{27} steroid) to pregnenolone (a C_{21} steroid). Pregnenolone (or alternatively, progesterone) is subsequently converted to 17α-hydrdoxypregnenolone (or 17α-hydroxyprogesterone) by P45017α. The lyase activity of P45017α then converts these to the corresponding C_{19} steroids dehydroepiandrosterone and androstenedione. Expression of full-length cDNA clones encoding P45017α from human, bovine, and rat cells has revealed that lyase activity of the former two will preferentially form dehydroepiandrosterone as distinct from androstenedione, suggesting that the Δ^5 pathway is the preferred route in these species.

On the other hand, rat P45017α will more readily convert 17α-hydroxyprogesterone to androstenedione, compared with other species. The electron donor for P45017α is the flavoprotein nicotinamide adenine dinucleotide phosphate (NADPH)-cytochrome P450 reductase, which is localized to the endoplasmic reticulum, as is P45017α. The enzyme 3β-hydroxysteroid dehydrogenase/$\Delta^{5\rightarrow4}$ isomerase (3β-HSD) catalyzes the formation of a 3-ketone and the subsequent isomerization of the Δ^5 double bond to a Δ^4 double bond; it may use pregnenolone, 17α-hydroxypregnenolone, or dehydroepiandrosterone (DHEA) as substrates. In the ovary, 17-oxidoreductase catalyzes the reduction of the 17-keto moiety of androstenedione (or estrone) to testosterone (or estradiol). The conversion of C_{19} steroids (androstenedione and testosterone) to estrogens is catalyzed by another microsomal P450, P450arom, which like P45017α, uses NADPH-cytochrome P450 reductase as its electron donor.

Studies on the origin of the cholesterol used in steroid biosynthesis in humans suggest that cholesterol used in progesterone synthesis by the corpus luteum is derived primarily from circulating serum lipoproteins, rather than from *de novo* synthesis from acetate (Tureck and Strauss, 1982; Carr *et al.*, 1981). The reliance on low-density lipoprotein (LDL)-derived cholesterol has been demonstrated by the observation that LDL is required for optimal progesterone formation in cultured cells. Furthermore, in the same studies, high-density lipoproteins (HDL) did not support human ovarian progesterone biosynthesis (Tureck and Strauss, 1982; Carr *et al.*, 1981). Conversely, in cattle (Carroll *et al.*, 1992) and in sheep (Wiltbank *et al.*, 1990), the addition of either HDL or LDL increased progesterone production by luteal cells *in vitro*, although the stimulation was greater with HDL in ovine cells. In human follicular fluid, there appears to be no detectable LDL, thereby conceivably limiting the ability of the granulosa cells of preovulatory follicles to synthesize progesterone (Simpson *et al.*, 1980). On the other hand, LDL is available to theca interna cells which are well vascularized.

A. Side Chain Cleavage Cytochrome P450

The first and rate-limiting step in steroidogenesis is the conversion of cholesterol to pregnenolone, and this reaction is catalyzed by P450scc. P450scc has been immunolocalized in follicles and CL from several different species [humans (Sasano *et al.*, 1989), cattle (Rodgers *et al.*, 1986a,b,c) sheep (Rodgers *et al.*, 1988), and rats (Hedin *et al.*, 1987; Zlotkin *et al.*, 1986; Farkash *et al.*, 1986; LeGoascogne *et al.*, 1989)]. In human (Sasano *et al.*, 1989), bovine (Rodgers *et al.*, 1986a), and rat follicles (Zlotkin *et al.*, 1986; Farkash *et al.*, 1986; LeGoascogne *et al.*, 1989), cells of the theca interna were the major site for positive staining for P450scc. Granulosa cells from women showed no staining for P450scc (Sasano *et al.*, 1989). However, in cattle, weak staining was observed for P450scc in granulosa cells from the larger preovulatory follicles (Rodgers *et al.*, 1986a). In addition, ovarian interstitial cells in immature rats stained positive for P450scc (Zlotkin *et al.*, 1986; Farkash *et al.*, 1986). The CL in women, cattle, and rats displayed the most intense staining for P450scc compared with follicles or interstial cells. Even though the theca interna is the major site of P450scc in the follicle, both the thecal and granulosa-derived luteal cells from humans (Sasano *et al.*, 1989) and the large and small cells from the bovine CL (Rodgers *et al.*, 1986b) stained positive for P450scc. As in women and in cattle, immunolocalization of P450scc in the rat follicle was limited to the theca interna until after ovulation, when luteinized granulosa cells displayed positive staining for P450scc (LeGoascogne *et al.*, 1989). Small and large cell types separated from the CL of pregnant rats also were positive for P450scc (Gibori *et al.*, 1988). These results help explain the comparison of different stages of ovarian development: the content of P450scc was lowest in small and medium follicles, increased in large follicles, and was greatest in tissue from CL in both cattle (Rodgers *et al.*, 1986c) and rats (Hedin *et al.*, 1987). Not only do the levels of P450scc protein change throughout the ovarian cycle, but the content of its electron donor adrenodoxin varies in a similar fashion (Rodgers *et al.*, 1986c). So the amount of P450scc as well as adrenodoxin seems to reflect the overall capacity for C_{21} steroid production at any given stage of the ovarian cycle.

It appears that the major control of production of P450scc is likely transcriptional, since changes in amounts of hybridizable transcript mirror the shifts in the quantity of P450scc protein throughout the ovarian cycle. In most species, the size of the P450scc transcript is approximately 2.0 kb. In women, low levels of P450scc mRNA were detected in all follicles measured (Days 7–13 of the menstrual cycle), with peak amounts observed in CL from midcycle and then declining until luteolysis (Doody *et al.*, 1990).

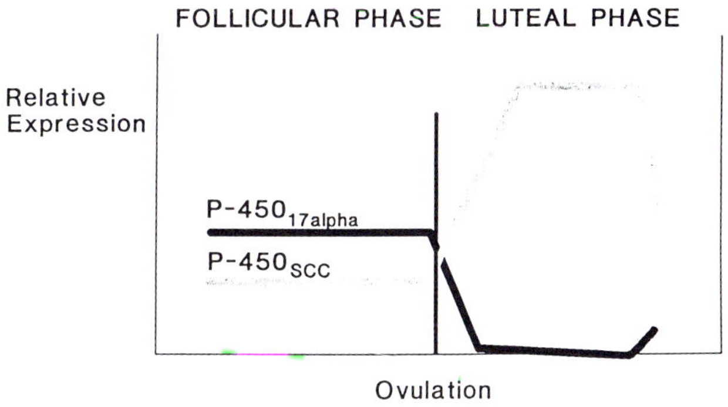

FIGURE 2 Diagrammatic representation of the expression of transcripts for P450scc and P45017α throughout the bovine ovarian cycle.

Analogous to the results found in women, the amount of hybridizable transcript was low in follicles from rats at different stages of the cycle, with the bulk of hybridizable transcript seen in CL from pregnant animals (Doody *et al.*, 1991). In a study performed with mRNA prepared from bovine ovaries obtained from a slaughterhouse, a similar situation was observed for both P450scc and adrenodoxin (Rodgers *et al.*, 1987) (Figure 2). Even though mRNA levels of P450scc seem to change in direct relationship to the amount of progesterone production, there are some exceptions. In cynomolgus macaques, the rise and fall in progesterone production caused by growth and regression of the CL was not reflected by a dramatic change in the amount of hybridizable transcript for P450scc until after the onset of menses (Bassett *et al.*, 1991). During pregnancy in rats, circulating levels of progesterone fall at parturition; however, levels of P450scc mRNA were no different at parturition than earlier in gestation (Goldring *et al.*, 1987). The changes in progesterone production observed throughout the luteal lifespan may therefore reflect in part differences in available substrate, rather than just alterations in the levels of P450scc.

Many factors known to increase progesterone production, such as human and equine chorionic gonadotropin (hCG and eCG), also increased the expression of P450scc in rat ovaries when administered *in vivo* (Doody *et al.*, 1991; Oonk *et al.*, 1989; Tuckey and Atkinson, 1989). Treatment with prolactin *in vitro* augmented the effect of hCG on 450scc transcript levels in these granulosa and luteal cells (Oonk *et al.*, 1989). Conversely, treatment with a gonadotropin-releasing hormone (GnRH) agonist not only decreased the content of progesterone in rat CL, but caused a decline in P450scc protein and mRNA (Smith *et al.*, 1991). In cattle, luteal cells in culture displayed a greater abundance of hybridizable transcript after LH was added, and this rise in levels of mRNA could be blocked by concurrent treatment with angiotensin II (Stirling *et al.*, 1990). Even though the amount of expression of P450scc appears to be low in follicles compared with CL, both human granulosa (McAllister *et al.*, 1990) and thecal cells (McAllister *et al.*, 1989) in culture did express hybridizable transcripts for P450scc, the levels of which were increased after the addition of forskolin, an analogue of cyclic adenosine monophosphate (cAMP).

Another hormone important in ovarian function known to act through cAMP is follicle-stimulating hormone (FSH). The addition of FSH to bovine (Funkenstein *et al.*, 1983), porcine (Veldhuis *et al.*, 1986; Veldhuis and Rogers, 1987), and rat (Trzeciak *et al.*, 1986) granulosa cells increased the activity of P450scc. This rise in activity after FSH treatment was a result of an increase in the amount of both P450scc and adrenodoxin protein (Funkenstein *et al.*, 1983; Trzeciak *et al.*, 1986). In porcine granulosa cells, as in granulosa cells from other species, the levels of transcripts for P450scc were greater after FSH was added (Urban *et al.*, 1991). Tetradecanoyl phorbol acetate (TPA), a stimulator of protein kinase C, blocked the FSH-induced increase in both the level of P450scc protein and its transcripts in rat granulosa in culture (Trzeciak *et al.*, 1987a). Several growth factors found in the ovary also appear to modulate the expression of the P450scc complex. Epidermal growth factor (EGF) increased the amounts of both P450scc and adrenodoxin protein in rat granulosa cells in either the presence or the absence of FSH (Trzeciak *et al.*, 1987b). Similar results were obtained with insulin-like growth factor I (IGF-I); however, not only did IGF-I cause a rise in levels of the components of the P450scc complex, but there was a synergy between FSH and IGF-I in inducing the synthesis of P450scc and adrenodoxin in porcine granulosa cells (Veldhuis *et al.*, 1986; Veldhuis and Rogers, 1987). Thus there are many factors in the ovary that have the ability to regulate the expression and hence activity of the P450scc complex.

One other study which may implicate a local growth factor in the molecular mechanisms of steroidogenesis is that of Stirling *et al.* (1991). They reported that transcripts for basic fibroblast growth factor (bFGF) could be detected in corpora lutea but not in follicles of the bovine ovary. The most abundant transcripts could be found in the regressing corpora lutea. In further studies they found that angiotensin II stimulated the expression of bFGF (Stirling *et al.*, 1991) but decreased the expression of cholesterol side chain cleavage mRNA (Stirling *et al.*, 1990) in bovine luteal cells in culture. It is interesting to speculate about the role of angiotensin II and bFGF in the process of luteinization. One possibility is that angiotensin II plays an indirect role in the process of angiogenesis which accompanies luteinization by inducing the expression of bFGF, a known angiogenic factor.

Of the components of the side chain cleavage electron transport system, the levels of transcripts for adrenodoxin have been studied in cultured human granulosa cells, and shown to increase following treatment with hCG, FSH, cAMP, or cholera toxin (Voutilainen *et al.*, 1988).

B. 17α-Hydroxylase Cytochrome P450

The 17α-hydroxylation of both pregnenolone and progesterone, as well as their conversion to C_{19} steroids, is accomplished by the same enzyme, P45017α, and its electron donor NADPH-cytochrome P450 reductase (Nakajin *et al.*, 1981). In ovaries from women, P45017α was immunolocalized to the cells of the theca interna, as well as to the luteinized thecal cells of the corpus luteum (Sasano *et al.*, 1989; Tamura *et al.*, 1992). These results agree with the ability of both follicles and CL to produce aromatizable substrate in the human. However, in cattle (Rodgers *et al.*, 1986a,b) and pigs (Suzuki *et al.*, 1992), species whose CL do not produce estrogen, there is no immunoreactive P45017α enzyme present in luteinized thecal cells. The ovulatory surge of LH may be the triggering factor in the decrease in P45017α, since there is no P45017α present even in CL from very early in the luteal phase in cattle (Rodgers *et al.*, 1986c) and injection of an ovulatory dose of hCG reduced the amount of P45017α within 7 hours in rats (Hedin *et al.*, 1987).

Another possible factor causing the drop in P45017α levels could be the preovulatory rise in estrogen, since Banks *et al.* (1991) reported that P45017α activity declined during proestrus in the rat. However, it should be noted that the process of luteinization that starts immediately with the surge of LH may not be irreversible, because thecal cell explants collected from rats 7 hours after hCG injection were able to reestablish androstenedione production *in vitro* (Richards *et al.*, 1986). This ability to reverse the process may help to explain the induction of P45017α in both thecal and interstitial cells when immature rats were primed with eCG followed by hCG (Ishimura *et al.*, 1990).

In adult animals, other factors may come into play to keep P45017α levels low after ovulation. Basic fibroblast growth factor, a proposed angiogenic factor in the corpus luteum (Gospodarowicz *et al.*, 1986), inhibited the hCG-induced increase in P45017α protein in rat ovarian cells in culture (Hurwitz *et al.*, 1990). However, in bovine thecal cells in culture that were not luteinized, bFGF did not inhibit 17α-hydroxylase activity, whereas transforming growth factor-β (TGFβ) was markedly effective in decreasing activity (Demeter-Arlotto *et al.*, 1993). So there may be many possible intraovarian and other interactions that may take place after ovulation in order to keep P45017α activity at low levels, and these effects may be species-specific.

Again, as in the case of P450scc, the amount of immunoreactive P45017α present in ovarian tissue seems directly related to the levels of P45017α mRNA in humans (Doody *et al.*, 1990), cattle (Rodgers *et al.*, 1987) (Figure 2), rats (Hedin *et al.*, 1987; Doody *et al.*, 1991), and pigs (Suzuki *et al.*, 1992). The amount of mRNA species approximately 1.8 kb in size was greatest in large follicles and midcycle CL from women (Doody *et al.*, 1990),

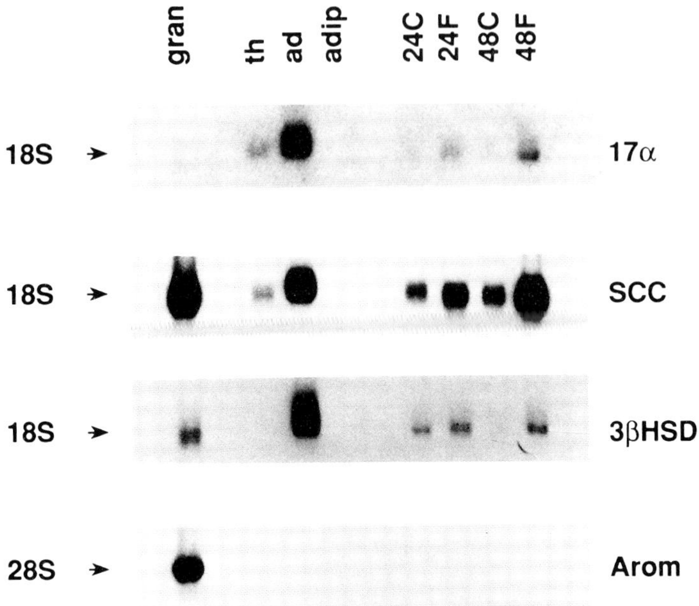

FIGURE 3 Northern blot analysis of poly(A)+ RNA obtained from freshly isolated bovine granulosa cells (gran) (3.2 μg), freshly isolated bovine theca interna (th) (3.2 μg), bovine adrenocortical cells (ad) (2 μg), and bovine adipose stromal cells (adip) (2 μg). Poly(A)+ RNA (11 μg) was also isolated from primary cultures of bovine theca interna cells grown for 5 days and then stimulated with forskolin (F) for 24 to 48 hours; C indicates control. Freshly harvested cells and tissue were taken from the same follicles from which the bovine theca interna cell cultures were prepared.

which is consistent with the maintenance of estrogen synthesis by the CL. The amount of hybridizable transcript for P45017α increased with increasing follicle size in cattle (Rodgers *et al.*, 1987), or as rats approached ovulation (Hedin *et al.*, 1987; Doody *et al.*, 1991), and then levels declined after ovulation (Figure 2). These changes in amount of transcript could be mimicked with exogenous hormone, or by adding forskolin to bovine thecal cells in culture (Demeter-Arlotto *et al.*, 1993; Figure 3). Injection of rats with eCG, a hormone with primarily FSH activity, increased total hybridizable transcript for P45017α, while injection with hCG caused a decrease (Doody *et al.*, 1991). Although P45017α mRNA levels dropped after ovulation in rats and were still low on Day 15 of pregnancy (Hedin *et al.*, 1987), the levels of hybridizable transcripts increased swiftly by Day 21 (Doody *et al.*, 1991). Since CL of rats synthesizes estrogen throughout pregnancy, this would indicate a role for placental androgen precursors in estrogen production by the CL of pregnancy (Warshaw *et al.*, 1986).

C. Aromatase Cytochrome P450

The last in the trio of cytochrome P450 enzymes involved in ovarian steroidogenesis is the aromatase cytochrome P450 complex (P450arom). P450arom is responsible for catalyzing the conversion of C_{19} steroids to C_{18} steroids. Although there have been many biochemical studies that indicate that granulosa cells are the source of P450arom activity in the follicle (Gore-Langton and Armstrong, 1988), there have been few studies that used immunolocalization techniques to demonstrate the cell types that contain P450arom. In women, P450arom was immunolocalized in granulosa and luteinized granulosa cells (Sasano *et al.*, 1989; Tamura *et al.*, 1992). However, these studies conflict with others that found P450arom was also localized in the theca interna (Matsuda *et al.*, 1984; Inkster and Brodie, 1991). The staining in the granulosa layer was always greater in larger rather than smaller follicles (Sasano *et al.*, 1989; Tamura *et al.*, 1992; Matsuda *et al.*, 1984; Inkster and Brodie, 1991). It was suggested by Inkster and Brodie (1991) that the different results obtained in the immunolocalization studies could be due to differences in fixation techniques, which altered the antigenic properties of P450arom. It is of interest to note that in studies of human ovarian cells in long-term culture, only granulosa cells showed hybridizable transcripts for P450arom (McAllister *et al.*, 1990), while cells of the theca interna did not (McAllister *et al.*, 1989). In rats and mice, P450arom was immunolocalized primarily in the granulosa cells of the preovulatory follicles (Ishimura *et al.*, 1989; Yoshinaga-Hirabayashi *et al.*, 1990; Sanghera *et al.*, 1991). Cells of some CL showed weak staining for P450arom, but the majority were negative (Yoshinaga-Hirabayashi *et al.*, 1990; Sanghera *et al.*, 1991). The theca interna and interstitial cells were also negative for P450arom.

In a study employing Northern analysis of mRNA prepared from rat granulosa cells, higher levels of P450arom transcripts were found in granulosa cells from preovulatory follicles compared with small follicles (Hickey *et al.*, 1988). This is in agreement with another study in rats that showed a greater abundance of P450arom mRNA levels during diestrus and proestrus compared with estrus and metestrus (Doody *et al.*, 1991). The greatest amounts of hybridizable transcripts in follicles from women were also found closest to the time of the preovulatory rise in estradiol, on Day 13 of the cycle (Doody *et al.*, 1990). Northern analysis of RNA prepared from human CL revealed a high abundance of P450arom transcripts during midcycle and then a decline with regression of the CL (Doody *et al.*, 1990). Therefore, based on the results of immunocytochemical and Northern analyses, there appears to be a correspondence between P450arom mRNA levels and the amount of enzyme protein throughout the ovarian cycle.

The number and sizes of P450arom transcripts appear to vary in different species. In human tissues, two sizes of P450arom transcripts of 3.4 and 2.9 kb appear to be functional, and the difference in size is apparently due to use of alternative polyadenylation sites (Doody *et al.*, 1990). Conversely, in rats, three species of approximately 2.7, 2.2, and 1.7 kb are found for P450arom mRNA (Hickey *et al.*, 1988; Doody *et al.*, 1991; Lephart *et al.*, 1990). However, further analysis employing specific probes revealed that the two smaller mRNA species contain an unspliced intron and lack the highly conserved heme-binding region, so that the transcripts encode nonfunctional proteins. It was concluded that the functional rat P450arom in ovarian tissue appears to be encoded by the 2.7-kb mRNA species (Lephart *et al.*, 1990, 1992b).

Even though the rat P450arom transcripts differ functionally, all three mRNA species seem to be expressed in the same fashion in the ovary regardless of age or hormonal status (Doody *et al.*, 1991; Hickey *et al.*, 1988; Lephart *et al.*, 1992b). This is especially true during pregnancy, where high P450arom levels are expressed during late gestation (Hickey *et al.*, 1988; Lephart *et al.*, 1992b). There appears to be a correlation between an increase in P450arom mRNA levels and enzymatic activity (Hickey *et al.*, 1988; Lephart *et al.*, 1992b). However, there clearly is not a direct relationship between the two, presumably because of the difference in time between the appearance of mRNA and the active P450arom enzyme (Hickey *et al.*, 1988; Lephart *et al.*, 1992a,b). In addition, the differences previously reported between the appearance of P450arom mRNA and enzyme activity may be explained, in part, by use of isolated preovulatory follicles (Hickey *et al.*, 1988) to measure activity in contrast to tissue slices from the whole ovary (Lephart *et al.*, 1992b).

As with the expression of P45017α mRNA, the levels of hybridizable transcripts for P450arom could be altered by the administration of hormones known to affect ovarian steroidogenesis. Immature rats injected with eCG had increased levels of P450arom mRNA, while a followup treatment with 10 IU of hCG negated this effect (Doody *et al.*, 1991). In another experiment, P450arom mRNA and activity were increased in granulosa cells from immature rats after treatment with FSH alone, and this effect was enhanced by addition of estrogen both *in vivo* (Hickey *et al.*, 1988; Fitzpatrick and Richards, 1991) and *in vitro* (Fitzpatrick and Richards, 1991). As in the previous study, treatment with LH or forskolin caused a rapid decline in P450arom mRNA levels. It appears that treatments that stimulate follicular growth enhance expression of P450arom in granulosa cells, while treatments that induce ovulation decrease expression of P450arom in preovulatory follicles. However, granulosa cells from preovulatory follicles induced to luteinize *in vivo* with hCG maintained a high expression of P450arom *in vitro,* independent of cAMP, when the cells were collected 5 to 7 hours after hCG was injected (Hickey *et al.*, 1990). The amount of P450arom mRNA observed during this treatment was lower than that observed in mRNA obtained from granulosa cells from preovulatory follicles (Hickey *et al.*, 1990). These results do, however, mirror the changes that occur in estrogen production during the transition from follicle to corpus luteum.

Another hormone, prolactin (PRL), known to be a luteotropic agent in the rat, blocked the increase in levels of P450arom transcripts normally observed after treatment with FSH in nonluteinized, preovulatory, rat granulosa cells in culture (Krasnow *et al.*, 1990). Similar results were obtained with granulosa cells that had just started to luteinize; PRL decreased the levels of P450arom transcripts and protein (Krasnow *et al.*, 1990). However, in luteal cells obtained from Days 8 to 12 of gestation, PRL increased the amount of P450arom mRNA and protein. So the effect of PRL on expression of P450arom is variable, depending on the physiological state of the animal, and appears to be related to estrogen secretion.

As indicated previously, the role of growth factors in ovarian steroidogenesis is under intense investigation. Although numerous studies have addressed their effects on production of steroids *in vitro* (for review, see Reed and James, 1989), there are few published studies on the actions of growth factors on the expression of steroidogenic enzymes. In one study, Steinkampf *et al.* (1988) examined the action of EGF on P450arom transcript levels in human granulosa cells in primary culture. They found that the inhibition of FSH-stimulated P450arom expression by EGF mirrored the inhibition of aromatase activity. In contrast to EGF, IGF-I stimulated aromatase activity and the levels of P450arom transcripts, both by itself and when added with FSH. It appears that growth factors are capable of modulating not only activity, but expression of P450arom.

D. 5α-Reductase

Another enzyme that has been studied in relation to estrogen production throughout the cycle is 5α-reductase. This enzyme provides an alternative to the P450arom enzyme as a pathway for androgen metabolism. Changes in both enzymatic activity and mRNA content for cytochrome P450arom and 5α-reductase (transcript size, 2.5 kb) were mapped throughout the rat estrous cycle (Lephart *et al.*, 1992a). Of interest in this study was the inverse relationship observed in both activity and levels of hybridizable transcripts between the two enzymes. At times when P450arom activity was greatest, at proestrus, 5α-reductase activity was at a nadir. Conversely, at diestrus and metestrus when activity of 5α-reductase was highest, P450arom activity was at its lowest. Changes in the amount of mRNA were similar for P450arom and 5α-reductase. The levels of transcripts for P450arom were greatest during diestrus and proestrus, while the amount of transcripts for 5α-reductase was greatest during estrus and metestrus (Lephart *et al.*, 1992a). The authors suggested that the inverse relationship in expression of the two enzymes may, at least in part, be a consequence of their localization in different cell types of the follicle, since the P450arom enzyme is in granulosa cells and 5α-reductase has been localized in the theca interna (Aono *et al.*, 1981). These differences in expression and activity of P450arom and 5α-reductase provide alternative pathways for testosterone metabolism and may play a role in controlling steroidogenesis throughout follicular development.

E. 3β-Hydroxysteroid Dehydrogenase/$\Delta^{5\rightarrow4}$ Isomerase

The enzyme responsible for the oxidation and isomerization of Δ^5-3β-hydroxysteroid precursors into Δ^4-3-ketosteroids is 3β-hydroxysteroid dehydrogenase/$\Delta^{5\rightarrow4}$ isomerase (3β-HSD (Luu-The *et al.*, 1989b; Lorence *et al.*, 1990). This enzyme has been immunolocalized in the human ovary, and levels vary with stage of follicular development; small follicles demonstrated immunoreactivity only in the theca interna, while immunoreactivity became apparent in the granulosa cells in larger follicles (Sasano *et al.*, 1990; Dupont *et al.*, 1992). The immunoreactivity was always less in the granulosa than in the theca interna. However, Sasano *et al.* (1990) reported that in large preovulatory follicles, the levels of 3β-HSD protein appeared nearly equal when the granulosa and thecal cell layers were compared. Both theca and granulosa-derived cells of the CL showed positive staining for 3β-HSD of equal intensity (Sasano *et al.*, 1990; Dupont *et al.*, 1992). In menopausal and postmenopausal women, some scattered interstitial cells stained positive for 3β-HSD (Dupont *et al.*, 1992). In rats, only the theca interna and cells of the CL showed positive staining for 3β-HSD (Dupont *et al.*, 1990).

When 3β-HSD activities were quantified with respect to the cycle, it was found in several species that there was much greater activity in the luteal versus follicular phase, in both women (Sano *et al.*, 1981) and cattle (Couet *et al.*, 1990), due to the presence of the CL. These changes in 3β-HSD activity throughout the lifespan of the CL from cattle were mirrored by changes in protein levels (Couet *et al.*, 1990). The difference in protein levels and activity of 3β-HSD does seem to be a direct effect of changes in the amounts of hybridizable transcripts for 3β-HSD since there was a high correlation ($R > 0.70$) between levels of 3β-HSD transcripts and activity using various substrates throughout the estrous cycle (Couet *et al.*, 1990). Transcripts for 3β-HSD were also much higher in midluteal phase CL from women than at other stages of the menstrual cycle (Doody *et al.*, 1990). However, it is of interest to note that in macaques, there was a trend for a decrease in mRNA levels for 3β-HSD from early in the luteal phase (Days 3–5) until luteolysis (Bassett *et al.*, 1991), which is in contrast to the results obtained in cattle (Couet *et al.*, 1990). The size of the 3β-HSD transcript in both humans and rats is 1.7 kb. In bovine CL, there was an increase in transcript levels for 3β-HSD after ovulation until Day 5 of the estrous cycle, and then levels remained fairly constant until about Days 12–13, when they started to decline (Couet *et al.*, 1990). When the amount of hybridizable transcripts for P450scc versus 3β-HSD in CL from rats (Doody *et al.*, 1991) and macaques (Bassett *et al.*, 1991) was compared, the levels of 3β-HSD declined more rapidly than those of P450scc as the CL aged. Even though levels of both 3β-HSD and P450scc mRNAs and activities change throughout the lifespan of the macaque CL, serum progesterone concentrations do not always mirror activity levels, suggesting that substrate availability and P450scc activity also play important roles in determining final progesterone levels.

As with the other steroidogenic enzymes, treatment with various hormones known to affect steroid production influenced not only 3β-HSD activity, but also mRNA levels. Treatment with hCG implants increased both the activity and levels of hybridizable transcripts for 3β-HSD in rat ovaries (Martel *et al.*, 1990a), and this effect was potentiated by the addition of PRL-containing implants. Conversely, when injected, hCG stimulated 3β-HSD activity, protein content, and mRNA only slightly; FSH had no influence; and PRL had a profound inhibitory effect (Martel *et al.*, 1990b). In this instance, the method of administration played a role in the effects of various pituitary hormones on 3β-HSD.

F. 17β-Hydroxysteroid Dehydrogenase

The enzyme responsible for the interconversion of 17-ketosteroids and their 17β-hydroxysteroid counterparts is 17β-hydroxysteroid dehydrogenase (17β-HSD). The presence of this enzyme has been demonstrated in a number of species and tissues (Martel *et al.*, 1992). However, so far, the cDNA and the gene have been isolated only in human tissues (Luu-The *et al.*, 1989a, 1990). The major interconversions for 17β-HSD in the ovary are between estrone (E_1) and estradiol (E_2), as well as androstenedione (A) and testosterone (T). It is interesting that in different species, opposing pathways of conversion may be favored. For example, while conversion of E_2 to E_1, namely the oxidative reaction, is the preferred route in the rat ovary, both the oxidative and reductive reactions appear to occur in the human ovary (Martel *et al.*, 1992).

Two mRNA species of 2.2 and 1.3 kb encoding 17β-HSD can be found in different human tissues, including the ovary. The dominant species is that of 1.3 kb. The difference between the two mRNA species lies in the 5′-untranslated region, the 2.2-kb species extending at least 814 base pairs further upstream than the 1.3-kb form (Luu-The *et al.*, 1989a).

III. Regulation of Expression of the Genes Encoding Steroidogenic Enzymes in Ovarian Cells

Studies on the levels of transcripts encoding steroidogenic enzymes in ovarian cells, although of great significance, do not address the issue of the nature of the regulatory mechanisms involved, since the steady-state levels of mRNA can be controlled at a number of sites, including both synthesis and turn-over. Very few, if any, studies employing techniques such as nuclear run-on to study transcript synthesis, or else mRNA half-lives, have applied these techniques to expression of steroidogenic enzymes in ovarian cells, in large part because of the technical difficulties involved. However, several laboratories have begun to study directly the expression of genes encoding steroidogenic enzymes, largely by means of transfection experiments employing chimeric DNA constructs, in which the appropriate promoter regions of steroidogenic genes are fused upstream of suitable reporter genes, such as chloramphenicol acetyl transferase (CAT) or β-globin.

The nucleotide sequence of part of the 5′-flanking region of the bovine P450scc (CYP11A) gene was recently reported by Ahlgren *et al.* (1990). In addition, the complete structure of the bovine P45017α (CYP17) gene was elucidated by Bhasker and co-workers (1989). In order to characterize the 5′-regulatory regions of these two genes, deletion DNA constructs using two different reporter genes were generated. Transfection of such chimeric DNA constructs into Y1 mouse adrenal tumor cells allowed the identification of specific DNA sequences involved in basal and cAMP-responsive gene expression (Ahlgren *et al.*, 1990; Lund *et al.*, 1990). Surprisingly, none of the P450scc (−186/−101 bp) and P45017α (−243/−225 and −80/−40 bp) specific DNA sequences showed homology to the classical cAMP-response element as described by Montiminy and co-workers (1986). Thus, it is likely that the mechanisms controlling cAMP-dependent expression of cytochrome P450 genes differ in important aspects from the classical cAMP-responsive element/cAMP response element-binding protein model.

In order to evaluate mechanisms involved in cAMP-dependent regulation of P450 gene expression in ovarian cells, specific reporter gene constructs were transfected into bovine luteal cells in primary culture (Lauber *et al.*, 1991) (Figure 4). The results of these experiments provided evidence that both basal and cAMP-dependent regulation of P450scc gene expression are controlled by a *cis*-acting element located at −186 to −101 bp upstream of the transcriptional start site. The apparent increase in basal expression observed with the P450scc-specific CAT constructs (−896/+12, −564/+12, and −240/+12 bp) was also found when these constructs were transfected into Y1 cells (Ahlgren *et al.*, 1990). Since the DNA sequence located at −186/−101 bp of the P450scc gene has previously been shown to function in Y1 and bovine adrenocortical cells, the transcriptional regula-

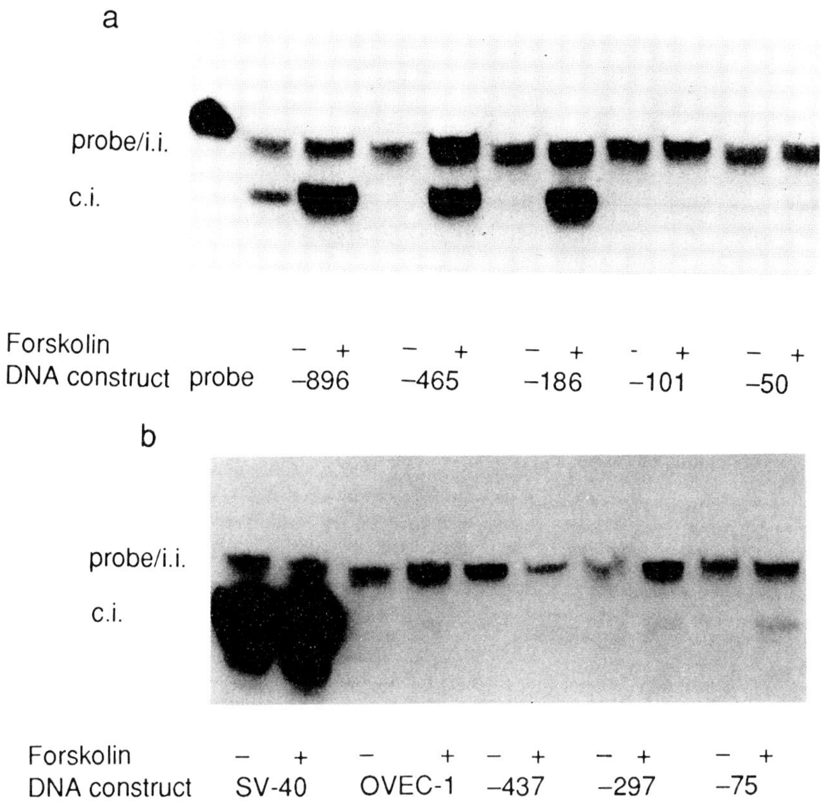

FIGURE 4(A) Expression of several P450scc-specific OVEC reporter gene constructs (−896/−32, −465/−32, −186/−32, −101/−32, and −50/−32 bp) after transfection into bovine luteal cells in primary culture in the presence (+) or absence (−) of forskolin (25 μM). The upper band (probe/i.i.) represents undigested probe and incorrectly initiated β-globin transcripts. The lower band (c.i.) indicates correctly initiated β-globin transcripts and directly reflects reporter gene expression. Positive (SV-40 OVEC) and negative (OVEC-1) controls are those shown in B. (B) Expression of several P45017α-specific OVEC reporter gene constructs (−437/−40, −297/−40, −75/40 bp) after transfection into bovine luteal cells in primary culture in the presence (+) or absence (−) of forskolin (25 μM). The upper band (probe/i.i.) represents undigested probe and incorrectly initiated β-globin transcripts. The lower band (c.i.) indicates correctly initiated β-globin transcripts and directly reflects reporter gene expression. The SV-40 OVEC construct was used as a positive control, whereas the OVEC-1 construct reflects nonspecific β-globin expression.

tion of this gene in luteal cells appears to be similar to that observed in adrenal cells. Further experiments involving gel-shift analysis will prove useful in comparing the DNA binding ability of nuclear extracts isolated from adrenal or luteal cells, respectively.

Whereas transcriptional regulation of the P450scc gene appears to be similar in adrenal and luteal cells, the regulation of P45017α gene expression is completely different in these two cell types. Transfection of reporter gene constructs containing sequences of the P45017α 5′-regulatory region into bovine luteal cells in primary culture revealed that none of these DNA constructs were expressed (Lauber *et al.*, 1991). Moreover, forskolin treatment could not alleviate possible repression of basal expression nor positively influence reporter gene expression. An interesting feature of Y1 cells, in contrast to luteal cells, is the fact that reporter gene constructs containing sequences of the P45017α regulatory region are expressed (Ahlgren *et al.*, 1990). However, the endogenous gene is not expressed in either cell type. Thus the lack of P45017α gene expression in Y1 and in bovine luteal cells must be the consequence of different mechanisms of cell-specific repression. The inability of the P45017α sequences to promote reporter gene expression in cultured bovine luteal cells could be explained either by the loss of activating transcriptional factor(s) or by the induction of specific repressor proteins during luteinization. Clearly, elucidation of the mechanisms responsible for these changes in transcription will be an important key to understanding the mechanism of luteinization.

Studies to examine the expression of transfected chimeric constructs containing regulatory elements of the human P450arom gene have been complicated by the realization that, uniquely among cytochrome P450 isoforms so far described, tissue-specific regulation of its expression involves use of different promoters by means of alternative splicing (Mahendroo *et al.*, 1991). Two unique untranslated first exons (exons I.1 and I.2) have been identified in mRNA specific for P450arom in human placenta (Means *et al.*, 1989). Although the proportion of transcripts encoding exon I.2 is very small, genomic clones encoding the sequences of both exons I.1 and I.2 have recently been isolated (Kilgore *et al.*, 1992). Exon I.1 is located at least 48 kb upstream from the translation start site, whereas exon I.2 is 9 kb upstream. The corpus luteum of human ovary differs in that promoters I.1 and I.2 are completely inactive (Means *et al.*, 1991). Sequence analysis of the DNA immediately 5′ of exon II (which contains the start site of translation) demonstrates the presence of a TATAA sequence beginning 149 base pairs 5′ of the ATG initiation codon identified in placental exon II. Using a combination of primer extension and S1 nuclease protection analysis, it appears that the initiation site of ovarian P450arom transcripts aligns 26 base pairs downstream of this TATAA sequence. It appears, therefore, that the expression of P450arom-specific mRNA in corpus luteum is regulated by an additional promoter (promoter II), which is located just 5′ of exon II. In studies of rat (Fitzpatrick and Richards, 1991) and chicken (Matsumine *et al.*, 1991) P450arom expression in the ovary, it appears that a proximal promoter is also used, consistent with the situation in the human. In addition to placenta, JEG-3 cells, and hydatid moles appear to use promoter I.1 and, to a limited extent, promoter I.2. Thus the proximal promoter,

promoter II, appears to be the primordial promoter, and the one used in the ovary. When, later in evolution, the human placenta acquired the need to make estrogens, it employed a distal element, promoter I.1, perhaps because estrogen biosynthesis by the placenta is continuous, in contrast to the episodic synthesis in the ovary.

IV Conclusions

Although much is known about the pattern of steroid secretion throughout the ovarian cycle, as well as the levels of steroidogenic enzymes and their transcripts, studies on the regulation of the genes encoding steroidogenic enzymes are still in their infancy. However, the stage is now set for major advances in this field. Analysis of the transcription factors involved in regulating the expression of these genes should in turn lead to discovery of the factors involved in that most mysterious of events, the dramatic process of cell differentiation and development that occurs with incredible regularity during every ovarian cycle in every female mammal of reproductive age.

Acknowledgments

This work was supported, in part, by USPHS Grant HD13234. M. Hinshelwood, M. Demeter-Arlotto and G. Means were supported, in part, by USPHS Training Grant 5-T32-HD07190. The authors gratefully acknowledge the skilled editorial assistance of Melissa Meister.

References

Ahlgren, R., Simpson, E. R., Waterman, M. R., and Lund, J. (1990). Characterization of the promoter-regulatory region of the bovine CYP11A (P450scc) expression. *J. Biol. Chem.* 265, 3313–3319.

Aono, T., Kitamura, Y., Fukuda, S., and Matsumoto, K. (1981). Localization of 4 ene-5α-reductase, 17β-ol-dehydrogenase and aromatase in immature rat ovary. *J. Steroid Biochem.* 14, 1369–1377.

Banks, P. K., Meyer, K., and Brodie, A. M. (1991). Regulation of ovarian steroid biosynthesis by estrogen during proestrus in the rat. *Endocrinology* 129, 1295–1304.

Bassett, S. G., Little-Ihrig, L. L., Mason, J. I., and Zeleznik, A. J. (1991). Expression of messenger ribonucleic acids that encode for 3β-hydroxysteroid dehydrogenase and cholesterol side-chain cleavage enzyme throughout the luteal phase of the macaque menstrual cycle. *J. Clin. Endocrinol. Metab.* 72, 362–366.

Bhasker, C. R., Adler, B. S., Dee, A., John, M. E., Kagimoto, M., Zuber, M. X., and Ahlgren, X.-D. (1989). Structural characterization of the bovine CYP17 (17α-hydroxylase) gene. *Arch. Biochem. Biophys.* 271, 479–487.

Carr, B. R., Sadler, R. K., Rochelle, D. B., Stamalch, M. A., MacDonald, P. C., and Simpson,

E. R. (1981). Plasma lipoprotein regulation of progesterone biosynthesis by human corpus luteum tissue in organ culture. *J. Clin. Endocrinol. Metab.* 52, 875–881.

Carroll, D. J., Grummer, R. R., and Mao, F. C. (1992). Progesterone production by cultured luteal cells in the presence of bovine low- and high-density lipoproteins purified by heparin affinity chromatography. *J. Anim. Sci.* 70, 2516–2526.

Couet, J., Martel, C., Dupont, E., Luu-The, V., Sirad, M-A., Shao, H-F., Pelletier, G., and Labrie, F. (1990). Changes in 3β-hydroxysteroid dehydrogenase/Δ^5-Δ^4 isomerase messenger ribonucleic acid activity and protein levels during the estrous cycle in the bovine ovary. *Endocrinology* 127, 2141–2148.

Demeter-Arlotto, M., Rainey, W., and Simpson, E. R. (1993). Maintenance and regulation of 17α-hydroxylase expression by bovine thecal cells in primary culture. *Endocrinology* 132, 1353–1358.

Doody, K. J., Lorence, M. C., Mason, J. I., and Simpson, E. R. (1990). Expression of messenger ribonucleic acid species encoding steroidogenic enzymes in human follicles and corpora lutea throughout the menstrual cycle. *J. Clin. Metab.* 70, 1041–1045.

Doody, K. J., Lephart, E. D., Stirling, D., Lorence, M. C., Magness, R. R., McPhaul, M. J., and Simpson, E. R. (1991). Expression of mRNA species encoding steroidogenic enzymes in the rat ovary. *J. Mol. Biol.* 6, 153–162.

Dupont, E., Zhao, H-F., Rheaume, E., Simard, J., Luu-The, V., Labrie, F., and Pelletier, G. (1990). Localization of 3β-hydroxysteroid dehydrogenase/Δ^5-Δ^4-isomerase in rat gonads and adrenal glands by immunocytochemistry and *in situ* hybridization. *Endocrinology* 127, 1394–1403.

Dupont, E., Labrie, F., Luu-The, V., and Pelletier, G. (1992). Immunocytochemical localization of 3β-hydroxysteroid dehydrogenase/Δ^5-Δ^4-isomerase in human ovary. *J. Clin. Endocrinol. Metab.* 74, 994–998.

Estabrook, R. W. (1989). Cytochrome P450 Viennese style. *In* "Cytochrome P450: Biochemistry and Biophysics" (I. Schuster, ed.), pp. 1–15. Taylor and Francis, Philadelphia, Pa.

Farkash, Y., Timberg, R., and Orly, J. (1986). Preparation of antiserum to rat cytochrome P-450 cholesterol side chain cleavage, and its use for ultrastructural localization of the immunoreactive enzyme by protein A-gold technique. *Endocrinology* 118, 1353–1365.

Fitzpatrick, S. L., and Richards, J. S. (1991). Regulation of cytochrome P450 aromatase messenger ribonucleic acid and activity by steroids and gonadotropins in rat granulosa cells. *Endocrinology* 129, 1452–1462.

Funkenstein, B., Waterman, M. R., Masters, B. S., and Simpson, E. R. (1983). Evidence for the presence of cholesterol side chain cleavage cytochrome P-450 and adrenodoxin in fresh granulosa cells: effects of follicle-stimulating hormone and cyclic AMP on cholesterol side chain cleavage cytochrome P-450 synthesis and activity. *J. Biol. Chem.* 258, 10187–10191.

Gibori, G., Khan, I., Warshwa, M. L., McLean, M. P., Puryear, T. K., Nelson, S., Durkee, T. J., Azhar, S., Steinschneider, A., and Rao, M. C. (1988). Placental-derived regulators and the complex control of luteal cell function. *Recent Prog. Horm. Res.* 44, 377–429.

Gore-Langton, R. E., and Armstrong, D. T. (1988). Follicular steroidogenesis and its control. *In* "The Physiology of Reproduction" (E. Knobil and J. Neill, eds.), pp. 331–352. Raven Press, New York.

Goldring, N. B., Durica, J. M., Lifka, J., Hedin, L., Ratoosh, S. L., Miller, W. L., Orly, J., and Richards, J. S. (1987). Cholesterol side-chain cleavage P450 messenger ribonucleic acid: evidence for hormonal regulation in rat ovarian follicles and constitutive expression in corpora lutea. *Endocrinology* 120, 1942–1950.

Gospodarowicz, D., Massoglia, S., Cheng, J., and Fujii, D. K. (1986). Effect of fibroblast growth factor and lipoproteins on the proliferation of endothelial cells derived from bovine adrenal cortex, brain cortex and corpus luteum capillaries. *J. Cell. Physiol.* 127, 121–136.

Hedin, L., Rodgers, R. J., Simpson, E. R., and Richards, J. S. (1987). Changes in content of cytochrome P45017α, cytochrome P450scc, and 3-hydroxy-3-methylglutaryl CoA reductase in developing rat ovarian follicles and corpora lutea: correlation with thecal cell steroidogenesis. *Biol. Reprod.* 37, 211–223.

Hickey, G. J., Chen, S., Besman, M. J., Shively, J. E., Hall, P. F., Gaddy-Kurten, D., and Richards, J. S. (1988). Hormonal regulation, tissue distribution, and content of aromatase cytochrome P450 messenger ribonucleic acid and enzyme in rat ovarian follicles and corpora lutea: relationship to estradiol biosynthesis. *Endocrinology* 122, 1426–1436.

Hickey, G. J., Krasnow, J. S., Beattie, W. G., and Richards, J. S. (1990). Aromatase cytochrome P450 in rat ovarian granulosa cells before and after luteinization: adenosine 3′,5′-monophosphate-dependent and independent regulation. Cloning and sequencing of rat aromatase cDNA and 5′ genomic DNA. *Mol. Endocrinol.* 4, 3–12.

Hurwitz, A., Hernandez, E. R., Resnick, C. E., Packman, J. M., Payne, D. W., and Adashi, E. Y. (1990). Basic fibroblast growth factor inhibits gonadotropin-supported ovarian androgen biosynthesis: mechanism(s) and possible site(s) of action. *Endocrinology* 126, 3089–3095.

Inkster, S. E., and Brodie, A.M.H. (1991). Expression of aromatase cytochrome P450 in premenopausal and postmenopausal human ovaries: an immunocytochemical study. *J. Clin. Endocrinol. Metab.* 73, 717–726.

Ishimura, K., Yoshinaga-Hirabayashi, T., Tsuri, H., Fujita, H., and Osawa, Y. (1989). Further immunocytochemical study on the localization of aromatase in the ovary of rats and mice. *Histochemistry* 90, 413–416.

Ishimura, K., Yoshinaga-Hirabayashi, T., Tsuri, H., Kominami, S., Takemori, S., and Fujita, H. (1990). Immunocytochemical and biochemical studies on the localization and changes of 17alpha-hydroxylase/C17-C20 lyase activity in immature rat ovary treated with PMSG and hCG. *Histochemistry* 94, 225–229.

Kilgore, M. W., Means, G. D., Mahendroo, M. S., Mendelson, C. R., and Simpson, E. R. (1992). Alternative promotion of aromatase cytochrome P450 expression in human fetal tissues. *Mol. Cell. Endocrinol.* 83, R9–R16.

Krasnow, J. S., Hickey, G. J., and Richards, J. S. (1990). Regulation of aromatase mRNA and estradiol biosynthesis in rat ovarian granulosa and luteal cells by prolactin. *Mol. Endocrinol.* 4, 13–21.

Lauber, M. E., Bengston, T., Waterman, M. R., and Simpson, E. R. (1991). Regulation of CYP11A (P450scc) and CYP17 (P45017α) gene expression in bovine luteal cells in primary culture. *J. Biol. Chem.* 266, 11170–11175.

Le Goascogne, C., Sananes, N., Gouizou, M., Baulieu, E. E., and Robel, P. (1989). Cell-specific variations and hormonal regulation of immunoreactive cytochrome P-450scc in the rat ovary. *J. Reprod. Fertil.* 85, 61–72.

Lephart, E. D., Peterson, K. G., Noble, J. F., George, F. W., and McPhaul, M. J. (1990). The structure of cDNA clones encoding the aromatase P-450 isolated from a rat Leydig cell tumor line demonstrates differential processing of aromatase mRNA in rat ovary and a neoplastic cell line. *Mol. Cell. Endocrinol.* 70, 31–40.

Lephart, E. D., Doody, K. J., McPhaul, M. J., and Simpson, E. R. (1992a). Inverse relationship between ovarian aromatase cytochrome P450 and 5α-reductase enzyme activities and mRNA levels during the estrous cycle in the rat. *J. Steroid Biochem. Mol. Biol.* 42, 439–447.

Lephart, E. D., Simpson, E. R., and McPhaul, M. J. (1992b). Ovarian aromatase cytochrome P-450 mRNA levels correlate with enzyme activity and serum estradiol levels in anestrous, pregnant and lactating rats. *Mol. Cell. Endocrinol.* 85, 205–214.

Lorence, M. C., Corbin, C. J., Kamimura, N., Mahendroo, M. S., and Mason, J. I. (1990). Structural analysis of the gene encoding human 3β-hydroxysteroid dehydrogenase/$\Delta^{5\rightarrow4}$ isomerase. *Mol. Endocrinol.* 4, 1850–1855.

Lund, J., Ahlgren, R., Wu, D., Kagimoto, M., Simpson, E. R., and Waterman, M. R. (1990). Transcriptional regulation of the bovine CYP17 (P45017α) gene: identification of two cAMP regulatory regions lacking the consenses CRE. *J. Biol. Chem.* 265, 3304–3312.

Luu-The, V., Labrie, C., Zhao, H.-F., Couet, J., Lachance, Y., Simard, J., Leblanc, G., Cote, J., Berube, D., Gagne, R., and Labrie, F. (1989a). Characterization of cDNAs for human estradiol 17β-hydrogenase and assignment of the gene to chromosome 17: evidence of two mRNA species with distinct 5′-termini in human placenta. *Mol. Endocrinol.* 3, 1301–1309.

Luu-The, V., Lachance, Y., Labrie, C., Leblanc, G., Thomas, J. L., Strickler, R. C., and Labrie, F. (1989b). Full length cDNA structure and deduced amino acid sequence of human 3β-hydroxy-5-ene steroid dehydrogenase. *Mol. Endocrinol.* 3, 1310–1319.

Luu-The, V., Labrie, C., Simard, J., Lachance, Y., Zhao, H.-F., Couet, J., Leblanc, G., and Labrie, F. (1990). Structure of two in tandem human 17β-hydroxysteroid dehydrogenase genes. *Mol. Endocrinol.* 4, 268–275.

Mahendroo, M. S., Means, G. D., Mendelson, C. R., and Simpson, E. R. (1991). Tissue-specific expression of human $P450_{arom}$: the promoter responsible for expression in adipose tissue is different from that utilized in placenta. *J. Biol. Chem.* 266, 11276–11281.

Martel, C., Labrie, C., Couet, J., Dupont, E., Trudel, C., Luu-The, V., Takahashi, M., Pelletier, G., and Labrie, F. (1990a). Effects of human chorionic gonadotropin (hCG) and prolactin (PRL) on 3β-hydroxy-5-ene-steroid dehydrogenase/Δ^5-Δ^4 isomerase (3β-HSD) expression and activity in the rat ovary. *Mol. Cell. Endocrinol.* 72, R7–R13.

Martel, C., Labrie, C., Dupont, E., Couet, J., Trudel, C., Rheaume, E., Simard, J., Luu-The, V., Pelletier, G., and Labrie, F. (1990b). Regulation of 3β-hydroxysteroid dehydrogenase/Δ^5-Δ^4 isomerase expression and activity in the hypophysectomized rat ovary: interactions between the stimulatory effect of human chorionic gonadotropin and the luteolytic effect of prolactin. *Endocrinology* 127, 2726–2737.

Martel, C., Rheaume, E., Takahashi, M., Trudel, C., Couet, J., Luu-The, V., Simard, J., and Labrie, F. (1992). Distribution of 17β-hydroxysteroid dehydrogenase gene expression and activity in rat and human tissues. *J. Steroid Biochem. Mol. Biol.* 41, 597–603.

Matsuda, H., Fujita, H., Ishimura, K., and Osawa, Y. (1984). Immunocytochemical localization of aromatase in ovaries of some rodents, cow and human. *Acta. Histochem. Cytochem.* 17, 311–322.

Matsumine, H., Herbst, M. A., Ignatius Oh, S-H., Wilson, J. D., and McPhaul, M. J. (1991). Aromatase in RNA in the extragonadal tissues of chickens with the henny-feathering trait is derived from a distinctive promoter structure that contains a segment of a retroviral long terminal repeat. *J. Biol. Chem.* 266, 19900–19907.

McAllister, J. M., Kerin, J.F.P., Trant, J. M., Estabrook, R. W., Mason, J. I., Waterman, M. R., and Simpson, E. R. (1989). Regulation of cholesterol side-chain cleavage and 17α-hydroxylase/lyase activities in proliferating human theca interna cells in long term monolayer culture. *Endocrinology* 125, 1959–1966.

McAllister, J. M., Mason, J. I., Byrd, W., Trant, J. M., Waterman, M. R., and Simpson, E. R. (1990). Proliferating human granulosa-lutein cells in long-term monolayer culture: expression of aromatase, cholesterol side-chain cleavage, and 3β-hydroxysteroid dehydrogenase. *J. Clin. Endocrinol. Metab.* 71, 26–33.

Means, G. D., Mahendroo, M., Corbin, C. J., Mathis, J. M., Powell, F. E., Mendelson, C. R., and Simpson, E. R. (1989). Structural analysis of the gene encoding aromatase cytochrome P-450, the enzyme responsible for estrogen biosynthesis. *J. Biol. Chem.* 264, 19385–19391.

Means, G. D., Kilgore, M. W., Mahendroo, M. S., Mendelson, C. R., and Simpson, E. R. (1991). Tissue-specific promoters regulate aromatase cytochrome P450 gene expression in human ovary and fetal tissues. *Mol. Endocrinol.* 5, 2005–2013.

Montminy, M. R., Sevarino, K. A., Wagner, J. A., Mandel, G., and Goodman, R. H. (1986). Identification of a cyclic-AMP-responsive element within the rat somatostatin gene. *Proc. Natl. Acad. Sci. USA* 83, 6682–6686.

Nakajin, S., Shively, J. E., Yuan, P., and Hall, P. F. (1981). Microsomal cytochrome P-450 from neonatal pig testis: two enzymatic activities (17α-hydroxylase and C17,21 lyase) associated with one protein. *Biochemistry* 20, 4037–4042.

Nebert, D. W., Nelson, D. R., Coon, M. J., Estabrook, R. W., Feyersisen, R., Fujii-Kuriyama, Y., Gonzalez, F. J., Guengerich, F. P., Gunsalus, I. C., Johnson, E. F., Loper, J. C., Sato, R., Waterman, M. R., and Waxman, D. J. (1991). The P450 superfamily: update on new sequences, gene mapping and recommended nomenclature. *DNA Cell Biol.* 10, 1–14.

Oonk, R. B., Krasnow, J. S., Beattie, W. G., and Richards, J. S. (1989). Cyclic AMP-dependent and -independent regulation of cholesterol side chain cleavage cytochrome P-450 (P-450scc) in rat ovarian granulosa cells and corpora lutea. *J. Biol. Chem.* 264, 21934–21942.

Reed, M. J., and James, V. H. (1989). Regulation of steroid synthesis and metabolism by growth factors. *Clin. Endocrinol.* 31, 511–525.

Richards, J. S., Hedin, L., and Caston, L. (1986). Differentiation of rat ovarian theca cells: evidence for functional luteinization. *Endocrinology* 118, 1660–1668.

Rodgers, R. J., Rodgers, H. F., Hall, P. F., Waterman, M. R., and Simpson, E. R. (1986a). Immunolocalization of cholesterol side-chain-cleavage cytochrome P-450 and 17α-hydroxylase cytochrome P-450 in bovine follicles. *J. Reprod. Fertil.* 78, 627–638.

Rodgers, R. J., Rodgers, H. F., Waterman, M. R., and Simpson, E. R. (1986b). Immunolocalization of cholesterol side-chain-cleavage cytochrome P-450 and ultrastructural studies of bovine corpora lutea. *J. Reprod. Fertil.* 78, 639–652.

Rodgers, R. J., Waterman, M. R., and Simpson, E. R. (1986c). Cytochromes P-450scc, P-45017α, adrenodoxin, and reduced nicotinamide adenine dinucleotide phosphate-cytochrome P-450 reductase in bovine follicles and corpora lutea: changes in specific contents during the ovarian cycle. *Endocrinology* 118, 1366–1374.

Rodgers, R. J., Waterman, M. R., and Simpson, E. R. (1987). Levels of messenger ribonucleic acid encoding cholesterol side-chain cleavage cytochrome P-450, 17α-hydroxylase cytochrome P-450, adrenodoxin, and low density lipoprotein receptor in bovine follicles and corpora lutea throughout the ovarian cycle. *Mol. Endocrinol.* 1, 274–279.

Rodgers, R. J., Waterman, M. R., Simpson, E. R., and Magness, R. R. (1988). Immunoblot analysis of cholesterol side-chain cleavage cytochrome P-450 and adrenodoxin in corpora lutea of cyclic and late-pregnant sheep. *J. Reprod. Fertil.* 83, 843–850.

Sanghera, M. K., Simpson, E. R., McPhaul, M. J., Kuzlowski, G., Conley, A. J., and Lephart, E. D. (1991). Immunocytochemical distribution of aromatase cytochrome P450 in the rat brain using peptide-generated polyclonal antibodies. *Endocrinology* 129, 2834–2844.

Sano, Y., Suzuki, K., Kiyoshi, A., Okinaga, S., and Tamaoki, B-I. (1981). Changes in enzyme activities related to steroidogenesis in human ovaries during the menstrual cycle. *J. Clin. Endocrinol. Metab.* 52, 994–1001.

Sasano, H., Okamoto, M., Mason, J. I., Simpson, E. R., Mendelson, C. R., Sasano, N., and Liverberg, S. G. (1989). Immunolocalization of aromatase, 17α-hydroxylase and side-chain-cleavage cytochromes P-450 in the human ovary. *J. Reprod. Fertil.* 85, 163–169.

Sasano, H., Mori, T., Sasano, N., Nagura, H., and Mason, J. I. (1990). Immunolocalization of 3β-hydroxysteroid dehydrogenase in human ovary. *J. Reprod. Fertil.* 89, 743–751.

Simpson, E. R., Rochelle, D. B., Carr, B. R., and MacDonald, P. C. (1980). Plasma lipoproteins in follicular fluid of human ovaries. *J. Clin. Endocrinol. Metab.* 51, 1469–1471.

Smith, C. J., Richards, J. S., Yasin, K., Sangster, J. N., and Sridaran, R. (1991). Changes in rat luteal cell ultrastructure and P450scc mRNA and protein content after in vivo treatment with a gonadotropin-releasing hormone agonist. *Biol. Reprod.* 44, 382–391.

Steinkampf, M. P., Mendelson, C. R., and Simpson, E. R. (1988). Effects of epidermal growth factor and insulin-like growth factor I on the levels of mRNA encoding aromatase cytochrome P-450 of human ovarian granulosa cells. *Mol. Cell. Endocrinol.* 59, 93–99.

Stirling, D., Magness, R. R., Stone, R., Waterman, M. R., and Simpson, E. R. (1990). Angiotensin II inhibits luteinizing hormone-stimulated cholesterol side chain cleavage expression and stimulates basic fibroblast growth factor expression in bovine luteal cells in primary culture. *J. Biol. Chem.* 265, 5–8.

Stirling, D., Waterman, M. R., and Simpson, E. R. (1991). Expression of mRNA encoding basic fibroblast growth factor (bFGF) in bovine corpora lutea and cultured luteal cells. *J. Reprod. Fertil.* 91, 1–8.

Strauss, J. F., Schuler, L. A., Rosenblum, M. F., and Tanaka, T. (1981). Cholesterol metabolism by ovarian tissue. *Adv. Lipid Res.* 18, 99–157.

Suzuki, T., Sasano, H., Sawai, T., Mason, J. I., and Nagura, H. (1992). Immunohistochemistry and in situ hybridization of P-45017 alpha (17alpha-hydroxylase/17,20-lyase). *J. Histochem. Cytochem.* 40, 903–908.

Tamura, T., Kitawaki, J., Yamamoto, T., Osawa, Y., Kominami, S., Kakemori, S., and Okada, H. (1992). Immunohistochemical localization of 17α-hydroxylase/c17-20 lyase and aromatase cytochrome P450 in the human ovary during the menstrual cycle. *J. Endocrinol.* 135, 589–595.

Trzeciak, W. H., Waterman, M. R., and Simpson, E. R. (1986). Synthesis of the cholesterol side-chain cleavage enzymes in cultured rat ovarian granulosa cells: induction by follicle-stimulating hormone and dibutryl adenosine 3′,5′-monophosphate. *Endocrinology* 119, 323–330.

Trzeciak, W. H., Duda, T., Waterman, M. R., and Simpson, E. R. (1987a). Tetradecanoyl phorbol acetate suppresses follicle-stimulating hormone-induced synthesis of the cholesterol side-chain cleavage enzyme complex in rat ovarian granulosa cells. *J. Biol. Chem.* 262, 15246–15250.

Trzeciak, W. H., Duda, T., Waterman, M. R., and Simpson, E. R. (1987b). Effects of epidermal growth factor on the synthesis of the cholesterol side-chain cleavage enzyme complex in rat ovarian granulosa cells in primary culture. *Mol. Cell. Endocrinol.* 52, 43–50.

Tuckey, R. C., and Atkinson, H. C. (1989). Pregnenolone synthesis from cholesterol and hydroxycholesterols by mitochondria from ovaries following the stimulation of immature rats with pregnant mare's serum gonadotropin and human choriogonadotropin. *Eur. J. Biochem.* 186, 255–259.

Tureck, R. W., and Strauss, J. F., III (1982). Progesterone synthesis by luteinized human granulosa cells in culture: the role of de novo sterol synthesis and lipoprotein-carried sterol. *J. Clin. Endocrinol. Metab.* 54, 367–373.

Urban, R. J., Garmey, J. C., Shupnik, M. A., and Veldhuis, J. D. (1991). Follicle-stimulating hormone increases concentrations of messenger ribonucleic acid encoding cytochrome P450 cholesterol side-chain cleavage enzyme in primary cultures of porcine granulosa cells. *Endocrinology* 128, 2000–2007.

Veldhuis, J. D., Rodgers, R. J., Dee, A., and Simpson, E. R. (1986). The insulin-like growth factor, somatomedin C, induces the synthesis of cholesterol side-chain cleavage cytochrome P450 and adrenodoxin in ovarian cells. *J. Biol. Chem.* 261, 2499–2502.

Veldhuis, J. D., and Rodgers, R. J. (1987). Mechanisms subserving the steroidogenic synergism between follicle-stimulating hormone and insulin-like growth factor I (somatomedin C). Alterations in cellular sterol metabolism in swine granulosa cells. *J. Biol. Chem.* 262, 7658–7664.

Voutilainen, R., Picado-Leonard, J., DiBlasio, A. M., and Miller, W. L. (1988). Hormonal and developmental regulation of adrenodoxin messenger RNA in steroidogenic tissues. *J. Clin. Endocrinol. Metab.* 66, 383–388.

Warshaw, M. L., Johnson, D. C., Khan, I., Eckstein, B., and Gibori, G. (1986). Placental secretion of androgens in the rat. *Endocrinology* 119, 2642–2648.

Wiltbank, M. C., Diskin, M. G., Flores, J. A., and Niswender, G. D. (1990). Regulation of the corpus luteum by protein kinase C. II. Inhibition of lipoprotein-stimulated steroidogenesis by prostaglandin F2α. *Biol. Reprod.* 42, 239–245.

Yoshinaga-Hirabayashi, T., Ishimura, K., Fujita, H., Kitawaki, J., Osawa, Y. (1990). Immunocytochemical localization of aromatase in immature rat ovaries treated with PMSG and hCG, and in pregnant rat ovaries. *Histochemistry* 93, 223–228.

Zlotkin, T., Farkash, Y., Orly, J. (1986). Cell specific expression of immunoreactive cholesterol side-chain cleavage cytochrome P450 during follicular development in the rat ovary. *Endocrinology* 119, 2809–2820.

7

Molecular Biology of Endocrine Receptors in the Ovary

OK-KYONG PARK-SARGE & KELLY E. MAYO

I. Introduction

In sexually mature mammals the ovaries undergo cyclic functional and morphological changes that result in the periodic discharge of female gametes, the oocytes. These changes are best described by considering the development of the ovarian follicle, which following its recruitment from the quiescent to the growing pool of follicles, progresses through a complex series of maturational stages involving cell proliferation and differentiation. It temporarily exists as a fully functional preovulatory follicle responsive to the signals that will culminate in its ovulation and subsequent luteinization. These processes are exquisitely coordinated by a growing array of autocrine, paracrine, and endocrine factors of ovarian and peripheral sources. Many of these factors have been classically established as ovarian regulators, for example, the pituitary gonadotropins, follicle-stimulating hormone and luteinizing hormone. Other factors, such as the neurotropic factors, were unsuspected and are only now receiving attention as important participants in this process. A key challenge in reproductive endocrinology is to gain an understanding of the interactions that subserve the effects of these regulatory factors on ovarian target cells.

The biological action of hormones in their target cells is generally mediated by receptors that transduce extracellular signaling into intracellular biochemical pathways that ultimately lead to the appropriate regulation of physiological processes. The nature of these receptors varies with the need to respond to different types of chemical signals (for example, a polypep-

Molecular Biology of the Female Reproductive System

tide versus a steroid) and with the need to trigger different types of effector systems (for example, the activation of a protein kinase or the opening of an ion channel). Tremendous progress has been made in elucidating the structural features of these receptors, investigating their interaction with the ligand, and in deciphering the components of the downstream signaling cascades. Several common themes have emerged that in a simplistic

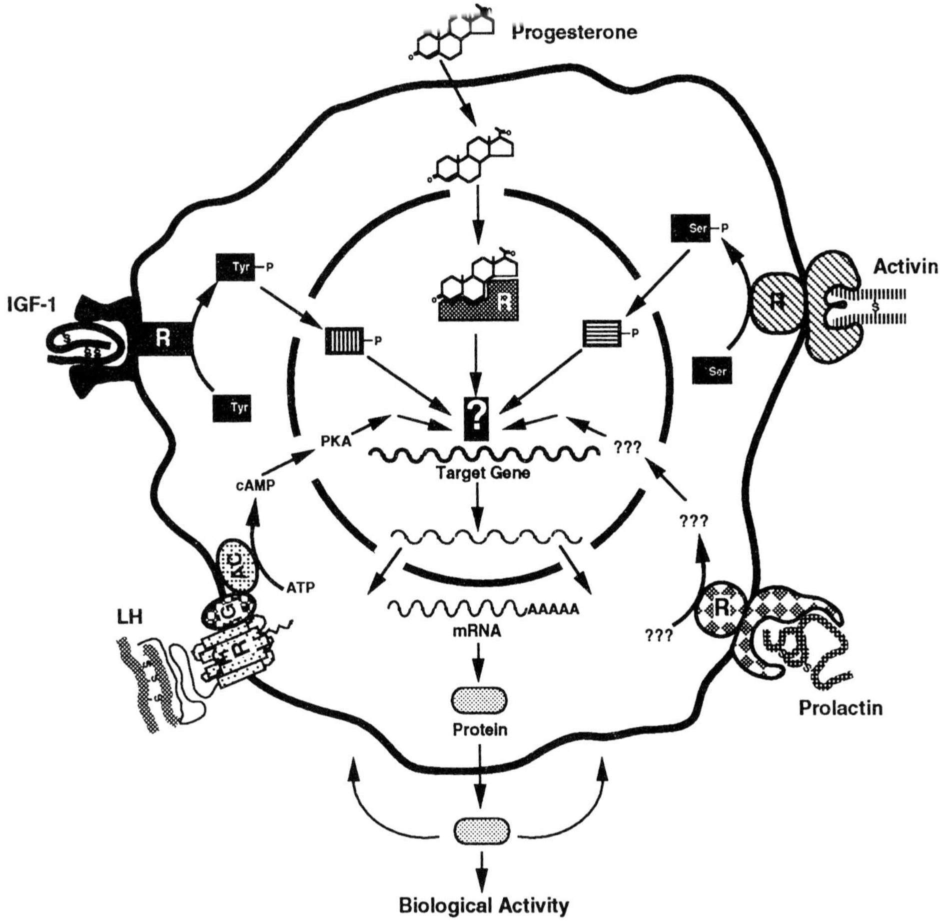

FIGURE 1 Overview of several prominent classes of receptors that are important in hormonal signaling within the ovary. The diagram shows a composite ovarian cell expressing representatives of the five receptor classes discussed in this chapter. These include the receptors for LH (a G protein-coupled receptor), progesterone (a DNA-binding transcription factor), activin (a protein serine and threonine kinase), IGF-1 (a protein tyrosine kinase), and prolactin (a protein with unknown signaling properties). As indicated, all initiate signaling events that eventually impinge on gene expression within the target cell, ultimately resulting in appropriate changes in its biological activities.

way allow these receptors to be categorized with respect to their primary structure, their ligand recognition properties, and their ability to interact with other signaling proteins.

This chapter discusses the roles of several of these receptor types in hormonal signaling within the ovary. We consider receptors in four well-characterized categories, as well as receptors that are not yet well enough understood to be easily classified. In each section, we begin with a broad overview of the known features of these receptors as a class, and then consider in more detail several examples of interest in ovarian physiology. Given the pace of advances in identifying both known and novel ligands and receptors, a comprehensive survey of all ovarian receptors is beyond the scope of this chapter, and numerous additional reviews in each of the topic areas are included for further consideration by the reader.

To introduce the receptor types discussed here, Figure 1 presents a composite ovarian cell expressing five receptor types; it briefly and schematically shows the downstream signaling events that are the basis of the classification. The groups to be considered are (1) the G protein-coupled receptors, shown as the luteinizing hormone (LH) receptor; (2) the transcription factor receptors, shown as the progesterone receptor; (3) the serine and threonine kinase receptors, shown as the activin receptor; (4) the tyrosine kinase receptors, shown as the insulin-like growth factor-1 receptor; and (5) the receptors with uncharacterized signaling mechanisms, shown as the prolactin receptor.

II. G Protein-Coupled Receptors

Numerous cell surface receptors use intracellular signaling pathways that employ G proteins as transducers. The G proteins are heterotrimers and are composed of three distinct subunits: a guanosine triphosphate (GTP)-binding α (39–46 kDa), as well as β (37 kDa) and γ (8 kDa) subunits (Birnbaumer, 1990; Taylor, 1990). Their characteristics have recently been reviewed elsewhere (Hepler and Gilman, 1992; Simon *et al.*, 1991). The G proteins regulate several hormone-induced intracellular signaling systems that play significant roles in ovarian function, including those that use adenylate cyclase, phosphodiesterase, and phospholipases A_2 and C (Simon *et al.*, 1991; Leung and Steele, 1992).

The members of the large family of transmembrane receptor proteins that interact with G proteins share a characteristic opsin-like (seven membrane-spanning domain) topological structure that has been revealed by molecular cloning and biochemical studies. The family includes receptors for neurotransmitters such as the β-adrenergic, muscarinic acetylcholine, substance P, dopamine, and serotonin receptors (Dixon *et al.*, 1986; Kubo *et al.*, 1986; Yokota *et al.*, 1989; Bunzow *et al.*, 1988; Julius *et al.*, 1988);

for polypeptide hormones such as the gonadotropin and thyrotropin receptors (McFarland *et al.*, 1989; Sprengel *et al.*, 1990; Parmentier *et al.*, 1989); and for peptide hormones such as the calcitonin, secretin, vasoactive intestinal peptide (VIP), and growth hormone-releasing hormone (GHRH) receptors (Lin *et al.*, 1991; Ishihara *et al.*, 1991, 1992; Mayo, 1992; C. Lin *et al.*, 1992).

Figure 2 shows a schematic structure of a G protein-coupled receptor, along with two of the potential signal transduction pathways these receptors are known to activate. While most G protein-coupled receptors are similar to the prototype rhodopsin in size and overall structure, there is also considerable variation among them. The general topology of these receptors has been inferred largely from studies of the β-adrenergic receptors. The receptors contain an amino-terminal extracellular domain, often containing potential N-linked glycosylation sites, that varies among receptors from small (neurotransmitter receptors) to quite large (gonadotropin receptors). They all have seven characteristic membrane-spanning domains that are connected by three short extracellular loops, two short cationic intracellular loops, and a somewhat longer intracellular loop between the fifth and sixth transmembrane domains. The carboxyl-terminal intracellular tail is of variable length, and it and the intracellular loops often contain serine and/or threonine residues that provide likely sites for phosphorylation. In most cases, it is not clear which parts of the receptor are required for its interaction with the ligand. The receptors for biogenic amines and neurotransmitters, which have short amino-terminal extracellular domains, probably require the transmembrane domains themselves for ligand–receptor interaction (Kobilka *et al.*, 1988; Catterall, 1988; Strader *et al.*, 1987). In contrast, the larger extracellular domain of the gonadotropin receptors appears to be sufficient for glycoprotein hormone binding (Tsai-Morris *et al.*, 1990; Xie *et al.*, 1990; Ji and Ji, 1991).

The ligand-induced interaction between the receptor and the G protein is also not fully understood. It appears that the base of the third intracellular loop in the region preceding the sixth transmembrane domain is important for G protein activation (Franke *et al.*, 1988; Kubo *et al.*, 1988; O'Dowd *et al.*, 1988). Many classes of G proteins have been characterized, and these regulate the activity of adenylate cyclases, phosphodiesterases, phospholipases, and a variety of ion channels in ovarian target cells. Although the activated α subunit of the trimeric G protein has classically been thought to interact with these effectors (for review, see Hepler and Gilman, 1992; Simon *et al.*, 1991), it has been suggested that ligand binding to receptor can stimulate one effector system through the G protein-associated α-subunit as well as a different effector system through the G protein-associated βγ complex (Katz *et al.*, 1992; Camps *et al.*, 1992). The flexibility of G protein-coupled transmembrane signaling is tremendous, given that the already characterized α, β, and γ subunits produce nearly 1000 possible

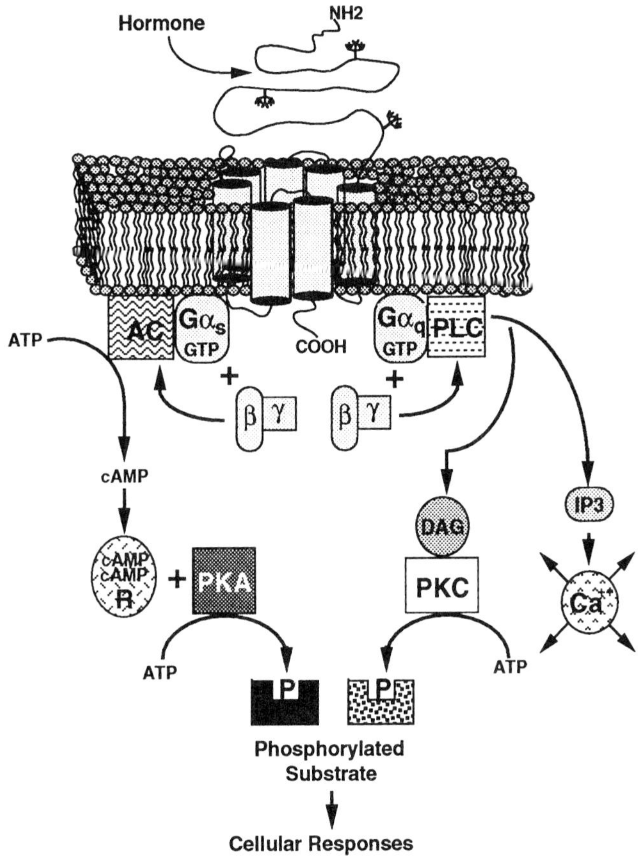

FIGURE 2 A schematic representation of the structural features of a G protein-coupled receptor and some of its potential signaling pathways. The example is the LH receptor, but the features have been generalized. This receptor has a large extracellular domain with several potential N-linked glycosylation sites (Ψ), and seven membrane-spanning domains. Binding of hormone to the extracellular domain of the receptor activates the associated G protein and results in the dissociation of Gα-subunits from the βγ complex. $G\alpha_s$ is shown to activate adenylate cyclase, leading to the initiation of a phosphorylation cascade controlled by the second messenger, cAMP. In an alternative pathway, $G\alpha_q$ is shown to activate phospholipase C, resulting in the breakdown of membrane lipids and the mobilization of intracellular Ca^{2+} that leads to activation of protein kinase C. The Gβγ complex is also thought to play a role in activating effector systems.

oligomeric combinations (Hepler and Gilman, 1992). Two classes of G protein-coupled ovarian receptors are discussed in the next section as representative examples: the large gonadotropin receptors and the smaller receptors for peptides such as gonadotropin-releasing hormone (GnRH) and growth hormone-releasing hormone (GHRH).

A. Gonadotropin Receptors

The importance of gonadotropins (LH and follicle-stimulating hormone (FSH) for ovarian steroidogenesis and oogenesis has been repeatedly documented (for reviews, see Hsueh *et al.*, 1984; Leung and Steele, 1992; Richards, 1980; Tonetta and DiZerega, 1989). Gonadotropin action is regulated through modulation of both circulating hormone levels and receptor expression by target cells. Because the LH-chorionic gonadotropin (CG) and FSH proteins display similar structural and functional characteristics, it has long been assumed that their receptor proteins would likewise be related. It has been recognized for some time that the gonadotropin receptors associate with G proteins and are coupled to the stimulation of adenylate cyclase (Hunzicker-Dunn and Birnbaumer, 1985). Further characterization of these receptors required the isolation and purification of the receptor protein (for review, see Roche and Ryan, 1985; Ascoli and Segaloff, 1989). From cross-linking, metabolic labeling, and ligand blotting studies using purified LH receptor protein (Kim *et al.*, 1987; Kusuda and Dufau, 1986; Minegish *et al.*, 1987), a consensus view emerged of the LH-CG receptor as a glycopolypeptide with a transmembrane anchor and a domain that served as a substrate for cAMP-dependent kinase (Kusuda and Dufau, 1986; Minegish *et al.*, 1987; Kim *et al.*, 1987; Wimalasena *et al.*, 1985; Dufau *et al.*, 1989).

The eventual isolation and characterization of the LH-CG receptor complementary deoxyribonucleic acid (cDNA) (McFarland *et al.*, 1989; Loosfelt *et al.*, 1989) predicted that this receptor cDNA encodes a single polypeptide chain consisting of a short signal peptide, a large amino-terminal domain, and a carboxyl-terminal domain capable of spanning the plasma membrane seven times and ending in a short cytoplasmic tail. This cDNA, when expressed in cultured mammalian cells, produced a protein that specifically bound human chorionic gonadotropin (hCG) and mediated a hormone-induced accumulation of intracellular cyclic adenosine monophosphate (cAMP). The LH-CG receptor has been shown to increase intracellular Ca^{2+} concentrations independently from its effects on cAMP formation (Gudermann *et al.*, 1992), indicating an apparent ability of the receptor to interact with distinct G proteins and effector systems. The subsequent cloning of the FSH receptor cDNA (Sprengel *et al.*, 1990; Minegish *et al.*, 1990) revealed it to be structurally similar to the LH-CG receptor; the two sequences are highly related, particularly within the 7 transmembrane domains (80% identity) and to a lesser extent within the amino-terminal domain (50%), which has been shown to be extracellular (Rodriguez and Segaloff, 1990). Although the gonadotropin receptors share the 7 transmembrane domains that are characteristic features of the G protein-coupled receptor family, they display several unique features. The gonadotropin receptors, as well as the related thyroid-stimulating hormone (TSH) receptor (Parmentier *et al.*, 1989), contain a large extracellular domain which has been suggested to be sufficient for hormone recognition and binding (Nagayama *et al.*, 1991; Tsai-Morris *et al.*, 1990; Xie *et al.*, 1990; Ji and Ji, 1991; Braun *et al.*, 1991; Tilly *et al.*, 1992a). It is within this large extracellular domain that a number of potential glycosylation sites as well as a highly repeated leucine-rich motif (Patthy, 1987) occur in all three receptors, defining the glycoprotein hormone receptors as a distinct subfamily within the larger G protein-coupled receptor family. The leucine-rich repeat motif is of particular interest because it is known to interact with both hydrophilic and hydrophobic protein surfaces (Lopez *et al.*, 1987). In addition, it contains sequences that are divergent among the three receptors and may provide for hormone-specific recognition by the receptor. Indeed, this motif has been shown to be important for hormone selectivity (Braun *et al.*, 1991). An additional feature of the glycoprotein hormone receptors is a fairly short third cytoplasmic loop, a region suggested as playing a key role in receptor interaction with the G protein.

Like other members of the G protein-coupled receptor family, the gonadotropin receptors contain within their intracellular domains numerous serines, threonines, and tyrosines, which are potential sites for phosphorylations that might modulate receptor activity. Although it is not completely understood which portion of the intracellular domain is involved in signaling, it has been suggested for other G protein-coupled receptors that amino acids within the third intracellular loop and proximal portion of the cytoplasmic tail of the receptors are involved in its coupling to and activation of G proteins (O'Dowd *et al.*, 1988; Chazenbalk *et al.*, 1990; Liggett *et al.*, 1991; Wong *et al.*, 1990). An LH-CG mutant receptor with a shorter cytoplasmic tail shows enhanced stimulation of cAMP accumulation (Rodriguez *et al.*, 1992), suggesting that the cytoplasmic tail has an inhibitory role in receptor coupling to or activation of G proteins; this could be through steric inhibition. The cytoplasmic tail of the receptor may be also involved in the uncoupling of the receptor from the G protein, or in desensitization of the receptor.

Both the LH-CG and FSH receptor genes give rise to multiple ribonucleic acid (RNA) transcripts in the ovary. There are at least four predominant LH-CG receptor transcripts (approximately 7.0, 4.2, 2.5, and 1.8 kb) as determined by RNA hybridization analysis (LaPolt *et al.*, 1990; K. Nakamura *et al.*, 1990; Segaloff *et al.*, 1990; Camp *et al.*, 1991). Although it is tempting to speculate that each of these transcripts might be produced in response to different hormonal stimuli, all transcripts were found to be induced in the immature rat ovary by gonadotropins [pregnant mare's serum gonadotropin (PMSG) or FSH] (LaPolt *et al.*, 1990; K. Nakamura *et al.*, 1990; Camp *et al.*, 1991), and subsequently downregulated by additional hCG (Camp *et al.*, 1991). Cellular localization of the LH receptor transcripts was in agreement with previous hormone autoradiography for receptor protein localization. The LH receptor mRNA has been shown to be present in the corpus luteum (McFarland *et al.*, 1989; Peng *et al.*, 1991; Camp *et al.*, 1991), in thecal and interstitial cells (Peng *et al.*, 1991; Camp *et al.*, 1991), and in granulosa cells of preovulatory follicles (Segaloff *et al.*, 1990; Peng *et al.*, 1991; Camp *et al.*, 1991). The FSH receptor RNA transcripts in the ovary are much less abundant and are also present in multiple forms (7.0, 4.2, 2.5, and 1.8 kb). The regulation of FSH receptor mRNA levels during gonadotropin-induced ovarian development has also been analyzed (LaPolt *et al.*, 1992; K. Nakamura *et al.*, 1991; Camp *et al.*, 1991). Expression of all four FSH receptor transcripts is increased during follicular development, as demonstrated by PMSG stimulation. In cultured granulosa cells, activation of adenylate cyclase with forskolin increases FSH receptor mRNA levels (Tilly *et al.*, 1992b). FSH receptor mRNA can be found in granulosa cells of small antral follicles, as demonstrated by *in situ* hybridization (Camp *et al.*, 1991). The expression of FSH receptor mRNA is quite high in granulosa cells of the preovulatory follicle, and decreases somewhat following the primary gonadotropin surges.

The human LH receptor gene has been shown to be a single copy and localized to chromosomal region 2p21 (Rousseau-Merk *et al.*, 1990). The structures of the rat LH-CG receptor gene (Koo *et al.*, 1991; Tsai-Morris *et al.*, 1991) and the rat FSH receptor gene (Heckert *et al.*, 1992) have revealed that the C-terminal domain (transmembrane regions and intracellular C-tail) of the gonadotropin receptors is derived from a single exon, while the amino-terminal extracellular domain is composed of nine (FSH receptor) or ten (LH receptor) exons. For the LH receptor, exons 1–10 appear to contain a high-affinity hormone binding site, while exon 11 contains both G protein modulatory information and a low-affinity hormone binding site (Ji and Ji, 1991). Both gonadotropin receptor genes contain multiple transcriptional start sites, and the promoter contains neither conventional TATA or CCAAT box elements nor a conventional cAMP response element (Koo *et al.*, 1990; Tsai-Morris *et al.*, 1991; Wang *et al.*, 1992; Heckert *et al.*, 1992). The LH-CG receptor promoter is cAMP responsive in Leydig cell transfection assays (H. Wang *et al.*, 1992), and it will be important to determine the molecular mechanisms by which these genes are regulated in ovarian cells.

B. GnRH and GHRH Receptors

Gonadotropin-releasing hormone, best known as the hypothalamic neuropeptide controlling anterior pituitary gonadotropin secretion, has been suggested to have actions in extrapituitary tissues, including the ovary (for review, see Hsueh and Jones, 1981). The ovary was found to contain a GnRH-immunoreactive protein (Aten *et al.*, 1986) and mRNA (Oikawa *et al.*, 1990; Clayton *et al.*, 1992), suggesting the local synthesis of this peptide.

Ovarian binding sites for GnRH have also been demonstrated in granulosa cells and thecal cells (Hsueh and Jones, 1981). In the human ovary, GnRH receptors appear to be localized to granulosa cells of the preovulatory follicles, as determined by ligand autoradiography (Latouche *et al.*, 1989). GnRH receptors in the pig ovary were localized to granulosa cells and mast cells of some but not all follicles (Dement-Liebenow, 1991).

Substantial evidence suggests that the GnRH receptor is G protein-coupled and molecular cloning of the mouse receptor (Tsutsumi *et al.*, 1992; Reinhart *et al.*, 1992) has verified that this is indeed a 7 transmembrane domain receptor. The receptor has a short amino-terminal domain with potential glycosylation sites, but unlike other receptors in the family, the GnRH receptor lacks the carboxyl-terminal cytoplasmic tail. The GnRH receptor has been suggested to be coupled to the activation of phospholipase C and subsequently protein kinase C during intracellular signaling (for review, see Huckle and Conn, 1988). Interestingly, Gregory and Taylor (1982) have suggested that this receptor dimerizes when it is occupied. The GnRH receptor mRNA is expressed as a single transcript (4.6 kb) (Tsutsumi *et al.*, 1992), which was originally suggested by a functional assay using size-fractionated mRNA (Sealfon *et al.*, 1992). The anterior pituitary has been shown to express the GnRH receptor mRNA as expected (Tsutsumi *et al.*, 1992) and the mRNA is also found in the ovary (Reinhart *et al.*, 1992), although its cellular localization remains to be determined.

Another class of G protein-coupled receptor is that for growth hormone-releasing hormone and related peptides. GHRH and the related hormones, secretin and vasoactive intestinal peptide, are members of the glucagon superfamily (Bell, 1986; Miyata *et al.*, 1989). VIP is produced in and acts on various tissues, including the brain, intestine, and lung (Said, 1986; Gozes and Brenneman, 1989), while GHRH is produced predominantly in restricted nuclei of the hypothalamus and acts on the anterior pituitary (Frohman and Jansson, 1986; Gelato and Merriam, 1986). Both of these peptides have been shown to be locally produced in the ovary and to function as intraovarian regulators. VIP immunoreactivity, mRNA, and binding sites have been demonstrated in the ovary (Huang *et al.*, 1984; Gozes and Tsafriri, 1986; Ahmed *et al.*, 1986). VIP can modulate steroid production, potentially by regulating steroidogenic enzymatic activity in the ovary (Ahmed *et al.*, 1986; Trzeciak *et al.*, 1986; George and Ojeda, 1987). Similarly, GHRH peptide and mRNA are present in the ovary (Bagnato *et al.*, 1992). Gonadotropins increase GHRH binding to ovarian cell membranes (Bagnato *et al.*, 1991) and this peptide may affect granulosa cell proliferation and differentiation (Spicer *et al.*, 1992; Moretti *et al.*, 1990).

Because both VIP and GHRH appear to act through intracellular cAMP accumulation (Labrie *et al.*, 1983; Bilezikjian and Vale, 1983; Trzeciak *et al.*, 1986; George and Ojeda, 1987), the receptor molecules for these peptides had been predicted to be coupled to a G protein-regulated adenylate cyclase. The recent cloning and characterization of receptors for several of the glucagon family peptides, including secretin (Ishihara *et al.*, 1991), VIP (Ishihara *et al.*, 1992), and GHRH (Mayo, 1992; C. Lin *et al.*, 1992; Gaylinn *et al.*, 1993) confirm these to be members of the G protein-coupled receptor superfamily capable of activating adenylate cyclase. These receptors have substantial homology, and they have amino-terminal extracellular domains that are smaller than those found in the gonadotropin receptors, but substantially larger than those found in the neurotransmitter receptors or the GnRH receptor. There are six cysteine residues in the extracellular domain of these receptors that are completely conserved, indicating that they are likely to be important for appropriate folding of the extracellular domain and therefore for ligand–receptor interactions. It is not yet known whether this extracellular domain is sufficient for ligand binding. VIP at high concentrations is able to bind to and activate the GHRH receptor, which suggests that some of the biological actions of locally produced VIP and GHRH could be mediated through a single receptor (Mayo, 1992).

The VIP receptor gene appears to be differentially regulated in a tissue-specific manner. Ishihara and colleagues (1992) have observed multiple VIP receptor mRNAs of 5.5, 2.3, and 1.3 kb. The 5.5-kb transcript appears to be present in all VIP-receptor expressing tissues, while 2.3- and 1.3-kb transcripts are present in a subset of tissues. The GHRH receptor gene has been partially characterized (Mayo, 1992). Unlike the gonadotropin receptor genes described earlier, the carboxyl-terminal domain of the GHRH receptor gene contains introns. Alternative RNA splicing of this gene generates an in-frame insertion into the critical third cytoplasmic loop connecting the fifth and sixth transmembrane domain. Similar alternative RNA processing has been observed in the dopamine D_2 receptor (Giros *et al.*, 1989; Monsma *et al.*, 1989) and repeats of variable length are found in this region of the dopamine D_4 receptor (O'Dowd, 1992); the significance of these isoforms remains to be determined. It is not known whether the VIP or GHRH receptors cloned thus far are expressed in the ovary, or whether additional receptors for these peptides might exist, as has been shown for the somatostatin receptors (Yamada *et al.*, 1992).

C. Other Receptors

Among other G protein-coupled receptors likely to be found in the ovary are the oxytocin receptor, substance P receptor, and neuropeptide Y receptor. All have now been cloned, but expression of these receptor genes in the ovary has yet to be studied. Oxytocin was considered a neuron-originated peptide primarily acting on the uterus, but it has now been shown to be locally synthesized and hormonally regulated within the ovary (Voss and Fortune, 1992) and to exert ovarian actions (for review, see Wathes, 1989; Ackland *et al.*, 1992). The action of oxytocin is mediated by a G protein-coupled receptor with all of the expected structural features

(Kimura *et al.*, 1992). The signaling pathway this receptor uses is not known. The oxytocin receptor gene is expressed as a single mRNA transcript of 4.4 kb in the ovary as well as in other tissues.

Substance P is a neuropeptide with a variety of functions in the central nervous system as well as in peripheral tissues (Pernow, 1983). Substance P fibers innervate the ovary (Ojeda *et al.*, 1985) and might modulate some ovarian functions, including steroidogenesis (Pitzel *et al.*, 1991). The action of substance P is mediated by a classical G protein-coupled receptor that appears to be coupled to phospholipase C, catalyzing the hydrolysis of inositol phospholipids with a concomitant mobilization of Ca^{2+} (Hershey *et al.*, 1991; Boyd *et al.*, 1991).

Neuropeptide Y, peptide YY, and pancreatic polypeptide are homologous peptides that function as gastrointestinal hormones and neurotransmitters (McDonald, 1988). Two subclasses of neuropeptide Y receptors have been described in the periphery as well as in the brain: Y_1 and Y_2 receptors (Wahlestedt *et al.*, 1990). cDNAs encoding the mammalian Y1 receptor (Larhammar *et al.*, 1992) and the *Drosophila* Y2 receptor (Li *et al.*, 1992) have been cloned, and these indicate that the neuropeptide Y receptors are members of the G protein-coupled receptor superfamily. An additional cDNA encoding a protein capable of binding to neuropeptide Y (Rimland *et al.*, 1992) represents a Y3 receptor, whose pharmacology is distinct from the Y1 or Y2 receptors; the presence of additional subtypes of neuropeptide Y receptors has been suggested (Inui *et al.*, 1992; Rimland *et al.*, 1992). Neuropeptide Y has been shown to be locally produced within the ovary (McDonald *et al.*, 1987) and to have intraovarian actions (Pitzel *et al.*, 1991). It is not known what receptor classes might mediate the actions of neuropeptide Y on the ovary.

III. Transcription Factor Receptors

In contrast to the membrane-anchored receptors that use various intracellular effectors to transduce extracellular hormonal signals, numerous hormone receptors are intracellularly located and act as DNA-binding transcription factors to activate or repress target genes. This family has been divided on the basis of the characteristics of DNA binding domains to include three subgroups of receptors: the steroid receptors, the thyroid hormone and retinoic acid receptors, and orphan receptors without known ligands (Laudet *et al.*, 1992). Our discussion is limited to the steroid receptors and focuses on the receptors for two key ovarian steroids, estrogen and progesterone.

Steroid receptors are ligand-inducible transcription factors that interact with specific *cis*-acting enhancer sequences in target genes termed hormone-response elements (HREs) (Evans, 1988; Beato, 1989; O'Malley,

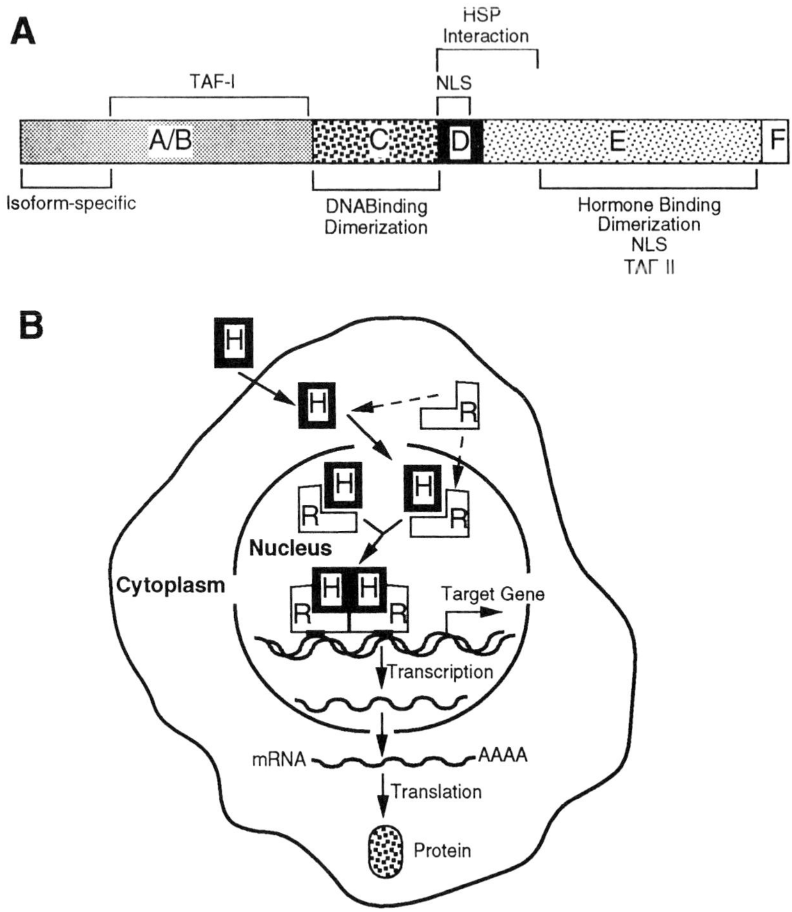

FIGURE 3 A schematic representation of the structural features of a steroid receptor and some of its potential signaling pathways. The example is the progesterone receptor, but the features have been generalized. (A) Structural domains of the receptor are indicated as A–F. The A/B region contains isoform-specific sequences as well as a constitutively active transactivation domain (TAF-I). The C region contains the zinc-finger DNA binding domain. The E region is important for the specificity of ligand binding as well as ligand-induced receptor dimerization, nuclear translocalization, and transactivation (TAF-II). Region F is found in some members of the family, such as the estrogen receptor. (B) In the simplified model, hormone-occupied receptors dimerize and tightly associate with hormone-response elements within target genes, resulting in transcriptional activation or repression and the production of an altered complement of cellular proteins.

1990). It is generally accepted that intracellular steroid receptors are associated with proteins of the heat shock family (Schlatter *et al.*, 1992; Smith *et al.*, 1992) and that they undergo ligand-induced dimerization and nuclear localization (Guiochon-Mantel *et al.*, 1989).

The structural and functional domains of many of the steroid receptors have been dissected in mutagenesis studies. Figure 3A schematically illustrates several key functional domains of the progesterone and estrogen receptors, and Figure 3B presents a simplified model of signal transduction by these receptors. Six characterized domains of these receptors are indicated as A–F in Figure 3A. The amino-terminal A/B region contains cell-specific and constitutive transactivation functions (TAF-1) (Gronemeyer *et al.*, 1987; Kumar *et al.*, 1987; Giguere *et al.*, 1986), and isoform-specific sequences for the progesterone receptor. The C region of the receptors includes the DNA-binding domain, a hallmark of these receptors as nuclear transcription factors. This basic and most conserved region of the C domain contains invariant cysteine repeats, originally observed in the amphibian transcription factor TFIIIA (Miller *et al.*, 1985; Brown *et al.*, 1985), which form two "zinc finger" structures, with four cysteines in each finger (for review, see Evans, 1988; Beato, 1989; O'Malley, 1990; Freedman, 1992).

Specific sequences within the DNA-binding domain are critical for DNA binding site specificity (Danielsen *et al.*, 1989; Mader *et al.*, 1989; Umesono and Evans, 1989), DNA-dependent dimerization (Umesono and Evans, 1989; Dahlman-Wright *et al.*, 1991), and positive control of transcription (Schena *et al.*, 1989). Recombinant steroid receptors have been expressed in various systems (for review, see Srinivasan, 1992) and the solution or crystal structure of the DNA-binding domain of the steroid receptors has been studied using these recombinant proteins (Schwabe *et al.*, 1990; Hard *et al.*, 1990; Luisi *et al.*, 1991). Extensive mutagenesis of the hormone-response element (HRE) suggests that the idealized DNA binding sequences for each steroid receptor are composed of two half-site palindromic or direct, repeated sequences with different spacing intervals (Truss *et al.*, 1991; Glass *et al.*, 1988). The region between the C and D domains is known to contain a constitutively active nuclear localization signal (Guiochon-Mantel *et al.*, 1989; Rusconi and Yamamoto, 1987). This region of the receptors, as well as the hormone-binding domain (E), is required for interaction between the receptor and heat shock proteins (Carson-Jurica *et al.*, 1989; Schlatter *et al.*, 1992). The carboxyl-terminal E domain is involved in multiple functions, including hormone binding (Kumar *et al.*, 1987), nuclear translocalization of the receptor (Picard and Yamamoto, 1987), hormone-dependent transactivation (Kumar *et al.*, 1987; Hollenberg and Evans, 1988), and receptor dimerization (Kumar and Chambon, 1988). The existence of an F domain is evident in the estrogen receptor, but not the progesterone receptor. As indicated in Figure 3B, upon hormone binding, there is a ligand-induced receptor dimerization and the receptor translocates to the nucleus (or if already there, is activated for DNA binding) and binds to HREs to stimulate or inhibit the transcription of target genes. Diamond and colleagues (1990) have suggested that some of the negative effects of the steroid hormones may be elicited predominantly by protein–protein interactions between the receptor and other transcription factors. Figure 3B shows a generalized model, and differences can be seen among the individual receptors, especially in association with heat shock proteins and initial cellular localization prior to hormone binding.

Steroid hormones are involved directly or indirectly in many aspects of gonadal development and function, and are of course major secretory products of the ovary. Both estrogen and progesterone have been implicated in ovarian follicular development, ovulation, and luteinization. The structure, function, and regulation of these two predominant ovarian steroid receptors are discussed in more detail in the following sections.

A. Progesterone Receptor

Ovarian functions that have been reported to be modulated by progesterone include follicular development (Goodman and Hodgen, 1982), estrogen production by granulosa cells (Schreiber *et al.*, 1980), FSH induction of LH receptors on granulosa cells (Schreiber *et al.*, 1982), the maintenance of the corpus luteum (Singh *et al.*, 1988), and ovulation (Swanson and Lipner, 1977; Mori *et al.*, 1977). The biological actions of this steroid in the ovary are likely to be mediated by the intracellular progesterone receptor, although nongenomic actions of progesterone have been reported (Schumacher, 1990). The presence of the progesterone receptor molecule in the ovary has been demonstrated by ligand-binding studies (Jacobs *et al.*, 1980; Jacobs and Smith, 1980; Schreiber and Hsueh, 1979; Naess, 1981), detection of immunoreactive protein (Korte and Isola, 1988; Press and Greene, 1988; Isola *et al.*, 1987; Yoshimura and Bahr, 1991), and mRNA analysis (Park and Mayo, 1991; Iwai *et al.*, 1991).

Biologically active progesterone receptors are phosphoproteins. Progesterone receptors undergo both hormone-independent and hormone-dependent phosphorylation (Denner *et al.*, 1987; Logeat *et al.*, 1985; Orti *et al.*, 1992; Wei *et al.*, 1987). Ligand-dependent phosphorylation is essential for receptor activation (Denner *et al.*, 1990) and can be bypassed by enhancing protein kinase or inhibiting protein phosphatase activity. It is generally thought that hormone-bound progesterone receptor dimers become tightly associated with DNA and subsequently interact with other components of the transcriptional apparatus (Bagchi *et al.*, 1992; Truss *et al.*, 1992). During these events, members of the heat shock protein family (in particular, hsp70 and hsp90) are thought to be involved as molecular chaperones (Schowalter *et al.*, 1991; DeMarzo *et al.*, 1991).

The progesterone receptor cDNA has been characterized in several species, including humans, chickens, rabbits, mice, and rats. Although Loosfelt and colleagues (1986) proposed that the rabbit progesterone receptor exists in only one form (termed B), the chicken, human, mouse, and rat progesterone receptors appear to exist as two isoforms (termed A and B) (Boyd and Spelsberg, 1979; Horwitz and Alexander, 1983; Conneely *et al.*, trol, at least in uterine epithelial cells and human breast cancer cells (Aronica and Katzenellenbogen, 1991; Katzenellenbogen and Norman, 1990; Sumida and Pasqualini, 1990). In addition to estrogen, progesterone, growth factors such as insulin-like growth factor (IGF) and EGF, and agents that increase intracellular cAMP levels all modulate progesterone receptor or receptor mRNA levels. Multiple progesterone receptor RNA transcripts are present in all target tissues or cells, and it is not clear if there are tissue-specific transcripts. Although it has been shown that some of these transcripts are specific for the A form of the receptor (Wei *et al.*, 1990), all of the transcripts appear to be increased or decreased in the same fashion in response to hormonal stimuli (Wei *et al.*, 1988, 1990). In the ovary, the multiple progesterone receptor mRNAs are induced in mature granulosa cells in response to the pituitary gonadotropins (Park and Mayo, 1991; Iwai *et al.*, 1991). During the rat estrous cycle, expression of the progesterone receptor gene is activated transiently by the preovulatory gonadotropin surges, and is rapidly suppressed, perhaps by elevated progesterone levels (Park and Mayo, 1991). Because the gonadotropin receptors are coupled to adenylate cyclase through a G protein as described earlier, and because cAMP analogues can mimic the expression of gonadotropin-induced progesterone receptor mRNA in ovarian granulosa cells (Park-Sarge and Mayo, unpublished observations; Iwai *et al.*, 1991), cAMP is likely to be a direct effector of the progesterone receptor gene in ovarian granulosa cells. The functional significance of progesterone receptors in ovarian granulosa cells remains to be determined; however, given the temporal pattern of receptor expression, it seems likely that ovarian progesterone receptors will have important functions in the periovulatory period.

B. Estrogen Receptor

It is believed that estrogen, in conjunction with the pituitary gonadotropins, plays a significant role in the growth and development of ovarian follicles (Richards, 1980; Erickson *et al.*, 1985; Hsueh *et al.*, 1984). Estrogen stimulates the proliferation of granulosa cells from small follicles (Goldenberg *et al.*, 1972), increases granulosa cell gonadotropin receptors and responsiveness to FSH and LH (Richards *et al.*, 1976; Rani *et al.*, 1981), regulates progesterone production (Welsh *et al.*, 1983), and enhances gap junction formation (Merk *et al.*, 1972). The presence of estrogen receptors in the rat ovary has been suggested by autoradiographic (Stumpf, 1969) and receptor binding studies (Richards, 1975; Kudolo *et al.*, 1984). Immunocytochemistry and RNA analyses have detected estrogen receptor or its mRNA in the mouse (Hillier *et al.*, 1989) and monkey ovary (Billiar *et al.*, 1992).

The overall structure of the estrogen receptor as predicted from estrogen receptor cDNAs (S. Green *et al.*, 1986; G. L. Greene *et al.*, 1986; Krust *et al.*, 1986; Koike *et al.*, 1987) is similar to that of the progesterone receptor, described in the preceding section. Unlike the progesterone receptor gene, the estrogen receptor gene structure has not been extensively studied. A putative promoter of the mouse estrogen receptor gene has been reported (White *et al.*, 1987; Ponglikitmongkol *et al.*, 1988), but it was recently suggested that there is an additional upstream exon (Keaveney *et al.*, 1991) preceded by the actual promoter (Piva *et al.*, 1992). It has been shown in both cases that transcription starts at a unique site (Green *et al.*, 1986; Piva *et al.*, 1992). It is possible that these two promoters may contribute to generation of the two isoforms of the estrogen receptors, in that the more recently characterized upstream initiation codon is in frame with the previously characterized initiation codon.

The estrogen receptor mRNA appears to be present as a single transcript of approximately 6.5 kb in estrogen target cells, including the human uterus, in which a genetic polymorphism generates a variant (B-variant) estrogen receptor mRNA that contains a point mutation in the A/B domain of the estrogen receptor (Garcia *et al.*, 1989). Several variant mRNAs of the estrogen receptor have been detected in human breast cancer cells (Dotzlaw *et al.*, 1992) and subsequently determined to encode unique proteins or proteins containing some functional domains of the normal estrogen receptor but missing others. The significance of these variant estrogen receptor proteins in human cancer cells remains to be determined. Hillier and colleagues (1989) reported a transcript of 1.5 kb in the human ovary that was related to the hormone-binding domain of the estrogen receptor. Since this would be too small to encode the entire protein and it lacks the DNA-binding domain, it is interesting to speculate that this mRNA could encode a novel estrogen-binding protein lacking normal receptor function, perhaps in some manner modulating the effects of estrogen within the ovary.

IV. Serine and Threonine Kinase Receptors

A large family of polypeptides that is exemplified by transforming growth factor (TGF) β and includes inhibin, activin, Müllerian-inhibiting substance, the bone morphogenic proteins, the *Drosophila decapentaplegic* gene product, the *Xenopus Vg-1* gene product, and the mammalian Vg-related proteins display wide-ranging effects on cell proliferation, differentiation, and organization (for review, see Roberts and Sporn, 1990; Vale *et al.*, 1990). Many members of this polypeptide family are expressed in the ovary or have been demonstrated to affect ovarian function (Ackland *et al.*, 1992). Although the biological actions of these peptides were suspected to be mediated by cell-surface receptor molecules, the molecular structure and signal transduction pathways of these receptors were unclear until the recent cloning and characterization of the receptors for activin (Mathews and Vale, 1991; Attisano *et al.*, 1992) and TGFβ (H. Lin *et al.*, 1992). It was found that these receptors are homologous, as are the ligands, and that the presumed intracellular domains of these receptors were related to a family of protein serine and threonine kinases, suggesting a possible signal transduction mechanisms involving protein phosphorylation.

Many types of cellular protein kinases have been identified, and they all contain conserved amino acid motifs within their catalytic domains, suggesting similar modes of activity or regulation among these enzymes (Hanks and Quinn, 1991). Conserved features of the catalytic domain include a nucleotide binding consensus sequence (Wierenga and Hol, 1983), an invariant lysine residue involved in phosphotransfer (Kamps and Sefton, 1986), several ATP binding sites, autophosphorylation sites, and two regions that are indicative of serine and threonine or tyrosine substrate specificity (Hanks and Quinn, 1991). These signature sequences suggest that the activin and TGFβ receptors are likely to be serine and threonine kinases. While numerous cytoplasmic serine and threonine kinases have been characterized, the activin and TGFβ receptors, along with the product of the *C. elegans daf*-1 gene (Georgi *et al.*, 1990), represent the initial examples of presumed receptor serine and threonine kinases.

Figure 4 shows a schematic illustration of the structure of the receptors in the class, and of their potential signaling pathways. The receptors have a cysteine-rich, hydrophilic amino-terminal domain that is presumed to be extracellular and contains consensus sequences for N-linked glycosylation. Sequences within the extracellular domain of this receptor family are somewhat conserved and can be aligned using the cysteine residues. There is no homology in this region between the serine and threonine kinase and the tyrosine kinase receptor families (see the next section). A single domain with strong hydrophobicity is thought to span the membrane once. The intracellular portion of the receptors is dominated by the kinase domain, which is followed by a short tail that is not highly conserved. As is discussed later, signaling by these receptors might involve at least two protein species, designated the type I and type II receptors in Figure 4. The following sections summarize information on the type II TGFβ and activin receptors, both of which are likely to have important roles in the ovary.

A. Type II Receptor for TGFβ

There is strong evidence that growth factors are involved in modulating the action of sex steroids and gonadotropins within the ovary (for review, see Adashi and Rohan, 1992; Ackland *et al.*, 1992). TGFβ affects many aspects of ovarian function and appears to be locally produced within the ovary. the TGFβ family includes three to five isoforms in various species. All of these but $TGF\beta_5$ have been cloned and their molecular structures have been reviewed elsewhere (Massague, 1990). The biolog-

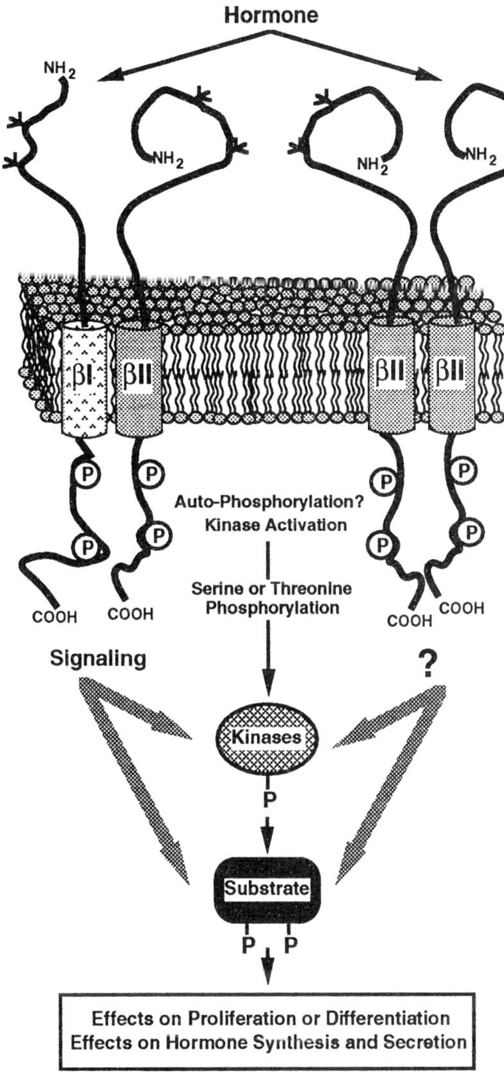

FIGURE 4 A schematic representation of the structural features of a serine and threonine kinase receptor and some of its potential signaling pathways. The example is the TGFβ receptor, but the features have been generalized. Type I and type II receptors are both transmembrane proteins with serine and threonine kinase domains. It is postulated that both the type I and type II receptors for TGFβ are necessary for complete activity, and thus a heterodimeric receptor is shown as one possible form, with the homodimeric form of the characterized type II receptor as an alternative possibility. (? indicates the uncertainty in whether this homodimer has signaling properties). Following hormone binding to the receptor, autophosphorylation, kinase activation, and association with cellular substrates are expected to occur, based upon more substantial information for the related tyrosine kinase receptors.

ically active hormone is a homodimer of subunits that arise by proteolytic processing of a larger precursor protein. Crystallographic studies of human recombinant TGFβ_2 have revealed important features of the monomeric structure and of dimer association that will surely improve our understanding of receptor-ligand interactions (Schlunegger and Grutter, 1992; Daopin *et al.*, 1992).

Three receptor proteins have been identified by their ability to bind to TGFβ_1 and TGFβ_2 with high affinity (Cheifetz *et al.*, 1987; Segarini *et al.*, 1987). Two are glycoproteins designated I (53 kDa) and II (75 kDa), the third is membrane proteoglycan designated as III (250 kDa). Essentially all cells express type I and type II receptors (Cheifetz *et al.*, 1987; Segarini *et al.*, 1987) and TGFβ-induced cellular signal transduction has been shown to be mediated through the type I and II receptors (Laiho *et al.*, 1991). With the recent identifications of a type I receptor cDNA, all three receptor types have now been cloned. As predicted from biochemical studies, the cDNA for the type III receptor encodes a membrane-anchored proteoglycan (Lopez-Casillas *et al.*, 1991; Wang *et al.*, 1991) with a small cytoplasmic domain containing no obvious signaling motifs. It has been suggested that the type III receptor might be involved in attracting and concentrating TGFβ on the cell surface for eventual transfer to closely situated signal-transducing receptors (X.-F. Wang *et al.*, 1991). The characterization of the type II receptor cDNA revealed the protein described earlier, with a protein serine and threonine kinase domain providing a likely signaling mechanism (H. Lin *et al.*, 1992). The presence of the type II receptor alone may not be sufficient for signaling the biological action of TGFβ since variant cell lines retaining the type II receptor but lacking the type I receptor were unable to respond to TGFβ (Laiho *et al.*, 1990). Subsequently, it has been suggested that TGFβ might act through a heteromeric complex of receptors I and II, and that the type I receptor might require the type II receptor for ligand binding, while the type II receptor might require the type I receptor for signaling (Wrana *et al.*, 1992). The recent cloning of a type-I receptor for TGF-β (Ebner *et al.*, 1993), which is structurally related to the type II receptor, further supports the notion of combinatorial interactions between the type I and type II receptors in TGF-β action.

Cellular substrates for this receptor kinase are not yet identified. The TGFβ receptor kinase domain, expressed in bacteria, is capable of auto- or cross-phosphorylation on serine and threonine residues (H. Lin *et al.*, 1992). Potential TGFβ receptor-mediated intracellular signaling pathways might include interactions with tumor-suppressor genes, as is the case for other growth factors (Cross and Dexter, 1991; Cantley *et al.*, 1991; Marshall, 1991). TGFβ has been shown to interfere with the phosphorylation of one tumor suppressor, the retinoblastoma protein RB (Laiho *et al.*, 1990) and to inhibit the transcription of the *c-myc* proto-oncogene (Pietenpol *et al.*, 1990), which perhaps explains some of its effects on cellular proliferation.

1989; Schrader and O'Malley, 1972; Schneider *et al.*, 1991; Ilenchuk and Walters, 1987; Gronemeyer *et al.*, 1987; Krett *et al.*, 1988). PR_A is an amino-terminally truncated variant of PR_B that appears to arise from alternative transcription or alternative translational initiation from an internal AUG codon (Conneely *et al.*, 1989; Kastner *et al.*, 1990a,b; Krett *et al.*, 1988; Wei *et al.*, 1990). Tora and co-workers (1988) have suggested that these two receptor proteins have different transactivating functions. The regulation of the progesterone receptor gene is not well understood. The gene is composed of multiple exons; exons 1 and 2 comprise the amino-terminal or regulatory domain, exons 2 and 3 the DNA-binding domain, and the remaining exons the ligand-binding domain. The first exon contains the translation initiation codons for both the B and A forms of the receptor (Misrahi *et al.*, 1988; Jeltsch *et al.*, 1990; Savouret *et al.*, 1991; Turcotte *et al.*, 1991; Fuchs-Young *et al.*, 1992; Kastner *et al.*, 1990b). Transcription of the PR_B mRNA starts at multiple sites downstream of the modified CCAAT and TATA consensus sequences found in many gene promoters; transcription of the PR_A mRNA also starts at multiple sites that fall within the coding region for PR_B. Kastner and co-workers (1990b) have suggested differential regulation of these two promoters in the human gene.

Estrogen is considered to be a predominant regulator of PR gene expression in many systems (Katzenellenbogen, 1980; Horwitz *et al.*, 1985). It is thought that the progesterone receptor gene is transcriptionally activated by estrogen and inhibited by progesterone. However, it has been difficult to definitively demonstrate DNA sequences that the estrogen receptor recognizes within the progesterone receptor gene, in that the promoter does not contain consensus DNA sequences known to mediate the actions of estrogen receptor. Savouret and colleagues (1991) however, have shown that the estrogen-induced transcription of the rabbit progesterone receptor gene is mediated by DNA sequences that are not classic estrogen-response elements (EREs) but can interact with estrogen receptors. In the human progesterone receptor gene, two promoters located between −711 and +31 basepairs and between +464 and +1105 basepairs have been shown to be estrogen inducible despite the lack of consensus ERE sequences (Kastner *et al.*, 1990).

An alternative model in examples where receptor binding sites have not been localized within target genes is to use transcription factors, present in limiting amounts, that are involved directly or indirectly in the action of estrogen or progesterone receptors on these genes (Meyer *et al.*, 1989; Shemshedin *et al.*, 1992). For example, progesterone rapidly stimulates the formation of AP-1 transcription complexes by activating the c-jun and jun-B genes (Alkhalaf and Murphy, 1992). Thus, indirect action of the steroid hormone would lead to eventual activation of genes that lack sites for direct steroid receptor binding.

The progesterone receptor gene appears to be under multifactorial con-

The general biological actions of the TGFβ peptides have been extensively discussed elsewhere (Massague, 1990). Within the ovary, TGFβ stimulates DNA synthesis (Bendell and Dorrington, 1988) and facilitates the ability of FSH to stimulate DNA synthesis in granulosa cells (Dorrington *et al.*, 1988). It antagonizes the action of epidermal growth factor (EGF) or platelet-derived growth factor (PDGF) on the growth of granulosa cells (Skinner *et al.*, 1987; Mondschein *et al.*, 1988) and modulates steroidogenesis in granulosa as well as thecal cells (Hutchinson *et al.*, 1987; Ying *et al.*, 1986). TGFβ alone or in combination with insulin stimulate production of FSH-induced LH receptor (Knecht *et al.*, 1986, 1987). Basal and FSH-stimulated secretion of inhibin in granulosa cells is also stimulated by TGFβ (Zhiwen *et al.*, 1988). These actions of TGFβ are considered to be autocrine or paracrine, and TGFβ peptides or mRNAs have been identified within the ovary (although there are species differences). TGFβ immunoreactivity or mRNA has been observed in rat granulosa cells as well as in theca-interstitial cells (Kim and Schomberg, 1989; Skinner *et al.*, 1987; Mulheron and Schomberg, 1990; Hernandez *et al.*, 1990). In the mouse, TGFβ immunoreactivity was localized to thecal-interstitial, but not to granulosa cells (Thompson *et al.*, 1989). In the human, however, most ovarian structures, including oocytes, granulosa-theca cells, corporal cells, and ovarian smooth muscles, were found to be positively stained using TGFβ immunohistochemistry (Chegini and Williams, 1992). It will be important to determine which ovarian cell types express which types of receptors for the TGFβ family of growth factors.

B. Type II Receptors for Activin

Activin is a member of the TGFβ superfamily that has a broad spectrum of biological actions but was first identified by its effects in the reproductive system. (for review, see Ying, 1988; deKretser and Robertson, 1989; Vale *et al.*, 1990). Activin is a dimer composed of β_A or β_B chains (activin A = $\beta_A\beta_A$, activin AB = $\beta_A\beta_B$, activin B = $\beta_B\beta_B$). These same β_A or β_B chains also contribute to the formation of the heterodimeric hormone inhibin, which acts to suppress pituitary FSH secretion by combining with the unique α chain (Ying, 1988; deKretser and Robertson, 1989; Vale *et al.*, 1990). The ovary is an important site of synthesis of activin and its counterpart inhibin, and the gonadotropins play an important role in regulating ovarian production of the inhibins and activins (Woodruff and Mayo, 1990). While the first description of inhibins and activins was based on their ability to act as endocrine regulators of pituitary FSH secretion, it is now recognized that they exert substantial autocrine or paracrine effects that regulate folliculogenesis within the ovary (Woodruff *et al.*, 1990; Findlay, 1993). Follistatin, a structurally unrelated protein, binds to the common β chains of inhibin and activin and inhibits the actions of activin both in the pituitary and in the ovary (Ying, 1988; Robertson, 1992; T. Nakamura *et al.*, 1990, 1992a).

Activin receptors have been identified by binding and cross-linking to iodinated activin A and B, and the 65 and 85 kDa forms designated type I and type II have been identified by analogy to the TGFβ receptors (Hino *et al.*, 1989; Centrella *et al.*, 1991). The mouse type II activin receptor (also called type IIA) was the first of the TGFβ receptor family to be isolated and characterized (Mathews and Vale, 1991). This receptor, when expressed in COS monkey kidney cells, was shown to bind to activin A or B, and to more weakly bind to inhibin, suggesting that the receptor for inhibin will be structurally similar to the activin receptor. Like the type II TGFβ receptor (H. Lin *et al.*, 1992), the activin receptor is predicted to be a serine and threonine-specific protein kinase (Mathews and Vale, 1991); however, biochemical analysis of a highly purified activin receptor suggests that it might also have tyrosine kinase activity (T. Nakamura *et al.*, 1992b). The type II activin receptor has been cloned from other species, including humans (Matzuk and Bradley, 1992; Donaldson *et al.*, 1992), chickens (Ohuchi *et al.*, 1992), rats (Brown-Schult *et al.*, 1992), and *Xenopus* (Nishimatsu *et al.*, 1992; Mathews *et al.*, 1992).

There appears to be a family of activin receptors. A type IIB activin receptor, similar to but distinct from the original type II receptor, was identified in the mouse (Attisano *et al.*, 1992) and rat (Legerski *et al.*, 1992; Brown-Schult *et al.*, 1992). The mouse cDNA contained two alternative splicing sites proximal to the transmembrane domain (one in the extracellular and the other in the intracellular domain), indicating that the type IIB activin receptor gene can generate four receptor isoforms. The two isoforms lacking the exon encoding a small proline-rich cluster just outside the membrane-spanning domain have a lower affinity for activin. The type II activin receptor gene produce two predominant transcripts of 3.0 and 6.0 kb which are found in many tissues but vary in abundance (Mathews and Vale, 1991). Likewise, the type IIB activin receptor mRNA exists as several species and varies among tissues; both the type II and type IIB receptor mRNA can be detected in the ovary (Brown-Schult *et al.*, 1992). Two forms of the activin receptor have also been identified in *Xenopus*. The important effects of the hormone in regulating cell fate (Green and Smith, 1990) can be seen in experiments that used a truncated receptor mutant lacking the kinase domain to interfere with the function of the endogenous receptor; the result was inhibition of mesoderm formation (Hemmati-Brivanlou and Melton, 1992).

V. Tyrosine Kinase Receptor Family

While knowledge of the serine and threonine kinase receptor family is just emerging, an analogous family of proteins with tyrosine kinase activity includes some of the best-characterized receptors for polypeptide hormones and growth factors (Ullrich and Schlessinger, 1990). The catalytic tyrosine kinase domains of these receptors are related to those of the serine and threonine kinases described earlier, but with clear differences that are indicative of amino acid substrate specificity (Hanks and Quinn, 1991). Many of the initial members of the tyrosine kinase family of receptors were identified as products of retrovirus-encoded oncogenes. Like the serine and threonine kinase family, the tyrosine kinase family includes members that are intracellular nonreceptor kinases (Cantley *et al.*, 1991) and members that are transmembrane receptor kinases (Ullrich and Schlessinger, 1990).

The tyrosine kinase receptors contain five major domains: an extracellular ligand-binding domain, a single transmembrane domain, the juxtamembrane domain, the tyrosine kinase domain, and a carboxyl-terminal tail. The structural characteristics of the extracellular domain subdivide these kinase receptors into three distinct classes, as shown in Figure 5A. The first class is exemplified by the EGF receptor, and is characterized by two cysteine-rich repeat sequences. The second class is exemplified by the IGF-1 receptor, and is characterized by a cysteine-rich and heterotetrameric structure generated by dissimilar disulfide-linked subunits. The third class is exemplified by the PDGF receptor, and is characterized by immunoglobulin-like repeats. The transmembrane domain appears to play a passive role by providing an anchor for these receptors, connecting extracellular binding to intracellular signaling. The tyrosine kinase receptors form dimers (for review, see Schlessinger, 1988; Williams, 1989; Ullrich and Schlessinger, 1990), and when these receptors bind to ligands, they transphosphorylate, which is thought to promote the active state (Honegger *et al.*, 1989; Kazlauskas and Cooper, 1989). All receptor tyrosine kinase signaling activities depend on a functional tyrosine kinase domain, which is capable of tyrosine phosphorylation of cellular substances (Ullrich and Schlessinger, 1990). The carboxyl-terminal tail appears to be involved in the negative control of tyrosine kinase receptor signaling (Roussel *et al.*, 1987).

Figure 5B shows a schematic illustration of the potential signal transduction pathways for a tyrosine kinase receptor of the PDGF class. Ligand binding promotes dimerization and transphosphorylation, which promotes in turn the active state in which the tyrosine kinase phosphorylates cellular substrates. The characterized substrates for the receptor tyrosine kinases include phosphatidylinositol (PI) kinase (Kaplan *et al.*, 1987), phosphatidylinositol-specific PLC-γ (Meisenhelder *et al.*, 1989), PLC-II (Wahl *et al.*, 1989), Raf-1 (Morrison *et al.*, 1989), and ras GTPase activating protein (GAP) (Kaplan *et al.*, 1990; Molloy *et al.*, 1989; Kazlauskas and Cooper, 1989). Representative receptors from each of the three subclasses of receptor tyrosine kinases are discussed in the following section with regard to their mechanisms of action and expression or functions in the ovary.

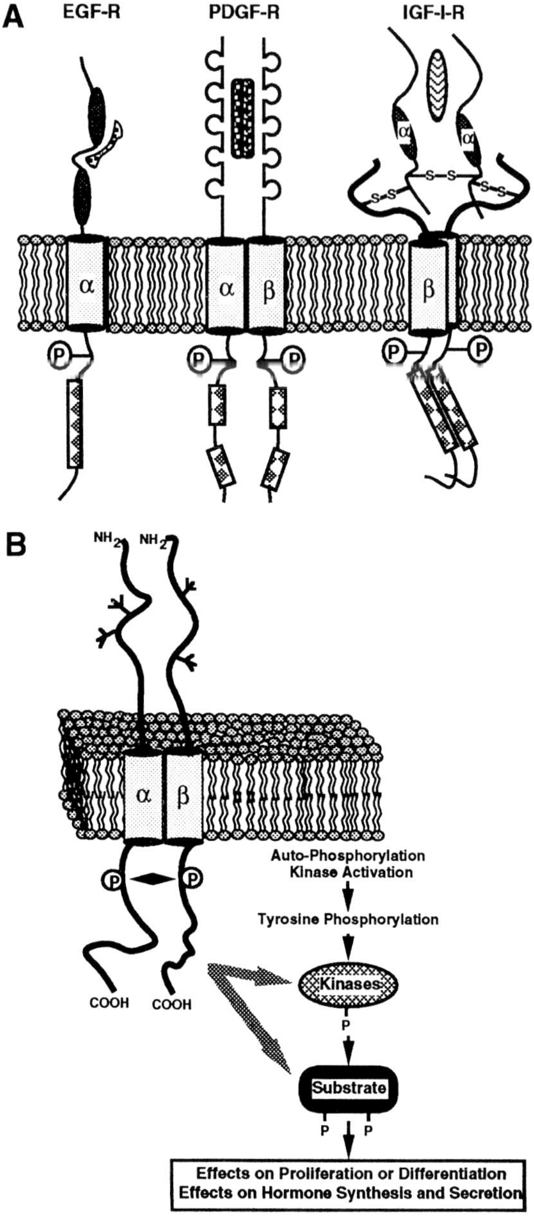

FIGURE 5 A schematic representation of the structural features of the tyrosine kinase receptors and some of their potential signaling pathways. The examples represent major subfamilies of tyrosine kinase receptors. (A) Three classes of tyrosine-kinase receptors distinguished by unique features in the extracellular domain; the EGF receptor contains two cysteine-rich ( ) subdomains; the PDGF receptor contains globulin-like ( ) structures, and disulfide bridges connect the subunits of the heterotetrameric IGF-I receptor. All these receptors undergo ligand-induced oligomerization following the binding of the hormone. The intracellular kinase domains ( ) are indicated; this domain is interrupted in the PDGF-type receptors. (B) The activated PDGF receptor dimerizes (a heterodimeric form is shown) and is autophosphorylated, stabilizing the tyrosine kinase activity and leading to the association of cellular substrates with the receptor.

A. PDGF and c-Kit Receptors

Although the functional roles of PDGF in ovarian function are unclear, it has distinct activities in this tissue; for example, PDGF increases mitotic activity, DNA synthesis, and IGF-1 secretion in the ovary (Knecht and Catt, 1983; Mondschein and Hammond, 1988). There is little information regarding the expression of the PDGF receptor in the ovary, although its activity in this tissue suggests that the receptor is present. The PDGF receptor is one of the best characterized of the tyrosine kinase receptors, and much information of general relevance to this class of signaling proteins has been obtained.

PDGF itself is a 30-kDa dimeric (homo- or heterodimeric) protein consisting of two different polypeptides, the A and B chains (Hannik and Donogue, 1989). The crystal structure of recombinant human PDGF-BB has been solved, and probable receptor recognition sites of the ligand have been identified (Oefner *et al.*, 1992). All forms of PDGF (AA, AB, and BB) are biologically active, but elicit distinct cellular responses (Nister *et al.*, 1988) by activating different receptor dimers (Hart *et al.*, 1988; Eriksson *et al.*, 1992). Like the ligand, PDGF receptors are dimers ($\alpha\alpha$, $\alpha\beta$, and $\beta\beta$). Each receptor subunit has been cloned and characterized (Yarden *et al.*, 1986; Matsui *et al.*, 1989). The two genes, located on human chromosomes 4 and 5, encode receptors that are structurally similar; each consists of an extracellular domain containing five immunoglobulin-like repeats and an intracellular domain with protein tyrosine kinase activity, including a characteristic insert that has been shown to be dispensable for kinase activity and mitogenic signaling (Taylor *et al.*, 1989). The kinase insert region does appear to be important for interaction between the kinase domain and intracellular substrates, as discussed later.

Receptor dimerization upon ligand binding leads to trans- or autophosphorylation of these receptors (Kelly *et al.*, 1991). The active receptor interacts predominantly with Src homology (SH)-2 domain-containing substrates such as phosphatidylinositol-3-kinase, GAP, phospholipase C-γ, and Src. Phosphorylation of the receptor within the insert region has been shown to mediate the interaction between receptor and GAP or PI-3-kinase (Fantl *et al.*, 1992; Kashishian *et al.*, 1992). Interaction between the receptor and PLC-γ, however, is mediated by phosphorylation of the receptor within the carboxyl-terminal tail (Ronnstrand *et al.*, 1992).

Another tyrosine kinase receptor that is structurally related to the PDGF receptor, and which may be important for oogenesis and embryogenesis, is the proto-oncogene c-kit, a cellular homologue of v-kit (the oncogene of the feline sarcoma virus) (Besmer *et al.*, 1986). The c-kit gene maps to the mouse dominant white-spotting (W) locus (Chabot *et al.*, 1988), the mutation of which affects the proliferation and migration of primordial germ cells (Mintz and Russell, 1957). Molecular cloning of c-kit (Yarden *et al.*, 1990; Qui *et al.*, 1990) demonstrated that this receptor is structurally related to the PDGF receptor, having the five immunoglobulin-like repeats, and the insert within the kinase domain. The ligand for the c-kit receptor has been identified as the product of the steel (Sl) locus in the mouse, and is a hematopoietic growth factor with a broad range of activities (for review, see Witte, 1990). Like the W locus, mutations in Sl exhibit many phenotypes, one of which is defective gonadal development.

The c-kit mRNA is expressed as a single transcript of 5.5 kb in various tissues, including the ovary (Manova *et al.*, 1990). Primordial germ cells appear to express c-kit mRNA during embryogenesis (Orr-Urtreger *et al.*, 1990). Throughout prenatal and postnatal development, oocytes of primordial and growing follicles express the c-kit mRNA (Manova *et al.*, 1990). Somatic expression of the c-kit gene was also observed in the mouse ovary at 14 days of age and thereafter; c-kit mRNA was expressed in thecal cells of healthy or early stage atretic follicles, and in interstitial and luteal cells. Contrary to these findings for mRNA, c-kit protein measured by immunoreactivity is present in oocytes, but not somatic cells (Horie *et al.*, 1991). Strong c-kit immunoreactivity was observed on the surface of oocytes, which is consistent with a signaling role for the oocyte receptor. Ovulated, unfertilized oocytes contained decreased amounts of c-kit protein. The c-kit protein was still observed in one-cell embryos but was undetectable in four-cell embryos (Horie *et al.*, 1991), suggesting a potential role for the Sl factor and the c-kit receptor during oocyte maturation and very early development.

B. EGF Receptor

Epidermal growth factor is a 53-amino acid, single-chain polypeptide, and its receptor is a transmembrane tyrosine kinase receptor (for review, see Ullrich and Schlessinger, 1990). The extracellular domain of this receptor contains two cysteine-rich repeats (Ullrich *et al.*, 1984). This receptor class is thought to form dimers due to a conformational change in the extracellular domain when monomeric ligands are bound to each receptor monomer (Greenfield *et al.*, 1989). Autophosphorylation and signal transduction mechanisms for the EGF receptor are thought to be similar to those described earlier for the PDGF receptor, and are not considered further here. The EGF receptor also mediates signaling by the related ligand, transforming growth factor-α (Derynck, 1986).

Epidermal growth factor can modulate ovarian function by inhibiting gonadotropin-stimulated granulosa cell differentiation (May and Schomberg, 1989) and by stimulating DNA synthesis (Bendell and Dorrington, 1990). There have been studies demonstrating the presence of EGF protein in hamster ovarian granulosa and theca cells (Roy and Greenwald, 1990) and EGF mRNA in the mouse ovary (Rall *et al.*, 1985), suggesting the local synthesis of this growth factor. In addition, TGFα is clearly produced in the ovary and has important biological actions there. Thecal cells appear to be the primary source of TGFα, which regulates follicular growth (Skinner and Coffey, 1988; Lobb *et al.*, 1989). TGFα induces aromatase activity and estradiol synthesis in ovarian cells, and this activity can be inhibited with the tyrosine kinase inhibitor, genistein, which is consistent with the action of a tyrosine kinase receptor (Gangrade *et al.*, 1991). EGF receptor binding sites have been demonstrated in ovarian granulosa (Jones *et al.*, 1982; Vlodavsky *et al.*, 1978; Chabot *et al.*, 1986), thecal (Chabot *et al.*, 1986), and luteal cells (Vlodavsky *et al.*, 1978; Chabot *et al.*, 1986). FSH increases EGF receptor numbers on the granulosa cell, while LH reduces EGF binding sites (Feng *et al.*, 1987).

C. IGF-I Receptor

Insulin-like growth factor-I is a 70-amino-acid hormone that has a variety of endocrine and metabolic roles; its biological action is mediated by the type I IGF receptor (IGF-I receptor), a transmembrane tyrosine kinase receptor (Ullrich *et al.*, 1986). The IGF-I receptor is an $\alpha_2\beta_2$ tetramer that is composed of two $\alpha\beta$ dimers linked by disulfide bonds (Soos and Siddle, 1989). The two α chains are completely extracellular and have a cysteine-rich domain similar to that in found in other tyrosine kinase receptors. They are connected to each other by a disulfide bond and are directly involved in ligand binding. The two β chains are transmembrane proteins with short extracellular domains linked to the α chain by a disulfide bond and with a cytoplasmic domain that contains the tyrosine kinase catalytic activity. The activation of this receptor involves an intracomplex conformational change subsequent to ligand binding to the disulfide-stabilized receptor tetramer. The activated β chain tyrosine kinase transphosphorylates other β chains, leading to a stable activated form that can act on additional substrates in the same way as the PDGF receptor. The receptor complex exhibits an affinity for the related peptide IGF-II that is only two to three times lower than that for IGF-I, but it has a low affinity for insulin (Steele-Perkins *et al.*, 1988, Germain-Lee *et al.*, 1992). The closely related insulin receptor is one of the most intensively studied receptor proteins, and its structure and activation are similar to those of IGF-I receptor; insulin receptors have been discussed elsewhere (Taylor *et al.*, 1992) and are not further considered in this chapter.

IGF-I is synthesized locally within the ovary, and acts directly on ovarian cells to affect many aspects of their function (Hammond *et al.*, 1991; Adashi and Rohan, 1992). FSH-stimulated LH receptor synthesis and steroid production in ovarian cells are enhanced by the actions of IGF-1. Granulosa cell production of the IGF-I peptide and its mRNA have been demonstrated (Oliver *et al.*, 1989; Hernandez *et al.*, 1989). In addition, IGF-1 binding sites as well as IGF-1 receptor mRNA have been demonstrated in ovarian tissues (Adashi *et al.*, 1990a; Davoren *et al.*, 1986; Cara *et al.*, 1990; Hernandez *et al.*, 1988, 1992; Monget *et al.*, 1989; Porestsky *et al.*, 1990). While binding sites appear to be modulated under some conditions (Monget *et al.*, 1989), the regulation of the receptor in the ovary is not well characterized. In addition to the IGF-1 receptor, an important contribution to IGF action in the ovary comes from a growing family of IGF-binding proteins that might modulate availability of the ligand in the local ovarian environment (Adashi *et al.*, 1990b; Nakatani *et al.*, 1991).

VI. Other Receptors

Thus far, we have discussed four categories of receptors with distinct and definable signal transduction pathways. There are, however, many other receptor proteins of importance to ovarian physiology that do not fit into these categories and for which signaling pathways have yet to be described or are poorly understood. In this section we briefly discuss the molecular structure and expression of three receptors of this type, using examples that are likely to be important for the ovary—the prolactin receptor, the nerve growth factor (NGF) receptor, and the urokinase-type plasminogen activator (uPA) receptor.

A. Prolactin Receptors

The prolactin and growth hormone receptor family is a branch of the cytokine receptor superfamily (for review, see Cosman *et al.*, 1990). GH is thought to act to stimulate the production of IGFs in its target cells. Jia and colleagues (1986) have described intraovarian actions of GH, and these are presumably mediated through an ovarian GH receptor (Lobie *et al.*, 1990). The structure of the GH receptor has been reviewed recently (Cosman *et al.*, 1990; Mathews, 1991). Interleukins, which are representative cytokines, have also been shown to exert intraovarian actions (Adashi and Rohan, 1992). The structures of the cytokine receptors and their potential signal transduction pathways have also been reviewed elsewhere (Foxwell *et al.*, 1992; Miyajima *et al.*, 1992). In this chapter, the discussion is limited to the prolactin receptors.

Prolactin is known to act on a broad spectrum of target tissues, including the ovary. It stimulates the synthesis of the LH receptor during the formation of the corpus luteum (Grinwich *et al.*, 1976) as well as during pregnancy (Gibori and Richards, 1978), and modulates steroidogenesis in the corpus luteum (Armstrong *et al.*, 1969; Gibori and Richards, 1978; Wang *et al.*, 1980; Dorrington and Gore-Langton, 1981; Wang and Chan, 1982; Krasnow *et al.*, 1990). Richards (1980) discussed the actions of prolactin on granulosa cells of mature ovarian follicles, theca-interstitial cells, and corpora lutea. Molecular cloning of the prolactin receptor revealed that there are at least three differentially spliced forms of the receptor (Boutin *et al.*, 1988, 1989; Davis and Linzer, 1989; Edery *et al.*, 1989; Kelly *et al.*, 1992). The primary structure of the prolactin receptor suggests that it is a transmembrane protein with a relatively large extracellular domain containing five cysteine residues and three potential N-linked glycosylation sites (Boutin *et al.*, 1988). Rozakis-Adcock and Kelly (1992) have shown that four of the cysteine residues in the extracellular domain are important for prolactin to bind to the receptor. The differences among the three prolactin receptors (long form, intermediate form, and short form) are in their intracellular domains, and all forms are predicted to bind to prolactin. The long form of the receptor presumably contains signaling information in that it has been shown to mediate prolactin-induced transcription of the milk protein gene (Lesueur *et al.*, 1990). Signaling by the remaining forms has not yet been demonstrated. As already discussed, the intracellular signaling pathways that are used by the prolactin receptor are currently unknown; however, the recent demonstration that the related growth hormone receptor associates with JAK2, a non-receptor tyrosine kinase, suggests the possibility that the prolactin receptor acts to activate JAK2 or a related tyrosine kinase (Argetsinger *et al.*, 1993).

The functional significance of multiple forms of the prolactin receptor has yet to be determined, and the existence of multiple forms becomes more complicated when one considers apparent species differences. The long-form prolactin receptor has been identified in the human, rabbit, mouse, and rat (Boutin *et al.*, 1988, 1989; Davis and Linzer, 1989; Shirota *et al.*, 1990). The intermediate-form prolactin receptor is present in the rat prolactin-responsive Nb2 lymphoma cell line (Ali *et al.*, 1991), and seems to arise from a deletion within the intracellular domain of the receptor. The short-form prolactin receptor has been identified in the rat (Boutin *et al.*, 1988) and in the mouse (Davis and Linzer, 1989). There are three different short-form prolactin receptors (I, II, and III) in the mouse (Davis and Linzer, 1989; Clarke and Linzer, 1993); they share common extracellular and transmembrane domains, but differ from each other in the most carboxyl-terminal portion of the intracellular domain. The regulation of prolactin receptor gene expression has been explored, but is complicated by the presence of multiple mRNA transcripts that most likely arise by differential RNA processing (Davis and Linzer, 1989). Buck and co-workers (1992) have shown that the prolactin receptor mRNAs are expressed in a tissue-specific fashion in the mouse. All four forms (the long form and three short forms) of the mouse prolactin receptor are expressed in the mouse ovary (Clarke and Linzer, 1993). Interestingly, the cellular location of each form of the prolactin receptor appears to be different (Clarke and Linzer, 1993); during early pregnancy, mouse luteal cells contain all four prolactin receptor mRNAs, while during late pregnancy, only the long form prolactin receptor mRNA is present in luteal cells. Granulosa cells of growing follicles also express the long-form prolactin receptor mRNA during late pregnancy. After parturition, granulosa cells of maturing preovulatory follicles express mRNAs for the long form and two short forms (II and III) of the receptor, while apparently atretic follicles express mRNA for the short form II. These results suggest that multiple forms of the prolactin receptor subserve different signaling functions in a cell-specific and temporally regulated fashion within the ovary.

B. Nerve Growth Factor Receptors

The bioactive form of nerve growth factor (NGF) is a homodimer (26 kDa) of two β subunits. βNGF is synthesized as a preprohormone that undergoes proteolytic cleavage to produce a mature hormone (for review, see Pintar, 1987). The actions and characteristics of NGF have been reviewed elsewhere (Levi-Montalcini, 1987; Thoenen and Bard, 1980). Nerve growth factor, produced locally in the ovary and transported to the sympathetic fibers in a retrograde fashion (Lara *et al.*, 1990a), has been suspected to play an important role in follicular development by supporting neuronal growth and differentiation (Lara *et al.*, 1990b). Measurements of NGF binding to target cells reveal two sites that can be distinguished by binding and kinetic measurements (10^{-9} *M* vs. 10^{-11} *M* affinity), the low- and high-affinity sites (for review, see Meakin and Shooter, 1992). Two NGF receptor types have been identified by molecular cloning, $P75^{NGFR}$ and $p140^{trkA}$. $p75^{NGFR}$ encodes a transmembrane protein with an extracellular domain containing potential N-linked glycosylation sites and four cysteine-rich repeat regions (Johnson *et al.*, 1986; Radeke *et al.*, 1987; Large *et al.*, 1989). The negative charges within the third and fourth cysteine-rich repeats are conserved among species, suggesting that these regions may participate in binding a highly positively charged NGF dimer (Angeletti and Bradshaw, 1971). $p75^{NGFR}$ binds not only to NGF but also to other neurotropins with a similar affinity (Rodriguez-Tebar *et al.*, 1990), which suggests that this receptor class plays a role as a tropic factor in a broad range of tissues. It exhibits low affinity binding of $\sim 10^{-9}$ *M*, and its signaling properties are unknown. Gene disruption experiments suggest a critical role for $p75^{NGFR}$ in neuronal survival.

$p140^{trkA}$ is a transmembrane tyrosine kinase receptor, and thus formally belongs to the receptor class discussed earlier (for review, see Chao, 1992). It is being discussed here because of uncertainty over the precise roles of the two NGF receptors that have been characterized. $p140^{trkA}$ is the product of the proto-oncogene trkA (Martin-Zanca *et al.*, 1989), which can bind NGF, resulting in stimulated tyrosine kinase activity and receptor autophosphorylation (Klein *et al.*, 1991b; Kaplan *et al.*, 1991a,b). The binding of NGF to this receptor also appears to be of low affinity, although this might reflect assay methods (Meakin and Shooter, 1992). Proteins related to p140^{trkA} include trkB and trkC (Klein *et al.*, 1991c; Lamballe *et al.*, 1991), which have been shown to bind to brain-derived neurotropic factor (BDNF) and neurotropin 3 (NT3), and are also receptor tyrosine kinases. The intracellular tyrosine kinase domains of these receptors, like the PDGF receptor, contain a kinase insert. The trkB gene produces not only full-length trkB but also a truncated trkB form lacking in the cytoplasmic domain (Klein *et al.*, 1991a). In the most extracellular domain, these receptors contain two cysteine-rich clusters divided by 24 leucine-rich repeats that might be important for protein-protein interactions. Near the transmembrane domain, they have two immunoglobulin-like domains, indicating the potential for interactions between cell matrix or cell surface proteins and the trk receptors. It has been suggested that both receptors are required for high-affinity binding to NGF (Hempstead *et al.*, 1991), but this is at present unclear. Current models of the roles of p75^{NGFR} and p140^{trkA} in NGF signaling have been reviewed recently (Meakin and Shooter, 1992; Chao, 1992).

As mentioned earlier, NGF appears to be locally synthesized and plays an important role in follicular development (Lara *et al.*, 1990a,b). Protein cross-linking, immunocytochemistry, and mRNA measurements have recently shown that the ovary produces p75^{NGFR} (Dissen *et al.*, 1991). p75^{NGFR} immunoreactivity was observed in nerve fibers surrounding blood vessels or associated with follicles at different stages of development. Most significantly, high p75^{NGFR} immunoreactivity as well as mRNA was observed in thecal cells of antral and preantral follicles; lesser expression was seen in granulosa cells. p75^{NGFR} mRNA levels were found to decrease shortly after ovulation in response to the gonadotropin surges (Dissen *et al.*, 1991). These studies suggest an involvement in this receptor in ovarian function, perhaps in mediating the actions of neurotropic substances in the ovary. Recent observations suggest that p140^{trkA} NGF receptor is also expressed in the ovary (Dissen *et al.*, 1992).

C. Urokinase-Type Plasminogen Activator Receptors

Plasminogen activators are involved in processes that require regulated extracellular proteolysis, and they catalyze the formation of the broad-spectrum protease, plasmin from the ubiquitous inactive plasminogen activator (Dano *et al.*, 1985). Strickland and Beers (1976) have shown that in the ovary, both tissue-type (tPA: 65 kDa) and urokinase-type (uPA: 55 kDa) plasminogen activators play a role in periovulatory processes that involve tissue remodeling. A specific surface receptor for uPA (Behrendt *et al.*, 1990) or tPA (Hajjar, 1991) appears to regulate cell surface proteolytic activity, localizing plasmin formation to the cell surface. While both uPA and tPA receptors have been purified, only the uPA receptor has been fully purified, characterized, and cloned (Roldan *et al.*, 1990). This cDNA, which is expressed in mouse LB6 cells, produces a membrane protein that can bind to human uPA to activate plasminogen. The sequence predicts a transmembrane protein with a large presumptive extracellular domain that includes a potential N-linked glycosylation site and a small intracellular domain. The uPA receptor recognizes 21 amino acid residues in the amino-terminal domain of uPA (Appella *et al.*, 1987). Although this receptor seems to play an important role in directing plasminogen-mediated extracellular proteolysis, it is not clear whether the uPA receptor mediates any subsequent intracellular signaling events important to uPA action

VII. Conclusions

In briefly considering some receptors that are fundamentally important in the ovary, several principles begin to emerge. First, it is apparent that a select number of themes have been used repeatedly, resulting in large families of related receptors that vary slightly in their hormone binding or signaling properties. This information has been extremely useful in designing strategies for identifying novel receptors (orphan receptors) and in devising experiments to understand the functions of these receptors. Second, it is clear that diversity at multiple steps of the signaling cascade, for example, receptor heterodimerization or coupling of single receptors to multiple G proteins, is used to dramatically expand the repertoire of possible responses. Finally, our knowledge of all but the initial signaling events is in most cases still rudimentary, and substantial progress needs to be made to obtain a fuller understanding of the complete signal transduction process in any of the known systems, and to uncover the novel signal transduction processes that surely await discovery.

An understanding of some of the relationships between structure and function for these endocrine receptors and their signaling partners has already led to important contributions to human health by providing scientific explanations for medically important human diseases or disorders. These explanations might ultimately be used for diagnosis or in the rational design of new therapies. For example, in ovarian tumors of both granulosa and thecal cell origin, mutations in the gene encoding the a_{i2} G protein subunit have been described and a new oncogene referred to as *gip*-2 has been defined (Lyons *et al.*, 1990). Mutations in the insulin receptor are responsible for the hyperandrogenism caused by overproduction of ovarian testosterone in women with type A insulin resistance syndrome (Taylor *et al.*, 1992). Furthermore, mutations in the hormone-binding domain of the androgen receptor account for the complete androgen insensitivity in genetic and gonadal males who have phenotypically developed as females (Imperato-McGinley and Canovatchel, 1992). Finally, mutations in the vasopressin V2 receptor result in congenital nephrogenic diabetes insipidus

(Rosenthal *et al.*, 1992). Further characterization of both receptors and the cellular signaling systems that they interact with promises to be a fruitful route to understanding ovarian hormone action, as well as many other cellular processes of fundamental biological importance.

References

Ackland, J. F., Schwartz, N. B., Mayo, K. E., and Dodson, R. D. (1992). Nonsteroidal signals originating in the gonads. *Physiol. Rev.* 72, 731–787.

Adashi, E. Y., and Rohan R. M. (1992). Intraovarian regulation: peptidergic signaling systems. *Trends Endocrinol. Metab.* 3, 243–248.

Adashi, E. Y., Resnick, C. E., and Rosenfeld, R. G. (1990a). Insulin like growth factor-I (IGF-I) and IGF-II hormonal action in cultured rat granulosa cells: mediation via type I but not type II IGF receptors. *Endocrinology* 126, 216–222.

Adashi, E. Y., Resnick, C. E., Hernandez, E. R., Hurwitz, A., and Rosenfeld, R. G. (1990b). Ovarian granulosa cell-derived insulin-like growth factor (IGF) binding proteins: release of low molecular weight, high-affinity IGF-selective species. *Mol. Cell. Endocrinol.* 74, 175–184.

Ahmed, C. E., Dees, W. L., and Ojeda, S. R. (1986). The immature rat ovary is innervated by vasoactive intestinal peptide (VIP)-containing fibers and responds to VIP with steroid secretion. *Endocrinology* 118, 1682–1689.

Ali, S., Pellegrini, I., and Kelly, P. A. (1991). A prolactin-dependent immune cell line (Nb2) expresses a mutant form of prolactin receptor. *J. Biol. Chem.* 266, 20110–20117.

Alkhalaf, M., and Murphy, L. C. (1992). Regulation of c-*jun* and *jun*-B by progestins in T-47D human breast cancer cells. *Mol. Endocrinol.* 6, 1625–1633.

Angeletti, R. H., and Bradshaw, R. A. (1971). Nerve growth factor from mouse submaxillary gland: amino acid sequence. *Proc. Natl. Acad. Sci. USA* 68, 2417–2420.

Appella, E., Robinson, E. A., Ullrich, S. J., Stoppelli, M. P., Corti, A., Cassani, G., and Blasi, F. (1987). The receptor-binding sequence of urokinase: biological function for the growth factor module of protease. *J. Biol. Chem.* 262, 4437–4440.

Argetsinger, L. S., Campbell, G. S., Yang, X., Witthuhn, B. A., Silrennoinen, O., Ihle, J. N., and Carter-Su, C. (1993). Identification of JAK2 as a growth hormone receptor-associated tyrosine kinase. *Cell* 47, 237–244.

Armstrong, D. T., Miller, L. S., and Knudsen, K. A. (1969). Regulation of lipid metabolism and progesterone production in rat corpora lutea and ovarian interstitial elements by prolactin and luteinizing hormone. *Endocrinology* 85, 393–401.

Aronica, S. M., and Katzenellenbogen, B. S. (1991). Progesterone receptor regulation in uterine cells: stimulation by estrogen, cyclic adenosine 3′,5′-monophosphate, and insulin-like growth factor I and suppression by antiestrogens and protein kinase inhibitors. *Endocrinology* 128, 2045–2052.

Ascoli, M., and Segaloff, D. L. (1989). On the structure of the luteinizing hormone/chorionic gonadotropin receptor. *Endocr. Rev.* 10, 27–44.

Aten, R. F., Williams, A. T., and Behrman, H. R. (1986). Ovarian gonadotropin-releasing hormone-like protein(s): demonstration and characterization. *Endocrinology* 118, 961–967.

Attisano, L., Wrana, J. L., Cheifetz, S., and Massague, J. (1992). Novel activin receptors: distinct genes and alternative mRNA splicing generate a repertoire of serine/threonine kinase receptors. *Cell* 68, 97–108.

Bagchi, M. K., Tsai, M.-J., O'Malley, B. W., and Tsai, S. Y. (1992). Analysis of the mechanism of steroid hormone receptor-dependent gene activation in cell-free systems. *Endocr. Rev.* 13, 525–535.

Bagnato, A., Moretti, C., Frajese, G., and Catt, K. (1991). Gonadotropin-induced expression of receptors for growth hormone releasing factor in cultured granulosa cells. *Endocrinology* 128, 2889–2894.

Bagnato, A., Moretti, C., Ohnishi, J., Frajese, G., and Catt, K. J. (1992). Expression of the growth hormone-releasing hormone gene and its peptide product in the rat ovary. *Endocrinology* 130, 1097–1102.

Beato, M. (1989). Gene regulation by steroid hormones. *Cell* 56, 335–344.

Behrendt, N., Ronne, E., Ploug, M., Petri, T., Lober, D., Nielsen, L. S., Schleuning, W.-D., Blasi, F., Appella, E., and Dano, K. (1990). The human receptor for urokinase plasminogen activator. *J. Biol. Chem.* 265, 6453–6460.

Bell, G. I. (1986). The glucagon superfamily: precursor structure and gene organization. *Peptides* 7 (Suppl.), 27–36.

Bendell, J. J., and Dorrington, J. H. (1988). Rat theca/interstitial cells secrete a transforming growth factor-β-like factor that promotes growth and differentiation in rat granulosa cell. *Endocrinology* 123, 941–948.

Bendell, J. J., and Dorrington, J. H. (1990). Epidermal growth factor influences growth and differentiation of rat granulosa cells. *Endocrinology* 127, 533–540.

Besmer, P., Murphy, J. E., George, P. C., Qui, F., Bergold, P. J., Lederman, L., Snyder, H. W., Brodeur, D., Zuckerman, E. E., and Hardy, W. D. (1986). A new acute transforming feline retrovirus and relationship of its oncogene v-*kit* with the protein kinase gene family. *Nature* 320, 415–421.

Bilezikjian, L., and Vale, W. (1983). Stimulation of adenosine 3′-5′-monophosphate production by growth hormone releasing factor and its inhibition by somatostatin in anterior pituitary cells *in vivo*. *Endocrinology* 113, 1726–1731.

Billiar, R. B., Loukides, J. A., and Miller, M. M. (1992). Evidence for the presence of the estrogen receptor in the ovary of the baboon (*Papio anubis*). *J. Clin. Endocrinol. Metab.* 75, 1159–1165.

Birnbaumer, L. (1990). G proteins in signal transduction. *Annu. Rev. Pharmacol. Toxicol.* 30, 675–705.

Boutin, J.-M., Jolicoeur, C., Okamura, H., Gagnon, J., Edery, M., Shirota, M., Banville, D., Dusanter-Fourt, I., Djiane, J., and Kelly, P. A. (1988). Cloning and expression of the rat prolactin receptor, a member of the growth hormone/prolactin receptor gene family. *Cell* 53, 69–77.

Boutin, J.-M., Edery, M., Shirota, M., Jolicoeur, C., Lesueur, L., Ali, S., Gould, D., Djiane, J., and Kelly, P. A. (1989). Identification of a cDNA encoding a long form of prolactin receptor in human hepatoma and breast cancer cells. *Mol. Endocrinol.* 3, 1455–1461.

Boyd, P. A., and Spelsberg, T. C. (1979). Seasonal changes in the molecular species and nuclear binding of the chick oviduct progesterone receptor. *Biochemistry* 18, 3685–3690.

Boyd, N. D., MacDonald, S. G., Kage, R., Luber-Narod, J., and Leeman, S. E. (1991). Substance P receptor: biochemical characterization and interactions with G proteins. *Ann. N.Y. Acad. Sci.* 632, 79–93.

Braun, T., Schofield, P. R., and Sprengel, R. (1991) Amino-terminal leucine–rich repeats in gonadotropin receptors determine hormone selectivity. *EMBO J.* 10, 1885–1890.

Brown, R. S., Sander, C., and Argos, P. (1985). The primary structure of transcription factor TFIIIA has 12 consecutive repeats. *FEBS Lett.* 186, 271–274.

Brown-Schult, N., Mordacq, J., and Mayo, K. E. (1992). Characterization of the rat activin receptor and a related receptor and expression of their mRNAs during development. *In* "74th Annu. Meeting of the Endocrine Society, Program and Abstracts," p. 421.

Buck, K., Vanek, M., Groner, B., and Ball, R. K. (1992). Multiple forms of prolactin receptor messenger ribonucleic acid are specifically expressed and regulated in murine tissues and the mammary cell line HC11. *Endocrinology* 130, 1108–1114.

Bunzow, J. R., Van Tol, H.H.M., Grandy, D. K., Albert, P., Salon, J., Christie, M., Marchida,

C. A., Neve, K. A., and Civelli, O. (1988). Cloning and expression of a rat D_2 dopamine receptor. *Nature* 336, 783–787.

Camp, T. A., Rahal, J. O., and Mayo, K. E. (1991). Cellular localization and hormonal regulation of follicle-stimulating hormone and luteinizing hormone receptor messenger RNAs in the rat ovary. *Mol. Endocrinol.* 5, 1405–1417.

Camps, M., Carozzi, A., Schnabel, P., Scheer, A., Parker, P. J., and Gierschik, P. (1992). Isozyme-selective stimulation of phospholipase C-β2 by G protein βγ-subunits. *Nature* 360, 684–686.

Cantley, L. C., Auger, K. R., Carpenter, C., Duckworth, B., Graziani, A., Kapeller, R., and Soltoff, S. (1991). Oncogenes and signal transduction. *Cell* 64, 281–302.

Cara, J. F., Fan, J., Azzarello, J., and Rosenfield, R. L. (1990). Insulin-like growth factor-I enhances luteinizing hormone binding to rat ovarian theca-interstitial cell. *J. Clin. Invest.* 86, 560–565.

Carson-Jurica, M. A., Lee, A. T., Dobson, A. W., Conneely, O. M., Schrader, W. T., and O'Malley, B. W. (1989). Interaction of the chicken progesterone receptor with heat shock protein (HSP) 90. *J. Steroid Biochem.* 34, 1–6.

Catterall, W. A. (1988). Analysis of ligand binding specificity of receptor chimeras. *Science* 243, 236–237.

Centrella, M., McCarthy, T. L., and Canalis, E. (1991). Activin-A binding and biochemical effects in osteoblast-enriched cultures from fetal-rat parietal bone. *Mol. Cell. Biol.* 11, 250–258.

Chabot, J.-G., St. Arnaud, R., Walker, P., and Pelletier, G. (1986). Distribution of epidermal growth factor receptors in the rat ovary. *Mol. Cell. Endocrinol.* 44, 99–108.

Chabot, B., Stephenson, D. A., Chapman, V. M., Besmer, P., and Bernstein, A. (1988). The protooncogene *c-kit* encoding a transmembrane tyrosine kinase receptor maps to the mouse W locus. *Nature* 335, 88–89.

Chao, M. V. (1992). Neutrophin receptors: a window into neuronal differentiation. *Neuron* 9, 583–593.

Chazenbalk, G. D., Nagayama, Y., Russo, D., Wadsworth, H. L., Rapoport, B. (1990). Functional analysis of the cytoplasmic domains of the human thyrotropin receptor by site-directed mutagenesis. *J. Biol. Chem.* 265, 20970–20975.

Chegini, N., and Williams, R. S. (1992). Immunocytochemical localization of transforming growth factors (TGFs) TGF-α and TGF-β in human ovarian tissues. *J. Clin. Endocrinol. Metab.* 74, 973–980.

Cheifetz, S., Weatherbee, J. A., Tsang, M. L.-S., Anderson, J. K., Mole, J. E., Lucas, R., and Massague, J. (1987). The transforming growth factor-β system, a complex pattern of cross-reactive ligands and receptors. *Cell* 48, 409–415.

Clarke, D. L., and Linzer, D. I. H. (1993). Changes in prolactin receptor expression during pregnancy in the mouse ovary. *Endocrinology* 133, 224–232.

Clayton, R. N., Eccleston, L., Gossard, F., Thalbard, J.-C., and Morel, G. (1992). Rat granulosa cells express the gonadotrophin-releasing hormone gene: evidence from *in situ* hybridization histochemistry. *J. Mol. Endocrinol.* 9, 189–195.

Conneely, O. M., Kettelberger, D. M., Tsai, M.-J., Schrader, W. T., and O'Malley, B. W. (1989). The chicken progesterone receptor A and B isoforms are products of an alternate translation initiation event. *J. Biol. Chem.* 264, 14062–14064.

Cosman, D., Lyman, S. D., Idzerda, R. L., Beckmann, M. P., Park, L. S., Goodwin, R. G., and March, C. J. (1990). A new cytokine receptor superfamily. *Trends Biochem. Sci.* 15, 265–270.

Cross, M., and Dexter, T. M. (1991). Growth factors in development, transformation, and tumorigenesis. *Cell* 64, 271–280.

Dahlman-Wright, K., Wright, A., Gustafsson, J.-A., and Carlstedt-Duke, J. (1991). Interaction of the glucocorticoid receptor DNA-binding domain with DNA as a dimer is mediated by a short segment of five amino acids. *J. Biol. Chem.* 266, 3107–3112.

Danielsen, M., Hinck, L., and Ringold, G. M. (1989). Two amino acids within the knuckle of the first zinc finger specify DNA response element activation by the glucocorticoid receptor. *Cell* 57, 1131–1138.

Dano, K., Andreasen, P. A., Grondahl-Hansen, J., Kristensen, P., Nielsen, L. S., and Skriver, L. (1985). Plasminogen activators, tissue degradation, and cancer. *Adv. Cancer Res.* 44, 139–266.

Daopin, S., Piez, K. A., Ogawa, Y., and Davies, D. R. (1992). Crystal structure of transforming growth factor-β2: an unusual fold for the superfamily. *Science* 257, 369–373.

Davis, J. A., and Linzer, D. I. H. (1989). Expression of multiple forms of the prolactin receptor in mouse liver. *Mol. Endocrinol.* 3, 674–680.

Davoren, J. B., Kasson, B. G., Li, C. H., and Hsueh, A. J. W. (1986). Specific insulin-like growth factor (IGF) I- and II-binding sites on rat granulosa cells: relation to IGF action. *Endocrinology* 119, 2155–2162.

deKretser, D. M., and Robertson, D. M. (1989). The isolation and physiology of inhibin and related proteins. *Biol. Reprod.* 40, 33–47.

DeMarzo, A. M., Beck, C. A., Onate, S. A., and Edwards, D. P. (1991). Dimerization of mammalian progesterone receptors occurs in the absence of DNA and is related to the release of the 90-kDa heat shock protein. *Proc. Natl. Acad. Sci. USA* 88, 72–76.

Dement-Liebenow, P. (1991). Localization of specific GnRH receptors on porcine granulosa cells and mast cells within ovarian follicles. *In* "Signaling Mechanisms and Gene Expression in the Ovary" (G. Gibori, ed.), pp. 405–410. Springer-Verlag, New York.

Denner, L. A., Bingman, W. E. III, Greene, G. L., and Weigel, N. L. (1987). Phosphorylation of the chicken progesterone receptor. *J. Steroid Biochem.* 27, 235–243.

Denner, L. A., Weigel, N. L., Maxwell, B. L., Schrader, W. T., and O'Malley, B. W. (1990). Regulation of progesterone receptor-mediated transcription by phosphorylation. *Science* 250, 1740–1743.

Derynck, R. (1986). Transforming growth factor α: structure and biological activities. *J. Cell Biochem.* 32, 294–304.

Diamond, M., Miner, J., Yoshinaga, S., and Yamamoto, K. (1990). Transcription factors: selectors of positive or negative regulation from a single DNA element. *Science* 249, 1266–1272.

Dissen, G. A., Hill, D. F., Costa, M. E., Ma, Y. J., and Ojeda, S. R. (1991). Nerve growth factor receptors in the peripubertal rat ovary. *Mol. Endocrinol.* 5, 1642–1650.

Dissen, G. A., Hill, D. F., and Ojeda, S. R. (1992). TrkA proto-oncogene expression in the developing rat ovary. *In* "74th Annu. Meeting Endocrine Society, Program and Abstracts," p. 58.

Dixon, A. R. A. F., Kobilke, B. K., Strader, D. J., Benovic, J. L., Dohlman, H. G., Frielle, T., Bolanowski, M. A., Bennett, C. D., Rands, E., Diehl, R. E., Mumford, R. A., Slater, E. E., Sigal, I. S., Carron, M. G., Lefkowitz, R. J., and Strader, C. D. (1986). Cloning of the gene and cDNA for mammalian β-adrenergic receptor and homology with rhodopsin. *Nature* 321, 75–79.

Donaldson, C. J., Mathews, L. S., and Vale, W. W. (1992). Molecular cloning and binding properties of the human type II activin receptor. *Biochem. Biophys. Res. Commun.* 184, 310–316.

Dorrington, J., and Gore-Langton, R. E. (1981). Prolactin inhibits oestrogen synthesis in the ovary. *Nature* 290, 600–602.

Dorrington, J., Chuma, A. V., and Bendell, J. J. (1988). Transforming growth factor-β and follicle-stimulating hormone promote rat granulosa cell proliferation. *Endocrinology* 123, 353–359.

Dotzlaw, H., Alkhalaf, M., and Murphy, L. C. (1992). Characterization of estrogen receptor variant mRNAs from human breast cancers. *Mol. Endocrinol.* 6, 773–785.

Dufau, M. L., Minegishi, T., Buczko, E. S., Delgado, C. J., and Zhang, R. (1989). Characterization and structure of ovarian and testicular LH/hCG receptors. *J. Steroid Biochem.* 33, 715–720.

Ebner, R., Chen, R.-H., Shum, L., Lawler, S., Zioncheck, T., Lee, A., Lopez, A. R., and Derynck, R. (1993). Cloning of a type I TGF-β receptor and its effects on TGF-β binding to the type II receptor. *Science* 260, 1344–1348.

Edery, M., Jolicoeur, C., Levi-Meyrueis, C., Dusanter-Fourt, I., Petridou, B., Boutin, J.-M., Lesueur, L., Kelly, P. A., and Djiane, J. (1989). Identification and sequence analysis of a second form of prolactin receptor by molecular cloning of complementary DNA from rabbit mammary gland. *Proc. Natl. Acad. Sci. USA* 86, 2112–2116.

Erickson, G. F., Magoffin, D. A., Dyer, C. A., and Hofeditz, C. (1985). The ovarian androgen producing cell: a review of structure/function relationships. *Endocr. Rev.* 6, 371–399.

Eriksson, A., Siegbahn, A., Westermark, B., Heldin, C.-H., and Claesson-Welsh, L. (1992). PDGF α- and β-receptors activate unique and common signal transduction pathways. *EMBO J.* 11, 543–550.

Evans, R. M. (1988). The steroid and thyroid hormone receptor superfamily. *Science* 240, 889–895.

Fantl, W. J., Escobedo, J. A., Martin, G. A., Turck, C. W., del Rosario, M., McCormick, F., and Williams, L. T. (1992). Distinct phosphotyrosines on a growth factor receptor bind to specific molecules that mediate different signaling pathway. *Cell* 69, 413–423.

Feng, P., Knecht, M., and Catt, K. (1987). Hormonal control of epidermal growth factor receptors by gonadotropins during granulosa cell differentiation. *Endocrinology* 120, 1121–1126.

Findlay, J. K. (1993). An update on the roles of inhibin, activin, and follistatin as local regulators of foliculogenesis. *Biol. Reprod.* 48, 15–23.

Foxwell, B. M. J., Barrett, K., and Feldmann, M. (1992). Cytokine receptors-structure and signal transduction. *Clin. Exp. Immunol.* 90, 161–169.

Franke, R. R., Sakmar, T. P., Oprian, D. D., and Khorana, H. G. (1988). A single amino acid substitution in rhodopsin (lysine 248 to leucine) prevents activation of transducin. *J. Biol. Chem.* 263, 2119–2122.

Freedman, L. P. (1992). Anatomy of the steroid receptor zinc finger region. *Endocr. Rev.* 13, 129–145.

Frohman, L. A., and Jansson, J.-O. (1986). Growth hormone-releasing hormone. *Endocr. Rev.* 7, 223–253.

Fuchs-Young, R., Robinson, A., Bettuzzi, S., Miles, R., Radigan, M., and Greene, G. L. (1992). Recognition and regulation of the 5′ flanking region of the human progesterone receptor gene by estrogen and progestin receptors. *In* "74th Annu. Meeting of the Endocrine Society, Program & Abstracts," p. 117.

Gangrade, B. K., Davis, J. S., and May, J. V. (1991). A novel mechanism for the induction of aromatase in ovarian cells *in vitro:* role of transforming growth factor alpha-induced protein tyrosine kinase. *Endocrinology* 129, 2790–2792.

Garcia, T., Sanchez, M., Cox, J. L., Shaw, P. A., Ross, J. B. A., Lehrer, S., and Schachter, B. (1989). Identification of a variant form of the human estrogen receptor with an amino acid replacement. *Nucleic Acids Res.* 17, 8364.

Gaylinn, B. D., Harrison, J. K., Zysk, J. R., Lyons, C. E., Lynch, K. R., and Thorner, M. O. (1993). Molecular cloning and expression of a human anterior pituitary receptor for growth hormone-releasing hormone. *Mol. Endocrinol.* 7, 77–84.

Gelato, M. C., and Merriam, C. R. (1986). Growth hormone-releasing hormone. *Annu. Rev. Physiol.* 48, 569–591.

George, F. W., and Ojeda, S. R. (1987). Vasoactive intestinal peptide enhances aromatase activity in the neonatal rat ovary before development of primary follicles or responsiveness to follicle-stimulating hormone. *Proc. Natl. Acad. Sci. USA* 84, 5803–5807.

Georgi, L. L., Albert, P. S., and Riddle, D. L. (1990). *daf*-1, a *C. elegans* gene controlling dauer larva development, encodes a novel receptor protein kinase. *Cell* 61, 635–645.

Germain-Lee, E., Janicot, M., Lammers, R., Ullrich, A., and Casella, S. J. (1992). Expression of a type I insulin-like growth factor receptor with low affinity for insulin-like growth factor II. *Biochem. J.* 281, 413–417.

Gibori, G., and Richards, J. S. (1978). Dissociation of two distinct luteotropin effects of prolactin: regulation of luteinizing hormone-receptor content and progesterone secretion during pregnancy. *Endocrinology* 102, 767–774.

Gick, G. G., Zeytin, F. N., Brazeau, P., Ling, N. C., Esch, F. S., and Bancroft, C. (1984). Growth hormone-releasing factor regulates growth hormone mRNA in primary cultures of rat pituitary cells. *Proc. Natl. Acad. Sci. USA* 81, 1553–1555.

Giros, B., Sokoloff, P., Martres, M.-P., Riou, J.-F., Emorine, L. J., and Schwartz, J.-C. (1989). Alternative splicing directs the expression of two D_2 dopamine receptor isoforms. *Nature* 342, 923–926.

Giguere, V., Hollenberg, S. M., and Rosenfeld, M. G., and Evans, R. M. (1986). Functional domains of the human glucocorticoid receptor. *Cell* 46, 645–652.

Glass, C. K., Holloway, J. M., Devary, O. V., and Rosenfeld, M. G. (1988). The thyroid hormone receptor binds with opposite transcriptional effects to common sequence motif in thyroid hormone and estrogen response elements. *Cell* 54, 313–323.

Goldenberg, R. L., Vaitukaitis, J. L., and Ross, G. T. (1972). Estrogen and follicle-stimulating hormone interactions on follicular growth in rats. *Endocrinology* 90, 1492–1498.

Goodman, A. L., and Hodgen, G. D. (1982). Antifolliculogenic action of progesterone despite hypersecretion of FSH in monkeys. *Am. J. Physiol.* 243, E387–E397.

Gozes, I., and Brenneman, D. E. (1989). VIP: molecular biology and neurobiological function. *Mol. Neurobiol.* 3, 201–236.

Gozes, I., and Tsafriri, A. (1986). Detection of vasoactive intestinal peptide-encoding messenger ribonucleic acid in the rat ovaries. *Endocrinology* 119, 2606–2610.

Green, J. B. A., and Smith, J. C. (1990). Graded changes in dose of a *Xenopus* activin A homologue elicit stepwise transitions in embryonic cell fate. *Nature* 347, 391–394.

Green, S., Walter, P., Kumar, V., Krust, A., Bornert, J.-M., Argos, P., and Chambon, P. (1986). Human oestrogen receptor cDNA: sequence, expression, and homology to v-*erb*-A. *Nature* 320, 134–139.

Greene, G. L., Gilna, P., Waterfield, M., Baker, A., Hort, Y., and Shine, J. (1986). Sequence and expression of human estrogen receptor complementary DNA. *Science* 231, 1150–1154.

Greenfield, C., Hiles, I., Waterfield, M. D., Federwisch, M., Wollmer, A., Blundell, T. L., and McDonald, N. (1989). Epidermal growth factor binding induces a conformational change in the external domain of its receptor. *EMBO J.* 8, 4115–4124.

Gregory, H., and Taylor, C. L. (1982). Luteinizing hormone release from dissociated pituitary cells by dimerization of occupied LHRH receptors. *Nature* 300, 269–271.

Grinwich, D. L., Hichens, M., and Behrman, H. R. (1976). Control of the LH receptor by prolactin and prostagland $F_{2\alpha}$ in rat corpora lutea. *Biol. Reprod.* 14, 212–218.

Gronemeyer, H., Turcotte, B., Quirin-Stricker, C., Bocquel, M. T., Meyer, M. E., Krozowski, Z., Jeltsch, J. M., Lerouge, T., Garnier, J. M., and Chambon, P. (1987). The chicken progesterone receptor: sequence, expression and functional analysis. *EMBO J.* 6, 3985–3994.

Gudermann, T., Nichols, C., Levy, F. O., Birnbaumer, M., and Birnbaumer, L. (1992). Ca^{2+} Mobilization by the LH receptor expressed in *Xenopus* oocytes independent of 3′,5′-cyclic adenosine monophosphate formation: evidence for parallel activation of two signaling pathways. *Mol. Endocrinol.* 6, 272–278.

Guiochon-Mantel, A., Loosfelt, H., Lescop, P., Sar, S., Atger, M., Perrot-Applanat, M., and Milgrom, E. (1989). Mechanisms of nuclear localization of the progesterone receptor: evidence for interaction between monomers. *Cell* 57, 1147–1154.

Hajjar, K. A. (1991). The endothelial cell tissue plasminogen activator receptor. Specific interaction with plasminogen. *J. Biol. Chem.* 266, 12962–21970.

Hammond, J. M., Mondschein, J. S., Samaras, S. E., and Canning, S. F. (1991). The ovarian insulin-like growth factors, a local amplification mechanism for steroidogenesis and hormone action. *J. Steroid Biochem. Mol. Biol.* 40, 411–416.

Hanks, S. K., and Quinn, A. M. (1991). Protein kinase catalytic domain sequence database: identification of conserved features of primary structure and classification of family members. *In* "Methods in Enzymology" (T. Hunter and B. M. Sefton, eds.), Vol. 200, pp. 38–62. Academic Press, San Diego, CA.

Hannik, M., and Donogue, D. J. (1989). Structure and function of platelet derived growth factor (PDGF) and releated proteins. *Biochim. Biophys. Acta* 989, 1–10.

Hard, T., Kellenbach, E., Boelens, R., Maler, B. A., Dahlman, K., Freedman, L. P., Carlstedt-Duke, J., Yamamoto, K. R., Gustafsson, J.-A., and Kaptein, R. (1990). Solution structure of the gulcocorticoid receptor DNA binding domain. *Science* 249, 157–160.

Hart, C. E., Forstrom, J. W., Kelly, J. D., Seifert, R. A., Smith, R. A., Ross, R., Murray, M. J., and Bowen-Pope, D. F. (1988). Two classes of PDGF receptor recognize different isoforms of PDGF. *Science* 240, 1529–1531.

Heckert, L. L., Daley, I. J., and Griswold, M. D. (1992). Structural organization of the follicle-stimulating hormone receptor gene. *Mol. Endocrinol.* 6, 70–80.

Hemmati-Brivanlou, A., and Melton, D. A. (1992). A truncated activin receptor inhibits mesoderm induction and formation of axial structures in *Xenopus* embryos. *Nature* 359, 609–614.

Hempstead, B. L., Martin-Zanca, D., Kaplan, D. R., Parada, L. F., and Chao, M. V. (1991). High-affinity NGF binding requires coexpression of the *trk* proto-oncogene and the low-affinity NGF receptor. *Nature* 350, 678–683.

Hepler, J. R., and Gilman, A. G. (1992). G proteins. *Trends Biochem. Sci.* 17, 383–387.

Hernandez, E. R., Resnick, C. E., Svoboda, M. E., Van Wyk, J. J., Payne, D. W., and Adashi, E. Y. (1988). Somatomedin-C/insulin-like growth factor-I as an enhancer of androgen biosynthesis by cultured rat ovarian cells. *Endocrinology* 122, 1603–1613.

Hernandez, E. R., Roberts, C. T. Jr., LeRoith, D., and Adashi, E. Y. (1989). Rat ovarian insulin-like growth factor I (IGF-I) gene expression is granulosa cell-selective: 5′-untranslated mRNA variant representation and hormonal regulation. *Endocrinology* 125, 572–574.

Hernandez, E. R., Hurwitz, A., Payne, D. W., Dhamarajan, A. M., Purchio, A. F., and Adashi, E. Y. (1990). Transforming growth factor-β1 inhibits ovarian androgen production: gene expression, cellular localization, mechanism(s), and site(s) for action. *Endocrinology* 127, 2804–2811.

Hernandez, E. R., Hurwitz, A., Vera, A., Pellicer, A., Adashi, E. Y., LeRoith, D., and Robert, C. T. Jr. (1992). Expression of the genes encoding the insulin-like growth factors and their receptors in the human ovary. *J. Clin. Endocrinol. Metab.* 74, 419–425.

Hershey, A. D., Polenzani, L., Woodward, R. M., Miledi, R., and Krause, J. E. (1991). Molecular and genetic characterization, functional expression, and mRNA expression patterns of a rat substance P receptor. *Ann. N.Y. Acad. Sci.* 632, 63–77.

Hillier, S. G., Saunders, P.T.K., White, R., and Parker, M. G. (1989). Oestrogen receptor mRNA and a related RNA transcript in mouse ovaries. *J. Mol. Endocrinol.* 2, 39–45.

Hino, M., Tojo, A., Miyazono, K., Miura, Y., Chiba, S., Eto, Y., Shibai, H., and Takaku, F. (1989). Characterization of cellular receptors for erythroid differentiation factor on murine erythroleukemia cells. *J. Biol. Chem.* 264, 10309–10314.

Hollenberg, S. M., and Evans, R. M. (1988). Multiple and cooperative *trans*-activation domains of the human glucocorticoid receptor. *Cell* 55, 899–906.

Honegger, A. M., Kris, R. M., Ullrich, A., and Schlessinger, J. (1989). Evidence that autophosphorylation of solubilized receptors for epidermal growth factor is mediated by intermolecular cross phosphorylation. *Proc. Natl. Acad. Sci. USA* 86, 925–929.

Horie, K., Takakura, K., Taii, S., Narimoto, K., Noda, Y., Nishikawa, S., Nakayama, H., Fujita, J., and Mori, T. (1991). The expression of c-*kit* protein during oogenesis and early embryonic development. *Biol. Reprod.* 45, 547–552.

Horwitz, K. B., and Alexander, P. S. (1983). *In situ* photolinked nuclear progesterone receptors of human breast cancer cells: subunit molecular weights after transformation and translocation. *Endocrinology* 113, 2195–2201.

Horwitz, K. B., Wei, L. L., Sedlacel, S. M., and D'Arville, C. N. (1985). Progestin action and progesterone receptor structure in human breast cancer. *Recent Prog. Horm. Res.* 41, 249–316.

Hsueh, A. J. W., and Jones, P. B. C. (1981). Extrapituitary actions of gonadotropin-releasing hormone. *Endocr. Rev.* 2, 437–461.

Hsueh, A. J. W., Adashi, E. Y., Jones, P. B. S., and Welsh, T. H. Jr. (1984). Hormonal regulation of the differentiation of cultured ovarian granulosa cells. *Endocr. Rev.* 5, 76–127.

Huang, W. M., Gu, J., Blank, M. A., Allen, J. M., Bloom, S. R., and Polak, J. M. (1984). Peptide-immunoreactive nerves in the mammalian female genital tract. *Histochem. J.* 16, 1297.

Huckle, W. R., and Conn, P. M. (1988). Molecular mechanism of gonadotropin-releasing hormone action. II. The effector system. *Endocr. Rev.* 9, 387–395.

Hunzicker-Dunn, M., and Birnbaumer, L. (1985). The involvement of adenylyl cyclase and cyclic AMP dependent protein kinases in luteinizing hormone actions. *In* "Luteinizing Hormone Action and Receptors." (M. Ascoli, ed.), pp. 57–134. CRC Press, Boca Raton, FL.

Hutchinson, L. A., Findlay, J. K., de Vos, F. L., and Robertson, D. M. (1987). Effects of bovine inhibin, transforming growth factor-β and bovine activin-A on granulosa cell differentiation. *Biochem. Biophys. Res. Commun.* 146, 1405–1412.

Ilenchuk, T. T., and Walters, M. R. (1987). Rat uterine progesterone receptor analyzed by [^{3}H]R5020 photoaffinity labeling: evidence that the A and B subunits are not equimolar. *Endocrinology* 120, 1449–1456.

Imperato-McGinley, J., and Canovatchel, W. J. (1992). Complete androgen insensitivity: pathophysiology, diagnosis, and management. *Trends Endocrinol. Metab.* 75–81.

Inui, A., Sato, K., Mura, M., Hirosue, Y., Nakajima, M., Oita, M., Baba, S., and Kasuga, M. (1992). Evidence for further heterogeneity of the receptors for neuropeptide-Y and peptide-YY in tumor cell lines derived from neural crest. *Endocrinology* 131, 2090–2096.

Ishihara, T., Nakamura, S., Kaziro, Y., Takahashi, T., Takahashi, K., and Nagata, S. (1991). Molecular cloning and expression of a cDNA encoding the secretin receptor. *EMBO J.* 7, 1635–1641.

Ishihara, T., Shigemoto, R., Mori, K., Takahashi, K., and Nagata, S. (1992). Functional expression and tissue distribution of a novel receptor for vasoactive intestinal polypeptide. *Neuron* 8, 811–819.

Isola, J., Korte, J. M., and Tuohimaa, P. (1987). Immunocytochemical localization of progesterone receptor in the chick ovary. *Endocrinology* 121, 1034–1040.

Iwai, M., Yasuda, K., Fukuoka, M., Iwai, T., Takakura, K., Taii, S., Nakanishi, S., and Mori, T. (1991). Luteinizing hormone induces progesterone receptor gene expression in cultured porcine granulosa cells. *Endocrinology* 129, 1621–1627.

Jacobs, B. R., and Smith, R. G. (1980). Evidence for a receptor-like protein for progesterone in bovine ovarian cytosol. *Endocrinology* 106, 1276–1282.

Jacobs, B. R., Suchock, S., and Smith, R. G. (1980). Evidence for human ovarian progesterone receptor. *Am. J. Obstet. Gynecol.* 138, 332–336.

Jeltsch, J.-M., Turcotte, B., Garnier, J.-M., Lerouge, T., Krozowski, Z., Gronemeyer, H., and Chambon, P. (1990). Characterization of multiple messenger RNAs originating from the chicken progesterone-receptor gene: evidence for a specific transcript encoding form A. *J. Biol. Chem.* 265, 3967–3974.

Ji, I., and Ji, T. H. (1991). Exons 1–10 of the rat LH receptor encode a high affinity hormone binding site and exon 11 encodes G-protein modulation and a potential second hormone binding site. *Endocrinology* 128, 2648–2650.

Jia, X. C., Kalmijn, J., and Hsueh, A. J. W. (1986). Growth hormone enhances follicle stimulating hormone-induced differentiation of cultured rat granulosa cells. *Endocrinology* 118, 1401–1409.

Johnson, D., Lanahan, A., Buck, C. R., Sehgal, A., Morgan, C., Mercer, E., Bothwell, M., and

Chao, M. (1986). Expression and structure of the human NGF receptor. *Cell* 47, 545–554.

Jones, P. B. C., Welsh, T. H. Jr., and Hsueh, A. J. W. (1982). Regulation of ovarian progestin production by epidermal growth factor in cultured rat granulosa cells. *J. Biol. Chem.* 257, 11268–11273.

Julius, D., MacDermott, A. B., Axel, R., and Jessell, T. (1988). Molecular characterization of a functional cDNA encoding the serotonin 1c receptor. *Science* 241, 558–564.

Kamps, M. P., and Sefton, B. M. (1986). Neither arginine nor histidine can carry out the function of lysine 295 in the ATP-binding site of $p60^{src}$. *Mol. Cell. Biol.* 6, 751–757.

Kaplan, D. R., Whitman, M., Schaffhausen, B., Pallas, D. C., White, M., Cantley, L., and Roberts, T. M. (1987). Common elements in growth factor stimulation and oncogenic transformation: 85 kd phosphoprotein and phosphatidylinositol kinase activity. *Cell* 50, 1021–1029.

Kaplan, D. R., Morrison, D. K., Wong, G., McCormick, F., and Williams, L. T. (1990). PDGF β receptor stimulates tyrosine phsophorylation of GAP and association of GAP with a signaling complex. *Cell* 61, 125–133.

Kaplan, D. R., Hempstead, B. L., Martin-Zanca, D., Chao, M. V., and Parada, L. F. (1991a). The *trk* proto-oncongene product: a signal transducing receptor for nerve growth factor. *Science* 252, 554–558.

Kaplan, D. R., Martin-Zanca, D., and Parada, L. F. (1991b). Tyrosine phosphorylation and tyrosine kinase activity of the *trk* proto-oncogene product induced by NGF. *Nature* 350, 158–162.

Kashishian, A., Kazlauskas, A., and Cooper, J. A. (1992). Phosphorylation sites in the PDGF receptor with different specificities for binding GAP and PI3 kinase *in vivo*. *EMBO J.* 11, 1373–1382.

Kastner, P., Bocquel, M.-T., Turcotte, B., Garnier, J.-M., Horwitz, K. B., Chambon, P., and Gronemeyer, H. (1990a). Transient expression of human and chicken progesterone receptors does not support alternative translational initiation from a single mRNA as the mechanism generating two isoforms. *J. Biol. Chem.* 265, 12163–12167.

Kastner, P., Krust, A., Turcotte, B., Stropp, U., Tora, L., Gronemeyer, H., and Chambon, P. (1990b). Two distinct estrogen-regulated promoters generate transcripts encoding the two functionally different human progesterone receptor forms A and B. *EMBO J.* 9, 1603–1614.

Katz, A., Wu, D., and Simon, M. I. (1992). Subunits βγ heterotrimeric G protein activate β2 isoform of phospholipase C. *Nature* 360, 686–689.

Katzenellenbogen, B. S. (1980). Dynamics of steroid hormone receptor action. *Annu. Rev. Physiol.* 42, 17–35.

Katzenellenbogen, B. S., and Norman, M. J. (1990). Multi-hormonal regulation of the progesterone receptor in MCF-7 human breast cancer cells: interrelationships among insulin/insulin-like growth factor-I, serum, and estrogen. *Endocrinology* 126, 891–898.

Kazlauskas, A., and Cooper, J. A. (1989). Autophosphorylation of the PDGF receptor in the kinase insert region regulates interactions with cell proteins. *Cell* 58, 1121–1133.

Keaveney, M., Klug, J., Dawson, M. T., Nestor, P. V., Neilan, J. G., Frode, R. C., and Gannon, F. (1991). Evidence for a previously unidentified upstream exon in the human oestrogen receptor gene. *J. Mol. Endocrinol.* 6, 111–115.

Kelly, J. D., Haldeman, B. A., Grant, F. J., Murray, M. J., Seifert, R. A., Bowen-Pope, D. F., Cooper, J. A., and Kazlauskas, A. (1991). Platelet-derived growth factor (PDGF) stimulates PDGF receptor subunit demerization and intersubunit transphosphorylation. *J. Biol. Chem.* 266, 8987–8992.

Kelly, P. A., Dijiane, J., and Edery, M. (1992). Different forms of the prolactin receptor: insights into the mechanisms of prolactin action. *Trends Endocrinol. Metab.* 3, 54–59.

Kim, I.-C., and Schomberg, D. W. (1989). The production of transforming growth factor-β activity by rat granulosa cell cultures. *Endocrinology* 124, 1345–1351.

Kim, I.-C., Ascoli, M., and Segaloff, D. L. (1987). Immunoprecipitation of the lutropin/choriogonadotropin receptor from biosynthetically labeled Leydig tumor cells: a 92 kDa protein. *J. Biol. Chem.* 262, 470–477.

Kimura, T., Tanizawa, O., Mori, K., Brownstein, M. J., and Okayama, H. (1992). Structure and expression of a human oxytocin receptor. *Nature* 356, 526–529.

Klein, R., Conway, D., Parada, L. F., and Barbacid, M. (1991a). The *trk* B tyrosine kinase gene codes for a second neurogenic receptor that lacks the catalytic kinase domain. *Cell* 61, 647–656.

Klein, R., Jing, S., Nanduri, V., O'Rourke, E., and Barbacid, M. (1991b). The *trk* proto-oncogene encodes a receptor for nerve growth factor. *Cell* 65, 189–197.

Klein, R., Nanduri, V., Jing, S., Lamballe, F., Tapley, P., Bryant, S., Cordon-Cardo, C., Jones, K. R., Reichardt, L. F., and Barbacid, M. (1991c). The *trk* B tyrosine protein kinase is a receptor for brain-derived neurotrophic factor and neurotrophin-3. *Cell* 66, 393–403.

Knecht, M. P., and Catt, K. J. (1983). Modulation of cAMP-mediated differentiation in ovarian granulosa cells by epidermal growth factor and platelet-derived growth factor. *J. Biol. Chem.* 258, 2789–2794.

Knecht, M. P., Feng, P., and Catt, K. J. (1986). Transforming growth factor-beta regulates the expression of luteinizing hormone receptors in ovarian granulosa cell. *Biochem. Biophys. Res. Commun.* 139, 800–807.

Knecht, M., Feng, P., and Catt, K. (1987). Bifunctional role of transforming growth factor-β during granulosa cell development. *Endocrinology* 120, 1243–1249.

Kobilka, B. K., Kobilka, T. S., Daniel, D., Regan, J. W., Caron, M. G., and Lefkowitz, R. J. (1988). Chimeric α_2-,β_2-adrenergic receptors: delineation of domains involved in effector coupling and ligand binding specificity. *Science* 240, 1310–1316.

Koike, S., Sakai, M., and Muramatsu, M. (1987). Molecular cloning and characterization of rat estrogen receptor cDNA. *Nucleic Acids Res.* 15, 2499–2513.

Koo, Y. B., Ji, I., Slaughter, R. G., and Ji, T. H. (1991). Structure of the luteinizing hormone receptor gene and multiple exons of the coding sequence. *Endocrinology* 128, 2297–2308.

Korte, J. M., and Isola, J. J. (1988). An immunocytochemical study of the progesterone receptor in rabbit ovary. *Mol. Cell. Endocrinol.* 58, 93–101.

Krasnow, J. S., Hickey, G. J., and Richards, J. S. (1990). Regulation of aromatase mRNA and estradiol biosynthesis in rat ovarian granulosa and luteal cells by prolactin. *Mol. Endocrinol.* 4, 13–21.

Krett, N. L., Wei, L. L., Francis, M. D., Nordeen, S. K., Gordon, D. F., Wood, W. M., and Hortwitz, K. B. (1988). Human progesterone A-receptors can be synthesized intracellularly and are biologically functional. *Biochem. Biophys. Res. Commun.* 157, 278–285.

Krust, A., Green, S., Argos, P., Kumar, V., Walter, P., Bornert, J.-M., and Chambon, P. (1986). The chicken oestrogen receptor sequence: homology with v-*erb* A and the human oestrogen and glucocorticoid receptors. *EMBO J.* 5, 891–897.

Kubo, T., Fukuda, K., Mikami, A., Maeda, A., Takahashi, H., Mishina, M., Haga, T., Haga, K., Ichiyama, A., Kangawa, K., Kojima, M., Matsuo, H., Hirose, T., and Numa, S. (1986). Cloning, sequencing and expression of complementary DNA encoding the muscarinic acetylcholine receptor. *Nature* 323, 411–416.

Kubo, T., Bujo, H., Akiba, I., Nakai, J., Mishina, M., and Numa, S. (1988). Location of a region of the muscarinic acetylcholine receptor involved in selective effector coupling. *FEBS Lett.* 241, 119–125.

Kudolo, G. B., Elder, M. G., and Myatt, L. (1984). A novel oestrogen-binding species in rat granulosa cells. *J. Endocrinol.* 102, 83–91.

Kumar, V., and Chambon, P. (1988). The estrogen receptor binds tightly to its responsive element as a ligand-induced homodimer. *Cell* 55, 145–156.

Kumar, V., Green, S., Stack, G., Berry, M., Jin, J.-R., and Chambon, P. (1987). Functional domains of the human estrogen receptor. *Cell* 51, 941–951.

Kusuda, S., and Dufau, M. L. (1986). Purification and characterization of the rat ovarian receptor for luteinizing hormone. Structural studies of subunit interaction. *J. Biol. Chem.* 261, 16161–16168.

Labrie, F., Gagne, B., and Lefevre, G. (1983). Growth hormone-releasing factor stimulates adenylate cyclase activity in the anterior pituitary gland. *Life Sci.* 33, 2229–2233.

Laiho, M., DeCaprio, J. A., Ludlow, J. W., Livingston, D. M., and Massague, J. (1990). Growth inhibition by TGF-β linked to suppression of retinoblastoma protein phosphorylation. *Cell* 62, 175–185.

Laiho, M., Weis, F. M. B., Boyd, F. T., Ignotz, R. A., and Massague, J. (1991). Responsiveness to transforming growth factor-β (TGF-β) restored by genetic complementation between cells defective in TGF-β receptors I and II. *J. Biol. Chem.* 266, 9108–9112.

Lamballe, F., Klein, R., and Barbacid, M. (1991). *trk* C, a new member of the *trk* family of tyrosine protein kinases, is a receptor for neurotrophin-3. *Cell* 66, 967–979.

LaPolt, P. S., Oikawa, M., Jia, X. C., Dargan, C., Hsueh, A. J. W. (1990). Gonadotropin induced up- and down-regulation of rat ovarian LH receptor message levels during follicular growth, ovulation and luteinization. *Endocrinology* 126, 3277–3279.

LaPolt, P. S., Tilly, J. L., Aihara, T., Nishimori, K., and Hsueh, A. J. W. (1992). Gonadotropin-induced up- and down-regeulation of ovarian follicle-stimulating hormone (FSH) receptor gene expression in immature rats: effects of pregnant mare's seerum gonadotropin, human chorionic gonadotropin, and recombinant FSH. *Endocrinology* 130, 1289–1295.

Lara, H. E., Hill, D. F., Katz, K. H., and Ojeda, S. R. (1990a). The gene encoding nerve growth factor is expressed in the immature rat ovary: effect of denervation and hormonal treatment. *Endocrinology* 126, 357–363.

Lara, H. E., McDonald, J. K., and Ojeda, S. R. (1990b). Involvement of nerve growth factor in female sexual development. *Endocrinology* 126, 364–375.

Large, T. H., Weskamp, G., Helder, J. C., Radeke, M. J., Misko, T. P., Shooter, E. M., and Reichardt, L. F. (1989). Structure and developmental expression of the nerve growth factor receptor in the chicken central nervous system. *Neuron* 2, 1123–1134.

Larhammar, D., Blomqvist, A. G., Yee, F., Jazin, E., Yoo, H., and Wahlested, C. (1992). Cloning and functional expression of a human neuropeptide Y: peptide YY receptor of the Y1 type. *J. Biol. Chem.* 267, 10935–10938.

Latouche, J., Crumeyrolle-Arias, M., Jordan, D., Kopp, N., Augendre-Ferrante, B., Cedard, L., and Haour, F. (1989). GnRH receptors in human granulosa cells: anatomical localization and characterization by autoradiography. *Endocrinology* 125, 1739–1741.

Laudet, V., Hanni, C., Coll, J., Catzeflis, F., and Stehelin, D. (1992). Evolution of the nuclear receptor superfamily. *EMBO J.* 3, 1003–1013.

Lee, K.-F., Li, E., Huber, L. J., Landis, S. C., Sharpe, A. H., Chao, M. V., and Jaenisch, R. (1992). Targeted mutation of the gene encoding the low affinity NGF receptor p75 leads to deficits in the peripheral sensory nervous system. *Cell* 69, 737–749.

Legerski, R., Zhou, X., Dresback, J., Eberspaecher, H., McKinney, S., Segarini, P., and de Crombrugghe, B. (1992). Molecular cloning and characterization of a novel rat activin receptor. *Biochem. Biophys. Res. Commun.* 183, 672–679.

Lesueur, L., Edery, M., Paly, J., Clark, J., Kelly, P. A., and Djiane, J. (1990). Prolactin stimulates milk protein promoter in CHO cells cotransfected with prolactin receptor cDNA. *Mol. Cell. Endocrinol.* 71, R7–R12.

Leung, P. K., and Steele, G. L. (1992). Intracellular signaling in the gonads. *Endocr. Rev.* 13, 476–498.

Levi-Montalcini, R. (1987). The nerve growth factor 35 years later. *Science* 237, 1154–1162.

Li, X.-J., Wu, Y.-N., North, A., and Forte, M. (1992). Cloning, functional expression, and developmental regulation of a neuropeptide Y receptor from *Drosophila melanogaster*. *J. Biol. Chem.* 267, 7–12.

Liggett, S. B., Caron, M. G., Lefkowitz, R. J., and Hnatowich, M. (1991). Coupling of a mutated form of the human β_2-adrenergic receptor to G_i and G_s. *J. Biol. Chem.* 266, 4816–4821.

Lin, C., Lin, S.-C., Chang, C.-P., and Rosenfeld, M. G. (1992). Pit-1-dependent expression of the receptor for growth hormone releasing factor mediates pituitary cell growth. *Nature* 360, 765–768.

Lin, H. Y., Harris, T. L., Flannery, M. S., Aruffo, A., Kaji, E. H., Gorn, A., Kolakowski, L. F. Jr., Lodish, H. F., and Goldring, S. R. (1991). Expression cloning of an adenylate cyclase-coupled calcitonin receptor. *Science* 254, 1022–1024.

Lin, H. Y., Wang, X.-F., Ng-Eaton, E., Weinberg, R. A., and Lodish, H. F. (1992). Expression cloning of the TGF-β type II receptor, a functional transmembrane serine/threonine kinase. *Cell* 68, 775–785.

Lobb, D. K., Kobrin, M. S., Kudlow, J. E., and Dorrington, J. H. (1989). Transforming growth factor-alpha in the adult bovine ovary: identification in growing ovarian follicles. *Biol. Reprod.* 40, 1089–1093.

Lobie, P. E., Breipohl, W., Aragon, J. A., and Waters, M. J. (1990). Cellular localization of the growth hormone receptor/binding protein in the male and female reproductive system. *Endocrinology* 126, 2214–2221.

Logeat, F., LeCunff, M., Pamphile, R., and Milgrom, E. (1985). The nuclear bound form of the progesterone receptor is generated through a hormone-dependent phosphorylation *Biochem. Biophys. Res. Commun.* 131, 421–427.

Loosfelt, H., Atger, M., Misrahi, M., Guichon-Mantel, A., Meriel, C., Logeat, F., Benarous, R., and Milgrom, E. (1986). Cloning and sequence analysis of rabbit progesterone receptor complementary DNA. *Proc. Natl. Acad. Sci. USA* 83, 9045–9049.

Loosfelt, H., Misrahi, M., Atger, M., Salesse, R., Thi, M. T. V. H.-L., Jolivet, A., Guiochon-Mantel, A., Sar, S., Jallal, B., Garnier, J., and Milgrom, E. (1989). Cloning and sequencing of porcine LH-hCG receptor cDNA: variants lacking transmembrane domain. *Science* 245, 525–527.

Lopez, J. A., Chung, D. W., Fujikawa, K., Hagen, F. S., Papayannopoulou, T., and Roth, G. J. (1987). Cloning of the a chain of human platelt glycoprotein Ic: a transmembrane protein with homology to leucine-rich α_2-glycoprotein. *Proc. Natl. Acad. Sci. USA* 84, 5615–5619.

Lopez-Casillas, F., Cheifetz, S., Doody, J., Andres, J. L., Lane, W. S., and Massague, J. (1991). Structure and expression of the membrane proteoglycan betaglycan, a component of the TGF-β receptor system. *Cell* 67, 785–795.

Luisi, B. F., Xu, W. X., Otwinowski, Z., Freedman, L. P., Yamamoto, K. R., and Sigler, P. B. (1991). Crystallographic analysis of the interaction of the glucocorticoid receptor with DNA. *Nature* 352, 497–505.

Lyons, J., Landis, C. A., Harsh, G., Vallar, L., Grunewald, K., Feichtinger, H., Duh, Q.-Y., Clark, O. H., Kawasaki, E., Bourne, H. R., and McCormick, F. (1991). Two G protein oncogenes in human endocrine tumors. *Science* 249, 655–659.

Mader, S., Kumar, V., deVereneuil, H., and Champon, P. (1989). Three amino acids of the oestrogen receptor are essential to its ability to distinguish an oestrogen from a glucocorticoid-responsive element. *Nature* 338, 271–274.

Manova, K., Nocka, K., Besmer, P., and Bachvarova, R. F. (1990). Gonadal expression of *c-kit* encoded at the W locus of the mouse. *Development* 110, 1057–1069.

Marshall, C. J. (1991). Tumor suppressor genes. *Cell* 64, 313–326.

Martin-Zanca, D., Oskam, R., Mitra, G., Copeland, T., and Barbacid, M. (1989). Molecular and biochemical characterization of the human *trk* proto-oncogene. *Mol. Cell. Biol.* 9, 24–33.

Massague, J. (1990). The transforming growth factor-β family. *Annu. Rev. Cell Biol.* 6, 597–641.

Mathews, L. S. (1991). Molecular biology of growth hormone receptors. *Trends Endocrinol. Metab.* 2, 176–180.

Mathews, L. S., and Vale, W. W. (1991). Expression cloning of an activin receptor, a predicted transmembrane serine kinase. *Cell* 65, 973–982.

Mathews, L. S., Vale, W. W., and Kintner, C. R. (1992). Cloning of a second type of activin receptor and functional characterization in *Xenopus* embryos. *Science* 255, 1702–1705.

Matsui, T., Heidaran, M., Miki, T., Popescu, N., La Rochelle, W., Kraus, M., Pierce, J., and Aaronson, S. (1989). Isolation of a novel receptor cDNA established the existence of two PDGF receptor genes. *Science* 243, 800–804.

Matzuk, M. M., and Bradley, A. (1992). Cloning of the human activin receptor cDNA reveals high evolutionary conservation. *Biochim. Biophys. Acta* 1130, 105–108.

May, J. V., and Schomberg, D. W. (1989). The potential relevance of epidermal growth factor and transforming growth factor-alpha to ovarian physiology. *Sem. Reprod. Endocrinol.* 7, 1–11.

Mayo, K. E. (1992). Molecular cloning and expression of a pituitary-specific receptor for growth hormone-releasing hormone. *Mol. Endocrinol.* 6, 1734–1744.

McDonald, J. K. (1988). NPY and related substance. *In* "CRC Critical Reviews in Neurobiology," (J. Nelson, ed.), Vol. 4, pp. 97–135. CRC Press, Boca Raton, FL.

McDonald, J. K., Dees, W. L., Ahmed, C. E., Noe, B. D., and Ojeda, S. R. (1987). Biochemical and immunocytochemical characterization of neuropeptide Y in the immature rat ovary. *Endocrinology* 120, 1703–1710.

McFarland, K. C., Sprengel, R., Phillips, H. S., Kohler, M., Rosemblit, N., Nikolics, K., Segaloff, D. L., and Seeburg, P. H. (1989). Lutropin-choriogonadotropin receptor: an unusual member of the G-protein-coupled receptor family. *Science* 245, 494–499.

Meakin, S. O., and Shooter, E. M. (1992). The nerve growth factor family of receptors. *Trends Neurosci.* 15, 323–331.

Meisenhelder, J., Suh, P.-G., Rhee, S. G., and Hunter, T. (1989). Phospholipase C-γ is a substrate for the PDGF and EGF receptor protein tyrosine kinases *in vivo* and *in vitro*. *Cell* 57, 1109–1122.

Merk, F. B., Botticelli, C. R., and Albright, J. T. (1972). An intercellular response to estrogen by granulosa cells in the rat ovary: an electron microscope study. *Endocrinology* 90, 992–1007.

Meyer, M.-E., Gronemeyer, H., Turcotte, B., Bocquel, M.-T., Tasset, D., and Chambon, P. (1989). Steroid hormone receptors compete for factors that mediate their enhancer function. *Cell* 57, 433–442.

Miller, J., McLachlan, A. D., and Klug, A. (1985). Repetitive zinc-binding domains in the protein transcription factor IIIA from *Xenopus* oocytes. *EMBO J.* 1609–1614.

Minegish, T., Kusuda, S., and Dufau, M. L. (1987). Purification and characterization of Leydig cell luteinizing hormone receptor. *J. Biol. Chem.* 262, 17138–17143.

Minegish, T., Nakamura, K., Takakura, Y., Miyamoto, K., Hasegawa, Y., Ibuki, Y., and Igarashi, M. (1990). Cloning and sequencing of human LH/hCG receptor cDNA. *Biochem. Biophys. Res. Commun.* 175, 1125–1130.

Mintz, B., and Russell, E. S. (1957). Gene induced embryological modifications of primordial germ cells in the mouse. *J. Exp. Zool.* 134, 207–237.

Misrahi, M., Loosfelt, H., Atger, M., Meriel, C., Zerah, V., Dessen, P., and Milgrom, E. (1988). Organization of the entire rabbit progesterone receptor mRNA and of the promoter and 5′ flanking region of the gene. *Nucleic Acids Res.* 16, 5459–5472.

Miyajima, A., Hara, T., and Kitamura, T. (1992). Common subunits of cytokine receptors and the functional redundancy of cytokines. *Trends Biochem. Sci.* 17, 378–382.

Miyata, A., Arimura, A., Dahl, R. R., Minamino, N., Uehara, A., Jiang, L., Culler, M. D., and Coy, D. H. (1989). Isolation of a novel 38 residue-hypothalamic polypeptide which stimulates adenylate cyclase in pituitary cells. *Biochem. Biophys. Res. Commun.* 164, 567–574.

Molloy, C. J., Bottaro, D. P., Fleming, T. P., Marshall, M. S., Gibbs, J. B., and Aaronson, S. A.

(1989). PDGF induction of tyrosine phosphorylation of GTPase activating protein. *Nature* 342, 711–714.

Mondschein, J. S., and Hammond, J. M. (1988). Growth factors regulate immunoreactive insulin-like growth factor-1 production by cultured porcine granulosa cells. *Endocrinology* 123, 463–468.

Mondschein, J. S., Canning, S. F., and Hammond, J. M. (1988). Effects of transforming growth factor-β on the production of immunoreactive insulin-like growth factor I and progesterone and on [³H]-thymidine incorporation in porcine granulosa cell cultures. *Endocrinology* 123, 1970–1976.

Monget, P., Monniauz, D., and Durand, P. (1989). Localization, characterization, and quantification of insulin-like growth factor-I binding sites in the ewe ovary. *Endocrinology* 125, 2486–2493.

Monsma, F. J., Jr., McVittie, L. D., Gerfen, C. R., Mahan, L. C., and Sibley, D. R. (1989). Multiple D_2 dopamine receptors produced by alternative RNA splicing. *Nature* 342, 926–929.

Moretti, C., Bagnato, A., Solan, N., Frajese, G., and Catt, K. J. (1990). Receptor-mediated actions of growth hormone releasing factor on granulosa cell differentiation. *Endocrinology* 127, 2117–2126.

Mori, T., Suzuki, A., Nishimura, T., and Kambegawa, A. (1977). Inhibition of ovulation in immature rats by antiprogesterone antiserum. *J. Endocrinol.* 73, 185–186.

Morrison, D. K., Kaplan, D. R., Escobedo, J. A., Rapp, U. R., Roberts, T. M., and Williams, L. T. (1989). Direct activation of the serine/threonine kinase activity of Raf-1 through tyrosine phosphorylation by the PDGFβ-receptor. *Cell* 58, 649–657.

Mulheron, G. W., and Schomberg, D. W. (1990). Rat granulosa cells express transforming growth factor-β type 2 messenger ribonucleic acid which is regulatable by follicle-stimulating hormone *in vitro*. *Endocrinology* 126, 1777–1779.

Naess, O. (1981). Characterization of cytoplasmic progesterone receptors in rat granulosa cells: evidence for nuclear translocation. *Acta Endocrinol. (Copenhagen)* 98, 288–294.

Nagayama, Y., Wadsworth, H. L., Chazenbalk, G. D., Russo, D., Seto, P., and Rapoport, B. (1991). Thyrotropin-luteinizing hormone/chorionic gonadotropin receptor extracellular domain chimeras as probes for thyrotropin receptor function. *Proc. Natl. Acad. Sci. USA* 88, 902–905.

Nakamura, K., Minegish, T., Takakura, Y., Miyamoto, K., Hasegawa, Y., Ibuki, Y., and Igarashi, M. (1990). Regulation of LH/hCG receptor by gonadotropins in rat ovary. *biochem. Biophys. Res. Commun.* 172, 786–792.

Nakamura, K., Minegishi, T., Takakura, Y., Miyamoto, K., Hasegawa, H., Ibuki, Y., and Igarashi, M. (1991). Hormonal regulation of gonadotropin receptor mRNA in rat ovary during follicular growth and luteinization. *Mol. Cell. Endocrinol.* 82, 259–263.

Nakamura, T., Takio, K., Eto, Y., Shibai, H., Titani, K., and Sugino, H. (1990). Activin-binding protein from rat ovary is follistatin. *Science* 247, 836–838.

Nakamura, T., Hasegawa, Y., Sugino, K., Kogawa, K., Titani, K., and Sugino, H. (1992a). Follistatin inhibits activin-induced differentiation of rat follicular granulosa cells *in vitro*. *Biochim. Biophys. Acta* 1135, 103–109.

Nakamura, T., Sugino, K., Kurosawa, N., Sawai, M., Takio, K., Eto, Y., Iwashita, S., Muramatsu, M., Titani, K., and Sugino, H. (1992b). Isolation and characterization of activin receptor from mouse embryonal carcinoma cells. *J. Biol. Chem.* 267, 18924–18928.

Nakatani, A., Shimasaki, S., Erickson, G. F., and Ling, N. (1991). Tissue-specific expression of four insulin-like growth factor-binding proteins (1, 2, 3, and 4) in the rat ovary. *Endocrinology* 129, 1521–1529.

Nishimatsu, S.-I., Oda, S., Murakami, K., and Ueno, N. (1992). Multiple genes for *Xenopus* activin receptor expressed during early embryogenesis. *FEBS Lett.* 303, 81–84.

Nister, M., Hammacher, A., Mellstrom, K., Siegbahn, A., Ronnstrand, L., Westermark, B., and Heldin, C.-H. (1988). A glioma-derived PDGF A chain homodimer has different functional activities from a PDGF AB heterodimer purified from human platelet. *Cell* 52, 791–799.

O'Dowd, B. F., (1992). Repeat sequences in the gene encoding the human D_4 dopamine receptor. *Gene* 118, 301–302.

O'Dowd, B. F., Hnatowich, M., Regan, J. W., Leader, W. M., Caron, M. G., and Lefkowitz, R. J. (1988). Site-directed mutagenesis of the cytoplasmic domains of the human β_2-adrenergic receptor: localization of regions involved in G protein-receptor coupling. *J. Biol. Chem.* 263, 15985–15992.

Oefner, C., D'Arcy, A., Winkler, F. K., Eggimann, B., and Hosang, M. (1992). Crystal structure of human platelet-derived growth factor BB. *EMBO J.* 11, 3921–3926.

Ohuchi, H., Noji, S., Koyama, E., Myokai, F., Nishikawa, K., Nohno, T., Tashiro, K., Shiokawa, K., Matsuo, N., and Taniguchi, S. (1992). Expression pattern of the activin receptor type IIA gene during differentiation of chick neural tissues, muscle, and skin. *FEBS Lett.* 303, 185–189.

Oikawa, M., Dargan, C., Ny, T., and Hsueh, A. J. W. (1990). Expression of gonadotropin-releasing hormone and prothymosin-α messenger ribonucleic acid in the ovary. *Endocrinology* 127, 2350–2356.

Ojeda, S. R., Costa, M. E., Katz, K. H., and Hersh, L. B. (1985). Evidence for the existence of substance P in the prepubertal ovary. I. Biochemical and physiological studies. *Biol. Reprod.* 33, 286–295.

Oliver, J. E., Aitman, T. J., Powell, J. F., Wilson, C. A., and Clayton, R. N. (1989). Insulin-like growth factor I gene expression in the rat ovary is confined to the granulosa cells of developing follicles. *Endocrinology* 124, 2671–2679.

O'Malley, B. (1990). The steroid receptor superfamily: more excitement predicted for the future. *Mol. Endocrinol.* 4, 363–369.

Orr-Urtreger, A., Avivi, A., Zimmer, Y., Givol, D., Yarden, Y., and Lonai, P. (1990). Developmental expression of c-kit, a proto-oncogene encoded by the W locus. *Development* 109, 911–923.

Orti, E., Bodwell, J. E., and Munck, A. (1992). Phosphorylation of steroid hormone receptors. *Endocr. Rev.* 13, 105–128.

Park, O.-K., and Mayo, K. E. (1991). Transient expression of progesterone receptor messenger RNA in ovarian granulosa cells after the preovulatory luteinizing hormone surge. *Mol. Endocrinol.* 5, 967–978.

Parmentier, M., Libert, F., Maenhaut, C., Lefort, A., Gerard, C., Perret, J., Van Sande, J., Dumont, J. E., and Vassart, G. (1989). Molecular cloning of the thyrotropin receptor. *Science* 246, 1620–1622.

Patthy, L. (1987). Detecting homology of distantly related proteins with consensus sequences. *J. Mol. Biol.* 198, 567–577.

Peng, X.-R., Hsueh, A. J. W., LaPolt, P. S., Bjersing, L., and Ny, T. (1991). Localization of luteinizing hormone receptor messenger ribonucleic acid expression in ovarian cell types during follicle development and ovulation. *Endocrinology* 129, 3200–3207.

Pernow, B. (1983). Substance P. *Pharmacol. Rev.* 35, 85–141.

Picard, D., and Yamamoto, K. R. (1987). Two signals mediate hormone dependent nuclear localization of the glucocorticoid receptor. *EMBO J.* 6, 3333–3340.

Pietenpol, J. A., Stein, R. W., Moran, E., Yaciuk, P., Schlegel, R., Lyons, R. M., Pittelkow, M. R., Munger, K., Howley, P. M., and Moses, H. L. (1990). TGF-β1 inhibition of *c-myc* transcription and growth in keratinocytes is abrogated by viral transforming proteins with pRB binding domains. *Cell* 61, 777–785.

Pintar, J. E. (1987). Structure and expression of the nerve growth factor gene. *In* "Oncogenes, Genes and Growth Factors" (G. Guroff, ed.), pp. 103–131. Wiley, New York.

Pitzel, L., Jarry, H., and Wuttke, W. (1991). Effects of substance-P and neuropeptide-Y on *in vitro* steroid release by porcine granulosa and luteal cells. *Endocrinology* 129, 1059–1065.

Piva, R., Gambari, R., Zorzato, F., Kumar, L., and del Senno, L. (1992). Analysis of upstream sequences of the human estrogen receptor gene. *Biochem. Biophys. Res. Commun.* 183, 996–1002.

Ponglikitmongkol, M., Green, S., and Chambon, P. (1988). Genomic organization of the human oestrogen receptor gene. *EMBO J.* 7, 3385–3388.

Poretsky, L., Bhargava, G., and Levitan, E. (1990). Type I insulin-like growth factor receptors in human ovarian stroma. *Horm. Res.* 33, 22–26.

Press, M. F., and Greene, G. L. (1988). Localization of progesterone receptor with monoclonal antibodies to the human progestin receptor. *Endocrinology* 122, 1165–1175.

Qiu, F., Ray, P., Brown, K., Barker, P. E., Jhanwar, S., Ruddle, F. H., and Besmer, P. (1990). Primary structure of c-*kit*:; relationship with the CSF-1/PDGF receptor kinase family-oncogenic activation of v-*kit* involves deletion of extracellular domain and C terminus. *EMBO J.* 7, 1003–1011.

Radeke, M. J., Misko, T. P., Hsu, C., Herzenberg, L. A., and Shooter, E. M. (1987). Gene transfer and molecular cloning of the rat nerve growth factor receptor. *Nature* 325, 593–597.

Rall, L. B., Scott, J., Bell, G. I., Crawford, R. J., Penschow, J. D., Niall, H. D., and Coghlan, J. P. (1985). Mouse prepro-epidermal growth factor synthesis by the kidney and other tissues. *Nature* 313, 228–230.

Rani, C., Salhanik, A., and Armstrong, D. T. (1981). Follicle-stimulating hormone induction of luteinizing hormone receptor in cultured granulosa cells: an example of the need for steroids in the induction process. *Endocrinology* 108, 1379–1385.

Reinhart, J., Mertz, L. M., and Catt, K. J. (1992). Molecular cloning and expression of cDNA encoding the morine gonadotropin-releasing hormone receptor. *J. Biol. Chem.* 267, 2181–2184.

Richards, J. S. (1975). Estrogen receptor content in rat granulosa cells during follicular development, modification by estradiol and gonadotropins. *Endocrinology* 97, 1174–1184.

Richards, J. S. (1980). Maturation of ovarian follicles: actions and interactions of pituitary and ovarian hormones on follicular cell differentiation. *Physiol. Rev.* 60, 51–89.

Richards, J. S., Ireland, J. J., Rao, M. C., Bernath, G. A., Midgley, A. R. Jr., and Reichert, L. E. Jr. (1976). Ovarian follicular development in the rat: hormone receptor regulation by estradiol, follicle-stimulating hormone, and luteinizing hormone. *Endocrinology* 99, 1562–1570.

Rimland, J., Xin, W., Sweetnam, P., Saijoh, K., Nestler, E. J., and Duman, R. S. (1992). Sequence and expression of a neuropeptide Y receptor cDNA. *Mol. Pharmacol.* 40, 869–879.

Roberts, A. B., and Sporn, M. B. (1990). The transforming growth factor-betas. *In* "Peptide Growth Factors and Their Receptors" (M. Sporn and A. B. Roberts, eds.), pp. 421–472. Springer-Verlag, Heidelberg.

Robertson, D. M. (1992). Follistatin/activin binding protein. *Trends Endocrinol. Metab.* 3, 65–68.

Roche, P. C., and Ryan, R. J. (1985). The LH/hCG receptor. *In* "Luteinizing Hormone Action and Receptors" (M. Ascoli, ed.), pp. 17–56. CRC Press, Boca Raton, FL.

Rodriguez, M. C., and Segaloff, D. L. (1990). The orientation of the lutropin/choriogonadotropin receptor in rat luteal cells as revealed by site-specific antibodies. *Endocrinology* 127, 674–681.

Rodriguez, M. C., Xie, Y.-B., Wang, H., Collison, K., and Segaloff, D. L. (1992). Effects of truncations of the cytoplasmic tail of the luteinizing hormone/chorinoic gonadotropin receptor on receptor-mediated hormone internalization. *Mol. Endocrinol.* 6, 327–336.

Rodriguez-Tebar, A., Dechant, G., and Barde, Y.-A. (1990). Binding of brain-derived neurotrophic factor to the nerve growth factor receptor. *Neuron* 4, 487–492.

Roldan, A. L., Cubellis, M. V., Masucci, M. T., Behrendt, N., Lund, L. R., Dano, K., Appella, E., and Blasi, F. (1990). Cloning and expression of the receptor for human urokinase plasminogen activator, a central molecule in cell surface, plasmin dependent proteolysis. *EMBO J.* 9, 467–474.

Ronnstrand, L., Mori, S., Arridsson, A.-K., Erikson, A., Wernstedt, C., Hellman, U.,

Calesson-Welsh, L., and Heldin, C.-H. (1992). Identification of two c-terminal autophosphorylation sites in the PDGF β-receptor: involvement in the interaction with phospholipase C-γ. *EMBO J.* 11, 3911–3919.

Rosenthal, W., Seibold, A., Antaramian, A., Lonergan, M., Arthus, M.-F., Hendy, G. N., Birnbaumer, M., and Bichet, D. G. (1992). Molecular identification of the gene responsible for congenital nephrogenic diabetes insipidus. *Nature* 359, 233–235.

Rousseau-Merck, M. F., Misrahi, M., Atger, M., Loosfelt, H., Milgrom, E., and Berger, R. (1990). Localization of the human luteinizing hormone/choriogonadotropin receptor gene (LHCGR) to chromosome 2p21. *Cytogenet. Cell Genet.* 54, 77–79.

Roussel, M. F., Dull, T. J., Rettenmier, C. W., Ralph, P., Ullrich, A., and Sherr, C. J. (1987). Transforming potential of the c-fms proto-oncogene (CSF-1 receptor). *Nature* 325, 545–552.

Roy, S. K., and Greenwald, G. S. (1990). Immunohistochemical localization of epidermal growth factor-like activity in the hamster ovary with a polyclonal antibody. *Endocrinology* 126, 1309–1317.

Rozakis-Adcock, M., and Kelly, P. A. (1992). Mutational analysis of the ligand-binding domain of the prolactin receptor. *J. Biol. Chem.* 266, 16472–16477.

Rusconi, S., and Yamamoto, K. R. (1987). Functional dissection of the hormone and DNA binding activities of the glucocorticoid receptor. *EMBO J.* 6, 1309–1315.

Said, S. I. (1986). Vasoactive intestinal peptide. *J. Clin. Invest.* 9, 191–200.

Savouret, J. F., Bailly, A., Misrahi, M., Rauch, C., Redeuilh, G., Chauchereau, A., and Milgrom, E. (1991). Characterization of the hormone responsive element involved in the regulation of the progesterone receptor gene. *EMBO J.* 10, 1875–1883.

Schena, M., Freedman, L. P., and Yamamoto, K. R. (1989). Mutations in the glucocorticoid receptor zinc finger region that distinguish interdigitated DNA binding and transcriptional enhancement activities. *Gene Dev.* 3, 1590–1601.

Schlatter, L. K., Howard, K. J., Parker, M. G., and Distelhorst, C. W. (1992). Comparison of the 90-kilodalton heat shock protein interaction with *in vitro* translated glucocorticoid and estrogen receptors. *Mol. Endocrinol.* 6, 132–140.

Schlessinger, J. (1988). Singla transduction by allosteric receptor oligomerization. *Trends Biochem. Sci.* 13, 443–447.

Schlunegger, M. P., and Grutter, M. G. (1992). An unusual feature revealed by the crystal structure at 2.2 A resolution of human transforming growth factor-β2. *Nature* 358, 430–434.

Schneider, W., Ramachandran, C., Satyaswaroop, P. G., and Shyamala, G. (1991). Murine progesterone receptor exists predominantly as the 83-kilodalton 'A' form. *J. Steroid. Biochem. Mol. Biol.* 38, 285–291.

Schowalter, D. B., Sullivan, W. P., Maihle, N. J., Dodson, A. D. W., Conneely, O. M., O'Malley, B. W., and Toft, D. O. (1991). Characterization of progesterone receptor binding to the 90- and 70-kDa heat shock proteins. *J. Biol. Chem.* 266, 21165–21173.

Schrader, W. T., and O'Malley, B. W. (1972). Progesterone-binding components of chick oviducts: characterizationn of purified subunits. *J. Biol. Chem.* 247, 51–59.

Schreiber, J. R., and Hsueh, A. J. W. (1979). Progesterone receptor in rat ovary. *Endocrinology* 105, 915–919.

Schreiber, J. R., Nakamura, K., and Erickson, G. F. (1980). Progestins inhibit FSH-stimulated steroidogenesis in cultured rat granulosa cells. *Mol. Cell. Endocrinol.* 19, 165–173.

Schreiber, J. R., Nakamura, K., Truscello, A. M., and Erickson, G. F. (1982). Progesterone inhibits FSH-induced functional LH receptors in cultured rat granulosa cells. *Mol. Cell. Endocrinol.* 25, 113–124.

Schumacher, M. (1990). Rapid membrane effects of steroid hormones: an emerging concept in neuroendocrinology. *Trends Neurosci.* 13, 359–361.

Schwabe, J. W. R., Neuhaus, D., and Rhodes, D. (1990). Solution structure of the DNA-binding domain of the oestrogen receptor. *Nature* 348, 458–461.

Sealfon, S. C., Gillo, B., Mundamattom, S., Mellon, P. L., Windle, J. J., Landau, E., and

Roberts, J. L. (1992). Gonadotropin-releasing hormone receptor expression in *Xenopus* oocytes. *Mol. Endocrinol.* 4, 119–124.

Segaloff, D. L., Wang, H., and Richards, J. S. (1990). Hormonal regulation of luteinizing hormone/chorionic gonadotropin receptor mRNA in rat ovarian cells during follicular development and luteinization. *Mol. Endocrinol.* 4, 1856–1865.

Segarini, P. R., Roberts, A. B., Rosen, D. M., and Seyedin, S. M. (1987). Membrane binding characteristics of two forms of transforming growth factor-β. *J. Biol. Chem.* 262, 14655–14662.

Segarini, P. R., Rosen, D. M., and Seyedin, S. M. (1989). Binding of transforming growth factor-β to cell surface proteins varies with cell type. *Mol. Endocrinol.* 3, 261–272.

Shemshedin, L., Ji, J., Brou, C., Chambon, P., and Gronemeyer, H. (1992). *In vitro* activity of the transcription activation functions of the progesterone receptor. *J. Biol. Chem.* 267, 1834–1839.

Shirota, M., Banville, D., Ali, S., Jolicoeur, C., Boutin, J.-M., Edery, M., Djiane, J., and Kelly, P. A. (1990). Expression of two forms of prolactin receptor in rat ovary and liver. *Mol. Endocrinol.* 4, 1136–1143.

Simon, M. I., Strathmann, M. P., and Gautam, N. (1991). Diversity of G proteins in signal transduction. *Science* 252, 802–808.

Singh, G., Singh, M. M., Maitra, S. C., Elger, W., Kalra, V., Upadhyay, S. N., Chowdhury, S. R., and Kamboj, V. P. (1988). Luteolytic action of two antiprogestational agents (RU-38486 and ZK-98734) in the rat. *J. Reprod. Fertil.* 83, 73–83.

Skinner, M. K., and Coffey, R. J., Jr. (1988). Regulation of ovarian cell growth through the local production of transforming growth factor-α by theca cells. *Endocrinology* 123, 2632–2638.

Skinner, M. K., Keski-Oja, J., Osteen, K. G., and Moses, H. L. (1987). Ovarian theca cells produce transforming growth factor-β which can regulate granulosa cell growth. *Endocrinology* 121, 786–792.

Smith, D. F., Stensgard, B. A., Welch, W. J., and Toft, D. O. (1992). Assembly of progesterone receptor with heat shock proteins and receptor activation are ATP mediated events. *J. Biol. Chem.* 267, 1350–1356.

Soos, M. A., and Siddle, K. (1989). Immunological relationships between receptors for insulin and insulin-like growth factor I: evidence for structural heterogeneity of insulin-like growth factor I receptors involving hybrids with insulin receptors. *Biochem. J.* 263, 553–563.

Spicer, L. J., Langhout, D. J., Alpizar, E., Williams, S. K., Campbell, R. M., Mowles, T. F., and Enright, W. J. (1992). Effects of growth hormone-releasing factor and vasoactive intestinal peptide on proliferation and steroidogenesis of bovine granulosa cells. *Mol. Cell. Endocrinol.* 83, 73–78.

Sprengel, R., Braun, T., Nikolics, K., Segaloff, D. L., and Seeburg, P. H. (1990). The testicular receptor for follicle stimulating hormone: structure and functional expression of cloned cDNA. *Mol. Endocrinol.* 4, 525–530.

Srinivasan, G. (1992). Overexpression of receptors of the steroid/thyroid family. *Mol. Endocrinol.* 6, 857–860.

Steele-Perkins, G., Turner, J., Edman, J. C., Hari, J., Pierce, S. B., Stover, C., Rutter, W. J., and Roth, R. A. (1988). Expression and characterization of a functional human insulin-like growth factor I receptor. *J. Biol. Chem.* 263, 11486–11492.

Strader, C. D., Sigal, I. S., Register, R. B., Candelore, M. R., Rands, E., and Dixon, R. A. F. (1987). Identification of residues required for ligand binding to the β-adrenergic receptor. *Proc. Natl. Acad. Sci. USA* 84, 4384–4388.

Strickland, S., and Beers, W. H. (1976). Studies on the role of plasminogen activator in ovulation. *J. Biol. Chem.* 251, 5694–5702.

Stumpf, W. E. (1969). Nuclear concentration of ^{3}H-estradiol in target tissues: dry-mount autoradiography of vagina, oviduct, ovary, testis, mammary tumor, liver, and adrenal. *Endocrinology* 85, 31–37.

Sumida, C., and Pasqualini, J. R. (1990). Stimulation of progesterone receptors by phorbol ester and cyclic AMP in fetal uterine cells in culture. *Mol. Cell. Endocrinol.* 69, 207–215.

Swanson, R. J., and Lipneer, H. (1977). Mechanism of ovulation: effect of intrafollicular progesterone antiserum. *Fed. Proc.* 36, 390.

Taylor, C. W. (1990). The role of G proteins in transmembrane signaling. *Biochem. J.* 272, 1–13.

Taylor, G. R., Reedijk, M., Rothwell, V., Rohrschneider, L., and Pawson, T. (1989). The unique insert of cellular and viral *fms* protein tyrosine-kinase domains is dispensable for enzymatic and transforming activities. *EMBO J.* 8, 2029–2037.

Taylor, S. I., Cama, A., Accili, D., Barbetti, F., Quon, M. J., DeLa Luz Sierra, M., Suzuki, Y., Koller, E., Levy-Toledano, R., Wertheimer, E., Moncada, V. Y., Kadowaki, H., and Kadowaki, T. (1992). Mutations in the insulin receptor gene. *Endocr. Rev.* 13, 566–595.

Thoenen, H., and Bard, Y.-A. (1980). Physiology of nerve growth factor. *Physiol. Rev.* 60, 1284–1335.

Thompson, N. L., Flanders, K. C., Smith, J. M., Ellingsworth, L. R., Roberts, A. R., and Sporn, M. B. (1989). Expression of transforming growth factor-β1 in specific cells and tissues of adult and neonatal mice. *J. Cell Biol.* 108, 661–669.

Tilly, J. L., Aihara, T., Nishimori, K., Jia, X.-C., Billig, H., Kowalski, K. I., Perlas, E. A., and Hsueh, A. J. W. (1992a). Expression of recombinant human follicle-stimulating receptor: species-specific ligand binding signal transduction, and identification of multiple ovarian mRNA transcripts. *Endocrinology* 131, 799–806.

Tilly, J. L., LaPolt, P. S., and Hsuch, A. J. W. (1992b). Hormonal regulation of follicle-stimulating hormone (FSH) receptor messenger ribonucleic acid levels in cultured rat granulosa cells. *Endocrinology* 130, 1296–1302.

Tonetta, S. A., and DiZerega, G. S. (1989). Intragonadal regulation of follicular maturation. *Endocr. Rev.* 10, 205–229.

Tora, L., Gronemeyer, H., Turcotte, B., Gaub, M.-P., and Chambon, P. (1988). The N-terminal region of the chicken progesterone receptor specifics target gene activation. *Nature* 333, 185–188.

Truss, M., Chalepakis, G., Slater, E. P., Mader, S., and Beato, M. (1991). Functional interaction of hybrid response elements with wild type and mutant steroid hormone receptors. *Mol. Cell. Biol.* 11, 3247–3258.

Truss, M., Chalepakis, G., Pina, B., Barettino, D., Bruggemeier, U., Kalff, M., Slater, E. P., and Beato, M. (1992). Transcriptional control by steroid hormones. *J. Steroid. Biochem. Mol. Biol.* 41, 241–248.

Treciak, W. H., Ahmed, C. E., Simpson, E. R., and Ojeda, S. R. (1986). Vasoactive intestinal peptide induces the synthesis of the cholesterol side-chain cleavage enzyme complex in cultured rat ovarian granulosa cells. *Proc. Natl. Acad. Sci. USA* 83, 7490–7494.

Tsai-Morris, C. H., Buczko, E., Wang, W., and Dufau, M. L. (1990). Intronic nature of the rat luteinizing hormone receptor gene defines a soluble receptor subspecies with hormone binding activity. *J. Biol. Chem.* 265, 19385–19388.

Tsai-Morris, C. H., Buczko, E., Wang, W., Xie, X.-Z., and Dufau, M. L. (1991). Structural organization of the rat luteinizing hormone (LH) receptor gene. *J. Biol. Chem.* 266, 11355–11359.

Tsutsumi, M., Zhou, W., Millar, R. P., Mellon, P. L., Roberts, J. L., Flanagan, C. A., Dong, K., Gillo, B., and Sealfon, S. C. (1992). Cloning and functional expression of a mouse gonadotropin-releasing hormone receptor. *Mol. Endocrinol.* 6, 1163–1169.

Turcotte, B., Meyer, M. E., Bellard, M., Dretzen, G., Gronemeyer, H., and Chambon, P. (1991). Control of transcription of the chicken progesterone receptor gene: *in vitro* and *in vivo* studies. *J. Biol. Chem.* 266, 2582–2589.

Ullrich, A., and Schlessinger, J. (1990). Signal transduction by receptors with tyrosine kinase activity. *Cell* 61, 203–212.

Ullrich, A., Coussens, L., Hayflick, J. S., Dull, T. J., Gray, A., Tam, A. W., Lee, J., Yarden, Y., Libermann, T. A., Schlessinger J., Downward, J., Mayes, E. L. V., Whittle, N., Waterfield, M. D., and Seeburg, P. H. (1984). Human epidermic growth factor receptor cDNA sequence and aberrant expression of the amplified gene in A431 epidermal carcinoma cells. *Nature* 309, 418–425.

Ullrich, A., Gray, A., Tam, A. W., Yang-Feng, T., Tsubokawa, M., Collins, C., Henzel, W., Le Bon, T., Kathuria, S., Chen, E., Jacobs, S., Francke, U., Ramachandran, J., and Fujita-Yamaguchi, Y. (1986). Insulin-like growth factor I receptor primary structure; comparison with insulin receptor suggests structural determinants that define functional specificity. *EMBO J.* 5, 2503–2512.

Umesono, K., and Evans, R. M. (1989). Determinants of target gene specificity for steroid/thyroid hormone receptors. *Cell* 57, 1139–1146.

Vale, W., Hsueh, A. J. W., River, C., and Yu, J. (1990). The inhibin/activin family of hormones and growth factors. *In* "Peptide Growth Factors and Their Receptors" (M. Sporn and A. B. Roberts, eds.), pp. 221–248. Springer-Verlag, Heidelberg

Vlodavsky, I., Brown, K. D., and Gospodarowicz, D. (1978). A comparison of the binding of epidermal growth factor to cultured granulosa and luteal cells. *J. Biol. Chem.* 253, 3744–3750.

Voss, A. K., and Fortune, J. E. (1992). Oxytocin/neurophysin-I messenger ribonucleic acid in bovine granulosa cells increases after the luteinizing hormone (LH) surge and is stimulated by LH *in vitro*. *Endocrinology* 131, 2755–2762.

Wahl, M. I., Nishibe, S., Suh, P.-G., Rhee, S. G., and Carpenter, G. (1989). Epidermal growth factor stimulates tyrosine phosphorylation of phospholipase C-II independently of receptor internalization and extracellular calcium. *Proc. Natl. Acad. Sci. USA* 86, 1568–1572.

Wahlstedt, C., Grundemar, L., Hakanson, R., Hailio, M., Shen, G. H., Zukowskaoroiec, Z., and Reis, D. J. (1990). Neuropeptide-Y receptor subtypes: Y_1 and Y_2. *Ann. N.Y. Acad. Sci.* 611, 7–26.

Wang, C., and Chan, V. (1982). Divergent effects of prolactin on estrogen and progesterone production by granulosa cells of rat Graafian follicles. *Endocrinology* 110, 1085–1093.

Wang, C., Hsueh, A. J. W., and Erickson, G. F. (1980). Prolactin inhibition of estrogen production by cultured rat granulosa cells. *Mol. Cell. Endocrinol.* 20, 135–144.

Wang, H., Nelson, S., Ascoli, M., and Segaloff, D. L. (1992). The 5′-flanking region of the rat luteinizing hormone/chorionic gonadotropin receptor gene confers Leydig cell expression and negative regulation of gene transcription by 3′,5′-cyclic adenosine monophosphate. *Mol. Endocrinol.* 6, 320–326.

Wang, X.-F., Lin, H. Y., Ng-Eaton, E., Downward, J., Lodish, H. F., and Weinberg, R. A. (1991). Expression cloning and characterization of the TGF-β type III receptor. *Cell* 67, 797–805.

Wathes, D. C. (1989). Oxytocin and vasopressin in the gonads. *In* "Oxford Reviews of Reproductive Biology" (S. R. Milligan, ed.), Vol. II, pp. 226–283. Oxford University Press, Oxford.

Wei, L. L., Sheridan, P. L., Krett, N. L., Francis, M. D., Toft, D. O., Edwards, D. P., and Horwitz, K. B. (1987). Immunologic analysis of human breast cancer progesterone receptors. 2. Structure, phosphorylation, and processing. *Biochemistry* 26, 6262–6272.

Wei, L. L., Krett, N. L., Francis, M. D., Gordon, D. F., Wood, W. M., O'Malley, B. W., and Horwitz, K. B. (1988). Multiple human progesterone receptor messenger ribonucleic acids and their autoregulation by progtestin agonists and antagonists in breast cancer cells. *Mol. Endocrinol.* 2, 62–72.

Wei, L. L., Gonzalez-Aller, G., Wood, W. M., Miller, L. A., and Horwitz, K. B. (1990). 5′-heterogeneity in human progesterone receptor transcripts predicts a new amino-terminal truncated "C"-receptor and unique A-receptor messages. *Mol. Endocrinol.* 4, 1833–1840.

Welsh, T. H. Jr., Zhuang, L.-Z., and Hsueh, A. J. W. (1983). Estrogen augmentation of gonadotropin-stimulated progestin biosynthesis in cultured granulosa cells. *Endocrinology* 112, 1916–1924.

White, R., Lees, J. A., Needham, M., Ham, J., and Parker, M. (1987). Structural organization and expression of the mouse estrogen receptor. *Mol. Endocrinol.* 1, 735–744.

Wierenga, R. K., and Hol, W. G. J. (1983). Predicted nucleotide-binding properties of p21 protein and its cancer-associated variant. *Nature* 302, 842–844.

Williams, L. T. (1989). Signal transduction by the platelet-derived growth factor receptor. *Science* 243, 1564–1570.

Wimalasena, J., Moore, P., Wiebe, J. P., Abel, J. Jr., and Chen, T. T. (1985). The porcine LH/hCG receptor. *J. Biol. Chem.* 260, 10689–10697.

Witte, O. (1990). Steel locus defines new multipotent growth factor. *Cell* 63, 5–6.

Woodruff, T. K., and Mayo, K. E. (1990). Regulation of inhibin synthesis in the rat ovary. *Annu. Rev. Physiol.* 52, 807–821.

Woodruff, T. K., Lyon, R. J., Hansen, S. E., Rice, G. C., Mather, J. P. (1990). Inhibin and activin locally regulate ovarian folliculogenesis. *Endocrinology* 127, 3196–3205.

Wong, S. K.-F., Parker, E. M., and Ross, E. M. (1990). Chimeric muscarinic cholinergic: β-adrenergic receptors that activate G_s in response to muscarinic agonists. *J. Biol. Chem.* 265, 6219–6224.

Wrana, J. L., Attisano, L., Carcamo, J., Zentella, A., Doody, J., Laiho, M., Wang, X.-F., and Massague, J. (1992). TGFβ signals through a heteromic protein kinase receptor complex. *Cell* 71, 1003–1014.

Xie, Y.-B., Wang, H., and Segaloff, D. L. (1990). Extracellular domain of lutropin/choriogonadotropin receptor expressed in transfected cells binds choriogonadotropin with high affinity. *J. Biol. Chem.* 265, 21411–21414.

Yamada, Y., Reisine, T., Las, S. F., Ihara, Y., Kubota, A., Kagimoto, S., Seino, M., Seino, Y., Bell, G. I., and Seino, S. (1992). Somatostatin receptors, an expanding gene family: cloning and functional characterization of human SSTR3, a protein coupled to adenylate cyclase. *Mol. Endocrinol.* 6, 2136–2142.

Yarden, Y., Escobedo, J. A., Kuang, W.-J., Yang-Feng, T. L., Daniel, T. O., Tremble, P. M., Chen, E. Y., Ando, M. E., Harkins, R. N., Francke, U., Fried, V. A., Ullrich, A., and Williams, L. T. (1986). Structure of the receptor for platelet-derived growth factor helps define a family of closely related growth factor receptors. *Nature* 323, 226–232.

Yarden, Y., Kuang, W.-J., Yang-Feng, T., Coussens, L., Munemitsu, S., Dull, T. J., Chen, E., Schlessinger, J., Francke, U., and Ullrich, A. (1990). Human proto-oncogene c-*kit:* a new cell surface receptor tyrosine kinase for an unidentified ligand. *EMBO J.* 6, 3341–3351.

Ying, S.-Y. (1988). Inhibins, activins, and follistatins: gonadal proteins modulating the secretion of follicle-stimulating hormone. *Endocr. Rev.* 9, 267–293.

Ying, S.-Y., Becker, A., Ling, N., Ueno, N., and Guillemin, R. (1986). Inhibin and beta type transforming growth factor (TGFβ) have opposite modulating effects on the follicle stimulating hormone (FSH)-induced aromatase activity of cultured rat granulosa cells. *Biochem. Biophys. Res. Commun.* 136, 969–975.

Yokota, Y., Sasai, Y., Tanaka, K., Fujiwara, T., Tsuchida, K., Shigemoto, R., Kakizuka, A., Ohkubo, H., and Nakanishi, S. (1989). Molecular characterization of a functional cDNA for rat substance P receptor. *J. Biol. Chem.* 264, 17649–17652.

Yoshimura, Y., and Bahr, J. M. (1991). Localization of progesterone receptors in pre- and post-ovulatory follicles of the domestic hen. *Endocrinology* 128, 323–330.

Zhiwen, Z., Findlay, J. K., Carson, R. S., Herington, A. C., and Burger, H. G. (1988). Transforming growth factor-β enhances basal and FSH-stimulated inhibin production by rat granulosa cells *in vitro*. *Mol. Cell. Endocrinol.* 58, 161–166.

8

Molecular Mechanisms in Ovulation

ALEX TSAFRIRI & NAVA DEKEL

I. Ovulatory Changes

The preovulatory surge of gonadotropins induces a series of changes in various follicular compartments that culminate in the release of a fertilizable ovum and the transformation of the follicle into a corpus luteum. Ovulation, in its strictest etymologic sense, denotes the release of the ovum. Nevertheless, the rupture of the follicular wall during ovulation is only one of the many follicular responses initiated by the gonadotropin surge. Therefore we prefer the use of the term "ovulation" for the entire sequence of follicular responses to gonadotropins. The more restricted term, "follicular rupture" will be used in conjunction with changes directly associated with the formation of the stigma and the breaching of the follicular wall. Here we describe two of the biological components of the ovulatory response—resumption of oocyte maturation and the rupture of the follicular wall.

II. Regulation of Oocyte Maturation

A. Life History of the Mammalian Germ Cell

The production of a fertilizable ovum is a protracted process spanning the embryonic, postnatal, prepubertal, and sexually mature life of the mammalian female. In Figure 1 the life history of the rat ovum is depicted

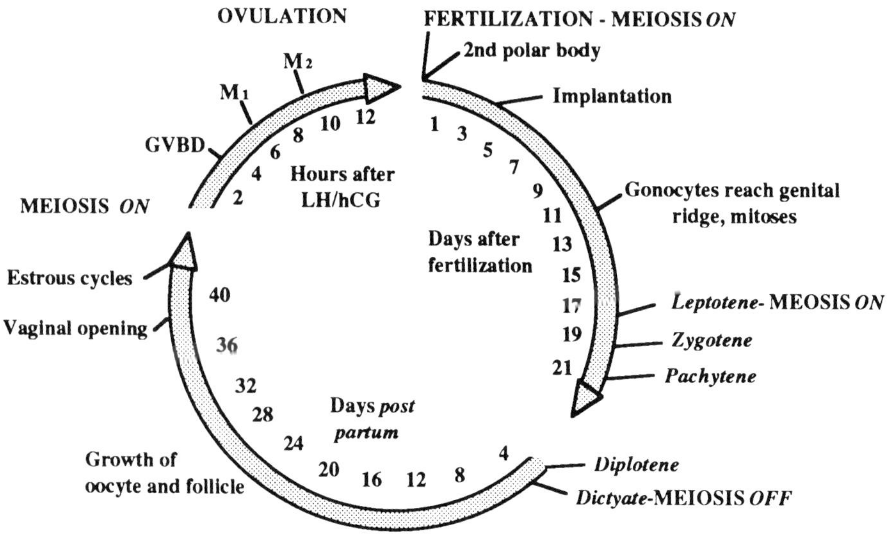

FIGURE 1 Schematic presentation of oogenesis and its multiple stop–go controls in the rat. Reproduced with permission from Lindner *et al.* (1980). Copyright © 1980 Raven Press.

schematically on a background of ontogenetic development. In brief, a short period of mitotic activity by oogonia is followed by their transformation into primary oocytes which embark on meiosis. The meiotic process is subject to multiple stop–go controls. The term "oogenesis" has been used in the literature to describe (1) the transformation of oogonia into oocytes (Franchi *et al.*, 1962; Peters and McNatty, 1980; Zuckerman and Baker, 1977) and (2) the entire process of the formation and maturation of the female germ cell (Baker, 1982). This wider definition is comparable to the term "spermatogenesis" and is at present used more frequently.

A close association with somatic cells of the ovary is established in early stages of germ cell life (Mitchell and Burghardt, 1986). Oocytes which do not become enclosed within granulosa cells, and thus incorporated into primordial follicles, usually degenerate (Ohno and Smith, 1964). Furthermore, the growth of the oocyte and of the follicle are closely coordinated and the ovulatory stimulus induces changes both in the oocyte and in the somatic compartment of the follicle. Therefore, normal development of the ovum, especially the later stages of its growth and maturation, can be properly discussed only within the framework of follicular physiology.

The meiotic process is initiated in most mammalian species during prenatal life or shortly after birth (Figure 1). The oocyte reaches the diplotene stage of prophase just before or immediately after birth. At this stage, by a mechanism yet not fully understood, the meiotic process is arrested (first meiotic arrest). In murid rodents, the chromosomes decondense, so that they disappear from view under light microscope, and resume their transcriptive activity. The oocyte, having a prominent nucleus, referred to as germinal vesicle (GV), enters the dictyate or diffuse diplotene stage, in which it may persist throughout infancy and a variable period beyond the onset of puberty. During fertile life, at each estrous cycle a number of oocytes, characteristic of the species, resume their meiotic process in response to the preovulatory surge of gonadotropins (Ayalon *et al.*, 1972). In many species meiosis is also resumed in oocytes of follicles undergoing atresia (reviewed by Tsafriri and Braw, 1984). Resumption of meiosis (Figure 2) includes chromatin condensation, disintegration of the nuclear

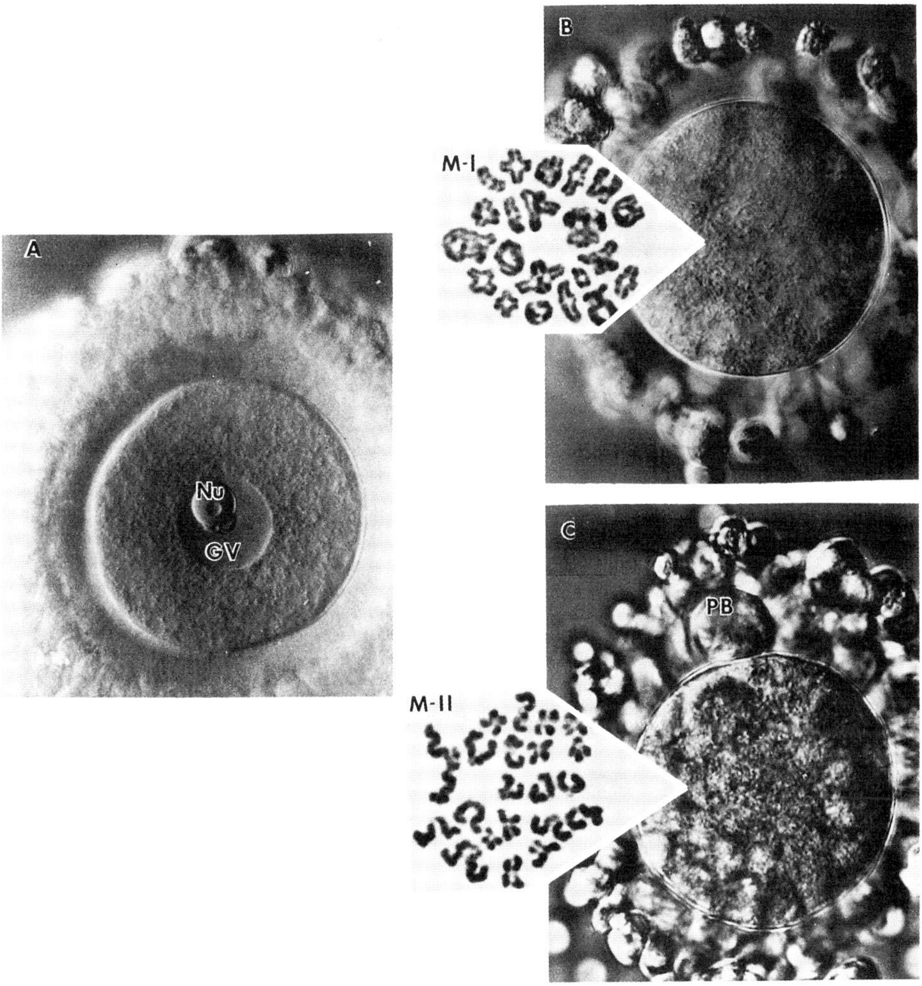

FIGURE 2 Resumption of meiosis in rat oocytes viewed with Nomarski interference contrast optics. (A) Dictyate oocyte. GV, germinal vesicle; Nu, nucleolus. (B) Germinal vesicle breakdown. Inset: bivalents at metaphase I. (C) Secondary oocyte. PB, first polar body. Inset: chromosomes seen at metaphase II.

membrane, usually referred to as germinal vesicle breakdown (GVB), chromosome segregation, and the abstriction of the first polar body shortly before ovulation. At ovulation, a fertilizable secondary oocyte arrested at the metaphase of the second meiotic division (second meiotic arrest) is released. In most mammals, fertilization occurs at the metaphase of the second meiotic division. In these species, oocytes that did not reach this stage usually cannot be penetrated by the spermatozoa or, if penetration does occur, the sperm nucleus fails to be transformed into a sperm pronucleus (Barros and Munoz, 1973; Iwamatsu and Chang, 1972; Niwa and Chang, 1975; Thibault and Gerard, 1973; Usui and Yanagimachi, 1976).

Several reviews on meiotic maturation are available (Dekel, 1986, 1988; Eppig and Downs, 1984; Eppig, 1991; Racowsky, 1991; Schuetz, 1985; Schultz, 1991; Thibault, 1977; Thibault *et al.*, 1987; Tsafriri, 1978; Tsafriri and Pomerantz, 1984, 1986; Tsafriri *et al.*, 1991b). Some of the mechanisms involved in the control of meiotic maturation in mammals are reviewed here.

B. Oocyte Maturation *in Vitro:* Experimental Model Systems

Two dissimilar *in vitro* models have been developed for investigating meiotic maturation in mammals (Figure 3). The simplest model, established by Pincus and Enzmann (1935), demonstrated that rabbit oocytes liberated from their follicles undergo spontaneous maturation in culture even in the absence of hormonal stimulation. Such spontaneous maturation of oocytes, either separated from their enveloping cumulus cells or incubated within their cumulus mass, has been described in all mammalian species examined (reviewed by Biggers, 1972; Tsafriri, 1978; Dekel and Beers, 1978, 1980). Subsequent studies demonstrated that the ability of oocytes to resume meiosis spontaneously, usually referred to as "meiotic competence," is acquired only at later stages of oocyte growth and is related to advanced phases of follicular development (Bar-Ami and Tsafriri, 1981; Erickson and Sorenson, 1974; Szybek, 1972). The development of meiotic competence appears to be regulated by follicle-stimulating hormone (FSH), an action which is at least partially mediated by follicular estrogen production (Bar-Ami *et al.*, 1983).

Isolated oocytes, cultured within or without their cumulus cells, are widely employed for studying certain physiological aspects of oocyte maturation. However, this model does not adequately represent the physiological events since: (1) *in vivo* oocyte maturation is clearly dependent on the preovulatory surge of gonadotropins (reviewed by Lindner *et al.*, 1974; Tsafriri, 1978); (2) with the exception of murids, only a very low percentage of oocytes that mature in this way are capable of normal fertilization and subsequent embryonic development (Mattioli *et al.*, 1988a,b; Staigmiller and Moor, 1984; Thibault, 1977). This model is, therefore, very useful for

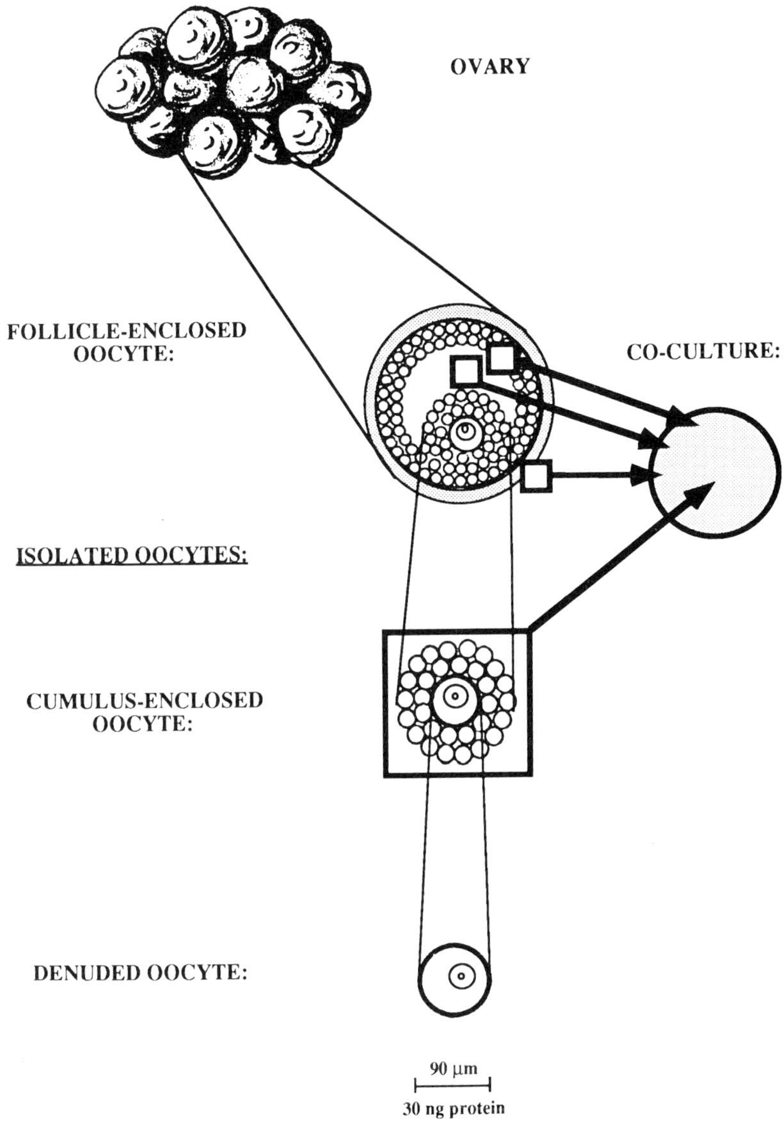

FIGURE 3 Model systems for studying the regulation of meiosis *in vitro*.

analyzing inhibitory effects on oocyte maturation, but is inadequate for studying the induction of the process.

In the second model, which consists of explanted preovulatory follicles, resumption of meiosis depends on hormonal stimulation (Mattioli *et al.*, 1988a,b; Meinecke and Meinecke-Tillman, 1981; Moor and Trounson,

1977; Neal and Baker, 1973; Thibault, 1977; Thibault and Gerard, 1973; Tsafriri *et al.*, 1972a; Dekel *et al.*, 1981). Exposure of explanted follicles to the relevant hormones was found to result in mature fertilizable oocytes. When these were transferred to foster mothers, a high percentage developed into normal young in the rabbit (Thibault *et al.*, 1975a,b), sheep (Moor and Trounson, 1977), and rat (Ben-Yosef *et al.*, 1992). This model permitted the analysis of some of the mechanisms involved in the hormonal induction of meiosis: the role of gonadotropins and other peptide hormones, steroids, prostaglandins, cyclic adenosine monophosphate (cAMP) and other second messengers such as protein kinase C (PKC) activators, intercellular communication, macromolecular synthesis, and follicular metabolism. However, being a multicompartmental model, it does not permit determination of the primary site of gonadotropin action on resumption of meiosis.

A third model, a coculture of isolated oocytes with various follicular components (Foote and Thibault, 1969; Leibfried and First, 1980a; Tsafriri and Channing, 1975a), was adopted to examine the relationship between the somatic compartment of the follicle and the oocyte. This approach may be compromised also, since the oocyte and follicle cells are removed from their natural environment and the normal association between the oocyte and the follicle cells, as well as between the follicle cells themselves is disrupted. Furthermore, this system is still too complex to allow the locality of specific cellular responses to be determined. Nevertheless, the isolated oocytes cultured with other follicular constituents provide an indispensable approach for examining mechanisms involved in maintaining meiotic arrest. Moreover, hormonal stimulation of granulosa cell-oocyte cocultures has been proven as a convenient method for obtaining mature and developmentally competent ovine (Crosby *et al.*, 1981; Staigmiller and Moor, 1984), bovine (Critser *et al.*, 1986; Gordon and Lu, 1990; Hensleigh and Hunter, 1985; Leibfried-Rutledge *et al.*, 1989), and porcine (Mattiolli *et al.*, 1988a,b) ova (reviewed by Racowsky, 1991).

C. Maintenance of Meiotic Arrest

1. Role of local regulators

a. Inhibition of meiosis by granulosa cells

The spontaneous maturation of isolated oocytes led Pincus and Enzmann (1935) to suggest that, within the follicle, oocyte maturation is inhibited by a follicular product. This suggestion was supported by the finding that meiosis was inhibited in porcine oocytes transferred to host follicles (Fleming *et al.*, 1985; Meinecke and Meinecke-Tillman, 1981), or cocultured with segments of follicle wall or with granulosa cells (Foote and Thibault, 1969; Tsafriri and Channing, 1975). In contrast, the theca layer only (Foote and Thibault, 1969; Tsafriri and Channing, 1975) or ovarian bursa (Leibfried and First, 1980a,b) failed to suppress spontaneous maturation. Coculture of porcine oocytes with porcine granulosa cells (Sato and Ishibashi, 1977; Tsafriri and Channing, 1975) or extroverted follicles (Mattioli *et al.*, 1986) inhibited their spontaneous maturation, the degree of inhibition depending upon granulosa cell number. The addition of rat oocytes to rat granulosa cell cultures suppressed the resumption of meiosis. Likewise, the spontaneous maturation of bovine oocytes was suppressed by coculture with granulosa cells (Sirard and Bilodeau, 1990a,b).

Extracts of granulosa cells, as well as medium in which granulosa cells had been previously cultured, inhibited the resumption of meiosis (Sato and Ishibashi, 1977; Tsafriri *et al.*, 1976a). Collectively, these results suggest that within the follicle, meiosis is kept in abeyance by an oocyte maturation inhibitor (OMI) produced by the granulosa cells.

b. Oocyte maturation inhibitor

Follicular fluid (FF) from ovaries of rabbit, sheep, cow, pig, hamster, and human origin was found to exert an inhibitory effect on the spontaneous maturation of isolated oocytes (Chang, 1955; Gwatkin and Andersen, 1976; Hillensjö *et al.*, 1978a; Jagiello *et al.*, 1977; Tsafriri and Channing, 1975; Tsafriri *et al.*, 1977). This inhibitory effect is not species-specific; thus human FF inhibited the maturation of pig (Hillensjö *et al.*, 1978a) and rat (Hillensjö *et al.*, 1981) oocytes; cow FF inhibited hamster oocytes (Gwatkin and Andersen, 1976); and pig FF inhibited the maturation of rat oocytes (Tsafriri *et al.*, 1977). The OMI activity in porcine FF appears to decline in the course of follicular growth (Stone *et al.*, 1978; Van de Wiel *et al.*, 1983).

The inhibitory action of a porcine OMI preparation can be reversed by transferring oocytes to an OMI-free medium (Hillensjö *et al.*, 1979a,b; Stone *et al.*, 1978) or by adding an antibody prepared against the OMI fraction (Tsafriri *et al.*, 1979). Thus, the inhibitory action of OMI on oocyte maturation appears not to be due to a toxic effect.

In vivo or in follicle-enclosed oocytes *in vitro*, resumption of meiosis is induced by LH. Hence the ability of LH to overcome the inhibition of meiosis exerted by FF, by granulosa cells, or by granulosa cell-conditioned medium, lends support to the view that OMI may have a physiological role in the regulation of meiosis (Gwatkin and Andersen, 1976; Mattioli *et al.*, 1988a,b; Tsafriri and Channing, 1975; Tsafriri *et al.*, 1977, 1979).

Oocyte maturation inhibitor from porcine FF inhibited the resumption of meiosis in oocytes cultured within their intact cumuli, but it did not interfere with the maturation of totally denuded pig (Hillensjö *et al.*, 1979b) or rat (Tsafriri and Bar-Ami, 1982) oocytes. It appears, therefore, that OMI exerts its inhibitory action on meiosis, not directly on the oocyte, but through the mediation of the cumulus cells.

Some investigators were unable to demonstrate OMI activity in follicular preparations. Granulosa cells did not inhibit maturation of cow, sow, or ewe oocytes, whereas FF showed such activity (Jagiello *et al.*, 1977). In contrast, Sato and coauthors (Sato and Ishibashi, 1977; Sato and Koide, 1984) found porcine follicular fluid to be inactive, whereas coculture with granulosa cells inhibited resumption of meiosis. Leibfried and First (1980a,b) found bovine and porcine FF as well as granulosa cells to be inactive, but coculture with follicle hemisections prevented meiosis and this was overcome by adding LH. However, in a later study, from the same laboratory, bovine FF inhibited the maturation of bovine oocytes (Sirard and First, 1988). Granulosa cells persistently inhibited meiosis in hamster oocytes, while the inhibitory effect of FF was only transitory (Racowsky and Baldwin, 1989). These discrepant findings may be related, at least in part, to the low OMI activity of follicular fluid and of granulosa cell cultures, the absence of OMI activity in follicular fluid and granulosa cells from large follicles, instability of OMI, and differences in the methods of oocyte collection and culture.

c. Chemical nature of OMI

The activity of OMI has been partially purified and characterized (for review, see Channing *et al.*, 1982; Channing and Segal, 1982). It appears to be a peptide of fewer than 2000 Da (Pomerantz *et al.*, 1979; Stone *et al.*, 1978; Tsafriri and Pomerantz, 1986; Tsafriri *et al.*, 1976a). OMI from bovine FF (Gwatkin and Andersen, 1976) appears to have similar properties. However, definitive isolation and chemical characterization of OMI are still to be accomplished.

Recent studies reported that inhibin (O *et al.*, 1989) or anti-Müllerian hormone (AMH; also called Müllerian-inhibiting factor, MIF) (Takahashi *et al.*, 1986; Ueno *et al.*, 1988) have OMI-like activity. In contrast, in our hands transforming growth factor β (TGFβ), inhibin, activin, and AMH did not inhibit the spontaneous maturation of rat oocytes *in vitro* (Tsafriri *et al.*, 1988, 1991b). It should be noted, however, that the concentrations of inhibin tested were lower than the ED_{50} (40 n*M* or 120 ng/ml) reported later to be effective (O *et al.*, 1989).

A promising clue for the identification of OMI was provided by the observations of Buscaglia and colleagues (1989). They reported that the 35-kDa form of porcine follistatin (FS, an ovarian glycoprotein suppressing pituitary cell FSH secretion), but not the C-terminal truncated 32-kDa form, exerted OMI-like activity on rat and mouse oocytes. Moreover, adding of an antibody against the C-terminal of 35-kDa FS (27-amino-acid peptide) prevented the OMI-like activity of FS and of porcine follicular fluid. In our hands the 35-kDa FS, as well as the 27-amino-acid C-terminal peptide, had no demonstrable OMI activity. Nevertheless, we did confirm the ability of the antiserum against the C-terminal peptide of FS to abolish the inhibitory action of partially purified OMI preparations of porcine follicular fluid on maturation of rat oocytes. While we cannot say whether follicular OMI is derived from the C-terminal of FS, the cross-reaction between the antiserum and the 27-amino-acid fragment of FS and OMI from porcine follicular fluid offers a unique opportunity to purify the elusive OMI (Tsafriri *et al.*, 1991b). Furthermore, in view of the recent demonstration that follistatin has activin- and inhibin-binding activity (Kogawa *et al.*, 1991; Nakamura *et al.*, 1990; Shimonaka *et al.*, 1991), it is possible that the stimulatory activity of high concentrations of activin (Itoh *et al.*, 1990) on meiotic maturation is related to its binding to follistatin.

d. Purine bases and regulation of oocyte maturation

On the basis of studies demonstrating that hypoxanthine exerts OMI-like activity on mouse oocytes, it was suggested that hypoxanthine, and possibly other purines are, therefore, responsible for OMI activity of porcine FF (Downs *et al.*, 1985, 1986; Eppig *et al.*, 1985). However, neither hypoxanthine nor adenosine were detected in bovine FF, despite its OMI activity. Furthermore, bovine oocytes showed only limited sensitivity to purines (Sirard and Bilodeau, 1990b; Sirard and First, 1988).

Purine bases were shown to modulate oocyte levels of cyclic adenosine monophosphate (cAMP), most probably by inhibiting cAMP-phosphodiesterase activity (Downs and Eppig, 1987; Downs *et al.*, 1989). Furthermore, microinjection of an inhibitor of cAMP-dependent protein kinase induced the resumption of meiosis in oocytes cultured in a medium containing hypoxanthine or guanosine (Downs *et al.*, 1989). It seems, therefore, that purines inhibit the resumption of meiosis by maintaining oocyte cAMP concentrations at a level that is inhibitory for meiosis (see Section II.C.2). Such action of purine bases does not exclude the possibility that other follicular factors may work in parallel to suppress the resumption of meiosis. Indeed, Downs *et al.* (1985) observed an additional inhibitory fraction in porcine FF which was not removed by charcoal extraction.

In conclusion, there seems to be little doubt that the somatic cells of the follicle exert control over the resumption of meiosis in mammals, but the detailed mechanism is still subject to further investigation. Whereas in preantral follicles the oocytes are still incompetent to resume meiosis (Bar-Ami and Tsafriri, 1981), it seems that in antral follicles, meiosis is prevented by a local paracrine factor produced by granulosa and cumulus cells. Both the yet-unidentified peptide OMI and hypoxanthine fulfill several of the criteria of physiological regulators of meiosis. Further studies are needed in order to determine their precise role in regulation of oocyte maturation, and these must await full chemical identification of OMI.

2. Role of cAMP in maintenance of meiotic arrest

Almost 40 years after the original observation of Pincus and Enzmann (1935), it was initially demonstrated that the spontaneous maturation in vitro of mouse oocytes released from their follicles can be reversibly blocked by adding a membrane-permeable derivative of cAMP or a phosphodiesterase inhibitor (Cho *et al.*, 1974). These findings were extended to include rat isolated oocytes and the complete dose and time dependency of the inhibitory effect of dibutyryl cAMP and the cyclic nucleotide phosphodiesterase inhibitor, methylisobutylxanthine (MIX) (Dekel and Beers, 1978, 1980). The inhibitory effect of cAMP on the spontaneous maturation of the oocyte was later confirmed in other species (Homa, 1988; Rice and McGaughey, 1981; Sirard and First, 1988), and led to the hypothesis that this cyclic nucleotide could serve as the physiological inhibitor involved in the maintenance of meiotic arrest. The observations that the continuous presence of cAMP derivatives or inhibitors of phosphodiesterase prevent LH-stimulated maturation in follicle-enclosed oocytes *in vitro* (Dekel *et al.*, 1981; Hillensjö *et al.*, 1978b; Lindner *et al.*, 1974) are in full accordance with this hypothesis.

To elucidate the possible regulatory role of intraoocyte concentrations of cAMP, a highly active adenylate cyclase preparation elaborated by the bacteria of the genus *Bordetella* and which can be internalized by mammalian cells (Confer and Eaton, 1982) has been employed. This "invasive" adenylate cyclase triggers the host cell to generate, from its own pool of adenosine triphosphate (ATP), high amounts of intracellular cAMP (Hanski and Farfel, 1985). It has been shown that in isolated cumulus-free rat oocytes subjected to the invasive adenylate cyclase, elevated levels of intraoocyte cAMP were associated with maintenance of meiotic arrest, while a drop in cAMP levels was followed by resumption of meiosis (Aberdam *et al.*, 1987). The physiological relevance of these findings is borne out by determinations of the intracellular levels of cAMP in immature compared with mature oocytes. Such measurements revealed that the postovulatory mature oocytes contain lower levels of cAMP than the follicular immature oocytes (Dekel, 1987). Moreover, they also demonstrated that the spontaneous maturation *in vitro* is preceded by a sharp drop in intraoocyte cAMP, while no decrease in cAMP levels is observed in oocytes maintained in meiotic arrest by a phosphodiesterase inhibitor (Aberdam *et al.*, 1987; Dekel, 1987). A clear correlation between the meiotic status and intraoocyte levels of cAMP was also demonstrated in the mouse (Schultz *et al.*, 1983; Vivarelli *et al.*, 1983), suggesting that cAMP levels within the oocyte negatively regulate its maturation.

Administration of the invasive enzyme enabled the oocyte to generate cAMP. This experiment, however, could not provide any indication of whether oocytes can generate inhibitory concentrations of cAMP using their own adenylate cyclase system. Other studies, however, suggest that rat oocytes do not have the potential to generate cAMP at levels that are sufficient to maintain meiotic arrest. Neither cholera toxin nor forskolin, which interact with the stimulatory, GTP-binding regulatory component and the catalytic subunit, respectively, to stimulate adenylate cyclase, were able to induce a rise in oocyte levels of cAMP (Dekel, 1987) and failed to inhibit the spontaneous maturation of oocytes *in vitro* (Dekel and Beers, 1980; Dekel *et al.*, 1984). On the other hand, mouse oocytes exposed to forskolin, but not cholera toxin, did generate cAMP levels sufficient to inhibit their spontaneous maturation (Sato and Koide, 1984; Schultz *et al.*, 1983; Urner *et al.*, 1983). This effect of forskolin is apparently of no physiological significance since a rapid decrease in cAMP concentrations is observed in both rat and mouse oocytes immediately after their isolation from the follicle, indicating a limited ability of the oocyte to produce cAMP at levels that are sufficient to maintain meiotic arrest (Aberdam *et al.*, 1987; Schultz *et al.*, 1983). This decrease in intraoocyte concentrations of cAMP probably triggers a series of events that result in an irreversible commitment of the oocyte to resume meiosis. Indeed, oocytes cultured for 1 hour in MIX-free medium and then exposed to MIX resumed meiotic maturation and were insensitive to the action of the phosphodiesterase inhibitor (Dekel and Beers, 1980; Schultz *et al.*, 1983).

3. Transfer of cAMP

As mentioned earlier, incubation of isolated oocytes in medium containing a phosphodiesterase inhibitor prevents both spontaneous maturation and a decrease in intraoocyte cAMP levels (Dekel and Beers, 1978, 1980; Schultz *et al.*, 1983; Dekel, 1987). These findings indicate that, regardless of its limited potential to generate cAMP, the follicular oocyte does contain inhibitory levels of this nucleotide. The possibility that the oocyte is affected by cAMP secreted into the follicular fluid should be ruled out since nonderivative cAMP cannot penetrate cell membranes and therefore does not exert an inhibitory action on the maturation of the isolated oocytes (Dekel and Beers, 1978, 1980). Alternatively, cAMP can be transmitted from follicle cells to the oocyte through cell-to-cell communication.

Observations of the intimate connections between the oocyte and the surrounding cumulus cells date back to 1890 (Paladino). Along with many other studies, we have used transmission electronmicroscopy to describe the long cytoplasmic projections of the cumulus cells which traverse the zona pellucida and contact the oolemma (Dekel *et al.*, 1976). However, the abundant number of cumulus cell projections, which intermingle with the microvilli on the oocyte surface, were apparent in cumulus–oocyte complexes examined by scanning electronmicroscopy following proteolytic digestion of the zona pellucida (Dekel *et al.*, 1978). Based on these close anatomical relationships, it was suggested that cumulus cell projections can serve as channels for communication of the follicular cAMP that maintains the oocyte in meiotic arrest (Dekel and Beers, 1978; Dekel *et al.*, 1981). Theoretically, because of its size, cAMP can easily be transferred through the gap junctions that are known to be present in the regions of contact between the cumulus cell projection and the oocyte surface (Szöllösi, 1975;

Amsterdam *et al.*, 1976; Anderson and Albertini, 1976; Gilula *et al.*, 1978). Furthermore, almost threefold higher levels of cAMP were found in oocytes that were derived from forskolin-stimulated cumulus-oocyte complexes compared with similarly stimulated cumulus-free oocytes (Dekel, 1987; Sherizly *et al.*, 1988). Transfer of cAMP from the cumulus to the oocyte was also demonstrated by other investigators (Racowsky, 1984; Bornslaeger *et al.*, 1986; Salustri *et al.*, 1985). These results are in full accordance with earlier findings that activators of adenylate cyclase could delay the spontaneous maturation of oocytes cultured within their cumuli but failed to affect the oocytes in the absence of the attached cumulus cells (Dekel *et al.*, 1984). Taken together, the results of these experiments suggest that cAMP generated by the cumulus cells is transferred to the oocyte to maintain it in meiotic arrest.

In conclusion, the studies reviewed support the communication of an inhibitory signal from the somatic cell compartment of the follicle to the oocyte. The rate of transfer of the signal may be modulated at several transition areas: oocyte–cumulus, cumulus–cumulus, cumulus–mural granulosa, or even granulosa–granulosa cell junctions. At present, there is sufficient evidence supporting the transfer of cAMP to the oocyte (Bornslaeger *et al.*, 1986; Sherizly *et al.*, 1988; Racowsky, 1984; Salustri *et al.*, 1985). Further studies are required to determine whether other regulators of meiosis are also transmitted to the oocyte or, alternatively, if their effect on oocyte maturation is exerted through the follicular somatic cell compartment, possibly by maintenance of basal, inhibitory levels of cAMP.

D. Resumption of Meiosis and Its Hormonal Control

1. Gonadotropins and resumption of oocyte maturation *in vivo*

Oocyte maturation, like other ovulatory processes, is triggered by pituitary gonadotropins. Thus, administration of human chorionic gonadotropin (hCG) induces meiosis in the mouse (Edwards and Gates, 1959), rat (Dekel *et al.*, 1979), and pig (Hunter and Polge, 1966). Blocking the preovulatory surge of gonadotropins in the rat by Nembutal or similar drugs (Ayalon *et al.*, 1972), hypophysectomy (Vermeiden and Zeilmaker, 1974), or administration of an antiserum to the β-subunit of luteinizing hormone (LH) (Tsafriri *et al.*, 1976b) all prevent oocyte maturation. Conversely, administration of exogenous LH to such Nembutal-treated rats results in blocking of the germinal vesicle breakdown (GVB) within 2.5–3.5 hours (Vermeiden and Zeilmaker, 1974; Magnusson *et al.*, 1977). The preovulatory surge of gonadotropins includes follicle-stimulating hormone (FSH) as well as LH. Therefore, it was of interest to test whether the FSH released during the preovulatory surge can trigger ovulation by itself. Indeed, immunopurified (Tsafriri *et al.*, 1976b) or human recombinant FSH (Galway *et al.*, 1990) can induce the ovulatory response. Nevertheless, since administration of an antibody directed toward the β-subunit of LH to proestrous rats prevented the ovulatory response, it appears that, at least in the rat, the amount of FSH normally secreted during the proestrous surge is inadequate to induce ovulation by itself (Tsafriri *et al.*, 1976b). Therefore, the action of FSH on ovulation in the above experiments must be regarded as merely a pharmacological effect.

In naturally ovulating mice, rats, and pigs, oocyte maturation and follicular rupture at ovulation are separated only rarely. Hormonal stimulation of ovulation increases the number of immature ova released (Donahue, 1972). Precocious treatment of pigs with hCG resulted in ovulation of only immature ova (Hunter *et al.*, 1976).

2. *In vitro* maturation of follicle-enclosed oocytes

When preovulatory follicles are explanted from adult cycling pregnant mare serum gonadotropin (PMSG)-treated immature rats on the day of proestrus into hormone-free medium, the behavior of the oocytes depends on the time of follicle isolation. Oocytes explanted before 1400 hours, i.e., before the preovulatory surge of gonadotropins, remain in the dictyate stage throughout a 24-hour culture period. On the other hand, in follicles explanted later on the afternoon of the same day, following the gonadotropin surge, the oocytes resume meiosis in culture (Ayalon *et al.*, 1972; Hillensjö, 1976). Follicles explanted prior to the endogenous surge of gonadotropins allowed to study the hormonal induction of meiosis *in vitro*. The gonadotropins LH, hCG, and FSH all triggered the resumption of meiosis in large Graafian follicles (Hillensjö, 1976; Tsafriri *et al.*, 1972a). On the other hand, in small antral follicles, in which the granulosa cells do not have as yet sufficient LH/hCG receptors, resumption of meiosis can be induced by FSH, but not LH (Dekel *et al.*, 1988a).

Likewise, resumption of meiosis was induced by gonadotropins in cultured follicle-enclosed oocytes of sheep (Hay and Moor, 1973; Miller and Jagiello, 1973; Moor and Trounson, 1977), cows (Thibault *et al.*, 1975a,b), pigs (Gerard *et al.*, 1979; Meinecke and Meinecke-Tillman, 1981; Mattiolli *et al.*, 1988a,b), rabbits (Thibault and Gerard, 1973) and hamsters (Gwatkin and Andersen, 1976).

3. Induction of oocyte maturation: Role of cAMP

Many of the agents inducing resumption of meiosis in follicle-enclosed oocytes *in vitro* also stimulate the production of cAMP (Hillensjö, 1976; Lindner *et al.*, 1974; Tsafriri *et al.*, 1972a). Indeed, injection of the cAMP derivative dibutyryl cAMP (dbcAMP) into the follicular antrum (Tsafriri *et al.*, 1972a), or a transient exposure of follicles to 8-bromo-cAMP (Hillensjö *et al.*, 1978b), dbcAMP, or isobutylmethylxanthine (IBMX) (Dekel *et al.*, 1981), all triggered the resumption of meiosis. Furthermore, forskolin was found to be a potent inducer of maturation in follicle-enclosed oocytes

(Dekel and Sherizly, 1983; Yoshimura *et al.*, 1992bc). Induction of oocyte maturation by forskolin was associated with elevation of cAMP in the follicle and was potentiated by a phosphodiesterase inhibitor (Dekel, 1986), suggesting that LH-induced oocyte maturation is a cAMP-mediated response. This last conclusion seems to present an apparent paradox. If cAMP mediates LH action to induce oocyte maturation, what mechanism allows the oocyte to mature under the influence of LH, since cAMP inhibits oocyte maturation?

Since both inhibition and induction of oocyte maturation are apparently elicited by cAMP, it can be assumed that each of these responses is stimulated by a different dose of the nucleotide. This assumption was tested by comparing the dose–response relationships of cAMP-mediated induction and inhibition of oocyte maturation. Oocyte maturation was induced in follicle-enclosed oocytes transiently exposed to either dibutryl cAMP or MIX. Inhibition of maturation was obtained by adding these agents to either follicle-enclosed oocytes incubated in the presence of LH or isolated cumulus-free oocytes that mature spontaneously *in vitro*. These experiments revealed that the concentrations of either dibutyryl cAMP or MIX needed to induce resumption of meiosis were fourfold higher than those required for inhibition of oocyte maturation (Dekel *et al.*, 1988b) suggesting that cAMP plays a dual role in regulating oocyte maturation. Basal levels of cAMP transferred to the oocyte result in maintenance of meiotic arrest, while elevated, LH-stimulated levels of the nucleotide mediate the induction of oocyte maturation. These contrasting responses are related not only to different thresholds to cAMP but also to different target cells for the nucleotide action. When the oocyte is the target for cAMP, maturation is inhibited. At higher LH-stimulated cAMP concentrations, it is the response of the cumulus cells to the nucleotide which results in breakdown of communication and leads to oocyte maturation (Dekel *et al.*, 1981). This hypothesis is in agreement with the reported absence of LH-hCG receptors on the oocyte and their presence on cumulus cells (Lawrence *et al.*, 1980).

Meiosis can also be induced in follicle-enclosed rat oocytes by factors which do not enhance follicular cAMP production. Thus, the hypothalamic gonadotropin-releasing hormone (GnRH) or its agonistic analogues induced resumption of meiosis both *in vivo* (Ekholm *et al.*, 1981) and *in vitro* (Hillensjö and Le Maire, 1980). Whereas a GnRH antagonist abolished the meiosis-inducing action of GnRH or its analogue, it did not block this effect of LH (Dekel *et al.*, 1983; Ekholm *et al.*, 1982). It appears, therefore, that GnRH or a similar ovarian factor does not mediate the effect of LH on meiosis. GnRH induces follicular prostaglandin synthesis (Clark *et al.*, 1980), and prostaglandin E_2 (PGE_2) was shown to induce resumption of meiosis (Tsafriri *et al.*, 1972a). However, indomethacin, a potent prostaglandin synthetase inhibitor, blocked the effect of GnRH analogue on ovulation but not on oocyte maturation (Ekholm *et al.*, 1982), suggesting that induction of meiosis by GnRH is not mediated by follicular prostaglandin synthesis. The demonstration that activators of protein kinase C, such as phorbol ester, synthetic diacylglycerol, and phospholipase C, also induce meiosis in follicle-enclosed oocytes (Aberdam and Dekel, 1985), suggests that GnRH-induced PKC activation may be involved in oocyte maturation. These findings may be of physiological significance, since Oikawa and coworkers (1990) have recently demonstrated expression of GnRH mRNA in rat ovaries. Nevertheless, the precise role of the mechanisms involved in regulation of oocyte maturation remains to be elucidated. Induction of oocyte maturation by epidermal growth factor (EGF) (Dekel and Sherizly, 1985), transforming growth factor-α (Tsafriri *et al.*, 1989b), and basic fibroblast growth factor (LaPolt *et al.*, 1990) may suggest that ligand-activated tyrosine kinases may participate in resumption of meiosis. Other drugs shown to induce resumption of meiosis in follicle-enclosed oocytes were clomiphene citrate (Laufer *et al.*, 1982) and tetrahydrocannabinol, the major psychoactive ingredient of marihuana, and cannabidiol, one of its nonpsychoactive constituents (Reich *et al.*, 1982).

4. LH-induced resumption of meiosis: Effects on oocyte–somatic cell communication

From the information presented earlier, it appears that elevated levels of cAMP in the oocyte are associated with arrest of meiosis. It was assumed, therefore, that any mechanism involved in LH-induced oocyte maturation should lead to a drop in cAMP levels within the oocyte. It was further predicted that if maintenance of meiotic arrest depends on communication of the inhibitory cAMP, then uncoupling the oocyte from the cumulus cells would terminate the supply of cAMP, providing the appropriate conditions for resumption of meiosis. It was previously demonstrated that coupling in the cumulus–oocyte complex gradually decreases after human chorionic gonadotropin is administered (Gilula *et al.*, 1978; Moor *et al.*, 1980). Subsequently, it has been shown that follicle-enclosed oocytes induced to mature *in vitro* by LH become uncoupled from the cumulus cells (Dekel *et al.*, 1981). A recent time analysis revealed that coupling in rat cumulus–oocyte complexes decreases to 50% of its initial level following 1 hour of incubation with LH, while almost all the oocytes seem still meiotically dormant by this point. Similar chronological relationships between uncoupling and reinitiation of meiosis were found in vivo in cumulus–oocyte complexes isolated from hCG-treated rats (Sherizly *et al.*, 1988). Uncoupling in the cumulus–oocyte complex was also correlated with resumption of meiosis when other inducing agents, such as epidermal growth factor, were employed (Dekel and Sherizly, 1985). A more recent study demonstrates that heptanol, a seven-carbon alcohol that blocks cell-to-cell communication in various experimental systems, including the cumulus–oocyte complex, reduces intraoocyte concentrations of cAMP and promotes maturation of rat follicle-enclosed oocytes (Dekel and Piontkewitz, 1991; Piontkewitz and Dekel, 1993). These findings suggest that a breakdown of communication can be sufficient to signal the induction of oocyte maturation.

A decrease in communication in the cumulus–oocyte complex during early stages of meiosis was also described in the hamster (Racowksy and Satterlie, 1985). However, the results reported by Moor *et al.* (1980) in the sheep and Eppig (1982) in the mouse seem to suggest that reinitiation of maturation precedes any reduction in cumulus cell–oocyte coupling. Motlik and colleagues (1986) have studied changes in intercellular coupling in pig follicles during oocyte maturation. These investigators reported that the contact point of the cumulus–oocyte complex with mural granulosa cells becomes distended before the oocyte is uncoupled from the cumulus cells. Larsen and colleagues (1986) have demonstrated that there is a significant loss in the net area of the cumulus gap junction membrane prior to breakdown of the germinal vesicle in the rat. Taken together, all these observations suggest that LH apparently interferes with the communication of cumulus cells with each other as well as their coupling with both the oocyte and the mural granulosa. Since the granulosa cells probably serve as the major source for cAMP, while the cumulus cells mainly provide the channels for transmitting this inhibitory signal, uncoupling of the oocyte from the cumulus cells could create conditions sufficient for resumption of meiosis. Alternatively, in the species in which decreases in coupling in the cumulus–oocyte complex could not be detected prior to onset of meiosis, downregulation of gap junctions at the interface between the cumulus stalk and mural granulosa cells, or even between neighboring cells within the membrana granulosa itself, could possibly lead to a substantial decrease in transfer of cAMP to the oocyte, resulting in maturation. Evidence from recent studies supports the idea that in some species, breakdown of communication within the membrana granulosa and/or between the granulosa and cumulus cells occurs soon after LH stimulation of ovulation, prior to uncoupling within the cumulus–oocyte complex (Larsen *et al.*, 1986; Motlik *et al.*, 1986; Racowsky *et al.*, 1989).

E. The Meiotic Cell Cycle

1. Maturation-promoting factor as M phase-promoting factor

Regulation of oocyte maturation by the somatic components of the ovarian follicle was discussed in detail in the previous sections of this chapter. This section focuses on the intraoocyte biochemical mechanisms involved in regulating the meiotic division of the female gamete.

Actively dividing eukaryotic cells proceed through an orderly sequence of events, termed "the cell cycle." The two major events in the cycle are replication of the genetic material in the S phase and segregation of the replicated DNA to daughter cells during the M phase. In most cells, two control points, the G1 and G2 phases, separate the S and the M phase. Meiosis is a special type of cell division that is characteristic of germ cells. As mentioned earlier in this chapter, meiosis in oocytes is a protracted process, with the fully grown oocyte arrested at the diplotene stage of the first prophase. This stage corresponds to the G2 phase of the cell cycle. Reinitiation of meiosis represents a transition from the G2 to the M phase and involves dissolution of the nuclear membrane (germinal vesicle breakdown), condensation of the interphase chromatin, and formation of spindles. The oocytes complete the first meiotic division and mature into unfertilized eggs that are arrested in metaphase.

Until quite recently, very little was known about the biochemical control of cell division. However, extensive recent studies have significantly advanced our understanding of this field of biology. This progress was based on the pioneering studies of Masui and Markert (1971) and, independently, of Smith and Ecker (1971) who discovered that maturing frog oocytes produce a factor that causes them to resume meiosis. This factor has been termed "maturation-promoting factor" or MPF. MPF was later found to be ubiquitous. It has been demonstrated in oocytes of the amphibians *Rana pipiens* (Masui and Markert, 1971), *Xenopus laevis* (Schorderet-Slatkine and Drury, 1973), and *Ambystoma mexicanun* (Reynhout and Smith, 1974), and in the starfish *Asterina pectinifera* (Kishimoto and Kanatani, 1976).

MPF activity has also been found in mitotically dividing cells, such as cleaving blastomers of starfish (Kishimoto *et al.*, 1982) and amphibians (Wasserman and Smith, 1978), mammalian cultured cells synchronized at M phase (Kishimoto *et al.*, 1982; Nelkin *et al.*, 1980; Sunkara *et al.*, 1979), and the temperature-sensitive cdc mutants of yeast arrested at M phase (Tachibana *et al.*, 1987; Weintraub *et al.*, 1982). MPF derived from this wide variety of eukaryotic cells exert the same maturation-inducing activity when injected into immature oocytes of starfish and amphibians. Thus, MPF acts in a nonspecies specific manner across the different phyla. Since MPF appears to control the transition from G2 to M phase in both meiosis and mitosis, the acronym MPF was proposed to mean M phase-promoting factor (Gerhart *et al.*, 1984). Murray and Kirschner (1989a) have recently established the central regulatory role of MPF at the onset of the M phase, particularly with respect to critical protein phosphorylation/dephosphorylation events that drive cells into the metaphase state.

The mechanism of activation of MPF involves interaction between the two components of this factor, one of which, a 34-kDa protein that is homologous to the product of the cdc2 gene of the fission yeast (Dunphy *et al.*, 1988; Gautier *et al.*, 1988), is termed $p34^{cdc2}$. This protein, which is highly conserved among eukaryotes, is a serine/threonine kinase. The level of $p34^{cdc2}$ is not altered during the cell cycle. The other component of

MPF belongs to a family of 42- to 45-kDa proteins known as cyclins. The cytoplasmic levels of cyclin fluctuate during the cell cycle; its concentrations rise just before cell division, turning on the MPF kinase activity. The cyclin concentrations then drop sharply and the kinase activity also declines, allowing the cells to complete the division cycle. Resynthesis of the cyclin can then trigger another round of division (Minshull *et al.*, 1989; Murray and Kirschner, 1989b). Binding to cyclin induces both tyrosine and threonine phosphorylation of the previously unphosphorylated $p34^{cdc2}$, rendering it inactivated. The transition into the M phase involves both a reduction in the rate of $p34^{cdc2}$ phosphorylation on tyrosine and an increase in the rate of dephosphorylation (Solomon *et al.*, 1990). Dephosphorylation of tyrosine and threonine residues of $p34^{cdc2}$ is accompanied by reciprocal phosphorylation of the cyclin (Meijer *et al.*, 1991). Destruction of cyclin is required for the exit of the cell from the M phase (Murray and Kirschner, 1989a,b).

2. Mammalian meiosis and MPF

The information presented earlier was obtained mostly from studies conducted on either somatic cells that undergo mitosis or invertebrate oocytes that resume meiotic division. Much less information is available on the regulatory mechanisms of meiosis in mammalian oocytes. Nevertheless, the results reported thus far allow mammalian oocytes to be included among the meiotic and mitotic cell types in which the transition from the G2 to the M phase is regulated by MPF (Figure 4). In this section, the studies on the meiotic cell cycle of mammalian oocytes are reviewed in detail.

The appearance of a cytoplasmic MPF in maturing mammalian oocytes was first suggested by Balakier and Czolowska (1977), who observed that enucleated fragments of immature mouse oocytes, when fused with interphase blastomeres from two-celled mouse embryos, induced premature dissolution of the nuclear membrane and premature chromatin condensation in the blastomeres when their nucleate sister fragments underwent GVB. Balakier (1978) further reported that young incompetent oocytes fused with maturing oocytes, underwent GVB, and progressed to metaphase of first meiotic division. Later, fusion experiments confirmed that, in addition to the mouse, MPF is also present in maturing oocytes of other mammalian species, such as the pig and the rabbit (Fulka, 1983; Fulka *et al.*, 1986).

Mammalian oocyte MPF is fully effective when introduced into oocytes of nonmammalian species. Starfish (Kishimoto *et al.*, 1984b), as well as *Xenopus laevis* oocytes (Sorensen *et al.*, 1985) underwent GVB following their injection with cytoplasm of maturing mouse oocytes. However, the significant difference in size between the *Xenopus* (1,300 μm diameter) and the mouse (72μm) oocyte made it necessary to use a minimum of 23 mouse

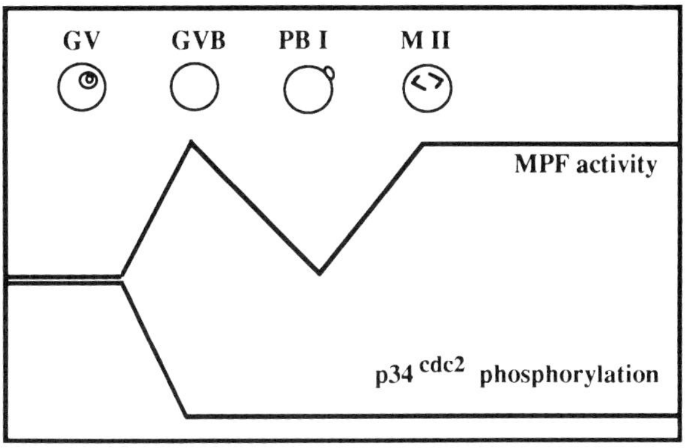

FIGURE 4 Schematic presentation of the changes in MPF activity and p34^{cdc2} tyrosine dephosphorylation during maturation of the mouse oocyte. Changes in MPF activity taken from Hashimoto and Kishimoto (1988) and in p34^{cdc2} tyrosine phosphorylation dephosphorylation from Choi *et al.* (1991). GV, germinal vesicle; GVB, GV breakdown; PBI, first polar body formation; MII, ovulated oocytes.

oocytes to obtain a positive response in the *Xenopus* oocyte (Sorensen *et al.*, 1985). A smaller number of mouse oocytes (at most 7) is needed to obtain MPF activity in one starfish oocyte (Hashimoto and Kishimoto, 1988).

The use of immature starfish oocytes as recipients of cytoplasmic transfer of maturing mouse oocytes was adopted by Hashimoto and Kishimoto (1988) for semiquantitation of changes in MPF activity during the course of meiotic maturation. These investigators demonstrated that MPF activity appears initially at GVB, reaches a high level at metaphase I and then disappears transiently when the first polar body is emitted. Thereafter, MPF activity reappears at metaphase II and remains at an elevated level until fertilization and activation. Similar dynamic changes in MPF activity were demonstrated by electrofusion of pig GV oocytes with homologous oocytes at different stages of maturation (Mattioli *et al.*, 1991). Thus, during oocyte maturation, MPF activity oscillates in exact correspondence with the meiotic cycles in mice and pigs, as well as in starfish (Doree *et al.*, 1983) and amphibians (Gerhart *et al.*, 1984).

Although MPF is highly conserved across the different phyla, its activation and the consequent resumption of meiosis seem to be regulated differently even among mammals. In mouse oocytes treated with inhibitors of protein synthesis, meiotic maturation is blocked before emission of the first polar body but GVB and chromosome condensation proceed (Wasserman and Letourneau, 1976; Golbus and Stern, 1976; Schultz and Wassarmann, 1977). On the other hand, in species with a longer interval between the stimulus and GVB, such as pigs, cows, and sheep, *de novo* protein synthesis is required for GVB (Fulka *et al.*, 1986; Hunter and Moor, 1987;

Moor and Crosby, 1986). These reports are in agreement with later studies demonstrating that in the pig both the first and the second rise in MPF activity require protein synthesis (Mattioli *et al.*, 1991), whereas in maturing mouse oocytes, the formation of MPF requires protein synthesis during the second, but not the first metaphase (Hashimoto and Kishimoto, 1988). However, protein synthesis inhibitors added to the culture medium during the first metaphase of mouse oocytes cause decondensation of the chromosomes and subsequent formation of the nuclear envelope (Clarke and Masui, 1983). These observations suggest that, in the mouse oocyte, newly synthesized proteins may control the activity of MPF soon after breakdown of the nuclear envelope. This conclusion is further supported by later studies demonstrating that, even in the absence of any effect on its initial appearance, MPF activity disappeared in the presence of cycloheximide, reappearing only after the protein synthesis inhibitors were removed (Hashimoto and Kishimoto, 1988). Removal of cycloheximide was followed by spindle formation and rearrangement of the chromosomes in the metaphase plate. These findings suggest that even in the mouse there is, in addition to the second metaphase, a critical period requiring protein synthesis shortly before the first metaphase.

3. Activation of MPF

The finding that protein synthesis inhibitors do not interfere with the initial phase of MPF activity in the mouse points toward a preexisting, latent form of MPF in these oocytes. These observations further suggest that activation of MPF is apparently dependent on some post-translational modifications of the protein. In agreement with this hypothesis are the demonstrations that the puromycin analogue, 6-dimethylaminopurine (6-DMAP), inhibits both GVB and the burst of the GVB-associated protein phosphorylation, without changing the rate of incorporation of [^{35}S]methionine into proteins (Rime *et al.*, 1989). When added after GVB and before emission of the first polar body, 6-DMAP decreased the level of protein phosphorylation and induced the decondensation of the chromosomes and formation of the nuclear envelope.

It is well established that protein tyrosine phosphorylation, which is associated with the binding of mitogenic ligands, plays an important role in the control of cell proliferation (Yarden and Ullrich, 1988). This information, combined with our earlier findings that epidermal growth factor promotes resumption of meiosis in rat oocytes (Dekel and Sherizly, 1985), led us to investigate the involvement of protein tyrosine kinases in regulation of oocyte maturation. We found that tyrphostins, a class of protein tyrosine kinase inhibitors (Yaish *et al.*, 1988), reversibly block maturation of rat oocytes (Dekel and Goren, 1993). These findings are in full agreement with an earlier demonstration that in mouse oocytes committed to resume meiosis the relative amounts of phosphotyrosine increase significantly (Bornslaeger *et al.*, 1988).

As mentioned earlier, the set of post-translational modifications occurring on transition from the G2 to M phase involve protein phosphorylation and dephosphorylation. The regulatory role of protein dephosphorylation in reinitiation of meiosis in mammalian oocytes was initially demonstrated by the use of okadaic acid, a specific inhibitor of protein phosphatase 1 and 2A (Cohen *et al.*, 1990). When microinjected into either *Xenopus* or starfish oocytes, it activates MPF (Goris *et al.*, 1989; Picard *et al.*, 1989), which suggests that protein dephosphorylation catalyzes the conversion of pre-MPF to MPF. Similarly, okadaic acid also induced arrested mouse oocytes to enter the M phase of the cell cycle (Rime and Ozon, 1990; Gavin *et al.*, 1991, 1992). Our recent finding that okadaic acid induces maturation in rat follicle-enclosed oocytes (Goren and Dekel, unpublished) supports the idea that phosphatases participate in reinitiation of meiosis and that MPF activation in mammalian oocytes involves protein dephosphorylation. Further support for this hypothesis is provided by experiments using vanadate, an inhibitor of tyrosine phosphatase (Klarlund, 1985; Tonks *et al.*, 1988). This agent, which prevents tyrosine dephosphorylation of $p34^{cdc2}$ and G2/M transition in mouse 3T3 fibroblasts (Morla *et al.*, 1989), also blocks spontaneous maturation of rat oocytes (Dekel and Goren, 1993).

MPF has been described as a protein kinase whose catalytic subunit, $p34^{cdc2}$, is activated through dephosphorylation of threonine and tyrosine residues (Lohka *et al.*, 1988). Changes in the phosphorylation state of $p34^{cdc2}$ associated with the cell cycle could only be demonstrated after purification of the protein and production of its specific antibodies (Guatier *et al.*, 1988). Using such antibodies, immunoblot analyses revealed that in maturing mouse oocytes the quantity of $p34^{cdc2}$ did not show appreciable changes (Choi *et al.*, 1991). The presence of similar quantities of $p34^{cdc2}$ in oocytes at different stages of their development has also been observed in the rat (Goren and Dekel, unpublished). However, the phosphorylated forms present in mouse oocytes decreased gradually within 2 hours after isolation from follicles, with no further change until metaphase II (Figure 4) (Choi *et al.*, 1991). These results suggest that in the mouse, dephosphorylation of $p34^{cdc2}$ triggers its activation at the first metaphase, but not inactivation and reactivation at the first and second metaphase, respectively. Similar dephosphorylation changes were associated with meiosis of *Xenopus* oocytes (Ferrel *et al.*, 1991).

At least two biochemical events have been established as necessary for $p34^{cdc2}$ kinase activation: the dephosphorylation of $p34^{cdc2}$ and its association with cyclin. Cyclin was first noted in eggs of marine invertebrates because of its pronounced synthesis during the interphase, followed by specific proteolysis at the end of mitosis (Evans *et al.*, 1983). Direct evidence that cyclin plays a role in meiotic regulation came from the finding that microinjections of clam and sea urchin cyclin mRNA induced meiotic maturation in *Xenopus* immature oocytes (Swenson *et al.*, 1986; Pines and Hunt, 1987), there is little information on the role of cyclin in regulation of meiosis in mammals. *In situ* hybridization analyses of adult mouse ovaries have detected the presence of cyclin B1 transcripts located in the ooplasm (Chapman and Wolgemuth, 1992). Although all of the oocytes in the adult mouse ovary are in the diplotene stage of meiosis, they exhibit different stages of growth. Cyclin B1 signal was clearly detected in growing as well as fully grown oocytes, and was totally absent in small oocytes that reside in primary follicles. No further information on a possible oscillatory pattern in the expression of cyclin during resumption of meiosis in either mouse or other mammalian oocytes is available at present.

4. MPF activation and cyclic AMP

As mentioned earlier, it has been demonstrated in several mammalian species that cAMP plays a dominant role in keeping the oocyte at the G2-arrested stage, and that a drop in intraoocyte concentrations of this cyclic nucleotide is associated with the transition from the G2 to the M phase (Aberdam *et al.*, 1987; Dekel, 1987; Schultz *et al.*, 1983). Membrane-permeable derivatives of cAMP, as well as phosphodiesterase inhibitors, are employed, therefore, to experimentally block GVB. Very limited information is available to date on the biochemical nature of cAMP control of the meiotic as well as mitotic cell cycle. Sorenson and colleagues (1985) have reported that extracts from mouse oocytes arrested with dibutyryl cAMP (dbcAMP) failed to induce GVBD after they were injected into *Xenopus* oocytes. These investigators suggested that a high cAMP concentration prevents MPF activation. On the other hand, fusion of metaphase II mouse oocytes rapidly induced GVB in homologous oocytes arrested with dbcAMP (Clarke and Masui, 1985). These latter results suggest that once cytoplasmic MPF has been activated, its ability to induce the transition of nuclei to metaphase is not affected by dbcAMP. Puromycin-treated metaphase I mouse oocytes were employed to explore the interrelationships between cAMP and MPF activation (Clarke and Masui, 1983). As mentioned earlier, treatment of metaphase I mouse oocytes with puromycin results in the formation of interphase nuclei containing decondensed chromosomes. The interphase–metaphase transition, which usually takes place in these oocytes when puromycin is removed, is prevented in the presence of dbcAMP (Clarke and Masui, 1985). These authors suggest that it is not the appearance of MPF, but rather some post-translational modification of this protein that is regulated by cAMP in both prophase (GV), as well as interphase-arrested oocytes.

Using okadaic acid, several studies have proposed that cAMP could possibly interfere with phosphorylation modification of MPF. Okadaic acid induces resumption of meiosis in mouse oocytes arrested in the first meiotic prophase by either dbcAMP or a phosphodiesterase inhibitor (Rime and Ozon, 1990; Gavin *et al.*, 1991, 1992). These studies suggest that the cAMP-regulated step in the G2 to M phase transition in the meiotic cell cycle is the protein dephosphorylation associated with MPF activation. Inhibition of dephosphorylation of $p34^{cdc2}$ by isobutylmethylxanthine, which has been demonstrated recently by immunoblot analysis of mouse oocytes, strongly supports this hypothesis (Choi *et al.*, 1991).

III. Follicular Rupture

Several reviews have been published on various aspects of follicular rupture during ovulation (Espey, 1974, 1978b, 1980, 1992; Espey and Lipner, 1993; Goetz *et al.*, 1991; LeMaire, 1989; LeMaire *et al.*, 1987; Lipner, 1973, 1988; Parr, 1975; Rondell, 1970, 1974; Thibault and Levasseur, 1988; Tsafriri and Reich, 1991; Tsafriri *et al.*, 1991a). The major developments which contributed to the recent advance in this field were the establishment of the *in vitro*-perfused ovary model (Ahrén *et al.*, 1975; Janson *et al.*, 1978, 1982; Koos *et al.*, 1984; LeMaire *et al.*, 1982; Löfman *et al.*, 1982; Wallach *et al.*, 1978; Yoshimura and Wallach, 1987); better understanding of the proteolytic processes associated with tissue remodeling and their molecular control in general; and, finally, the application of advanced biochemical and physiological approaches to the study of follicular rupture. Conceptually, the recognition of the attributes common to acute inflammation and follicle rupture (Parr, 1974, 1975; Espey, 1980) provided most helpful working hypotheses which were testable experimentally. It is obvious that follicular rupture, as well as other components of the ovulatory response, is the end result of many finely orchestrated biochemical changes which are regulated by multiple pathways and modulated by even larger numbers of factors and processes. Some of these may be obligatory and others may play merely an auxiliary role under varying physiological, pathophysiological, or experimental conditions. The relative role of each of these in ovulation remains to be elucidated. Here some of the structural, physiological, and molecular aspects of follicular rupture are overviewed. Our current understanding of the ovarian proteolytic cascade and the modulatory role of steroids and eicosanoids in ovulation is emphasized.

A. Structural Changes in Follicular Wall Associated with Ovulation

The mature preovulatory follicle is embedded within the connective tissue matrix of the ovary and only a small part of it protrudes from the ovarian surface. At its apex, the following layers are distinguished: (1) a single layer of surface epithelium; (2) collagenous connective tissue consisting of the tunica albuginea and theca externa, which contributes most of the tensile strength of the follicle wall (Espey, 1967a); (3) theca interna, a well-vascularized layer containing differentiated fibrocytes very active in steroidogenesis; and (4) granulosa cells which are separated from the theca interna by the basal lamina and enclose the follicular fluid-filled cavity.

The most important structural changes leading to ovulation are those of the connective tissue of the tunica albuginea and theca externa. As ovulation approaches, there is dissolution of the extracellular matrix and dissociation of thecal collagen (Espey, 1967a,b). These changes in the connective tissue are accompanied by increased permeability of the blood vessels, resulting in leakage of blood cells and edema of follicular tissue (Abisogun *et al.*, 1988; Bjersing and Cajander, 1974; Parr, 1974).

B. Biophysical–Mechanical Changes

An increase in intrafollicular pressure was for long assumed to be the cause of rupture and extrusion of the oocyte–cumulus complex. However, direct measurements of intrafollicular pressure did not demonstrate any significant increase in rat, rabbit, and sow follicles approaching ovulation (Blandau and Rumery, 1963; Bronson *et al.*, 1979; Espey and Lipner, 1963). Nevertheless, Janson (1975) demonstrated an increase in follicular volume that was related to the LH-induced increase in follicular blood flow and the decrease in vascular resistance, leading to hyperemia. Therefore, follicular rupture occurs at constant pressure and is preceded by a marked increase in follicle volume. This results in an increase in the distensibility of the follicular wall and a decrease in its tensile strength as ovulation approaches (Espey, 1967a; Rondell, 1964).

The demonstration of perifollicular muscle-like cells containing myoid proteins in the rat, mouse, rabbit, monkey, and human (Amsterdam *et al.*, 1977; Bjersing and Cajander, 1974; Fumagalli *et al.*, 1971; Okamura *et al.*, 1972; O'Shea, 1970; Osvaldo-Decima, 1970) led to the suggestion that ovarian contractions may be involved in follicular rupture. Ovarian contractions were demonstrated in several species, and these were enhanced by α-adrenergic (Rocerto *et al.*, 1969) and serotonergic stimulation, and by various peptides such as enkephalins, substance P, and vasoactive intestinal peptide (VIP) (Lipner, 1988). Furthermore, inhibitors of smooth muscle contraction prevented ovulation in the hamster (Martin and Talbot, 1981). However, the involvement of ovarian contractions in follicular rupture was questioned by others (Burden and Lawrence, 1980; Didio *et al.*, 1980; Espey, 1978a). It remains to be determined whether ovarian contractility has an obligatory role in follicular rupture and the expulsion of the oocyte–cumulus complex.

C. Vascular and Related Changes

Periovulatory changes in ovarian and follicular microcirculation have been described (Janson, 1975; Lee and Novy, 1978; Murdoch *et al.*, 1983).

These studies revealed an initial increase in blood flow, which in the sheep was followed by a decrease (Murdoch *et al.*, 1983). Eicosanoids are probably involved in ovulation by modulating follicular microcirculation. Indeed, in the rabbit, indomethacin reduced the LH-induced hyperemic response (Lee and Novy, 1978) and in the ewe it prevented the second-phase reduction in follicular blood flow (Murdoch *et al.*, 1983). In the rat, inhibitors of cyclo-oxygenase and of lipoxygenase prevented the hCG-induced increase in vascular permeability, as evidenced by uptake of labeled iodinated bovine serum albumin (Abisogun *et al.*, 1988).

Platelet-activating factor (PAF) (1-*O*-alkyl-2-acetyl-*sn*-glycero-3-phosphocholine) is a potent bioactive phospholipid. It elicits a wide spectrum of biological responses by platelets and leukocytes and induces marked changes in vascular permeability, blood pressure, and heart and lung function. Unilateral injection of a specific PAF antagonist BN52021, isolated from the *Ginkgo biloba* (Braquet *et al.*, 1985), or its analogue, BN52111, into the ovarian bursa resulted in a dose-dependent inhibition of follicle rupture in the treated ovary. The inhibition was prevented when the inhibitors and PAF were administered into the ovarian bursa simultaneously with hCG. On the other hand, administration of PAF to rats that were not stimulated by exogenous hCG and had their endogenous surge of LH blocked by Nembutal failed to induce ovulation by itself (Abisogun *et al.*, 1989; Daphna-Iken *et al.*, 1989). These results were confirmed in studies employing additional PAF antagonists: in the rat using Y24180 (Li *et al.*, 1991) and in the mouse CV3988 (Kikukawa *et al.*, 1991). PAF was detected in follicular fluid collected from women undergoing *in vitro* fertilization–embryo transfer (Amiel *et al.*, 1991) and the secretion of PAF from ovine follicles *in vitro* was stimulated after the LH surge (B. M. Alexander *et al.*, 1990). Conversely, a decrease in rat ovarian PAF was reported after hCG was used to stimulate ovulation (Espey *et al.*, 1989c). While these results support the participation of PAF in follicular rupture, the precise mechanisms involved must still be determined.

Additional vasoactive substances were implicated in ovulation. During the menstrual cycle, prorenin (a large molecular weight precursor of the active enzyme renin) rises following the LH surge (Sealey *et al.*, 1985, 1987) or hCG stimulation (Itskovitz *et al.*, 1987). Likewise, renin-like activity was located in follicular fluid (Fernandez *et al.*, 1985) and it is, therefore, possible that renin substrate, angiotensinogen, is converted into angiotensin II (Ang II) in the ovary. Ang II has profound hypertensive effects. The ovarian renin–angiotensin system has been implicated in ovulation by the use of an antagonist of Ang II receptor, saralasin, which blocked ovulation in approximately 50% of the hCG-treated rats (Pellicer, 1988). Later studies with saralasin *in vivo* produced variable results (Daud, 1989; Naftolin *et al.*, 1989), but in perfused rat (Peterson *et al.*, 1993) and rabbit (Kuo *et al.*, 1991; Yoshimura *et al.*, 1992a) ovaries, saralasin blocked ovulation and this was reversed by Ang II. Furthermore, angiotensin II by itself induced ovulation in perfused rabbit ovaries (Yoshimura *et al.*, 1992a).

Kinins, which cause local vasodilation and increases in vascular permeability, have likewise been implicated in ovulation. An increase in ovarian kinin-generating activity and in bradykinin has been demonstrated (Espey *et al.*, 1986, 1989b; Gao *et al.*, 1992). Bradykinin induced ovulation in perfused rabbit ovaries (Yoshimura *et al.*, 1988) and potentiated the action of LH in perfused rat ovaries (Brännström and Hellberg, 1989); this effect was prevented by a competitive antagonist of bradykinin (Hellberg *et al.*, 1991a).

The involvement of histamine in ovulation has been suggested on the basis of its ability to mimic the hyperemic action of LH (Wurtman, 1964) and the ability of antihistamines to block the hyperemic action of LH (Piacsek and Huth, 1971). Nevertheless, the role of histamine in ovulation remains obscure in view of the conflicting data obtained in various experimental models (see detailed reviews by Goetz *et al.*, 1991; Lipner, 1988).

Finally, the role of oxygen free radicals in ovulation was demonstrated by the use of the scavenging enzymes, superoxide dismutase (SOD) and catalase. SOD and the combination of SOD and catalase reduced the number of ovulating follicles and prolonged the time interval between the gonadotropic stimulation of ovulation and follicle rupture from rabbit perfused ovaries (Miyazaki *et al.*, 1991). Similarly, SOD reduced the number of ovulated ova in immature rats *in vivo* (Tsafriri *et al.*, unpublished observations). These results suggest a role for oxygen free radicals in follicle rupture.

Emigration of neutrophils and eosinophils from the vascular compartment in preovulatory ovine follicles, followed by extravasation of blood with numerous platelets adhering to damaged vascular endothelium, were observed as ovulation approached (Cavender and Murdoch, 1988). Mast cells, macrophages, and granulocytes were reported to accumulate in rat ovaries during the preovulatory period (Gaytan *et al.*, 1991; Brännstrom *et al.*, 1992). These morphological changes allow the blood cells to participate in the ovulatory response in a way that is analogous to acute inflammation. The ovarian accumulation of neutrophils prior to ovulation was confirmed by measuring ovarian levels of myeloperoxidase, a neutrophil-specific enzyme (Chun *et al.*,1993). Furthermore, supplementing the medium with leukocytes augmented the LH-stimulated ovulation in rat-perfused ovaries (Hellberg *et al.*, 1991b), thus supporting the suggested role of leukocytes in follicular rupture. Nevertheless, acute depletion of peripheral leukocytes with cyclophosphamide or neutrophils with vinblastine did not affect follicular rupture in PMSG-hCG treated immature rats (Chun *et al.*, 1993). Likewise, ovarian eosinopenia induced by prednisolone did not affect ovulation rate in the ewe (Murdoch and Steadman, 1991). The data available seem to indicate that blood-derived cells may facilitate follicular rupture, but it remains to be determined whether this action is obligatory.

D. Proteolysis and Follicular Rupture

Microscopic examination of the wall of ovulatory follicles revealed dissolution of the connective tissue matrix and of collagen fibers in the tunica albuginea and theca externa. These structural changes were correlated with an increase in distensibility of the follicular wall. It was suggested by Schochet (1916) that proteases may play a role in ovulation. Indeed, many proteolytic enzymes have been identified in follicular tissue or fluid (see Lipner, 1988; Goetz, 1991). Espey and Lipner (1965) showed that injection of proteolytic enzymes into the follicular antrum resulted in follicular rupture and application of enzyme solutions to follicular wall resulted in ovulatory changes (Espey, 1967a; Espey and Rondell, 1967; Rondell, 1970). Recent studies on follicle rupture focused on two proteolytic enzyme systems—the plasmin-activating system and interstitial collagenase. Inhibition of ovarian steroid and eicosanoid synthesis has been shown to block follicular rupture. The stimulation of ovarian collagen degradation seems to be regulated by ovarian eicosanoids. A simplified scheme depicting some of the factors implicated in regulation of the ovarian proteolysis required for follicle rupture is presented in Figure 5.

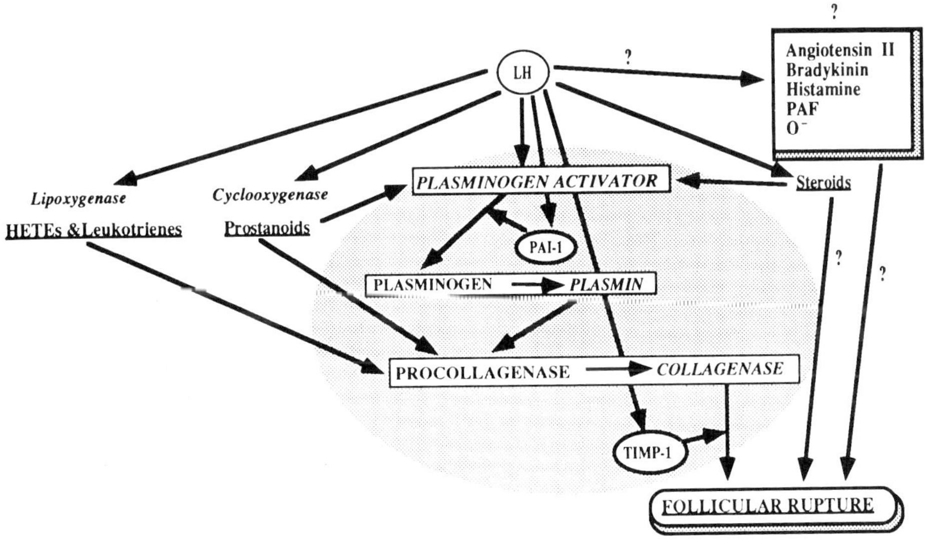

FIGURE 5 A simplified scheme of some of the factors involved in follicle rupture during ovulation. The role of a proteolytic cascade, including local regulators of proteinases, as well as eicosanoids is emphasized. Active enzymes are in italics. Further details in the text.

1. Plasminogen activator

The demonstration of a preovulatory increase in plasminogen activator (PA) (Beers, 1975; Beers and Strickland, 1978) and the ability of plasmin to decrease the tensile strength of Graafian follicle wall (Beers, 1975) led to the suggestion that PA may be involved in the ovulatory process. Beers and Strickland (1978) suggested that plasmin, the product of PA action on plasminogen, activates latent collagenase and thereby initiates the proteolytic processes that culminate in follicular rupture. *In vitro* studies revealed that granulosa cells contribute 80–90% of total follicular PA activity (Reich *et al.*, 1986). Most of granulosa cell PA is secreted into the culture media, whereas intact follicles or theca tissue retain the activity within the tissue (Reich *et al.*, 1985b; Reich *et al.*, 1986). Similarly, in cultured granulosa cells, most of the secreted PA is associated with extracellular matrix (Knecht, 1988). These findings suggest that the theca serves as a specific barrier which prevents the secretion of PA into the extrafollicular compartment and thereby allows localized action of PA within the follicular wall.

Two different molecular types of PA have been identified in mammalian tissues—the urokinase type (UPA) and the tissue type (TPA). These enzymes are products of different genes (Ny *et al.*, 1984; Riccio *et al.*, 1985; Suenson *et al.*, 1984). Both molecular types of PA were identified in rat, mouse, and human follicles on the basis of electrophoretic mobility and immunoabsorption. In the rat, both PAs were present in granulosa and theca compartments, and the granulosa cells contributed 80–90% of the total follicular activity. Reich and colleagues (1986) observed a significant increase in TPA in both granulosa and theca cells after stimulation with gonadotropin. A similar increase in human granulosa cells was observed by Reinthaller *et al.*, (1990). Nevertheless, in mice, urokinase rather than TPA seems to be the gonadotropin-stimulated type of PA (Canipari *et al.*, 1987). Treatment with inhibitors of serine proteases prevented ovulation in rats *in vivo* (Akazawa *et al.*, 1983a,b; Reich *et al.*, 1985b), in hamster ovaries *in vitro* (Ichikawa *et al.*, 1983), and in perfused rat ovaries (Woessner *et al.*, 1989). Furthermore, α_2-antiplasmin and antibodies raised against TPA suppressed ovulation rate in the rat (Tsafriri *et al.*, 1989a). A delayed administration of these inhibitors or antibodies (> 4 hours after hCG) was ineffective in blocking ovulation (Reich *et al.*, 1985b; Tsafriri *et al.*, 1989a). Collectively, these data support the obligatory role of PA and plasmin at an early step of follicular rupture, most probably in the activation of collagenase.

Ovarian steroids were found to synergize with LH in stimulating follicular PA secretion. Furthermore, the addition of inhibitors of steroidogenesis inhibited the LH-stimulated PA activity. This inhibition could be restored by progesterone, testosterone, and estradiol-17β, but not the nonaromatizable 5α-dihydrotestosterone (Reich *et al.*, 1985b). Glucocorticoids were found to stimulate granulosa cell TPA secretion and mRNA levels (Jia *et al.*, 1990; Wang and Leung, 1989).

PA activity is regulated by specific local inhibitors (Hekman and Loskutoff, 1988), with plasminogen activator inhibitor type-1 (PAI-1) being the primary physiological inhibitor of both TPA and UPA (Vassali *et al.*, 1991). Ny *et al.* (1985) demonstrated that granulosa cells secrete PAI-1 and that FSH suppressed this activity. Increased levels of PAI-1 activity were found in human preovulatory follicular fluid (Reinthaller *et al.*, 1990), in ovarian extracellular fluid of hCG-treated immature rats primed with PMSG (Liu *et al.*, 1991), and in preovulatory follicles of prepubertal gilts treated with eCG and hCG (Politis *et al.*, 1990). Liu *et al.* (1991), using Northern blot analysis, demonstrated that the expression of PAI-1 mRNA is regulated by gonadotropins differentially in granulosa and theca-interstitial cells during the preovulatory period. The expression of PAI-1 mRNA in theca-interstitial cells is maximal at 6 hours after LH-hCG stimulation, and in granulosa cells it reaches maximum only 24 hours after the stimulation (Liu *et al.*, 1991). With *in situ* hybridization, only negligible expression of PAI-1 mRNA limited to theca cells was observed before hCG stimulated ovulation. Following hCG administration, increased expression of PAI-1 mRNA was observed in theca cells of most antral follicles, whereas granulosa cell expression was limited to preovulatory follicles and only to areas where the basal membrane was dissociated (Chun *et al.*, 1992). Glucocorticoids were shown to suppress plasminogen-dependent fibrinogen activity in conditioned medium of a rat hepatoma cell line and this was later demonstrated to be due to an increase in PAI-1 (Gelehrter *et al.*, 1987; Heaton and Gelehrter, 1990). Curiously, in rat granulosa cells, glucocorticoids did not affect PAI-1 activity, but did stimulate PA (Wang and Leung, 1989). This effect of glucocorticoids on granulosa cell TPA activity probably underlies the observed increase in ovulation rate in the rat after dexamethasone treatment (de Greef and van der Schoot, 1987).

2. Collagenase

In view of the abundance of collagen fibers in ovarian tunica albuginea covering the apex of ovulatory follicles, and the extensive changes in follicular collagen as revealed by microscopic observation, it was assumed that collagenases play a major role in follicular rupture. Nevertheless, early experiments to extract follicular collagenase and to demonstrate its ability to degrade the follicular wall were not successful (Espey and Stacy, 1970), nor has a correlation between collagenase activity and ovulatory changes been demonstrated (Fukumoto *et al.*, 1981; Morales *et al.*, 1978, 1983). This failure should probably be attributed to the following characteristics of mammalian collagenase: (1) it is secreted as an inactive zymogen; (2) it is firmly bound to its natural substrate; and (3) its activity is tightly regulated by local inhibitors and activators (Alexander and Werb, 1989; Werb, 1989). To circumvent these difficulties, ovarian collagen was labeled by local administration of [^{3}H]proline, which in collagen is partly hydroxylated. Ovarian collagen metabolism and/or catabolism was followed by assessing [^{3}H]hydroxyproline levels in acid hydrolysates of the tissue. With this approach, a significant degradation of labeled ovarian and follicular collagen was detected following the gonadotropin surge (Reich *et al.*, 1985c). The use of special extraction procedures (Reich *et al.*, 1987) and rigorous *in vitro* activation (Curry *et al.*, 1985), made it possible to confirm an LH-hCG-induced rise in ovarian collagenase activity.

More recently, using cDNA probes for interstitial and type IV collagenases, a highly significant increase in the expression of interstitial collagenase mRNA at 3–6 hours following hCG administration to PMSG-treated immature rats was observed. The mRNA of collagenase IV, which is responsible for the degradation of basement membrane collagen IV, was stimulated later, at 9 hours following the treatment (Reich *et al.*, 1991). The role of collagenase in follicular rupture was also corroborated using inhibitors of collagenase, such as cysteine (Reich *et al.*, 1985c), the microbial metalloprotease inhibitor—talopeptine (Ichikawa *et al.*, 1983), and specific inhibitors of interstitial collagenases (Brännström *et al.*, 1988).

Metalloproteinases, the enzymes that include the collagenases, are locally regulated by specific serum and tissue inhibitors. A tissue inhibitor of metalloproteinases (TIMP) was recently identified in ovarian tissues (Curry *et al.*, 1989; Woessner *et al.*, 1989) and shown to be induced by gonadotropins (Curry *et al.*, 1989; Mann *et al.*, 1991; Reich *et al.*, 1991). *In situ* hybridization revealed low constitutive expression of TIMP-1 mRNA in theca cells, which became markedly stimulated within 9 hours after hCG stimulation of ovulation in antral and preovulatory follicles. TIMP-1 mRNA expression in granulosa cells was increased by hCG only in preovulatory follicles (Chun *et al.*, 1992).

LH/hCG seem to stimulate ovarian coexpression of proteolytic enzymes as TPA and collagenase, and their respective inhibitors, PAI-1 and TIMP-1. The preovulatory stimulation of thecal PAI-1 and TIMP-1 even in small antral follicles seems to protect them against the proteolytic activity diffusing from ovulating follicles and thus allows their development in subsequent cycles to the ovulatory stage.

E. Eicosanoids

An increase in follicular prostaglandin synthesis after gonadotropic stimulation was demonstrated in several studies (Ainsworth *et al.*, 1975, 1984; Armstrong and Zamecnik, 1975; Bauminger and Lindner, 1975; LeMaire *et al.*, 1973). This increase was associated in the rat with a dose- and time-dependent induction of cyclo-oxygenase enzyme (prostaglandin synthase-PGS) in the preovulatory follicles (Brännström *et al.*, 1989; Hedin *et al.*, 1987; Huslig *et al.*, 1987). Two variants of PGS were detected in the rat ovary, of which only the 72-kDa PGS was stimulated by LH in preovulatory granulosa cells in culture. The other isoform of 69 kDa was constitutively expressed, primarily in theca cells of antral and preovulatory follicles (Wong and Richards, 1991; Sirois and Richards, 1991). It was shown that the stimulation of PGE synthesis by LH is not dependent on steroid synthesis (Bauminger *et al.*, 1975), and that most follicular responses to LH, such as ovum maturation, activation of adenylate cyclase, steroidogenesis, and luteinization proceed undisturbed even when prostaglandin synthesis is inhibited by drugs (reviewed by Zor and Lamprecht, 1977). The exception was follicular rupture, which was prevented by administration of indomethacin (Armstrong and Grinwich, 1972; Orczyk and Behrman, 1972; Tsafriri *et al.*, 1972b; Wallach *et al.*, 1975a), or intrafollicular injection of $PGF_2\alpha$ antiserum in rabbits (Armstrong *et al.*, 1974). The inhibitory action of indomethacin was overcome by administration of PGE_2 to rats (Tsafriri *et al.*, 1972b), $PGF_2\alpha$ to rhesus monkeys (Wallach *et al.*, 1975b). Both PGE_2 and $PGF_2\alpha$ restored ovulation in the ewe (Murdoch *et al.*, 1986) and the perfused rat ovary (Sogn *et al.*, 1987). Furthermore, in the perfused rabbit ovary, $PGF_2\alpha$ was able to induce follicular rupture by itself, albeit somewhat earlier than LH (Holmes *et al.*, 1983). Collectively, these results suggest that prostaglandins have a physiological role in follicular rupture.

An additional group of arachidonic acid products was implicated in follicle rupture—the lipoxygenase-derived metabolites. Inhibitors of the lipoxygenase pathway were shown to block ovulation in a dose-dependent manner (Reich *et al.*, 1983). In a later study, it was shown that gonadotropins stimulated incorporation of radiolabeled substrates into lipoxygenase products and lipoxygenase activity was located within the preovulatory follicle (Reich *et al.*, 1985a, 1987).

An increase in the levels of several lipoxygenase products was observed by radioimmunoassay following hCG stimulation of ovulation in the rat (Espey *et al.*, 1989a; Tanaka *et al.*, 1989). These increases included an early rise in leukotriene $(LT)B_4$ and the peptidic LTs (LTC_4, D_4, and E_4, measured with an antibody recognizing them as a group), the products of 5-lipoxygenase, and a later rise in 15-hydroxyeicosatetraenoic acid (15-HETE). While these studies support the involvement of lipoxygenase products in ovulation, pharmacological inhibitors of lipoxygenases facilitated rather than suppressed LH-stimulated ovulation in perfused rabbit ovaries (Hellberg *et al.*, 1990). Likewise, the rise in sheep follicular LTB_4 occurred only after follicle rupture, and inhibitors of lipoxygenases did not affect ovulation in the sheep (Carvalho *et al.*, 1989). It remains to be determined whether these seemingly inconsistent results obtained with inhibitors of lipoxygenases are caused by species differences or the methods employed.

More recent studies suggest that eicosanoids mediate the LH-hCG-induced rise in ovarian collagenolytic activity (Reich *et al.*, 1985b; Tsafriri *et al.*, 1991a). Likewise, indomethacin doses which blocked ovulation in the sheep also suppressed degradation of follicular collagen (Murdoch and McCormick, 1991). Furthermore, a specific inhibitor of 5-lipoxygenase, MK 886, suppressed ovulation as well as collagenase activity, as revealed by zymography (Tsafriri *et al.*, 1992). Inhibitors of eicosanoid synthesis also suppressed the hCG-induced rise in expression of ovarian interstitial collagenase mRNA as revealed by Northern blot analysis, thus suggesting the involvement of eicosanoids in the regulation of collagenase synthesis. The reduced levels of collagenase mRNA may be due to reduced transcription or reduced stability of the message. However, since positive regulation of metalloproteinase gene expression seems to occur mainly at the level of transcription (Matrisian, 1990), it is likely that the increase in ovarian interstitial collagenase expression is similarly regulated. In contrast, the expression of collagenase IV or of the metalloproteinase inhibitor TIMP mRNA was not altered by the inhibitors of eicosanoid synthesis (Reich *et al.*, 1991).

Evidence has been obtained showing that human follicles or follicle cells can produce eicosanoids, which are products of both cyclo-oxygenase (Feldman *et al.*, 1986; Liedtke and Seifert, 1978; Patwardhan and Lanthier, 1981; Plunkett *et al.*, 1975) and lipoxygenase (Feldman *et al.*, 1986) pathways. Likewise, an increase in follicular fluid prostanoid levels was observed following hCG (Darling *et al.*, 1982; Lumsden *et al.*, 1986). In hCG-stimulated patients, in addition to prostanoids, LTB_4 and LTC_4, D_4 and E_4 were demonstrated (Priddy *et al.*, 1989). Finally, administration of the cyclo-oxygenase inhibitors, azapropazone and indomethacin, to healthy, cycling women markedly suppressed ovulation in response to hCG (Killick and Elstein, 1987). Thus, it seems that eicosanoids also play an essential role in follicle rupture in the human.

F. Steroids

The issue of whether ovarian steroidogenesis is essential for follicle rupture has been discussed at length over the past few years (reviewed by Espey, 1989; Tsafriri *et al.*, 1987). Early studies demonstrated that progesterone induces distensibility of sow follicles while neither FSH nor estrogen could do so. This effect of LH on the follicle wall could be mimicked by cAMP and prevented by an inhibitor of 3β-hydroxysteroid dehydrogenase, and the inhibition could be reversed by exogenous progesterone (Rondell, 1970). Administration of aminoglutethimide and cyanoketone, drugs that block the synthetic pathway to progesterone, inhibits follicle rupture (Lipner and Greep, 1971; Lipner and Wendelken, 1971). Hypophysectomy of rats on the afternoon of the day of proestrus blocked follicular rupture, and this was effectively reversed by three consecutive injections of progesterone (5–10 mg each), suggesting an essential role of steroids in ovulation. This conclusion was challenged by Bullock and Kappauf (1972), who could dissociate the effect of these inhibitors on plasma progesterone levels and on follicle rupture; thus they could induce ovulation by hCG when the rise in plasma progesterone was attenuated by aminoglutethimide, and cyanoketone prevented ovulation even when steroid plasma levels remained unaffected. Nevertheless, it is possible that plasma levels of progesterone are not a precise indicator of follicular levels of steroids, which are probably involved in follicular rupture through paracrine mechanisms. Recently, the ability of another inhibitor of 3β-hydroxysteroid dehydrogenase, epostane, to inhibit follicular rupture in the rat was demonstrated, and this inhibition was overcome by administration of progesterone (Snyder *et al.*, 1984). Ovulation in the ewe was blocked by isoxazol, an inhibitor of progesterone synthesis (Murdoch *et al.*, 1986). Furthermore, administration of the progesterone receptor antagonist RU 38486 partially blocked ovulation in mature and immature rats in which ovulation was triggered by hCG (Tsafriri *et al.*, 1987). This supports the paracrine role of progesterone, which is exerted through its interaction with cellular receptors, in the mediation of LH/hCG induction of follicular rupture.

The participation of ovarian steroids in ovulation was closely examined in an *in vitro* model of the perfused ovary (Janson *et al.*, 1982; Lambertsen *et al.*, 1976). In perfused rabbit ovaries, inhibition of the cholesterol chain cleavage enzyme by aminoglutethimide or of 3β-hydroxysteroid dehydrogenase did not affect follicle rupture induced by gonadotropin (Holmes *et al.*, 1985; Yoshimura *et al.*, 1986). In contrast, an inhibitor 3β-hydroxysteroid dehydrogenase significantly reduced ovulation in perfused rat ovaries stimulated by a combination of LH and methylisobutylxanthine (MIX) (Brännström and Janson, 1989). The preovulatory decline in follicular estrogen is not required for follicular rupture in the rabbit, since addition of estrogen (10 μg/ml) to the perfusate did not inhibit ovulation, despite a more than twofold increase in follicular estradiol-17β content (LeMaire *et al.*, 1982). Likewise, blockade of the early rise in estrogen by an inhibitor of aromatase did not affect follicular rupture in perfused rat ovaries (Koos *et al.*, 1984; Morioka *et al.*, 1988). While inhibition of steroidogenesis did not affect the ovulatory response of perfused ovaries in the rabbit, addition of progesterone into the perfusate modified it. Progesterone somewhat enhanced follicular rupture in one study (Hamada *et al.*, 1979) and it reduced the number of ovulating ovaries in another when a higher dose was used (Holmes *et al.*, 1985). Thus, a modulatory role for progesterone in follicular rupture *in vitro* cannot be excluded. Collectively, the data presented seem to support the involvement of steroids, most probably progesterone, in follicular rupture. Nevertheless, their specific role remains obscure and species differences may exist.

Acknowledgments

The meticulous secretarial help of Mrs. M. Kopelowitz is gratefully acknowledged. The studies of the authors were supported by grants from the German-Israel Foundation for Scientific Research and Development (GIF), Jerusalem, Israel; the Israel Science Foundation of The Israel Academy of Sciences and Humanities; and the United States–Israel Binational Science Foundation.

References

Aberdam, E., and Dekel, N. (1985). Activators of protein kinase C stimulate meiotic maturation of rat oocytes. *Biochem. Biophys. Res. Commun.* 132, 570–574.

Aberdam, E., Hanski, E., and Dekel, N. (1987). Maintenance of meiotic arrest in isolated rat oocytes by the invasive adenylate cyclase of *Bordetella pertussis*. *Biol. Reprod.* 36, 530–535.

Abisogun, A. O., Daphna-Iken, D., Reich, R., Kranzfelder, D., and Tsafriri, A. (1988). Modulatory role of eicosanoids in vascular changes during the preovulatory period in the rat. *Biol. Reprod.* 38, 756–762.

Abisogun, A. O., Braquet, P., and Tsafriri, A. (1989). The involvement of platelet activating factor in ovulation. *Science* 243, 381–383.

Ahrén, I., Janson, P. O., and Selstam, G. (1975). A technique for perfusion of rabbit ovaries *in vitro*. *In* "Methods in Enzymology" (J. G. Hardman and B. W. O'Malley, eds.), Vol. 39, pp. 1–9. Academic Press, San Diego, CA.

Ainsworth, L., Baker, R. D., and Armstrong, D. T. (1975). Preovulatory changes in follicular fluid prostaglandin F levels in swine. *Prostaglandins* 9, 915–925.

Ainsworth, L., Tsang, B. H., Marcus, G. J., and Downey, B. R. (1984). Prostaglandin production by dispersed granulosa and theca interna cells from porcine preovulatory follicles. *Biol. Reprod.* 31, 115–121.

Akazawa, K., Matsuo, O., Kosugi, T., Mihara, H., and Mori, N. (1983a). The role of plasminogen activator in ovulation. *Acta Physiol. Latino America* 33, 105–110.

Akazawa, K., Mori, N., Kosogi, T., Matsuo, O., and Mihara, H. (1983b). Localization of fibrinolytic activity in ovulation of the rat follicle as determined by the fibrin slide method. *Jpn. J. Physiol.* 33, 1011–1018.

Alexander, B. M., Van Kirk, E. A., and Murdoch, W. J. (1990). Secretion of platelet-activating factor by periovulatory ovine follicles. *Life Sci.* 47, 865–868.

Alexander, C. M., and Werb, Z. (1989). Proteinases and extracellular matrix remodeling. *Curr. Opin. Cell Biol.* 1, 974–982.

Amiel, M. L., Testart, J., and Benveniste, J. (1991). Platelet-activating factor-acether is a component of human follicular fluid. *Fertil. Steril.* 56, 62–65.

Amsterdam, A., Josephs, R., Lieberman, M. E., and Lindner, H. R. (1976). Organization of intermembrane particles in freeze-cleaved gap junctions of rat Graafian follicles: optical diffraction analysis. *J. Cell Sci.* 21, 93–105.

Amsterdam, A., Lindner, H. R., and Groschel-Stewart, V. (1977). Localization of actin and myosin in the rat oocyte and follicular wall by immunofluorescence. *Anat. Rec.* 187, 311–328.

Anderson, E., and Albertini, P. F. (1976). Gap junctions between the oocyte and the companion follicle cells in the mammalian ovary. *J. Cell Biol.* 71, 680–686.

Armstrong, D. T., and Grinwich, D. L. (1972). Blockade of spontaneous and LH-induced ovulation in rats by indomethacin, an inhibitor of prostaglandin biosynthesis. *Prostaglandins* 1, 21–28.

Armstrong, D. T., and Zamecnik, J. (1975). Preovulatory elevation of rat ovarian prostaglandin F and its blockade by indomethacin. *Mol. Cell. Endocrinol.* 2, 125–131.

Armstrong, D. T., Grinwich, D. L., Moon, Y. S., and Zamecnik, J. (1974). Inhibition of ovulation in rabbits by intrafollicular injection of indomethacin and $PGF_2\alpha$ antiserum. *Life Sci.* 14, 129–140.

Ayalon, D., Tsafriri, A., Lindner, H. R., Cordova, T., and Harrel, A. (1972). Serum gonadotropin levels in pro-estrous rats in relation to the resumption of meiosis by the oocytes. *J. Reprod. Fertil.* 31, 51–58.

Baker, T. G. (1982). Oogenesis and ovulation. *In* "Reproduction in Mammals. 1. Germ Cells and Fertilization" (C. R. Austin and R. V. Short, eds.), pp. 17–45, Cambridge University Press. Cambridge.

Balakier, H. (1978). Induction of maturation in small oocytes from sexually immature mice by fusion with meiotic and mitotic cells. *Exp. Cell Res.* 112, 137–141.

Balakier, H., and Czolowska, N. (1977). Cytoplasmic control of nuclear maturation in mouse oocytes. *Exp. Cell Res.* 110, 466–469.

Bar-Ami, S., and Tsafriri, A. (1981). Acquisition of meiotic competence in the rat: role of gonadotropin and estrogen. *Gamete Res.* 4, 463–472.

Bar-Ami, S., Nimrod, A., Brodie, A.M.H., and Tsafriri, A. (1983). Role of FSH and oestradiol 17-β in the development of meiotic competence in rat oocytes. *J. Steroid Biochem.* 19, 965–971.

Barros, C., and Munoz, G. (1973). Sperm–egg interaction in immature hamster oocytes. *J. Exp. Zool.* 186, 73–78.

Bauminger, S., and Lindner, H. R. (1975). Periovulatory changes in ovarian prostaglandin formation and their hormonal control in the rat. *Prostaglandins* 9, 737–751.

Bauminger, S., Lieberman, M. E., and Lindner, H. R. (1975). Steroid-independent effect in gonadotropins on prostaglandin synthesis in rat Graafian follicles *in vitro*. *Prostaglandins* 9, 753–764.

Beers, W. H. (1975). Follicular plasminogen and plasminogen activator and the effect of plasmin on ovarian follicular wall. *Cell* 6, 379–386.

Beers, W. H., and Strickland, S. (1978). A cell culture assay for follicle-stimulating hormone. *J. Biol. Chem.* 253, 3877–3881.

Beers, W. H., Strickland, S., and Reich, E. (1975). Ovarian plasminogen activator: relationship to ovulation and hormonal regulation. *Cell* 6, 387–394.

Ben-Yosef, D., Galiani, D., Dekel, N., and Shalgi, R. (1992). Rat oocytes induced to mature by epidermal growth factor are successfully fertilized. *Mol. Cell. Endocrinol.* 88, 135–141.

Biggers, J. D. (1972). Metabolism of the oocyte. *In* "Oogenesis" (J. D. Biggers and A. W. Schuetz, eds.), pp. 241–251, University Park Press, Baltimore, Md.

Bjersing, L., and Cajander, S. (1974). Ovulation and the mechanism of follicular rupture. VI. Ultrastructure of theca interna and the inner vascular network surrounding rabbit Graafian follicles prior to induced ovulation. *Cell Tissue Res.* 153, 31–44.

Blandau, R. J., and Rumery, R. E. (1963). Measurements of intrafollicular pressure in ovulatory and preovulatory follicles of rat. *Fertil. Steril.* 14, 330-341.

Bornslaeger, E. A., and Schultz, R. M. (1985). Regulation of mouse oocyte maturation: effect of elevating cumulus cell cAMP on oocyte cAMP levels. *Biol. Reprod.* 33, 698–704.

Bornslaeger, E. A., Mattei, P. M., and Schultz, R. M. (1986). Involvement of cAMP-dependent protein kinase and protein phosphorylation in the regulation of mouse oocyte maturation. *Dev. Biol.* 114, 453–462.

Bornslaeger, E. A., Mattei, P. M., and Schultz, R. M. (1988). Protein phosphorylation in meiotically competent and incompetent mouse oocytes. *Mol. Reprod. Dev.* 1, 19–25.

Brännström, M., and Hellberg, P. (1989). Bradykinin potentiates LH-induced follicular rupture in the rat ovary perfused *in vitro*. *Hum. Reprod.* 4, 475–481.

Brännström, M., and Janson, P. O. (1989). Progesterone is a mediator in the ovulatory process in the *in vitro* perfused rat ovary. *Biol. Reprod.* 40, 1170–1177.

Brännström, M., Woessner, J. F., Koos, R. D., Sear, C. H. J., and LeMaire, W. J. (1988). Inhibitors of mammalian tissue collagenase and metaloproteinases suppress ovulation in the perfused rat ovary. *Endocrinology* 122, 1715–1721.

Brännström, M., Larson, L., Basta, B., and Hedin, L. (1989). Regulation of prostaglandin endoperoxide synthase by cyclic adenosine 3′,5′-monophosphate in the *in vitro*-perfused rat ovary. *Biol. Reprod.* 41, 513–522.

Brännström, M., Mayrhofer, G., and Robertson, S. (1992). Distribution of leukocyte subsets in the rat ovary during ovulation. *Biol. Reprod.* 46 (Suppl 1), 91. [Abstract 162]

Braquet, P., Spinnewyn, B., Braquet, M., Bourgain, R. H., Taylor, J. E., Etienne, A., and Drien, K. (1985). BN 52021 and related compounds: a new series of highly specific PAF-acether receptor antagonists isolated from *Ginkgo biloba*. *Blood Vessels* 16, 559–572.

Bronson, R. A., Bryant, G., Balk, M., and Emanuele, N. (1979). Intrafollicular pressure within preovulatory follicles of the pig. *Fertil. Steril.* 31, 205–210.

Bullock, W., and Kappauf, B. H. (1972). Dissociation of gonadotropin-induced ovulation and steroidogenesis in immature rats. *Endocrinology* 982, 1625–1628.

Burden, H. W., and Lawrence, I. E. (1980). Nerve supply of the ovary. *In* "Biology of the Ovary" (P. M. Motta and E. S. E. Hafez, eds.), pp. 99–105. Martinus Nijhoff, The Hague.

Buscaglia, M., Fuller, F., Mazzola, T., Bacigalupi, K., Shimasaki, S., Ui, M., Simmons, D., Castillo, F., Schroeder, R., and Ling, N. (1989). A new intra-ovarian function for follistain: the inhibition of oocyte meiosis. *Proc. Endocrine Soc. USA* 71, [Abstract No. 883]

Canipari, R., O'Connell, M. L., Meyer, G., and Strickland, S. (1987). Mouse ovarian granulosa cells produce urokinase-type plasminogen activator, whereas the corresponding rat cells produce tissue type plasminogen activator. *J. Cell. Biol.* 205, 977–981.

Carvalho, C. B., Yeik, B. S., and Murdoch, W. J. (1989). Significance of follicular cyclooxygenase and lipoxygenase pathways of metabolism of arachidonate in sheep. *Prostaglandins* 37, 533–558.

Cavender, J. L., and Murdoch, W. J. (1988). Morphological studies of the microcirculatory system of periovulatory ovine follicles. *Biol. Reprod.* 39, 989–997.

Chang, M. C. (1955). The maturation of rabbit oocytes in culture and their maturation, activation, fertilization and subsequent development in Fallopian tube. *J. Exp. Zool.* 128, 378–399.

Channing, C. P., and Segal, S. (1982). Intraovarian control mechanisms. *Adv. Exp. Med. Biol.* 147, 392.

Channing, C. P., Anderson, L. D., Hoover, D. J., Kolena, J., Osteen, K. G., Pomerantz, S. H., and Tanabe, K. (1982). The role of non-steroidal regulators in control of oocyte and follicular maturation. *Recent Prog. Horm. Res.* 38, 331–408.

Chapman, D. L., and Wolgemuth, D. J. (1992). Identification of a mouse B-type cyclin which exhibits developmentally regulated expression in the germ line. *Mol. Reprod. Dev.* 33, 259–269.

Cho, W. K., Stern, S., and Biggers, J. D. (1974). Inhibitory effect of dibutyryl cAMP on mouse oocyte maturation *in vitro*. *J. Exp. Zool.* 187, 383–386.

Choi, T., Aoki, F., Mori, M., Yamashita, M., Nagahama, Y., and Kohmoto, K. (1991). Activation of $p34^{cdc2}$ protein kinase activity in meiotic and mitotic cell cycles in mouse oocytes and embryos. *Development* 113, 789–795.

Chun, S.-Y., Popliker, M., Reich, R., and Tsafriri, A. (1992). Localization of preovulatory expression of plasminogen activator inhibitor type-1 and tissue inhibitor of metalloproteinase type-1 in the rat ovary. *Biol. Reprod.* 47, 245–253.

Chun, S. Y., Daphna-Iken, D., Calman, D., and Tsafriri, A. (1993). Severe leukocyte depletion does not affect follicular rupture in the rat. *Biol. Reprod.* 48, 905–909.

Clark, M. R., Thibier, C., Marsh, J. M., and LeMaire, W. (1980). Stimulation of prostaglandin accumulation by luteinizing hormone-releasing hormone (LHRH) and LHRH analogs in rat granulosa cells *in vitro*. *Endocrinology* 107, 17–23.

Clarke, H. J., and Masui, Y. (1983). The induction of reversible and irreversible chromosome decondensation by protein synthesis inhibition during meiotic maturation of mouse oocytes. *Dev. Biol.* 97, 291–301.

Clarke, H. J., and Masui, Y. (1985). Inhibition by dibutyryl cyclic AMP of the transition to metaphase of mouse oocyte nuclei and its reversal by cell fusion to metaphase oocytes. *Dev. Biol.* 108, 32–37.

Cohen, F., Holmes, C.F.B., and Tsukitani, Y. (1990). Okadaic acid: a new probe for the study of cellular regulation. *TIBS* 15, 98–102.

Confer, D. L., and Eaton, J. W. (1982). Phagocyte impotence caused by an invasive bacterial adenylate cyclase. *Science* 217, 948–950.

Critser, E. S., Leibfried-Rutledge, M. L., Eyestone, W. H., Northey, D. L., and First, N. L. (1986). Acquisition of developmental competence during maturation *in vitro*. *Theriogenology* 25, 150.

Crosby, I. M., Osborn, J. C., and Moor, R. M. (1981). Follicle cell regulation of protein synthesis and developmental competence in sheep oocytes. *J. Reprod. Fertil.* 62, 575–582.

Curry, T. E., Dean, D. D., Woessner, J. F., and LeMaire, W. J. (1985). The extraction of a tissue collagenase associated with ovulation in the rat. *Biol. Reprod.* 33, 981–991.

Curry, T. E. J., Dean, D. D., Sanders, S. L., Pedigo, N. G., and Jones, P. B. C. (1989). The role of ovarian proteases and their inhibitors in ovulation. *Steroids* 54, 501–521.

Daphna-Iken, D., Chun, S. Y., Abisogun, A. O., and Tsafriri, A. (1989). Platelet-activating factor, ginkgolides and follicle rupture during ovulation. *In* "Ginkgolides–Chemistry, Biology, Pharmacology and Clinical Perspectives" (P. Braquet, ed.), Vol. 2, pp. 835–844. J. R. Prous, Barcelona.

Darling, M.R.N., Joyce, M., and Elder, M. G. (1982). Prostaglandin $F_2\alpha$ levels in the human ovarian follicle. *Prostaglandins* 23, 551–556.

Daud, A. I., Bumpus, F. M., and Husain, A. (1989). Angiotensin II: Does it have a direct obligate role in ovulation? *Science* 245, 870–871.

de Greef, W. J., and van der Schoot, P. (1987). Effects of dexamethosome on ovarian activity in the rat. *Acta Endocrinol.* 116, 350–356.

Dekel, N. (1986). Hormonal control of ovulation. *In* "Biochemical Actions of Hormones" (G. Litwack, ed.), Vol. XIII, pp. 57–90, Academic Press, Orlando, FL.

Dekel, N. (1987). Interaction between the oocyte and the granulosa cells in the preovulatory follicle. *In* "Endocrinology and Physiology of Reproduction" (D. T. Armstrong, H. G. Freisen, P.C.K. Leung, W. Moger, and K. B. Ruf, eds.), pp. 1997–209. Plenum, New York.

Dekel, N. (1988). Spatial relationship of follicular cells in the control of meiosis. *In* "Meiotic Inhibition: Molecular Control of Meiosis" (F. P. Haseltine, ed.), pp. 87–101. A. R. Liss, New York.

Dekel, N., and Beers, W. H. (1978). Rat oocyte maturation *in vitro:* relief of cyclic AMP inhibition by gonadotropins. *Proc. Natl. Acad. Sci. USA* 75, 3469–3473.

Dekel, N., and Beers, W. H. (1980). Development of the rat oocyte in vitro: inhibition and induction of maturation in the presence or absence of the cumulus oophorus. *Dev. Biol.* 75, 247–254.

Dekel, N., and Goren, S. (1993). Protein tyrosine phosphorylation/dephosphorylation in control of meiosis of rat oocytes. *In* "Gondal Development and Function" (S. Hillier, ed.), Vol. 94, pp. 99–106. Serono Symposia Reviews, Romen Press, New York.

Dekel, N., and Piontkewitz, Y. (1991). Induction of maturation of rat oocyte by interruption of communication in the cumulus oocyte complex. *Bull. Anat.* 75, 51–53.

Dekel, N., and Sherizly, I. (1983). Induction of maturation in rat follicle-enclosed oocyte by forskolin. *FEBS Lett.* 151, 153–155.

Dekel, N., and Sherizly, I. (1985). Epidermal growth factor induces maturation of rat follicle-enclosed oocytes. *Endocrinology* 116, 406–409.

Dekel, N., Dabush, S., and Messer, Y. (1976). Effect of LH on the ultrastructure of rat cumulus cells. *In* "Proc. 6th Europ. Congr. Electron Microscopy," pp. 596–598.

Dekel, N., Kraicer, P. F., Phillips, D. M., Ramon, S., and Segal, S. J. (1978). Cellular association in the rat oocyte-cumulus complex: morphology and ovulatory changes. *Gamete Res.* 1, 47–57.

Dekel, N., Hillensjö, T., and Kraicer, P. F. (1979). Maturational effects of gonadotropins on the cumulus-oocyte complex of the rat. *Biol. Reprod.* 20, 191–197.

Dekel, N., Lawrence, T. S., Gilula, N. B., and Beers, W. H. (1981). Modulation of cell-to-cell communication in the cumulus-oocyte complex and the regulation of oocyte maturation by LH. *Dev. Biol.* 86, 356–362.

Dekel, N., Sherizly, I., Tsafriri, A., and Naor, Z. (1983). A comparative study of the mechanism of action of luteinizing hormone and a gonadotropin releasing hormone analog on the ovary. *Biol. Reprod.* 28, 161–166.

Dekel, N., Aberdam, E., and Sherizly, I. (1984). Spontaneous maturation *in vitro* of cumulus-enclosed rat oocytes is inhibited by forskolin. *Biol. Reprod.* 31, 244–250.

Dekel, N., Galiani, D., and Beers, W. H. (1988a). Induction of maturation in follicle-enclosed oocytes: Specific response to gonadotropins at different stages of follicular development. *Biol. Reprod.* 38, 517–521.

Dekel, N., Galiani, D., and Sherizly, I. (1988b). Dissociation between the inhibitory and the stimulatory action of cAMP on maturation of rat oocytes. *Mol. Cell. Endocrinol.* 56, 115–121.

Didio, L. J. A., Allen, D. J., Correr, S., and Motta, P. M. (1980). Smooth muscle in the ovary. *In* "Biology of the Ovary" (P. M. Motta and E. S. E. Hafez, eds.), pp. 107–118. Martinus Nijhoff, The Hague.

Donahue, R. P. (1972). The relation of oocyte maturation to ovulation in mammals. *In* "Oogenesis" (J. D. Biggers and A. W. Schuetz, eds.), pp. 413–438. University Park Press, Baltimore, MD.

Doree, M., Peaucellier, G., and Picard, A. (1983). Activity of the maturation-promoting factor and the extent of protein phosphorylation oscillate simultaneously during meiotic maturation of star-fish oocytes. *Dev. Biol.* 99, 489–501.

Downs, S. M., and Eppig, J. J. (1987). Induction of mouse oocyte maturation *in vivo* by perturbants of purine metabolism. *Biol. Reprod.* 36, 431–437.

Downs, S. M., Coleman, D. L., Ward-Bailey, P. F., and Eppig, J. J. (1985). Hypoxanthine is the principal inhibitor of murine oocyte maturation in a low molecular weight fraction of porcine follicular fluid. *Proc. Natl. Acad. Sci. USA* 82, 454–458.

Downs, S. M., Coleman, D. L., and Eppig, J. J. (1986). Maintenance of murine oocyte meiotic arrest: uptake and metabolism of hypoxanthine and adenosine by cumulus cell-enclosed and denuded oocytes. *Dev. Biol.* 117, 174–183.

Downs, S. M., Daniel, S.A.J., Bornslaeger, E. A., Hoppe, P. C., and Eppig, J. J. (1989). Maintenance of meiotic arrest in mouse oocytes by purines. *Gam. Res.* 23, 323–334.

Dunphy, W. G., Brizuela, L., Beach, D., and Newport, J. (1988). The *Xenopus* cdc2 protein is a component of MPF, a cytoplasmic regulator of mitosis. *Cell* 54, 423–431.

Edwards, R. G., and Gates, A. H. (1959). Timing of the stages of the maturation, division, ovulation, fertilization and the first cleavage of eggs of adult mice treated with gonadotrophins. *J. Endocrinol.* 18, 292–304.

Ekholm, C., Hillensjö, T., and Isaksson, O. (1981). Gonadotropin releasing hormone agonists stimulate oocyte meiosis and ovulation in hypophysectomized rats. *Endocrinology* 108, 2022–2024.

Ekholm, C., Clark, M. R., Magnusson, C., Isaksson, O., and LeMaire, W. J. (1982). Ovulation induction by a GnRH analog in hypophysectomized rats involves prostaglandins. *Endocrinology* 110, 288–290.

Eppig, J. J. (1982). The relationship between cumulus cell-oocyte coupling, oocyte meiotic maturation, and cumulus expansion. *Dev. Biol.* 89, 268–272.

Eppig, J. J. (1991). Mammalian oocyte development *in vivo* and *in vitro*. *In* "Elements of Mammalian Fertilization" (P. M. Wassarman, ed.), Vol. I, pp. 57–76. CRC Press, Boca Raton, FL.

Eppig, J. J., and Downs, S. M. (1984). Chemical signals that regulate mammalian oocyte maturation. *Biol. Reprod.* 30, 1–12.

Eppig, J. J., Ward-Bailey, P. F., and Coleman, D. L. (1985). Hypoxanthine and adenosine in murine ovarian follicular fluid: concentrations and activity in maintaining oocyte meiotic arrest. *Biol. Reprod.* 33, 1041–1049.

Erickson, G. F., and Sorensen, R. A. (1974). In vitro maturation of mouse oocytes isolated from ovaries of immature hypophysectomized rats. *J. Exp. Zool.* 190, 123–127.

Espey, L. L. (1967a). Tenacity of porcine Graafian follicle as it approaches ovulation. *Am. J. Physiol.* 212, 1397–1401.

Espey, L. L. (1967b). Ultrastructure of the apex of the rabbit Graafian follicle during the ovulatory process. *Endocrinology* 81, 267–276.

Espey, L. L. (1974). Ovarian proteolytic enzymes and ovulation. *Biol. Reprod.* 10, 216–235.

Espey, L. L. (1978a). Ovulation. *In* "The Vertebrate Ovary" (R. E. Jones, ed.), pp. 503–532. Plenum, New York.

Espey, L. L. (1978b). Ovarian contractility and its relationship to ovulation: a review. *Biol. Reprod.* 19, 540–551.

Espey, L. L. (1980). Ovulation as an inflammatory reaction: a hypothesis. *Biol. Reprod.* 22, 73–106.

Espey, L. L. (1989). The role of steroids in mammalian ovulation. *In* "Follicular Development and the Ovulatory Response" (A. Tsafriri and N. Dekel, eds.), Vol. 23, pp. 189–198. Ares-Serono Symposia, Rome.

Espey, L. L. (1992). A review of factors that could influence membrane potentials of ovarian follicular cells during mammalian ovulation. *Acta Endocrinol.* 126, 1–31.

Espey, L. L., and Lipner, H. (1963). Measurements of intrafollicular pressures in the rabbit ovary. *Am. J. Physiol.* 205, 1067–1072.

Espey, L. L., and Lipner, H. (1965). Enzyme-induced rupture of rabbit Graafian follicle. *Am. J. Physiol.* 208, 208–213.

Espey, L. L., and Lipner, H. (1993). The mechanism of ovulation. *In* "Physiology of Reproduction," in press.

Espey, L. L., and Rondell, P. (1967). Estimation of mammalian collagenolytic activity with a synthetic substrate. *J. Appl. Physiol.* 23, 457–461.

Espey, L. L., and Stacy, S. (1970). Failure of an ovarian collagenolytic extract to decompose the connective tissue in the mature sow graafian follicle. *Fed. Proc.* 29, 833.

Espey, L. L., Miller, D. H., and Margolius, H. S. (1986). Ovarian increase in kinin-generating capacity in PMSG/hG-primed immature rat. *Am. J. Physiol.* 251, E361–E365.

Espey, L. L., Tanaka, N., and Okamura, H. (1989a). Increase in ovarian leukotrienes during hormonally induced ovulation in the rat. *Am. J. Physiol.* 256, E753–E759.

Espey, L. L., Tanaka, N., Winn, V., and Okamura, H. (1989b). Increase in ovarian kallikrein activity during ovulation in the gonadotrophin-primed immature rat. *J. Reprod. Fertil.* 87, 503–508.

Espey, L. L., Tanaka, N., Woodard, D. S., Harper, M. J. K., and Okamura, H. (1989c). Decrease in ovarian platelet-activating factor during ovulation in the gonadotropin-primed immature rat. *Biol. Reprod.* 40, 104–110.

Evans, T., Rosenthal, E., Youngblom, J., Distel, D., and Hunt, T. (1983). Cyclin: a protein specified by maternal mRNA in sea urchin eggs that is destroyed at each cleavage division. *Cell* 33, 389–396.

Feldman, E., Haberman, S., Abisogun, A. O., Reich, R., Levran, D., Maschiach, S., Zuckermann, H., Rudak, E., Dor, J., and Tsafriri, A. (1986). Arachidonic acid metabolism in human granulosa cells: evidence for cyclooxygenase and lipoxygenase activity *in vitro*. *Hum. Reprod.* 1, 353–356.

Fernandez, C. A., Tarlatzis, B. C., Rzasa, P. J., Caride, V. J., Laufer, N., Negro-Vilar, A. F., DeCherney, A. H., and Naftolin, F. (1985). Renin-like activity in ovarian follicular fluid. *Fertil. Steril.* 44, 119–222.

Ferrel, J. E., Jr., Wu, M., Gerhart, J. C., and Martin, S. G. (1991). Cell cycle tyrosine phosphorylation of $p34^{cdc2}$ and a microtubule-associated protein kinase homolog in *Xenopus* oocytes and eggs. *Mol. Cell. Biol.* 11, 1965–1971.

Fleming, A. D., Kuehl, T. J., and Armstrong, D. T. (1985). Maturation of pig and rat oocytes transplanted into surrogate pig follicles *in vitro*. *Gam. Res.* 11, 107–119.

Foote, W. D., and Thibault, C. (1969). Recherches experimentales sur la maturation *in vitro* des ovocytes de truie et de veau. *Ann. Biol. Anim. Biochem. Biophys.* 9, 329–349.

Franchi, L. L., Mandl, A. M., and Zuckerman, S. (1962). The development of the ovary and the process of oogenesis. *In* "The Ovary" (S. Zuckerman, ed.), Vol. 1, pp. 1–88. Academic Press, New York.

Fukumoto, M., Yajima, Y., Okamura, H., and Midorikawa, O. (1981). Collagenolytic enzyme activity in human ovary: an ovulatory enzyme system. *Fertil. Steril.* 36, 746–750.

Fulka Jr., J. (1983). Nuclear maturation in pig and rabbit oocytes after interspecific fusion. *Exp. Cell. Res.* 146, 212–218.

Fulka, Jr., J., Motlik, J., Fulka, J., and Jílek, F. (1986). Effect of cycloheximide on nuclear maturation of pig and mouse oocytes. *J. Reprod. Fertil.* 77, 281–285.

Fumagalli, Z., Motta, P., and Calvieri, S. (1971). The presence of smooth muscular cells in the ovary of several mammals as seen under the electron microscope. *Experientia* 27, 82–683.

Galway, A. B., LaPolt, P. S., Tsafriri, A., Dargan, C. M., Boime, I., and Hsueh, A. J. W. (1990). Recombinant FSH induces ovulation and tissue plasminogen activator expression in hypophysectomized rats. *Endocrinology* 127, 3023–3028.

Gao, X., Greenbaum, L. M., Mahesh, V. M., and Brann, D. W. (1992). Characterization of the kinin system in the ovary during ovulation in the rat. *Biol. Reprod.* 47, 945–951.

Gautier, J., Norbury, C., Lohka, M., Nurse, P., and Maller, J. (1988). Purified maturation-promoting factor contains the product of a *Xenopus* homolog of the fission yeast cell cycle control gene cdc^{2+}. *Cell* 54, 433–439.

Gavin, A. C., Tsukitani, Y., and Schorderet-Slatkine, S. (1991). Induction of M-phase entry of prophase-blocked mouse oocytes through microinjection of okadaic acid, a specific phosphatase inhibitor. *Exp. Cell Res.* 192, 75–81.

Gavin, A. C., Vassalli, J. D., Cavadore, J. C., and Schorderet-Slatkine, S. (1992). Okadaic acid and $p13^{sc1}$ modulate the reinitiation of meiosis in mouse oocytes. *Mol. Reprod. Dev.* 33, 287–296.

Gaytan, F., Aceitero, J., Bellido, C., Sanchez-Criado, J. E., and Aguilar, E. (1991). Estrous cycle-related changes in mast cell numbers in several ovarian compartments in the rat. *Biol. Reprod.* 45, 27–33.

Gelehrter, T. D., Sznycer-Lascuk, R., Zeheb, R., and Cwikel, B. J. (1987). Dexamethasone inhibition of tissue-type plasminogen activator (tPA) activity: paradoxical induction of both tPA antigen and plasminogen activator inhibitor. *Mol. Endocrinol.* 1, 97–101.

Gerard, M., Menezo, Y., Rombauts, P., Szöllösi, D., and Thibault, C. (1979). *In vitro* studies of oocyte maturation and follicular metabolism in the pig. *Ann. Biol. Anim. Biochem. Biophys.* 18, 1521–1535.

Gerhart, J., Wu, M., and Kirschner, M. (1984). Cell cycle dynamics of an M-phase-specific cytoplasmic factor in *Xenopus laevis* oocytes and eggs. *J. Cell Biol.* 98, 1247–1255.

Gilula, N. B., Epstein, M. L., and Beers, W. H. (1978). Cell-to-cell communication and ovulation: a study of the cumulus-occyte complex. *J. Cell Biol.* 78, 58–75.

Goetz, F. W., Berndtson, A. K., and Ranjan, M. (1991). Ovulation: mediators at the ovarian level. *In* "Vertebrate Endocrinology: Fundamentals and Biomedical Implications" (P. K. Pang, M. P. Schreibman and R. Jones, eds.) Vol. 4, Part A, pp. 127–204. Academic Press, San Diego, CA.

Golbus, M. S., and Stern, M. P. (1976). Qualitative patterns of protein synthesis in the mouse oocytes. *J. Exp. Zool.* 198, 337–342.

Gordon, I., and Lu, K. H. (1990). Production of embryos *in vitro* and its impact on livestock production. *Theriogenology* 33, 77–87.

Goris, J., Hermann, J., Hendrix, P., Ozon, R., and Merlevede, W. (1989). Okadaic acid, a specific protein phosphatase inhibitor, induces maturation and MPF formation in *Xenopus laevis* oocytes. *FEBS Lett.* 245, 91–94.

Guatier, J., Norbury, C. J., Lohka, M., Nurse, P., and Maller, J. (1988). Purified maturation-promoting factor contains the product of a *Xenopus* homolog of the fission yeast cell cycle control gene cdc2+. *Cell* 54, 33–439.

Gwatkin, R. B. L., and Andersen, O. F. (1976). Hamster oocyte maturation *in vitro* inhibition by follicular components. *Life Sci.* 19, 527–536.

Hamada, Y., Wright, K. H., and Wallach, E. E. (1979). The effect of progesterone and human chorionic gonadotrophin on ovulation in the *in vitro* perfused ovary. *Fertil. Steril.* 32, 335–339.

Hanski, E., and Farfel, Z. (1985). Bordetella pertussis invasive adenylate cyclase: partial resolution and properties of its cellular penetration. *J. Biochem.* 290, 5526–5532.

Hashimoto, N., and Kishimoto, T. (1988). Regulation of meiotic metaphase by a cytoplasmic maturation-promoting factor during mouse oocyte maturation. *Dev. Biol.* 126, 242–252.

Hay, M. F., and Moor, R. M. (1973). The Graafian follicle of the sheep: relationship between gonadotropins, steroid production morphology and oocyte maturation. *Ann. Biol. Anim. Biochem. Biophys.* 13, 241–249.

Heaton, J. H., and Gelehrter, T. D. (1990). Cyclic nucleotide regulation of plasminogen activator-inhibitor messenger RNAs in rat hepatoma cells. *Mol. Endocrinol.* 4, 171–178.

Hedin, L., Gaddy-Kurten, D., Kurten, R., DeWitt, D. L., Smith, W., and Richards, J. S. (1987). Prostaglandin endoperoxide synthase in rat ovarian follicles: content, cellular distribution, and evidence for hormonal induction preceeding ovulation. *Endocrinology* 121, 722–731.

Hekman, C. M., and Loskutoff, D. J. (1988). Bovine plasminogen activator inhibitor 1: specificity determinations and comparison of the active, latent, and guanidine-activated forms. *Biochemistry* 27, 2911–2918.

Hellberg, P., Holmes, P. V., Brannstrom, M., Olofsson, J., and Janson, P. O. (1990). Inhibitors of lipoxygenase increase the ovulation rate in the *in vitro* perfused luteinizing hormone-stimulated rabbit ovary. *Acta Physiol. Scand.* 138, 557–564.

Hellberg, P., Larson, L., Olofsson, J., Hedin, L., and Brännström, M. (1991a). Stimulatory effects of bradykinin on the ovulatory process in the *in vitro*-perfused rat ovary. *Biol. Reprod.* 44, 269–274.

Hellberg, P., Thomsen, P., Janson, P. O., and Brännström, M. (1991b). Leukocyte supplementation increases the luteinizing hormone-induced ovulation rate in the in vitro perfused rat ovary. *Biol. Reprod.* 44, 791–797.

Hensleigh, H. C., and Hunter, A. G. (1985). *In vitro* maturation of bovine cumulus enclosed primary oocytes and their subsequent *in vitro* fertilization and cleavage. *J. Dairy Sci.* 88, 1456–1462.

Hillensjö, T. (1976). Oocyte maturation and glycolysis in isolated pre-ovulatory follicles of PMS-injected immature rats. *Acta Endocr. (Copenhagen)* 82, 809–830.

Hillensjö, T., and LeMaire, W. J. (1980). Gonadotropin-releasing hormone agonists stimulate meiotic maturation of follicle-enclosed rat oocytes *in vitro*. *Nature (London)* 287, 145–146.

Hillensjö, T., Batta, S. K., Schwartz-Kripner, A., Wentz, A. C., Sulewski, J., and Channing, C. P. (1978a). Inhibitory effect of human follicular fluid upon the maturation of porcine oocytes in culture. *J. Clin. Endocrinol. Metab.* 47, 1332–1335.

Hillensjö, T., Ekholm, C., and Ahrén, K. (1978b). Role of cyclic AMP in oocyte maturation and glycolysis in the preovulatory rat follicle. *Acta Endocr. (Copenhagen)* 87, 377–388.

Hillensjö, T., Channing, C. P., Pomerantz, S. H., and Schwartz-Kripner, A. (1979a). Intrafollicular control of oocyte maturation in the pig. *In Vitro* 15, 32–40.

Hillensjö, T., Kripner, A. S., Pomerantz, S. H., and Channing, C. P. (1979b). Action of porcine follicular fluid oocyte maturation *in vitro:* possible role of the cumulus cells. *In* "Ovarian Follicular and Corpus Luteum Function" (C. P. Channing, J. M. Marsh, and W. A. Sadler, eds.), pp. 283–291. Plenum, New York.

Hillensjö, T., Chari, S., Magnusson, C., Daume, E., and Sturm, G. (1981). Inhibitory effects of low molecular weight fractions of human follicular fluid upon rat granulosa cells and oocyte *in vitro. In* "International Congress Series No. 551" (K. Semm and K. Mettler, eds.), pp. 458–462. (Proceedings II World Congress of Human Reproduction): Excerpta Medica, Berlin.

Holmes, P. V., Janson, P. O., Sogn, J., Källfelt, B., LeMaire, W. J., Ahrén, K., Cajander, S., and Bjersing, L. (1983). Effects of $PGF_2\alpha$ and indomethacin on ovulation and steroid production in the isolated perfused rabbit ovary. *Acta Endocrinol.* 104, 233–239.

Holmes, P. V., Sogn, J., Schillinger, E., and Janson, P. O. (1985). Effects of high and low preovulatory concentrations of progesterone on ovulation from the isolated perfused rabbit ovary. *J. Reprod. Fertil.* 76, 393–399.

Homa, S. T. (1988). Effects of cyclic AMP on the spontaneous meiotic maturation of cumulus-free bovine oocyte cultured in chemically defined medium. *J. Exp. Zool.* 248, 222–231.

Hunter, A. G., and Moor, R. M. (1987). Stage-dependent effects of inhibiting ribonucleic acids and protein synthesis on meiotic maturation of bovine oocytes *in vitro. J. Dairy Sci.* 70, 1646–1651.

Hunter, R. H. F., and Polge, C. (1966). Maturation of follicular oocytes in the pig after injection of human chorionic gonadotrophin. *J. Reprod. Fertil.* 12, 525–531.

Hunter, R. H. F., Cook, B., and Baker, T. G. (1976). Dissociation of response to injected gonadotropin between the Graafian follicle and oocyte in pigs. *Nature (London)* 260, 156–158.

Huslig, R. L., Malik, A., and Clark, M. R. (1987). Human chorionic gonadotropin stimulation of immunoreactive prostaglandin synthase in rat ovary. *Mol. Cell. Endocrinol.* 50, 237–246.

Ichikawa, S., Morioka, H., Oda, M., Oda, K., and Murao, S. (1983). Effects of various proteinase inhibitors on ovulation of explanted hamster ovaries. *J. Reprod. Fertil.* 68, 407–412.

Itoh, M., Igarashi, M., Yamada, K., Hasegawa, Y., Seki, M., Eto, Y., and Shibai, H. (1990). Activin A stimulated meiotic maturation of the rat oocyte *in vitro. Biochem. Biophys. Res. Commun.* 166, 1479–1484.

Itskovitz, J., Sealey, J. E., Glorioso, N., and Rosenwaks, Z. (1987). Plasma prorenin response to human chorionic gonadotropin in ovarian-hyperstimulated women: correlation with the number of ovarian follicles and steroid hormone concentrations. *Proc. Natl. Acad. Sci. USA* 84, 7285–7289.

Iwamatsu, T., and Chang, M. C. (1972). Sperm penetration *in vitro* of mouse oocytes at various times during maturation. *J. Reprod. Fertil.* 31, 237–247.

Jagiello, G., Graffeo, J., Ducayen, M., and Prosser, R. (1977). Further studies of inhibitors of *in vitro* mammalian oocyte maturation. *Fertil. Steril.* 28, 476–481.

Janson, P. O. (1975). Effects of luteinizing hormone on blood in the follicular rabbit ovary as measured by radioactive microspheres. *Acta Endocrinol.* 79, 122–123.

Janson, P. O., Amato, F., Weiss, T., Ralph, M., and Seamark, R. F. (1978). On the isolated perfused sheep ovary as a model for the study of ovarian function. *Fertil. Steril.* 30, 230–236.

Janson, P. O., LeMaire, W. J., Källfelt, B., Holmes, P. V., Cajander, S., Bjersing, L., Wiqvist, N., and Ahrén, K. (1982). The study of ovulation in the isolated perfused rabbit ovary. I. Methodology and patterns of steroidogenesis. *Biol. Reprod.* 26, 456–465.

Jia, X. C., Ny, T., and Hsueh, A. J. W. (1990). Synergistic effect of glucocorticoids and androgens on the hormonal induction of tissue plasminogen activator activity and messenger ribonucleic acid levels in granulosa cells. *Mol. Cell. Endocrinol.* 68, 143–151.

Kikukawa, Y., Ishikawa, M., Sengoku, K., Kasamo, M., and Shimizu, T. (1991). The effect of platelet activating factor on ovulation. *Prostaglandins* 42, 95–104.

Killick, S., and Elstein, M. (1987). Pharmacologic production of luteinized unruptured follicles by prostaglandin synthetase inhibitors. *Fertil. Steril.* 47, 773–777.

Kishimoto, T., and Kanatani, H. (1976). Cytoplasmic factor responsible for germinal vesicle breakdown and meiotic maturation in starfish oocyte. *Nature (London)* 260, 321–322.

Kishimoto, T., Kuriyama, R., Kondo, H., and Kanatani, H. (1982). Generality of the action of various maturation-promoting factors. *Exp. Cell Res.* 137, 121–126.

Kishimoto, T., Usui, N., and Kanatani, H. (1984a). Breakdown of starfish ovarian follicle induced by maturation-promoting factor. *Dev. Biol.* 101, 28–34.

Kishimoto, T., Yamazaki, K., Kato, Y., Koide, S. S., and Kanatani, H. (1984b). Induction of starfish oocyte maturation by maturation-promoting factor of mouse and surf clam oocytes. *J. Exp. Zool.* 231, 293–295.

Klarlund, J. K. (1985). Transformation of cells by an inhibitor of phosphatases acting on phosphotyrosine in proteins. *Cell* 41, 707–717.

Knecht, M. (1988). Plasminogen activator is associated with the extracellular matrix of ovarian granulosa cells. *Mol. Cell. Endocrinol.* 56, 1–9.

Kogawa, K., Nakamura, T., Sugino, K., Takio, K., Titani, K., and Sugino, H. (1991). Activin-binding protein is present in pituitary. *Endocrinology* 128, 1434–1440.

Koos, R. D., Feiertag, M. A., Brodie, A. M. H., and LeMaire, W. J. (1984). Inhibition of estrogen synthesis does not inhibit luteinizing hormone-induced ovulation. *Am. J. Obstet. Gynecol.* 148, 939–945.

Kuo, T. C., Endo, K., Dharmarajan, A. M., Miyazaki, T., Atlas, S. J., and Wallach, E. E. (1991). Direct effect of angiotensin II on *in vitro* perfused rabbit ovary. *J. Reprod. Fertil.* 92, 469–474.

Lambertsen, C. J., Greenbaum, D. F., Wright, K. H., and Wallach, E. E. (1976). *In vitro* studies of ovulation in the perfused rabbit ovary. *Fertil. Steril.* 27, 178–187.

LaPolt, P. S., Yamoto, M., Veljkovic, M., Sincich, C., Ny, T., Tsafriri, A., and Hsueh, A. J. W. (1990). Basic fibroblast growth factor induction of granulosa cell tissue-type plasminogen activator expression and oocyte maturation: potential role as a paracrine ovarian hormone. *Endocrinology* 127, 2357–2363.

Larsen, W. J., Wert, S. E., and Brunner, C. D. (1986) A dramatic loss of cumulus cell gap junctions is correlated with germinal vesicle breakdown in rat oocytes. *Dev. Biol.* 113, 517–521.

Laufer, N., Reich, R., Braw, R., Shenker, J. G., and Tsafriri, A. (1982). The effect of clomiphene citrate on preovulatory rat follicles in culture. *Biol. Reprod.* 27, 463–471.

Lawrence, T. S., Dekel, N., and Bers, W. H. (1980). Binding of hCG by rat cumuli oophori and granulosa cells: a comparative study. *Endocrinology* 106, 1114–1118.

Lee, W., and Novy, M. J. (1978). Effects of luteinizing hormone and indomethacin on blood flow and steroidogenesis in the rabbit ovary. *Biol. Reprod.* 18, 799–807.

Leibfried, L., and First, N. L. (1980a). Follicular control of meiosis in the porcine oocyte. *Biol. Reprod.* 23, 705–709.

Leibfried, L., and First, N. L. (1980b). Effect of bovine and porcine follicular fluid and granulosa cells on maturation of oocyte *in vitro*. *Biol. Reprod.* 23, 699–704.

Leibfried-Rutledge, M. L., Critser, E. S., Parrish, J. J., and First, N. L. (1989). *In vitro* maturation and fertilization of bovine oocytes. *Theriogenology* 31, 61–74.

LeMaire, W. J. (1989). Mechanism of mammalian ovulation. *Steroids* 54, 455–469.

LeMaire, W. J., Yang, N. S. T., Behrman, H. H., and Marsh, J. M. (1973). Preovulatory changes in the concentration of prostaglandins in rabbit Graafian follicles. *Prostaglandins* 3, 367–376.

LeMaire, W. J., Janson, P. O., Källfelt, B. J., Holmes, P. V., Cajander, S., Bjersing, L., and Ahrén, K. (1982). The preovulatory decline in follicular estradiol is not required for ovulation in the rabbit. *Acta Endocrinol. (Copenhagen)* 101, 452–457.

LeMaire, W. J., Curry, T. E., Morioka, N., Brännstöm, M., Clark, M. R., Woessner, J. F., and Koos, R. D. (1987). Regulation of ovulatory processes. *In* "The Primate Ovary" (R. L. Stouffer, ed.), pp. 91–111. Plenum, New York.

Li, X. M., Sagawa, N., Ihara, Y., Okagaki, A., Hasegawa, M., Inamori, K., Itoh, H., Mori, T., and Ban, C. (1991). The involvement of platelet-activating factor in thrombocytopenia and follicular rupture during gonadotropin-induced superovulation in immature rats. *Endocrinology* 129, 3132–3138.

Liedtke, M. P., and Seifert, B. (1978). Biosynthesis of prostaglandins in human ovarian tissues. *Prostaglandins* 16, 825–833.

Lindner, H. R., Bar-Ami, S., and Tsafriri, A. (1980). "Animal Models in Human Reproduction" (M. Serio and L. Martini, eds.) pp. 65–85. Raven Press, New York.

Lindner, H. R., Tsafriri, A., Lieberman, M. E., Zor, U., Koch, Y., Bauminger, S., and Barnea, A. (1974). Gonadotropin action on cultured Graafian follicles: Mechanism of induction of maturation of the mammalian oocyte. *Recent Progr. Horm. Res.* 30, 79–138.

Lipner, H. (1973). Mechanism of mammalian ovulation. *In* "Handbook of Physiology, Endocrinology" (R. Greep, ed.), Vol. II, Part 1, pp. 409–437. American Physiological Society, Washington, D.C.

Lipner, H. (1988). Mechanism of mammalian ovulation. *In* "The Physiology of Reproduction" (E. Knobil and J. Neill, eds.), pp. 447–488. Raven Press, New York.

Lipner, H., and Greep, R. O. (1971). Inhibition of steroidogenesis at various sites in the biosynthetic pathway in relation to induced ovulation. *Endocrinology* 88, 602–607.

Lipner, H., and Wendelken, L. (1971). Inhibition of ovulation by inhibition of steroidogenesis in immature rats. *Proc. Soc. Exp. Biol. Med.* 136, 1141–1145.

Liu, Y.-X., Peng, X.-R., and Ny, T. (1991). Tissue-specific and time-coordinated hormone regulation of plasminogen-activator-inhibitor type 1 and tissue-type plasminogen activator in the rat ovary during gonadotropin-induced ovulation. *Eur. J. Biochem.* 195, 549–555.

Löfman, C. O., Janson, P. O., Källfelt, B. J., Ahrén, K., and LeMaire, W. J. (1982). The study of ovulation in the isolated perfused rabbit ovary. II. Photographic and cinematographic observations. *Biol. Reprod.* 26, 467–473.

Lohka, M. J., Hayes, M. K., and Maller, J. L. (1988). Purification of maturation-promoting factor, an intracellular regulator of early mitotic events. *Proc. Natl. Acad. Sci. USA* 85, 3009–3013.

Lumsden, M. A., Kelly, R. W., Templeton, A. A., van Look, P. F. A., Swanston, I. A., and Baird, D. T. (1986). Changes in the concentration of prostaglandins in preovulatory human follicles after administration of hCG. *J. Reprod. Fertil.* 77, 119–124.

Magnusson, C., Hillensjö, T., Tsafriri, A., Hultborn, R., and Ahrén, K. (1977). Oxygen consumption of maturing rat oocytes. *Biol. Reprod.* 17, 9–15.

Mann, J. S., Kindy, M. S., Edwards, D. R., and Curry, J. T. E. (1991). Hormonal regulation of matrix metalloproteinase inhibitors in rat granulosa cells and ovaries. *Endocrinology* 128, 1825–1832.

Martin, G. G., and Talbot, P. (1981). The role of follicular smooth muscle cells in hamster ovulation. *J. Exp. Zool.* 216, 469–482.

Masui, Y., and Markert, C. L. (1971). Cytoplasmic control of nuclear behavior during maturation of frog oocytes. *J. Exp. Zool.* 177, 129–146.

Matrisian, L. M. (1990). Metalloproteinases and their inhibitors in matrix remodeling. *TIG* 6, 121–125.

Mattioli, M., Galeati, G., and Seren, E. (1986). Pig oocyte maturation in vitro and evaluation of their developmental capability by IVF. *Atti. Soc. It. Sci. Vet.* 40, 200–202.

Mattioli, M., Galeati, G., and Seren, E. (1988a). Effect of follicle somatic cells during pig oocyte maturation on egg penetrability and male pronucleus formation. *Gamete Res.* 20, 177–183.

Mattioli, M., Galeati, G., Bacci, M. L., and Seren, E. (1988b). Follicular factors influence oocyte fertilizability by modulating the intercellular cooperation between cumulus cells and oocyte. *Gamete Res.* 21, 223–232.

Mattioli, M., Galeati, G., Bacci, M. L., and Barboni, B. (1991). Changes in maturation-promoting activity in the cytoplasm of pig oocytes throughout maturation. *Mol. Reprod. Dev.* 30, 119–125.

Meijer, L., Azzi, L., and Wang, J. H. J. (1991). Cyclin B targets $p34^{cdc2}$ for tyrosine phsophorylation. *EMBO J.* 10, 1545–1554.

Meinecke, B., and Meinecke-Tillman, S. (1981). Induction and inhibition of meiotic maturation of follicle-enclosed porcine oocytes *in vivo. Theriogenology* 216, 205–209.

Miller, W. A., and Jagiello, G. (1973). Gonadotropin dependency of isolated ovine ovarian follicles cultured *in vitro. Fertil. Steril.* 24, 609–172.

Minshull, J., Blow, J. J., and Hunt, T. (1989). Translation of cyclin mRNA is necessary for extracts of activated *Xenopus* eggs to enter mitosis. *Cell* 56, 947–956.

Mitchell, P. A., and Burghardt, R. C. (1986). The ontogeny of nexuses (gap junctions) in the ovary of the fetal mouse. *Anat. Rec.* 214, 283–288.

Miyazaki, T., Sueoka, K., Dharmarajan, A. M., Atlas, S. J., Bulkley, G. B., and Wallach, E. E. (1991). Effect of inhibition of oxygen free radical on ovulation and progesterone production by the *in vitro* perfused rabbit ovary. *J. Reprod. Fertil.* 91, 207–212.

Moor, R. M., and Crosby, I. M. (1986). Protein requirements for germinal vesicle breakdown in ovine oocytes. *J. Embryol. Exp. Morphol.* 94, 207–220.

Moor, R. M., and Trounson, A. O. (1977). Hormonal and follicular factors affecting maturation of sheep oocytes *in vitro* and their subsequent developmental capacity. *J. Reprod. Fertil.* 49, 101–109.

Moor, R. M., Smith, M. W., and Dawson, R. M. C. (1980). Measurement of intercellular coupling between oocytes and cumulus cells using intracellular markers. *Exp. Cell. Res.* 126, 15–30.

Morales, T. I., Woessner, J. F., Howell, D. S., Marsh, J. M., and LeMaire, W. J. (1978). A microassay for the direct demonstration of collagenolytic activity in Graafian follicles of the rat. *Biochim. Biophys. Acta* 524, 428–434.

Morales, T. I., Woessner, J. F. J., Marsh, J. M., and LeMaire, W. J. (1983). Collagen, collagenase and collagenolytic activity in rat Graafian follicles during follicular growth and ovulation. *Biochim. Biophys. Acta* 756, 119–122.

Morla, A. O., Draetta, G., Beach, D., and Wang, J. Y. J. (1989). Reversible tyrosine phosphorylation of cdc2: Dephosphorylation accompanies activation during entry into mitosis. *Cell* 58, 193–203.

Morioka, N., Brännström, M., Koos, R. D., and LeMaire, W. J. (1988). Ovulation in the perfused rat ovary *in vitro:* further evidence that estrogen is not required. *Steroids* 51, 173–183.

Motlik, J., Fulka, J., and Flechon, J. E. (1986). Changes in intercellular coupling between pig oocytes and cumulus cells during *in vivo* and *in vitro* maturation. *J. Reprod. Fertil.* 76, 32–37.

Motlik, J., Nagai, T., and Kikuchi, K. (1991). Resumption of meiosis in pig oocytes cultured with cumulus and parietal granulosa cells: the effect of protein synthesis inhibition. *J. Exp. Zool.* 259, 386–391.

Murdoch, W. J., and McCormick, R. J. (1991). Dose-dependent effects of indomethacin on ovulation in the sheep: relationship to follicular prostaglandin production, steroidogenesis, collagenolysis, and leukocyte chemotaxis. *Biol. Reprod.* 45, 907–911.

Murdoch, W. J., and Steadman, L. E. (1991). Investigations concerning the relationship of ovarian eosinophilia to ovulation and luteal function in the sheep. *Am. J. Reprod. Immunol.* 25, 81–87.

Murdoch, W. J., Nix, K. J., and Dunn, T. G. (1983). Dynamics of ovarian blood supply to periovulatory follicles of the ewe. *Biol. Reprod.* 28, 1001–1006.

Murdoch, W. J., Peterson, T. A., Van Kirk, E. A., Vincent, D. L., and Inskeep, E. K. (1986). Interactive roles of progesterone, prostaglandin, and collagenase in the ovulatory mechanism of the ewe. *Biol. Reprod.* 35, 1187–1194.

Murray, A. W., and Kirschner, M. W. (1989a). Cyclin synthesis drives the early embryonic cell cycle. *Nature* 339, 275–280.

Murray, A. W., and Kirschner, M. W. (1989b). Dominoes and clocks: the union of two views of the cell cycle. *Science* 246, 614–621.

Naftolin, F., Andrade-Gordon, P., Pellicer, A., Palumbo, A., Apa, R., Zreik, T., Yoon, T. K., and DeCherney, A. H. (1989). (Response to) Angiotensin II: Does it have a direct obligate role in ovulation? *Science* 245, 870–871.

Nakamura, T., Takio, K., Eto, Y., Shibai, H., Titani, K., and Sugino, H. (1990). Activin-binding protein from rat ovary is follistatin. *Science* 247, 836–838.

Neal, P., and Baker, T. G. (1973). Response of mouse ovaries *in vivo* and in organ culture to pregnant mare's serum gonadotrophin and human chorionic gonadotrophin. I. Examination of critical time intervals. *J. Reprod. Fertil.* 33, 399–410.

Nelkin, B., Nichols, C., and Vogelstein, B. (1980). Protein factor(s) from mitotic CHO cells induces meiotic maturation in *Xenopus laevis* oocytes. *FEBS Lett.* 109, 233–238.

Niwa, K., and Chang, M. C. (1975). Fertilization of rat eggs *in vitro* at various times before and after ovulation with special reference to fertilization of ovarian oocytes matured in culture. *J. Reprod. Fert.* 43, 435–451.

Ny, T., Elgh, F., and Lund, B. (1984). The structure of the human tissue type plasminogen activator gene. Correlation of intron and exon structures to functional and structural domains. *Proc. Natl. Acad. Sci. USA* 81, 5355–5359.

Ny, T., Bjersing, L., Hsueh, A. J. W., and Loskutoff, D. J. (1985). Cultured granulosa cells produce two plasminogen activators and an antiactivator, each regulated differently by gonadotropins. *Endocrinology* 116, 1666–1668.

O, W.-S., Robertson, D. M., and de Kretser, D. M. (1989). Inhibin as an oocyte meiotic inhibitor. *Mol. Cell. Endocrinol.* 62, 307–311.

O'Shea, J. D. (1970). An ultrastructural study of smooth muscle-like cells in the theca externa of ovarian follicles in the rat. *Anat. Rec.* 167, 127–140.

Ohno, S., and Smith, J. B. (1964). Role of fetal follicular cells in meiosis of mammalian oocytes. *Cytogenetics* 3, 324–333.

Oikawa, M., Dargan, C., Tor, N. Y., and Hsueh, A. J. W. (1990). Expression of gonadotropin-releasing hormone and prothymosin-α messenger ribonucleic acid in the ovary. *Endocrinology* 127, 2350–2356.

Okamura, H., Virutamasen, P., Wright, K. H., and Wallach, E. E. (1972). Ovarian smooth muscle in the human being, rabbit, and cat. *Am. J. Obstet. Gynecol.* 122, 183–191.

Orczyk, G. P., and Behrman, H. R. (1972). Ovulation blockade by aspirin or indomethacin: *in vivo* evidence for a role of prostaglandin in gonadotrophin secretion. *Prostaglandins* 1, 3–20.

Osvaldo-Decima, L. (1970). Smooth muscle in the ovary of the rat and monkey. *J. Ultrastruct. Res.* 29, 218–237.

Parr, E. L. (1974). Histological examination of the rat ovarian follicle wall prior to ovulation. *Biol. Reprod.* 11, 483–503.

Parr, E. L. (1975). Rupture of ovarian follicles at ovulation. *J. Reprod. Fertil. (Suppl.)* 22, 1–22.

Patwardhan, V. V., and Lanthier, A. (1981). Prostaglandins PGE and PGF in human ovarian follicles: Endogenous contents and *in vitro* formation by theca and granulosa cells. *Acta Endocrinol.* 97, 543–550.

Pellicer, A., Palumbo, A., DeCherney, A. H., and Naftolin, F. (1988). Blockage of ovulation by an angiotensin antagonist. *Science* 240, 1660–1661.

Peters, H., and McNatty, K. P. (1980). "The Ovary." Granada Publishing, London.

Peterson, C. M., Zhu, C., Mukaida, T., Butler, T. A., Woessner, J. F., and LeMaire, W. J. (1993). The angiotensin II antagonist, saralasin, inhibits ovulation in the perfused rat ovary. *Am. J. Obstet. Gynecol.* 168, 242–245.

Piacsek, B. E., and Huth, J. F. (1971). Changes in ovarian venous blood flow following cannulation: effect of luteinizing hormone (LH) and antihistamine. *Proc. Soc. Exp. Biol. Med.* 138, 1022–1024.

Picard, A., Capony, J. P., Brautigan, D. L., and Doree, M. (1989). Involvement of protein phosphatase 1 and 2A in the control of M-phase-promoting factor activity in starfish. *J. Cell Biol.* 109, 3347–3354.

Pincus, G., and Enzmann, E. V. (1935). The comparative behaviour of mammalian eggs *in vivo* and *in vitro*. *J. Exp. Med.* 62, 655–675.

Pines, J., and Hunt, T. (1987). Molecular cloning and characterization of the mRNA for cyclin from sea urchin eggs. *EMBO J.* 6, 2987–2995.

Piontkewitz, Y., and Dekel, N. (1993). Heptanol, an alkanol that blocks gap junctions, induces oocytes maturation. *Endocrin. J.* (in press).

Plunkett, E. R., Moon, Y. S., Zamecnik, J., and Armstrong, D. T. (1975). Preliminary evidence of a role for prostaglandin F in human follicular function. *Am. J. Obstet. Gynecol.* 123, 392–397.

Politis, I., Srikandakumar, J., Turner, J. D., Tsang, B. K., Ainsworth, L., and Downey, B. R. (1990). Changes in and partial identification of the plasminogen activator and plasminogen activator inhibitor systems during ovarian follicular maturation in the pig. *Biol. Reprod.* 43, 636–642.

Pomerantz, S., Tsafriri, A., and Channing, C. P. (1979). Studies on purification and actions of an oocyte maturation inhibitor isolated from porcine follicular fluid. *In* "Peptides, Structure and Biological Function" (E. Gross and J. Meienhofer, eds.), pp. 765–774. Pierce Chemical Company, Rockford.

Priddy, A. R., Killick, S. R., Elstein, M., Morris, J., Sullivan, M., Patel, L., and Elder, M. G. (1989). Ovarian follicular fluid eicosanoid concentrations during the pre-ovulatory period in humans. *Prostaglandins* 38, 197–202.

Racowsky, C. (1984). Effect of forskolin on the spontaneous maturation and cyclic AMP content of rat oocyte cumulus complexes. *J. Reprod. Fertil.* 72, 107–116.

Racowsky, C. (1985). Effect of forskolin on maintenance of meiotic arrest and stimulation of cumulus expansion, progesterone and cyclic AMP production by pig oocyte-cumulus complexes. *J. Reprod. Fertil.* 74, 9–21.

Racowsky, C. (1991). Gamete resources: origin and production of oocytes. *In* "Animal Applications of Research in Mammalian Development" (R. A. Pedersen, A. McLaren, and N. L. First, eds.), pp. 23–82. Cold Spring Harbor Laboratory Press, Cold Spring Harbor, New York.

Racowsky, C., and Baldwin, K. V. (1989). *In vitro* and *in vivo* studies reveal that hamster oocyte meiotic arrest is maintained only transiently by follicular fluid, but persisently by membrana/cumulus granulosa cell contact. *Dev. Biol.* 134, 297–306.

Racowsky, C., and Satterlie, R. A. (1985). Metabolic, fluorescent dye and electrical coupling between hamster oocytes and cumulus cells during meiotic maturation *in vivo* and *in vitro*. *Dev. Biol.* 60, 318–325.

Racowsky, C., Baldwin, K. V., Larabell, C. A., DeMarais, A. A., and Kazilek, C. J. (1989). Down-regulation of membrana granulosa cell gap junctions is correlated with irreversible commitment to resume meiosis in gold Syrian hamster oocytes. *Eur. J. Cell Biol.* 49, 244–251.

Reich, R., Laufer, N., Lewysohn, O., Cordova, T., Ayalon, D., and Tsafriri, A. (1982). *In vitro* effects of cannabinoids on follicular function in the rat. *Biol. Reprod.* 27, 223–231.

Reich, R., Kohen, F., Naor, Z., and Tsafriri, A. (1983). Possible involvement of lipoxygenase products of arachidonic acid pathway in ovulation. *Prostaglandins* 26, 1011–1019.

Reich, R., Kohen, F., Slager, R., and Tsafriri, A. (1985a). Ovarian lipoxygenase activity and its regulation by gonadotropin in the rat. *Prostaglandins* 30, 581–590.

Reich, R., Miskin, R., and Tsafriri, A. (1985b). Follicular plasminogen activator: involvement in ovulation. *Endocrinology* 116, 516–521.

Reich, R., Tsafriri, A., and Mechanic, G. L. (1985c). The involvement of collagenolysis in ovulation in the rat. *Endocrinology* 116, 521–527.

Reich, R., Miskin, R., and Tsafriri, A. (1986). Intrafollicular distribution of plasminogen activators and their hormonal regulation *in vitro*. *Endocrinology* 119, 1588–1601.

Reich, R., Haberman, S., Abisogun, A. O., Sofer, Y., Grossman, S., Adelmann-Grill, B. C., and Tsafriri, A. (1987). Follicular rupture at ovulation: collagenase activity and metabolism of arachidonic acid. *In* "The Control of Follicle Development, Ovarian and Luteal Function: Lessons from in Vitro Fertilization" (F. Naftolin and A. H. deCherney, eds.), pp. 317–329. Raven Press, New York.

Reich, R., Daphna-Iken, D., Chun, S. Y., Popliker, M., Slager, R., Adelmann-Grill, B. C., and Tsafriri, A. (1991). Preovulatory changes in ovarian expression of collagenases and tissue metalloproteinase inhibitor mRNA: role of eicosanoids. *Endocrinology* 129, 1869–1875.

Reinthaller, A., Bieglmayer, C., Kirchheimer, J. C., Christ, G., Deutinger, J., and Binder, B. R. (1990). Plasminogen activators, plasminogen activator inhibitor, and fibronectin in human granulosa cells and follicular fluid related to oocyte maturation and intrafollicular gonadotropin levels. *Fertil. Steril.* 54, 1045–1051.

Reynhout, J. K., and Smith, L. D. (1974). Studies on the appearance and nature of maturation-inducing factor in the cytoplasm of amphibian oocytes exposed to progesterone. *Dev. Biol.* 38, 394–400.

Riccio, A., Grimaldi, G., Verde, P., Gebastio, G., Boast, S., and Blasi, F. (1985). The human urokinase plasminogen activator gene and its promoter. *Nucleic Acid Res.* 13, 2759–2771.

Rice, C., and McGaughey, R. W. (1981). Effect of testosterone and dibutyryl cAMP on the spontaneous maturation of pig oocytes. *J. Reprod. Fertil.* 62, 245–256.

Rime, H., and Ozon, R. (1990). Protein phosphatases are involved in the in vivo activation of histone H1 kinase in mouse oocyte. *Dev. Biol.* 141, 115–122.

Rime, H., Néant, I., Guerrier, P., and Ozon, R. (1989). 6-dimethylaminopurine (6-DMAP), reversible inhibitor of the transition to metaphase during the first meiotic cell division of the mouse oocyte. *Dev. Biol.* 133, 169–179.

Rocerto, T., Jacobowitz, D., and Wallach, E. (1969). Observations of spontaneous contractions of the cat ovary *in vitro*. *Endocrinology* 84, 1336–1341.

Rondell, P. (1964). Follicular pressure and distensibility. *Am. J. Physiol.* 207, 590–594.

Rondell, P. (1970). Biophysical aspects of ovulation. *Biol. Reprod. (Suppl.)* 2, 64–89.

Rondell, P. (1974). Role of steroid synthesis in the process of ovulation. *Biol. Reprod.* 10, 199–215.

Salustri, A., Petrungaro, S., de Felici, M., Conti, M., and Siracusa, D. (1985). Effect of follicle-stimulating hormone on cyclic adenosine monophosphate level and on meiotic maturation in mouse cumulus cell-enclosed oocyte cultured *in vitro*. *Biol. Reprod.* 33, 797–802.

Sato, E., and Ishibashi, T. (1977). Meiotic arresting action of the substance obtained from cell surface to porcine ovarian granulosa cells. *Jpn. J. Zootech. Sci.* 48, 22–26.

Sato, E., and Koide, S. S. (1984). A factor from bovine granulosa cells preventing oocyte maturation. *Differentiation* 26, 59–62.

Schochet, S. S. (1916). A suggestion as to the process of ovulation and ovarian cyst formation. *Anat. Rec.* 10, 447–457.

Schorderet-Sltakine, S., and Drury, K. C. (1973). Progesterone induced mature oocytes of *Xenopus laevis:* appearance of a "maturation-promoting factor" in entire oocytes. *Cell Differ.* 2, 247–254.

Schuetz, A. W. (1985). Local control mechanisms during oogenesis and folliculogenesis. *In* "Developmental Biology. A Comprehensive Synthesis" (L. W. Browder, ed.), Vol. 1, pp. 3–83. Plenum, New York.

Schultz, R. M. (1991). Meiotic maturation of mammalian oocytes. *In* "Elements of Mammalian Fertilization" (P. M. Wassarman, ed.), Vol. I, pp. 77–104. CRC Press, Boca Raton, FL.

Schultz, R. M., and Wassarman, P. M. (1977). Biochemical studies of mammalian oogenesis: protein synthesis during oocyte growth and meiotic maturation in the mouse. *J. Cell Sci.* 24, 167–194.

Schultz, R. M., Montgomery, R. R., and Belanoff, J. R. (1983). Regulation of mouse oocyte meiotic maturation: implication of a decrease in oocyte cAMP and protein dephosphorylation in commitment to resume meiosis. *Dev. Biol.* 97, 264–273.

Sealey, J. E., Atlas, S. A., Glorioso, N., Manapat, H., and Laragh, J. H. (1985). Cyclical secretion of prorenin during the menstrual cycle: synchronization with luteinizing hormone and progesterone. *Proc. Natl. Acad. Sci. USA* 82, 8705–8709.

Sealey, J. E., Cholst, I., Glorioso, N., Torffa, C., Weintraub, I. D., James, G., and Larach, J. H. (1987). Sequential changes in plasma luteinizing hormone and plasma prorenin during the menstrual cycle. *J. Clin. Endocrinol. Metab.* 65, 1–5.

Sherizly, I., Galiani, D., and Dekel, N. (1988). Regulation of oocyte maturation: communication in the rat cumululs-oocyte complex. *Hum. Reprod.* 3, 761–766.

Shimonaka, M., Inouye, S., Shimasaki, S., and Ling, N. (1991). Follistatin binds to both activin and inhibin through the common beta-subunit. *Endocrinology* 128, 3313–3315.

Sirard, M. A., and Bilodeau, S. (1990a). Granulosa cells inhibit the resumption of meiosis in bovine oocytes *in vitro*. *Biol. Reprod.* 43, 777–783.

Sirard, M. A., and Bilodeau, S. (1990b). Effects of granulosa cell co-culture on *in-vitro* meiotic resumption of bovine oocytes. *J. Reprod. Fertil.* 89, 459–465.

Sirard, M. A., and First, N. L. (1988). *In vitro* inhibition of oocyte nuclear maturation in the bovine. *Biol. Reprod.* 39, 229–234.

Sirois, J., and Richards, J. (1991). Identification and characterization of a novel, distinct isoform of prostaglandin endoperoxide synthase. *J. Biol. Chem.* 267, 6382–6388.

Smith, L. D., and Ecker, R. E. (1971). The interaction of steroids with *Rana pipiens* oocytes in the induction of maturation. *Dev. Biol.* 25, 232–247.

Snyder, B. W., Beecham, G. D., and Schane, H. P. (1984). Inhibition of ovulation in rats with epostane, an inhibitor of 3 β-hydroxysteroid dehydrogenase (41865). *Proc. Soc. Exp. Biol. Med.* 176, 238–242.

Sogn, J. H., Curry, T. E., Brännström, M., LeMaire, W. J., Koose, R. D., Papkoff, H., and Janson, P. O. (1987). Inhibition of follicle-stimulating hormone induced ovulation by indomethacin in the perfused rat ovary. *Biol. Reprod.* 36, 536–542.

Solomon, M. J., Glotzer, M., Lee, T. H., Philippe, M., and Kirschner, M. W. (1990). Cyclin activation of $p34^{cdc2}$. *Cell* 63, 1013–1024.

Sorensen, R. A., Cyert, M. S., and Pedersen, R. A. (1985). Active maturation-promoting factor is present in mature mouse oocytes. *J. Cell Biol.* 100, 1637–1640.

Staigmiller, R. B., and Moor, R. M. (1984). Effect of follicle cells on the maturation and developmental competence of ovine oocytes matured outside the follicle. *Gamete Res.* 9, 221–229.

Stone, S. L., Pomerantz, S. H., Schwartz-Kripner, A., and Channing, C. P. (1978). Inhibitor of oocyte maturation from porcine follicular fluid: further purification and evidence for reversible action. *Biol. Reprod.* 19, 585–592.

Suenson, E., Lötzen, O., and Thorsen, S. (1984). Initial plasmin degradation of fibrin as the basis of a positive feed-back mechanism in fibrinolysis. *Eur. J. Biochem.* 140, 513–522.

Sunkara, P. S., Wright, D. A., and Rao, P. N. (1979). Mitotic factors from mammalian cells induce germinal vesicle breakdown and chromosome condensation in amphibian oocytes. *Proc. Natl. Acad. Sci. USA* 76, 2799–2802.

Swenson, K. I., Farrell, K. M., and Ruderman, J. V. (1986). The clam embryo protein cyclin a induces entry into M phase and the resumption of meiosis in *Xenopus* oocytes. *Cell* 47, 861–870.

Szöllösi, D. (1975). Ultrastructural aspects of oocyte maturation and fertilization in mammals. *In* "La Fécondation" (C. Thibault, ed.), pp. 13–35. Masson et Cie, Paris.

Szöllösi, D., Gérard, M., Ménézo, Y., and Thibault, C. (1978). Permeability of ovarian follicle; corona cell-oocyte relationship in mammals. *Ann. Biol. Anim. Biochem. Biophys.* 18, 511–521.

Szybek, K. (1972). *In vitro* maturation of oocytes from sexually immature mice. *J. Endocrinol.* 54, 527–528.

Tachibana, K., Yanagishima, N., and Kishimoto, T. (1987). Preliminary characterization of maturation-promoting factor from yeast *Saccharomyces cerevisiae*. *J. Cell Sci.* 88, 273–281.

Takahashi, M., Koide, S. S., and Donahue, P. K. (1986). Müllerian inhibiting substance as oocyte meiosis inhibitor. *Mol. Cell. Endocrinol.* 47, 225–234.

Tanaka, N., Espey, L., and Okamura, H. (1989). Increase in ovarian 15-hydroxyeicosatetraenoic acid during ovulation in the gonadotropin-primed immature rat. *Endocrinology* 125, 1373–1377.

Thibault, C. (1977). Are follicular maturation and oocyte maturation independent processes? *J. Reprod. Fertil.* 51, 1–15.

Thibault, C., and Gerard, M. (1973). Cytoplasmic and nuclear maturation of rabbit oocytes *in vitro*. *Ann. Biol. Anim. Biochem. Biophys* 13, 145–155.

Thibault, C., and Levasseur, M. C. (1988). Ovulation. *Hum. Reprod.* 3, 513–523.

Thibault, C., Gerard, M., and Menezo, Y. (1975a). Acquisition par l'ovocyte de Lapine et de Veau du facteur de decondensation du noyau du spermatozoide fecondant (MPGF). *Ann. Biol. Anim. Biochem. Biophys.* 15, 705–714.

Thibault, C., Gerard, M., and Menezo, Y. (1975b). Preovulatory and ovulatory mechanisms in oocyte maturation. *J. Reprod. Fertil.* 45, 605–610.

Thibault, C., Szöllösi, D., and Gerard, M. (1987). Mammalian oocyte maturation. *Reprod. Nutr. Dev.* 27, 865–896.

Tonks, N. K., Diltz, C. D., and Fischer, E. H. (1988). Characterization of the major protein-tyrosine-phosphatases of human placenta. *J. Biol. Chem.* 263, 6731–6737.

Tsafriri, A. (1978). Oocyte maturation in mammals. *In* "The Vertebrate Ovary" (R. E. Jones, ed.), pp. 409–442. Plenum, New York.

Tsafriri, A., and Bar-Ami, S. (1982). Oocyte maturation inhibitor: A 1981 perspective. *In* "Intra-Ovarian Control Mechanisms" (C. P. Channing and S. J. Segal, eds.), Vol. 47, pp. 145–160. Advances in Experimental Medicine and Biology. Plenum, New York.

Tsafriri, A., and Braw, R. (1984a). Experimental approaches to atresia in mammals. *Oxford Rev. Reprod. Biol.* 6, 226–265.

Tsafriri, A., and Channing, C. P. (1975). An inhibitory influence of granulosa cells and follicular fluid upon porcine oocyte meiosis *in vitro*. *Endocrinology* 96, 922–927.

Tsafriri, A., and Pomerantz, S. H. (1984). The control of ovum maturation in mammals: Comparative aspects. *In* "Gonadal Peptides and Proteins" (S. H. Sairam and L. E. Atkinson, eds.), pp. 191–206. World Scientific Publishing, Singapore.

Tsafriri, A., and Pomerantz, S. H. (1986). Oocyte maturation inhibitor. *In* "Clinics in Endocrinology and Metabolism" (P. Franchimont, ed.), Vol. 15/1, pp. 157–170. W. B. Saunders, Philadelphia.

Tsafriri, A., and Reich, R. (1991). Plasminogen activators in the preovulatory follicle: role in ovulation. *In* "Plasminogen Activators: From Cloning to Therapy" (R. Abbate, T. Barni, and A. Tsafriri, eds.), pp. 81–94. Raven Press, New York.

Tsafriri, A., Lindner, H. R., Zor, U., and Lamprecht, S. A. (1972a). *In vitro* induction of meiotic division in follicle-enclosed rat oocytes by LH, cyclic AMP and prostaglandin E_2. *J. Reprod. Fertil.* 31, 39–50.

Tsafriri, A., Lindner, H. R., Zor, U., and Lamprecht, S. A. (1972b). Physiological role of prostaglandins in the induction of ovulation. *Prostaglandins* 2, 1–10.

Tsafriri, A., Pomerantz, S. H., and Channing, C. P. (1976a). Porcine follicular fluid inhibitor of oocyte meiosis: Partial characterization of the inhibitor. *Biol. Reprod.* 14, 511–516.

Tsafriri, A., Lieberman, M. E., Koch, Y., Bauminger, S., Chobsieng, P., Zor, U., and Lindner, H. R. (1976b). Capacity of immunologically purified FSH to stimulate cyclic AMP accumulation and steroidogenesis in Graafian follicles and to induce ovum maturation and ovulation in the rat. *Endocrinology* 98, 655–661.

Tsafriri, A., Channing, C. P., Pomerantz, S. H., and Lindner, H. R. (1977). Inhibition of maturation of isolated rat oocytes by porcine follicular fluid. *J. Endocrinol.* 75, 285–291.

Tsafriri, A., Weinstein, Y., Bar-Ami, S., Channing, C. P., Pomerantz, S. H., and Lindner, H. R. (1979). The control of meiotic maturation of the rat oocyte. *In* "Research on Steroids" (E. Conti, ed.), Vol. 8, pp. 193–198. Academic Press, New York.

Tsafriri, A., Abisogun, A. O., and Reich, R. (1987). Steroids and follicular rupture at ovulation. *J. Steroid Biochem.* 27, 359–363.

Tsafriri, A., Picard, Y.-Y., and Josso, N. (1988). Immunopurified anti-Müllerian hormone does not inhibit spontaneous maturation *in vitro* of rat oocytes. *Biol. Reprod.* 38, 481–485.

Tsafriri, A., Bicsak, T. A., Cajander, S. B., Ny, T., and Hsueh, A. J. W. (1989a). Suppression of ovulation rate by antibodies to tissue type plasminogen activator and α_2-antiplasmin. *Endocrinology* 124, 415–421.

Tsafriri, A., Vale, W., and Hsueh, A. J. W. (1989b). Effects of transforming growth factors and inhibin-related proteins on rat preovulatory Graafian follicles. *Endocrinology* 125, 1857–1862.

Tsafriri, A., Daphna-Iken, D., Abisogun, A. O., and Reich, R. (1991a). Follicular rupture during ovulation: Activation of collagenolysis. *In* "Advances in Assisted Reproductive Technologies" (S. Mashiach, Z. Ben-Rafael, N. Laufer, and J. G. Schenker, eds.), pp. 103–112. Plenum, New York.

Tsafriri, A., Veljkovic, M. V., Pomerantz, S. H., and Ling, N. (1991b). The action of transforming growth factors and inhibin-related proteins on oocyte maturation. *Bull. L'Assoc. Anatom.* 75, 109–113.

Tsafriri, A., Chun, S. Y., and Reich, R. (1992). Role of eicosanoids in the mediation of LH/hCG stimulation of ovarian collagenase and of ovulation. *In* "8th International Conference on Prostaglandins and Related Compounds," Montreal, Canada. [Abstract 510]

Ueno, S., Manganaro, T. F., and Donahoe, P. K. (1988). Human recombinant Müllerian inhibiting substance inhibition of rat oocyte meiosis is reversed by epidermal growth factor *in vitro*. *Endocrinology* 123, 1652–1659.

Urner, F., Herrman, W. L., Baulieu, E. E., and Schorderet-Slatkine, S. (1983). Inhibition of denuded mouse oocyte meiotic maturation by forskolin, an activator of adenylate cyclase. *Endocrinology* 113, 1170–1172.

Usui, N., and Yanagimachi, R. (1976). Behaviour of hamster sperm nuclei incorporated into eggs at various stages of maturation, fertilization and early development. *J. Ultrastruct. Res.* 57, 276–288.

Van de Wiel, D. F. M., Bar-Ami, S., Tsafriri, A., and de Jong, F. H. (1983). Oocyte maturation inhibitor, inhibin and steroid concentrations in porcine follicular fluid at various stages of the oestrous cycle. *J. Reprod. Fertil.* 68, 247–252.

Vassali, J. D., Sappino, A. P., and Belin, D. (1991). The plasminogen activator/plasmin system. *J. Clin. Invest.* 88, 1067–1072.

Vermeiden, J. P. W., and Zeilmaker, G.H. (1974). Relationship between maturation, division, ovulation and luteinizatioin in the female rat. *Endocrinology* 95, 341–351.

Vivarelli, E., Conti, M., DeFelici, M., and Siracusa, G. (1983). Meiotic resumption and intraclelular cAMP levels in mouse oocytes treated with compounds which act on cMP metabolism. *Cell Differ.* 12, 271–276.

Wallach, E. E., Bronson, R., Hamada, Y., Wright, K. H., and Stevens, V. C. (1975a). Effectiveness of prostaglandin $F_2\alpha$ in restoration of HMG-HCG induced ovulation in indomethacin-treated rhesus monkeys. *Prostaglandins* 10, 129–138.

Wallach, E. E., de Ila Cruz, A., Hunt, J., Wright, K. H., and Stevens, V. C. (1975b). The effect of indomethacin on HMG-HCG induced ovulation in the rhesus monkey. *Prostaglandins* 9, 645–658.

Wallach, E. E., Wright, K. H., and Hamada, Y. (1978). Investigation of mammalian ovulation with an *in vitro* perfused rabbit ovary preparation. *Am. J. Obstet. Gynecol.* 132, 728–738.

Wang, C., and Leung, A. (1989). Glucocorticoids stimulate plasminogen activator production by rat granulosa cells. *Endocrinology* 124, 1595–1601.

Wassarman, P. M., and Letourneau, G. E. (1976). Meiotic maturation of mouse oocytes *in vitro:* association of newly synthesized protein with condensing chromosomes. *J. Cell Sci.* 20, 549–568.

Wasserman, W. J., and Smith, L. D. (1978). The cyclic behavior of a cytoplasmic factor controlling nuclear membrane breakdown. *J. Cell Biol.* 78, R15–R22.

Weintraub, H., Buscaglia, M., Ferrez, M., Weiller, S., Boulet, A., Fabre, F., and Baulieu, E. E. (1982). Mise en evidence d'une activite "MPF" chez *Saccharomyces cerevisiae. C.R. Acad. Sci. Paris Ser. 3* 295, 787–790.

Werb, Z. (1989). Proteinases and matrix degradation. *In* "Textbook of Rheumatology" (W. N. Kelley, E. D., Harris, Jr., S. Ruddy, and C. B. Sledge, eds.) Chap. 18, pp. 300–321. W. B. Saunders, Philadelphia.

Woessner, J. J. F., Butler, W. J., LeMaire, W. J., Morioka, N., Mukaida, T., and Zhu, C. (1989). The role of collagenase in ovulation in the rat. *In* "Follicular Development and the Ovulatory Response" (A. Tsafriri and N. Dekel, eds.), Vol. 23, pp. 167–178. Ares-Serono Symposia, Rome.

Wong, W. Y. L., and Richards, J. S. (1991). Evidence for two antigenically distinct molecular weight variants of prostaglandin H synthase in the rat ovary. *Mol. Endocrinol.* 5, 1269–1279.

Wurtman, R. J. (1964). An effect of luteinizing hormone on the fractional perfusion of the rat ovary. *Endocrinology* 75, 927–933.

Yaish, P., Gazit, A., Gilon, C., and Levitzki, A. (1988). Blocking of EGF-dependent cell proliferation by EGF receptor kinase inhibitors. *Science* 242, 933–935.

Yarden, Y., and Ullrich, A. (1988). Growth facator receptor tyrosine kinases. *Annu. Rev. Biochem.* 57, 443–478.

Yoshimura, Y., and Wallach, E. E. (1987). Studies of the mechanism(s) of mammalian ovulation. *Fertil. Steril.* 47, 22–34.

Yoshimura, Y., Hosoi, Y., Atlas, S. J., Bongiovanni, A. M., Santulli, R., and Wallach, E. E. (1986). The effect of ovarian steroidogenesis on ovulation and fertilizability in the *in vitro* perfused rabbit ovary. *Biol. Reprod.* 35, 943–948.

Yoshimura, Y., Espey, L., Hosoi, Y., Adashi, T., Atlas, S. J., Ghodgoaonkar, R. B., Dubin, N. H., and Wallach, E. E. (1988). The effects of bradykinin on ovulation and prostaglandin production by the perfused rabbit ovary. *Endocrinology* 122, 2540–2546.

Yoshimura, Y., Karube, M., Koyama, N., Shoikawa, S., Nanno, T., and Nakamura, Y. (1992a). Angiotensin II directly induces follicle rupture and oocyte maturation in the rabbit. *FEBS Lett.* 307, 305–308.

Yoshimura, Y., Nakamura, Y., Ando, M., Jinno, M., Oda, T., Karube, M., Koyama, N., and Nanno, T. (1992b). Stimulatory role of cyclic adenosine monophosphate as a mediator of meiotic resumption in rabbit oocytes. *Endocrinology* 131, 351–356.

Yoshimura, Y., Nakamura, Y., Oda, T., Ando, M., Ubukata, Y., Karube, M., Koyama, N., and Yamada, H. (1992c). Induction of meiotic maturation of follicle-enclosed oocytes of rabbits by a transient increase followed by an abrupt decrease in cyclic AMP concentration. *J. Reprod. Fertil.* 95, 803–812.

Zor, U., and Lamprecht, S. A. (1977). Mechanism of prostaglandin action in endocrine glands. *In* "Biochemical Actions of Hormones" (G. Litwack, ed.), Vol. 4, pp. 88–133. Academic Press, New York.

Zuckerman, S., and Baker, T. G. (1977). The development of the ovary and the process of oogenesis. *In* "The Ovary" (S. Zuckerman and B. J. Weir, eds.), 2nd Ed., pp. 41–67. Academic Press, New York.

9

Hormonal Control of Luteal Function

TIMOTHY D. BRADEN, CAROL J. BELFIORE, & GORDON D. NISWENDER

I. Introduction

Formation of the Corpus Luteum

Luteinization and formation of the corpus luteum are hormonal responses to the preovulatory surge of luteinizing hormone (LH). Luteinization of ovarian follicular cells includes morphological and biochemical changes and overall reorganization of these cells into the corpus luteum. Following the preovulatory surge of LH, the basement membrane between the theca interna and membrana granulosa begins to break down. Blood vessels from the theca invade the follicular antral space and, under the influence of angiogenic factors, new and more extensive vascular structures develop. During luteinization, there is significant hypertrophy and hyperplasia of theca cells (O'Shea *et al.*, 1980). As luteinization progresses, theca cells become more loosely packed and migrate into the previous follicular cavity and become dispersed among the luteinizing granulosa cells. Throughout luteinization, granulosa cells accumulate smooth endoplasmic reticulum, mitochondria become rounded with tubulovescicular cristae (Niswender and Nett, 1988), and glycogen-containing granules accumulate. While there is significant mitotic activity in theca cells during luteinization, little if any mitotic activity has been observed in luteinizing granulosa cells (McClellan *et al.*, 1975).

In addition to the morphological changes that occur during luteinization, there is functional differentiation of ovarian cells. In response to the

Molecular Biology of the Female Reproductive System

preovulatory surge of LH, cytochrome P450 side chain cleavage (P450scc) enzyme activity increases with an associated decrease in 17α-hydroxylase cytochrome P450 activity (Rodgers *et al.*, 1986, 1987; Lauber *et al.*, 1991). These alterations allow enhanced breakdown of cholesterol esters to form pregnenolone with subsequent conversion to progesterone as the final product. There are also changes in the populations of receptors for gonadotropins during luteinization. Receptors for FSH on granulosa cells usually disappear following ovulation, probably from internalization of occupied receptors and reduced gene expression for these receptors (Nakamura *et al.*, 1991); however, FSH receptors have been demonstrated in corpora lutea from cows (Manns *et al.*, 1984) and hamsters (Oxberry and Greenwald, 1982). After exposure to the preovulatory surge of LH, receptors for LH are downregulated, most likely as a result of internalization of occupied receptors and reduced gene expression (Segaloff *et al.*, 1990b; LaPolt *et al.*, 1990; Nakamura *et al.*, 1991). Subsequently, enhanced expression of these receptors occurs (approximately 48 hours after exposure to gonadotropins). The remainder of this chapter is primarily concerned with regulation of the mature corpus luteum, given this background.

II. Morphological Aspects of Luteal Function

The corpus luteum of most species consists of a number of cell types. The majority of the cells in the parenchyma of this organ are progesterone-producing cells, referred to as luteal cells. Additional cells, assumed to have a support role, account for approximately 20% of the volume of the corpus luteum and include those associated with vascular areas (endothelial cells, pericytes), macrophages, smooth muscle cells, and fibroblasts (Rodgers *et al.*, 1984; Farin *et al.*, 1986). These populations of cells are assumed to be primarily involved with their conventional roles, although clearly the vascular compartment is involved in the response of the corpus luteum to prostaglandin $F_2\alpha$ ($PGF_2\alpha$) in those species in which it is luteolytic.

A. Steroidogenic Cell Types

1. Morphological properties

There are at least two morphologically and biochemically distinct steroidogenic cell types in the ewe (Fitz *et al.*, 1982; Rodgers and O'Shea, 1982), cow (Ursely and Leymarie, 1979; Koos and Hansel, 1981), pig (Corner, 1919; Lemon and Loir, 1977), goat (Gemmell *et al.*, 1977; Azmi and Bongso, 1985), rat (Wilkinson *et al.*, 1976; Miyauchi and Midgley, 1990; Nelson *et al.*, 1992), rabbit (Hoyer *et al.*, 1986), monkey (Gulyas *et al.*, 1979), and human (Ohara *et al.*, 1987). The two steroidogenic luteal cell types differ in a number of characteristics, with size being the most obvious.

Thus, they are usually designated small luteal cells and large luteal cells. Ovine small luteal cells are approximately 12–22 μm in diameter and usually spindle shaped. They are characterized by an abundance of smooth endoplasmic reticulum, numerous mitochondria, and lipid droplets within the cytoplasm (characteristics consistent with their role as steroid-secreting cells). In contrast, ovine large luteal cells are 22–40 μm in diameter and usually spherical in shape. They contain numerous mitochondria and an abundance of smooth endoplasmic reticulum. In addition, large luteal cells contain stacks of rough endoplasmic reticulum and electron-dense secretory granules that have been shown to contain oxytocin (Sawyer *et al.*, 1979) or relaxin (Fields, 1984), which is released by exocytosis (Gemmell and Stacy, 1979; Sawyer *et al.*, 1979). It seems likely that these granules also contain growth factors, depending on the reproductive state and species.

Several reports have described the relative numbers of each luteal cell type following enzymatic dissociation of the tissue, based on size and steroidogenic capacity (Gamboni *et al.*, 1984; Schwall *et al.*, 1986; Hansel *et al.*, 1987). It is difficult to reliably interpret these data since large and small steroidogenic luteal cells do not survive the dissociation procedure equally well (Rodgers *et al.*, 1984; Braden *et al.*, 1988) and the diameters of small and large luteal cells change during different reproductive states (Farin *et al.*, 1986; Braden *et al.*, 1988). Morphometrical analyses have provided more reliable data on numbers of steroidogenic luteal cells in the ewe and cow (Farin *et al.*, 1986; O'Shea *et al.*, 1986; O'Shea *et al.*, 1989). There are approximately 10–12 million large luteal cells in an ovine corpus luteum and this number remains relatively constant throughout the estrous cycle (Farin *et al.*, 1986; O'Shea *et al.*, 1986). However, in the ewe, the size of large luteal cells increased approximately 50% from the early to midluteal phases of the estrous cycle (Farin *et al.*, 1986). In contrast, there are approximately 10 million small luteal cells per corpus luteum early in the luteal phase, and this number reached a maximum of 50–60 million cells by the midluteal phase (Farin *et al.*, 1986; Braden *et al.*, 1988). Therefore, there are dynamic changes in both populations of steroidogenic luteal cells that are most likely controlled by hormonal factors that have yet to be identified.

2. Biochemical properties

Both types of steroidogenic luteal cells secrete progesterone; however, regulation of this process appears to be different in the two cell types. Under basal conditions, large luteal cells in culture secrete progesterone at a higher rate (2- to 40-fold) than small luteal cells (Lemon and Loir, 1977; Koos and Hansel, 1981; Fitz *et al.*, 1982; Alila *et al.*, 1988a; Hoyer and Kong, Nelson *et al.*, 1992). When the cells are supplemented with cholesterol substrate in the form of high-density lipoprotein (HDL), this difference becomes even more apparent (Wiltbank *et al.*, 1990). Thus, large luteal cells appear to produce over 80% of the progesterone secreted by the corpus luteum during the midluteal phase of the estrous cycle (Niswender *et al.*, 1985). Stimulation of small luteal cells with maximally effective doses of LH resulted in a large increase (up to 40-fold) in secretion of progesterone, while similar exposure of large luteal cells had little or no effect (Lemon and Loir, 1977; Koos and Hansel, 1981; Fitz *et al.*, 1982; Rodgers and O'Shea, 1982; Hoyer and Kong, 1989; Nelson *et al.*, 1992).

This difference between large and small luteal cells in responsiveness to LH may be related to the lack of functional receptors for LH on large cells. Data from initial studies using superovulated ewes to produce large and small steroidogenic luteal cells indicated that small luteal cells possess most of the receptors for LH (Fitz *et al.*, 1982), while the large luteal cells contain the majority of the receptors for estradiol (Glass *et al.*, 1984) and prostaglandins (Fitz *et al.*, 1982). However, in normally cycling animals, a similar number of LH receptors was observed on large and small luteal cells in the ewe (Harrison *et al.*, 1987), cow (Chegini *et al.*, 1991), and rat (Nelson *et al.*, 1992). Treatment of large luteal cells with LH does not result in enhanced intracellular levels of cyclic adenosine monophosphate (cAMP) or secretion of progesterone (Hoyer and Niswender, 1986). Small luteal cells in the bovine corpus luteum may also contain significant numbers of receptors for various prostaglandins (Alila *et al.*, 1988a; Chegini *et al.*, 1991), which was not observed in luteal cells from ewes (Fitz *et al.*, 1982). This disparity may be related to the reported presence of high-affinity sites for $PGF_2\alpha$ on large luteal cells and low-affinity binding sites on small luteal cells (Belapure *et al.*, 1989).

3. Follicular source of origin

Both granulosa and theca cells of the preovulatory follicle appear to contribute to the corpus luteum in most species except horses. Sequential observations of the developing and mature corpus luteum (McClellan *et al.*, 1975; O'Shea *et al.*, 1980) support the hypothesis that granulosa cells become large luteal cells. In addition, the number of large luteal cells generally approximates the number of granulosa cells present in the preovulatory follicle (McNatty *et al.*, 1982; Rodgers *et al.*, 1984; Farin *et al.*, 1986). Perhaps more direct evidence that large luteal cells are of granulosa origin has been obtained through the use of a monoclonal antibody specific for granulosa (vs. theca) cell surface antigens (Alila and Hansel, 1984). Large luteal cells during the early luteal phase bound an antibody specific for granulosa cell antigens. However, not all the large luteal cells bound this antibody; the proportion decreased from a maximum of 77% on Day 4 postestrus to 30% by Days 16–18. Some large luteal cells (10%) bound an antibody specific for antigens from theca cells; this fraction increased dramatically as the luteal phase progressed (35–45% on Days 10–18 postestrus).

Morphological studies by Donaldson and Hansel (1965), Deane *et al.*

(1966), and McClellan *et al.* (1975), support the origin of small luteal cells from theca cells. In addition, a majority (58–70%) of small bovine luteal cells bound theca antigen-specific antibody throughout the luteal phase (Alila and Hansel, 1984). Additional evidence that small luteal cells are derived from theca cells has been obtained through the use of alkaline phosphatase as a marker for theca cells (O'Shea *et al.*, 1980). Taken together, these data suggest that large luteal cells are initially derived from granulosa cells, and small luteal cells from theca cells. However, in the mature corpus luteum, the origin of these cell types is less clear because there are indications that some large luteal cells may be derived from small luteal and/or theca cells.

Given that there are few mitoses in luteal cells and that the number of small luteal cells increases during development of the corpus luteum, the possibility also exists that luteal cells arise from cells outside the follicle. The presence of cells with characteristics intermediate between fibroblasts and small luteal cells has been reported (O'Shea *et al.*, 1980; Farin *et al.*, 1986), indicating a potential differentiation of fibroblasts into small luteal cells. The possibility also exists that small luteal cells can differentiate into large ones. This was suggested by the histological observations of several researchers (Warbritton, 1934; Donaldson and Hansel, 1965). Further evidence for this type of conversion was obtained by Alila and Hansel (1984) using granulosa and theca antigen-specific antibodies to stain dispersed bovine luteal cells (see the discussion above). Treatment of ewes with high doses of LH for 5 days or with human chorionic gonadotropin (hCG) resulted in a significant increase in the number of large luteal cells and a concomitant decrease in the mean number of small luteal cells (Gamboni *et al.*, 1984; Farin *et al.*, 1988). Moreover, in LH-treated animals, several luteal cells were observed that appeared to be "intermediate" between small and large cells. These cells shared some characteristics of both luteal cell types and were not observed in untreated animals (Farin *et al.*, 1988). These data may indicate that small luteal cells are converted to large cells, but it is not known whether this occurs under normal conditions.

B. Luteolysis

Since the initial observations that $PGF_2\alpha$ could induce luteolysis (Phariss and Wyngarden, 1969), data have shown that $PGF_2\alpha$ is the uterine luteolytic factor responsible for normal regression of the corpus luteum (reviewed by Inskeep and Murdoch, 1980; McCracken *et al.*, 1981; Niswender and Nett, 1988). Morphological changes associated with luteal regression, whether natural or induced, have been reported for dogs (Abel *et al.*, 1975), guinea pigs (Crombie *et al*, 1971), hamsters (Leavitt *et al.*, 1973), humans (Van Lenny and Madden, 1965; Adams and Hertig, 1969; Christensen and Gillim, 1969; Crisp *et al.*, 1973), pigs (Cavazos *et al.*, 1969), rabbits (Blanchette, 1966; Koering and Kirton, 1973; Koering and Thor, 1978), rats (Enders and Lyons, 1964), sheep (McClellan *et al.*, 1975; Nett *et al.*, 1976), and cattle (Archbald *et al.*, 1981; Betts *et al.*, 1985). The most striking morphological feature of luteal cells during regression is accumulation of large lipid droplets; cells from nonregressing corpora lutea accumulate only small amounts of lipid within the cytoplasm.

During luteolysis, the number of electron-dense granules in large steroidogenic luteal cells decreases (Umo, 1975; Nett *et al.*, 1976), which suggests that these granules are released under the influence of $PGF_2\alpha$. The appearance of autophagosomes (Nett *et al.*, 1976; Stacy *et al.*, 1976) and an increase in the activity of lysosomal enzymes (McClellan *et al.*, 1975) occurs concurrently with the loss of cytoplasmic granules.

There are also specific changes in the appearance of the vascular compartment during luteal regression. During normal (Niswender *et al.*, 1976) and $PGF_2\alpha$-induced (Nett *et al.*, 1976; Braden *et al.*, 1988) luteolysis, the portion of the ovine corpus luteum occupied by vascular elements decreases. Capillary endothelial cells appear to bulge into the lumen of the capillary, thus reducing lumenal space (Warbritton, 1934; O'Shea *et al.*, 1977). In addition, cellular debris blocks the capillary lumen. Therefore, the characteristic decrease in blood flow to the corpus luteum during luteolysis (Niswender *et al.*, 1976) is likely due to the vasoconstrictor actions of $PGF_2\alpha$, combined with blocking of the capillaries by endothelial cell debris.

There is also a general decrease in the size of steroidogenic cells (McClellan *et al.*, 1975; Nett *et al.*, 1976; Stacy *et al.*, 1976). In the sheep, large luteal cells contain most of the high-affinity plasma membrane receptors for $PGF_2\alpha$ (Fitz *et al.*, 1982; Belapure *et al.*, 1989); therefore, it would be expected that during luteolysis, the large luteal cell populations would be primarily affected. Morphometrical analysis of corpora lutea at specific times after administration of a luteolytic dose of $PGF_2\alpha$ indicated that the number of large luteal cells does not change for at least 36 hours (Braden *et al.*, 1988), well after functional luteolysis as defined by a decrease in serum levels of progesterone. However, the diameter of large luteal cells did decrease as regression of the corpus luteum progressed (Braden *et al.*, 1988). Fewer large luteal cells survived enzymatic dissociation as regression progressed, which is indicative of increased fragility in this cell population. Carlson *et al.* (1989) reported changes in the plasma membrane fluidity of luteal cells after exposure to $PGF_2\alpha$. By 24 hours after $PGF_2\alpha$ treatment, there was a significant decrease in the number of small luteal cells, with no change in size (Braden *et al.*, 1988).

It has become apparent that the antisteroidogenic actions of $PGF_2\alpha$ appear to be mediated through activation of the protein kinase C (PKC) second messenger system, while its cytotoxic effects on large cells are thought to result from a sustained, $PGF_2\alpha$-induced increase in intracellular levels of calcium (Wiltbank *et al.*, 1992). The increased intracellular levels of calcium may induce the characteristic nuclear changes seen after $PGF_2\alpha$ administration (Sawyer *et al.*, 1990) and ultimately the biochemical changes associated with apoptosis (Juengel *et al.*, 1993). The details of these responses are discussed in Section III,A,2.

III. Hormonal Control of Luteal Steroidogenesis

Steroid Biosynthesis

During the estrous cycle, secretion of progesterone and maintenance of the corpus luteum are dependent on LH (Hansel *et al.*, 1973; Niswender *et al.*, 1980; Niswender and Nett, 1988). Notable exceptions are the rat and canine corpus luteum (which are also dependent on prolactin) (reviewed by Niswender *et al.*, 1985; Gibori *et al.*, 1988), and the rabbit corpus luteum (which is also dependent on estrogens) (Keys and Nalbandov, 1967; Keyes and Armstrong, 1968; Spies *et al.*, 1968). The fact that large luteal cells do not respond to LH with increased secretion of progesterone, yet are the primary source of this steroid makes these early observations more intriguing.

1. Effects of LH

a. Receptors for LH

With the recent identification of the structure of the LH receptor, and the development of probes to investigate functions of specific regions within this receptor, new information on the molecular characteristics of this receptor is quickly becoming available. The LH receptor is composed of a single polypeptide chain with a predicted sequence of 674 amino acid residues. This chain contains two major domains—a 341-residue amino terminal which forms the extracellular portion of the receptor (Rodriguez and Segaloff, 1990) and a 333-residue C terminus which contains seven hydrophobic regions thought to represent transmembrane regions characteristic of G protein-linked receptors (McFarland *et al.*, 1989). Compared with other receptors, the extracellular domain of the LH receptor is large and contains six potential glycosylation sites. Recent studies have indicated that elimination of potential glycosylation sites at Asn^{77}, Asn^{152}, or Asn^{173} results in significant decreases in hormone binding to these mutant receptors (Zhang *et al.*, 1991). The cytoplasmic domain contains numerous potential phosphorylation sites and two clusters of basic amino acids which may serve as cleavage sites. Studies using LH receptors that have truncations of the cytoplasmic tail have indicated that regions between residues 616 and 631 are required for proper insertion and/or targeting of the receptor into the plasma membrane, and that responsiveness to agonist binding with cAMP generation as well as internalization of agonist was not impaired in mutants lacking 43 residues at the C terminus (Rodriguez *et al.*, 1992). Specific regions of the LH receptor that mediate other intracellular effects of this receptor remain to be identified.

The rat has a single copy of the gene encoding the LH receptor in its genome (Koo *et al.*, 1991), which contains 11 exons. Exons 1–10 code for the N-terminal half of the molecule (hormone-binding region), and exon 11 codes for the C-terminal half (transmembrane regions and cytoplasmic tail; Koo *et al.*, 1991). Multiple species of mRNA have been observed (3–5) and are expressed in sizes of 1.1–4.7, 6.7 (abundant in testis), 5.8, 4.0, 2.6, and 1.2–1.4 kb (McFarland *et al.*, 1989; Loosfelt *et al.*, 1989). The LH receptor gene has been localized to chromosome 2p21 in the human (Rousseau-Merk *et al.*, 1990) and 3q2.2 → q2.3 in the pig (Yerle *et al.*, 1992).

b. Steroidogenic effects of LH

As described earlier, both types of steroidogenic luteal cells are capable of secreting progesterone; however, regulation of this process appears to differ. The steroidogenic pathway for a generic luteal cell is shown in Figure 1. Steroidogenic cells are capable of producing cholesterol from acetate (Gwynne and Strauss, 1982). However, the preferred substrate for progesterone synthesis consists of cholesterol esters or cholesterol delivered to steroidogenic cells by lipoprotein, rather than from cholesterol synthesized *de novo* (Gwynne and Strauss, 1982). Lipoproteins are present in blood primarily as high-density and low-density species. Uptake of available substrate carried in low-density lipoproteins is regulated through receptor-mediated endocytosis (Brown and Goldstein, 1986). High-density lipoprotein has been shown to bind a 58-kDa polypeptide membrane receptor (Ferreri and Menon, 1990). Following binding to the membrane, relatively less high-density lipoprotein than lipid is internalized by the cell (Schuler *et al.*, 1981; Paavola and Strauss, 1983), indicating delivery of substrate by a mechanism distinctly different from the receptor-mediated endocytic uptake of low-density lipoprotein.

The rate-limiting enzyme involved in biosynthesis of cholesterol from acetyl coenzyme A is 3-hydroxy-3-methyl glutaryl coenzyme A reductase, or HMG-CoA reductase (Luskey *et al.*, 1988). Cholesterol not immediately inserted into the plasma membrane or used as substrate for steroid hormones is esterified through the action of acyl coenzyme A:cholesterol acyltransferase, or ACAT (Schuler *et al.*, 1979), and stored in cytoplasmic lipid droplets (Flint and Armstrong, 1973). The stored cholesterol esters are deesterified by cholesterol esterase (cholesterol ester hydrolase (Caffrey *et al.*, 1979a). Cholesterol is then transported to the inner mitochondrial membrane, the site of cholesterol side chain cleavage. The cytochrome P450 side chain cleavage enzyme is associated with adrenodoxin and adrenodoxin reductase, through which electrons necessary for hydroxylation are acquired from NADPH (Kimura, 1981; Lambeth *et al.*, 1982). Hy-

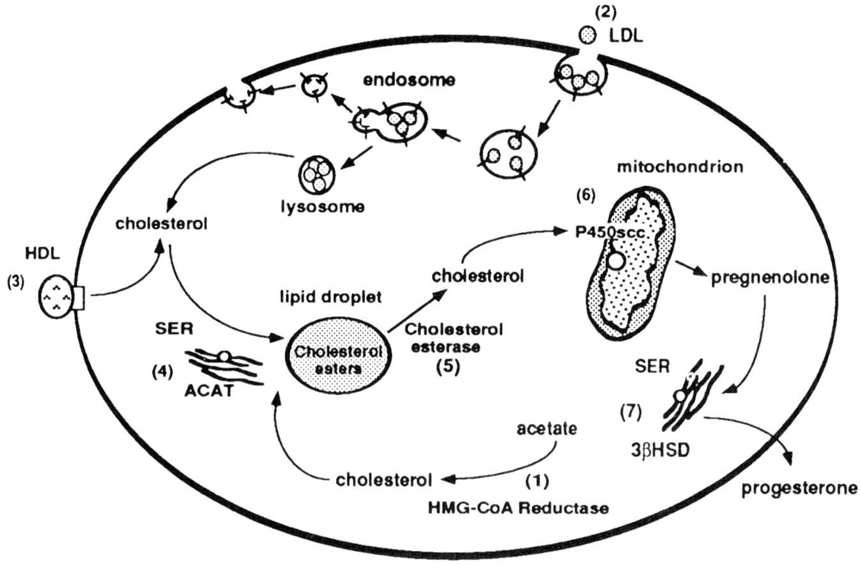

FIGURE 1 Steroidogenic luteal cell, illustrating enzymes and processes involved in synthesis of progesterone. *De novo* cholesterol synthesis (1) proceeds from acetate through action of, primarily, 3-hydroxy-3-methyl glutaryl coenzyme A reductase (HMG-CoA reductase). Cholesterol and cholesterol esters are also taken up (2) from low-density lipoprotein (LDL) via receptor-mediated endocytosis or (3) from high-density lipoprotein (HDL) through a nonendocytic process. Cholesterol is esterified (4) by acyl coenzyme A:cholesterol acyltransferase (ACAT) for storage in lipid droplets. Cholesterol esterase (CE) effects deesterification (5), and the resultant cholesterol is transported (6) to the inner mitochondrial membrane where cytochrome P450scc cleaves the side chain of cholesterol to produce pregnenolone. Pregnenolone proceeds (7) to the smooth endoplasmic reticulum (SER), where it is converted to progesterone by 3β-hydroxysteroid dehydrogenase/Δ^5-Δ^4 isomerase (3β-HSD).

droxylation at the C-22 and C-20 positions is followed by cleavage of the bond between the hydroxylated carbons to form pregnenolone and isocaproaldehyde (Shikita and Hall, 1974). This conversion of cholesterol to pregnenolone has been shown to be the rate-limiting step in steroid synthesis (Halkerston *et al.*, 1961). The final enzymatic transformation to progesterone occurs in the microsomal fraction of luteal cells (Caffrey *et al.*, 1979b). Here 3β-hydroxysteroid dehydrogenase/Δ^5-Δ^4 isomerase (3β-HSD) effects oxidation of the C-3 hydroxyl group and isomerization of the double bond from Δ^5 to Δ^4 (Samuels *et al.*, 1951).

Delivery of the hydrophobic cholesterol, cholesterol ester, and pregnenolone from aqueous cytosol to the enzymes responsible for their eventual conversion to progesterone requires mechanisms beyond simple diffusion. There is evidence for both lipid carrier protein and cytoskeletal involvement, neither mechanism being necessarily exclusive of the other.

Inhibition of protein synthesis blocks both mitochondrial intramembrane movement of cholesterol and steroid synthesis by a variety of steroidogenic tissues (Davis and Garren, 1968; Jefcoate *et al.*, 1974; Cooke *et al.*, 1975). Numerous proteins have been proposed as facilitators of steroidogenesis (Noland *et al.*, 1980; Pon and Orme-Johnson, 1986; Yanagibashi *et al.*, 1988). Cytoskeletal involvement in steroidogenesis has also been suggested by decreased steroid production following treatment with disruptors of microtubules or microfilaments (Sawyer *et al.*, 1979; Silavin *et al.*, 1980).

In small cells, LH induced secretion of progesterone is thought to occur through cAMP production by adenylate cyclase activation of protein kinase A (PKA) (Hoyer and Niswender, 1986). The stimulatory effects of LH can be mimicked by analogues of cAMP or hormone-independent activation of adenylate cyclase (Hoyer and Niswender, 1986). The presence of cAMP-binding sites (a regulatory subunit of PKA) has been demonstrated in both large and small luteal cells; however, increased intracellular levels of cAMP result in enhanced secretion of progesterone only in small luteal cells. While there is some evidence for LH-stimulated release of progesterone from large luteal cells (which was most consistently observed with bovine luteal cells) (Alila *et al.*, 1988a; Farin *et al.*, 1989; Hansel *et al.*, 1991), maximal effective doses of LH only double progesterone secretion from large luteal cells. This limited effect could be due to contamination of the preparation of large luteal cells used for these studies with small luteal cells.

Potential intracellular sites of PKA activation include synthesis and activity of steroidogenic enzymes, intracellular transport of substrate, and uptake of substrate from lipoproteins. The number of receptors for low-density lipoproteins was enhanced in human granulosa cells treated with hCG or 8-bromo-cAMP (Golos and Strauss, 1987). Luteinizing hormone or hCG treatment also increased binding of high-density lipoproteins to rat luteal tissue (Schuler *et al.*, 1981; Strauss *et al.*, 1982). Reduced negative feedback of cholesterol appears to be the mechanism for control of lipoprotein receptor synthesis (Brown and Goldstein, 1975; Talavera and Menon, 1989).

Effects of PKA activation on steroidogenic enzyme activity appear to be mediated directly through phosphorylation or alteration of cellular concentrations of cholesterol. Stimulation of steroidogenesis, leading to depletion of cholesterol, resulted in increased transcription of MHG-CoA reductase (Rodgers *et al.*, 1987; Golos and Strauss, 1988), while supplementation of cellular cholesterol by lipoproteins reduced *de novo* cholesterol synthesis (Pate and Condon, 1989). Esterification of cholesterol by ACAT occurred in the presence of adequate substrate (Schuler *et al.*, 1979). The activity of cholesterol esterase appears to be regulated directly through phosphorylation by PKA (Bisgaier *et al.*, 1979; Caffrey *et al.*, 1979a; Wiltbank *et al.*, 1993).

Synthesis of components of the side chain cleavage enzyme complex appears to be induced during luteinization (Rodgers *et al.*, 1986, 1987; Hedin *et al.*, 1987). Gonadotropic stimulation of granulosa cells increased adrenodoxin reductase, adrenodoxin, and P450scc protein, as well as mRNA encoding the p0roteins (Trzeciak *et al.*, 1986; Golos *et al.*, 1987). However, in luteal tissue, the levels of mRNA encoding P450scc enzyme appeared to be less responsive to hormonal regulation (Richards *et al.*, 1986; Oonk *et al.*, 1989; Belfiore *et al.*, 1993). Assessment of ovine luteal side chain cleavage activity using hydroxylated cholesterol as substrate, however, indicated no acute effect of PKA activation on the activity of the P450scc side chain cleavage enzyme (Wiltbank *et al.*, 1993; Belfiore, 1993).

Induction of mRNA encoding 3β-HSD by gonadotropin has been noted in granulosa cells (Chedrese *et al.*, 1991). In midcycle luteal tissue, expression of both 3β-HSD enzyme and its mRNA appear to be relatively constant (Couet *et al.*, 1990; Hawkins *et al.*, 1993). Activity of 3β-HSD has been viewed as unregulated because of its high cellular concentration compared with other steroidogenic enzymes (Caffrey *et al.*, 1979b). Acute stimulation of enzyme activity was not evident in ovine luteal cells treated with LH or forskolin (Wiltbank *et al.*, 1993).

Besides the direct effects of second messenger systems on the activity and synthesis of steroidogenic enzymes themselves, there is also ever-increasing evidence that intracellular transport of steroidogenic substrates is regulated. Activation of PKA affects the arrangement of cytoskeletal components (Gospodarowicz and Gospodarowicz, 1975; Soto *et al.*, 1986), altering the shape of cultured steroidogenic cells. Such rearrangement could result in translocation of intracellular organelles involved in steroid biosynthesis (Paavola *et al.*, 1985) and might bring substrate and enzyme into close proximity. Both protein synthesis and phosphorylation of proteins by PKA appear to be critical for progesterone biosynthesis. Synthesis of steroid carrier protein-2 (SCP2) (Noland *et al.*, 1980) responded to gonadotropic regulation in rat Leydig cells (van Noort *et al.*, 1986). Increases in mRNA encoding SCP2 which paralleled mRNA encoding P450scc enzyme occurred in gonadotropin-primed rat ovary (Billheimer *et al.*, 1990). Mitochondrial phosphoproteins 28 to 32 kDa in size have been detected by Pon and Orme-Johnson (1986) in rat corpus luteum following gonadotropic stimulation; they suggested that these proteins were involved in the transport of cholesterol. Direct evidence for enhancement of cholesterol transport to the side chain cleavage enzyme by gonadotropic stimulation has been indicated by increased association of cholesterol with P450scc enzyme in rat luteal tissue (Ghosh *et al.*, 1987).

c. Other effects of LH

In addition to the steroidogenic effects of LH on luteal cells, other effects have been proposed. A number of studies have shown desensitization to LH (Hunzicker-Dunn, 1976; Niswender *et al.*, 1985) and homologous downregulation of receptor numbers (Richards and Williams, 1976; Suter *et al.*, 1980). Homologous downregulation of receptors for LH is associated with decreased quantities of LH receptor mRNA (Segaloff *et al.*, 1990b). Studies evaluating the very early (10 minutes post-treatment) response of ovine luteal cells to administration of LH provided evidence that LH also caused upregulation of its own receptor (Suter *et al.*, 1980).

While it is generally held that LH stimulates release of progesterone from small luteal cells through generation of cAMP and activation of PKA, the results of recent studies have also suggested activation of PKC by LH-hCG. Davis *et al.* (1987) reported increased inositol phospholipid turnover and intracellular calcium in LH-treated dispersed bovine luteal cells, and noted simultaneous increases in dinositol triphosphate (IP3) and cAMP in bovine small luteal cells (Davis *et al.*, 1989). Benhaim *et al.* (1990) suggested that LH triggered a negative regulatory signal beyond cAMP synthesis, and that this signal might consist of phospholipase C activation. Allen *et al.* (1988), using porcine luteal cells, also suggested that gonadotropins stimulated inositol phospholipid hydrolysis. Stimulation of inositol phospholipid metabolism was not seen in studies of rat corpora lutea (Schuler *et al.*, 1978), human granulosa-luteal cells (Davis *et al.*, 1989), or ovine luteal cells (Jacobs *et al.*, 1987). It should be pointed out that activation of PKC was not directly measured in any of these studies.

Other investigators have reported increases in intracellular calcium in suspensions of small bovine luteal cells after exposure to LH (Alila *et al.*, 1989). However, LH did not influence calcium levels in ovine small luteal cells under conditions where the secretion of progesterone was dramatically increased (Wiltbank *et al.*, 1991). Thus, there is controversy as to whether LH induces calcium influx as part of its actions.

2. Effects of $PGF_2\alpha$

Many theories have been advanced to explain the mechanisms of $PGF_2\alpha$-induced luteolysis. These include an alteration in the number of receptors for LH (Hichens *et al.*, 1974), changes in luteal blood flow (Nett *et al.*, 1976; Niswender *et al.*, 1976), uncoupling of adenylate cyclase from LH receptors (Behrman *et al.*, 1979; Fletcher and Niswender, 1982), and a direct cytotoxic effect (Silvia *et al.*, 1984; Fitz *et al.*, 1984). The conclusions from these early studies must be reevaluated in light of recent data that large luteal cells appear to be the primary target for $PGF_2\alpha$ while small cells appear to be the physiological target for LH.

More recent studies have revealed that $PGF_2\alpha$ appears to influence the function of large luteal cells in sheep through two distinct mechanisms: (1) activation of PKC, which inhibits the secretion of progesterone (Wiltbank *et al.*, 1989b), and (2) binding that causes an influx of extracellular calcium (Wiltbank *et al.*, 1989a), which appears to be responsible for the cytotoxic effects of this hormone. Treatment of small luteal cells with PKC activators has yielded differing results with respect to basal progesterone production —no effect (Hansel *et al.*, 1987; Wiltbank *et al.*, 1989b) or stimulation (Alila *et al.*, 1988b; Benhaim *et al.*, 1990). However, a consistent attenuation of LH-stimulated progesterone release from small luteal cells has been observed (Alila *et al.*, 1988b; Hoyer and Marion, 1989; Wiltbank *et al.*, 1989b). PKC exerted this effect prior to cholesterol side chain cleavage (Wiltbank *et al.*, 1989b). Activation of PKC in large luteal cells by phorbol esters resulted in either no effect (Alila *et al.*, 1988b) or suppression of basal production of progesterone by these cells (Hoyer and Marion, 1989; Wiltbank *et al.*, 1989b). PKC activity has been shown to be present in large luteal cells (Wiltbank *et al.*, 1989b; Hoyer and Kong, 1989) and the enzyme activity translocated to the plasma membrane when activated by phorbol esters or $PGF_2\alpha$ (Wiltbank *et al.*, 1989b). Thus, it appears that $PGF_2\alpha$ activates PKC and that activation of this enzyme results in reduced secretion of progesterone from LH-stimulated small luteal cells and from large luteal cells. Interestingly, treatment of ewes with phorbol ester to pharmacologically activate the PKC system on Day 10 of the estrous cycle resulted in a rapid decrease in serum levels of progesterone without any evidence of luteolysis (McGuire *et al.*, 1992). Thus, the concepts developed using purified preparations of luteal cells *in vitro* appear to be correct *in vivo*.

Although it seems clear that at least some of the antisteroidogenic actions of $PGF_2\alpha$ are mediated through the PKC second messenger system, the precise cellular mechanisms are not clear. Activation of PKC potentially affects the same steroidogenic enzymes and processes (Figure 1) as does activation of PKA. $PGF_2\alpha$ treatment does not appear to affect binding or uptake of lipoprotein substrate in rat (Rajkumar *et al.*, 1985b) or bovine luteal cells (Grusenmeyer and Pate, 1992). Hepatic 3-hydroxyl 3-methylglutaryl coenzyme A reductase (HMG-CoA reductase) activity is inhibited through reversible phosphorylation of the enzyme (Ingebritsen *et al.*, 1979; Zammit and Easom, 1987). Pate and Condon (1989) observed decreased *de novo* cholesterol synthesis by bovine luteal cells following treatment with $PGF_2\alpha$. This effect might readily be attributed to phosphorylation activity by PKC because reductase kinase (which phosphorylates the HMG-CoA reductase) is active in the phosphorylated form (Ingebritsen *et al.*, 1979). Behrman *et al.* (1970, 1971) observed decreased activity of both ACAT and cholesterol esterase in rat luteal cells following $PGF_2\alpha$ treatment. However, it was concluded that this effect was not PKC mediated. PMA (phorbol 12-myristate 13-acetate) activation of PKA had no effect on ovine luteal cholesterol esterase activity (Wiltbank *et al.*, 1993).

Likewise, when lipoproteins were used to facilitate uptake of exogenous substrate, PMA had no effect on cholesterol side chain cleavage activity (Belfiore, 1993; Wiltbank *et al.*, 1993). *In vivo* treatment of ewes with $PGF_2\alpha$ did not influence the levels of mRNA encoding P450scc, despite significantly decreased production of progesterone (Belfiore *et al.*, 1992).

However, treatment of ewes with $PGF_2\alpha$ resulted in a decrease in message encoding 3β-HSD (Hawkins *et al.*, 1993). Treatment of ovine luteal cells with PMA had no acute effect on the activity of 3β-HSD (Wiltbank *et al.*, 1993).

Involvement of PKC in substrate movement is suggested by phosphorylation of SCP2 by PKC but not by PKA or calmodulin kinase II (Steinschneider *et al.*, 1989). In addition, increased mitochondrial cholesterol noted in bovine luteal cells treated with $PGF_2\alpha$ (Grusenmeyer and Pate, 1992) may indicate inhibition of a process involved in transporting cholesterol to P450scc.

While some antisteroidogenic effects of $PGF_2\alpha$ are mediated by PKC activation, not all the actions of $PGF_2\alpha$ appear to require activation of PKC. Administration of $PGF_2\alpha$ to large luteal cells which had been depleted of PKC activity resulted in cell damage and eventual cell death, although no acute reduction in progesterone was observed (Wiltbank *et al.*, unpublished observations). Moreover, hormone-independent activation of PKC by phorbol esters did not result in cell death (Hoyer and Marion, 1989; Wiltbank *et al.*, 1989b). Therefore, although the initial antisteroidogenic actions of $PGF_2\alpha$ appear to be mediated through PKC activation, the cytotoxic effects of $PGF_2\alpha$ appear to be mediated by another mechanism.

Exposure of large luteal cells to $PGF_2\alpha$ resulted in increased intracellular calcium levels in single large luteal cells (Wiltbank *et al.*, 1989a) or large luteal cell suspensions (Wegner *et al.*, 1990). No change in intracellular calcium concentrations was detected in small luteal cells exposed to $PGF_2\alpha$ (Wiltbank *et al.*, 1989a; Wegner *et al.*, 1990). After $PGF_2\alpha$ stimulation, increased intracellular calcium concentrations remained elevated for a relatively long time compared with ionophore-treated small luteal cells, indicating a less efficient mechanism for equilibration of calcium in large luteal cells (Wiltbank *et al.*, 1989a; Wegner *et al.*, 1990). Hoyer and Marion (1989) showed that sustained elevations of intracellular levels of calcium are toxic to ovine luteal cells. This may be an explanation for the direct cytotoxic effects of $PGF_2\alpha$ observed on large luteal cells (Fitz *et al.*, 1984; Silvia *et al.*, 1984). Recent evidence indicates that $PGF_2\alpha$ induces oligonucleosome formation (Juengel *et al.*, 1993), which may be involved in apoptosis during the luteolytic process.

3. Additional regulators of luteal function

Clearly, the primary function of the corpus luteum is production of progesterone for maintenance of early pregnancy. The primary regulators of progesterone production appear to be LH and $PGF_2\alpha$. However, there are also other significant products formed by the corpus luteum which may be involved in paracrine regulation of this gland. These additional products include oxytocin, vasopressin, and relaxin; growth factors such as insulin, insulin-like growth factors I and II; GnRH-like peptides, several eicosanoids in addition to $PGF_2\alpha$, and some biogenic amines such as epinephrine, serotonin, and dopamine. It seems likely that any or all of these factors might be involved in the cell-to-cell communication which must be a continuous activity among the different cell types in the corpus luteum. Table 1 contains a summary of these various factors and their effects on luteal function. Further research is needed to define the physiological significance of each of the potential paracrine and/or autocrine regulators of luteal function.

Oxytocin may have both a paracrine and an endocrine role in the regulation of luteal function, although its endocrine role is most clearly understood. Exogenous administration of oxytocin *in vivo* can result in luteolysis (Armstrong and Hansel, 1959; Cooke and Knifton, 1981; Auletta *et al.*, 1984); however, the luteolytic effect requires a uterus. The prevailing experimental data suggest that release of $PGF_2\alpha$ from the uterus stimulates the release of oxytocin from the corpus luteum, which in turn stimulates further $PGF_2\alpha$ release from the uterus (Flint *et al.*, 1990). This positive feedback loop appears to be responsible for the large pulses of circulating

TABLE 1 Effects of Various Agents on Luteal Function

Agent[a]	Species	Actions	Reference
ANP	Cow	↑ cGMP accumulation	Vollmar *et al.* (1988)
Biogenic amines	Cow	↑ P release	Alila and Dowd (1991)
Growth factors			
EGF	Rat	↓ Gonadotropin-stimulated P release	Arakawa *et al.* (1990)
IGF-I	Rat	↑ P release from large luteal cells	Dowd *et al.* (1990)
	Cow	↑ Oxytocin release; ↑ P release	McArdle and Holtorf (1989)
	Rat	↑ P release	Talavera and Menon (1991)
	Rabbit	↑ P release	Constantino *et al.* (1991)
IGF-II	Cow	NE-P release	McArdle and Holtorf (1989)
bFGF	Pig	↓ Relaxin release	Taylor and Clark (1992)
Insulin	Cow	↑ Oxytocin and P release	McArdle and Holtorf (1989)
Eicosanoids			
PGE_2, PGI_2	Sheep	↑ P release	Fitz *et al.* 1984
	Cow	↑ P release	Alila *et al.* 1988a
5-HETE	Cow	↓ Prelease	Milvae *et al.* (1986)
Interferon-γ	Cow	↑ PG; ↓ PG; ↓ gonadotropin-stimulated P release	Fairchild and Pate (1991)
IL-1β	Cow	↑ PG release; $NE\text{-}P_4$ release	Nothnick and Pate (1990)
TNF	Cow	↑ PG release; ↓ P release	Benyo and Pate (1992)
Neuropeptide Y	Pig	↓ Basal and gonadotropin-stimulated P release	Pitzel *et al.* (1991)
Substance P	Pig	↓ Basal and gonadotropin-stimulated P release	Pitzel *et al.* (1991)

[a] ANP, atrial natriuretic factor; EGF, epidermal growth factor; IGF, insulin-like growth factor; IL interleukin; NE, no effect; P, progesterone; PG, prostaglandins; TNF, tumor necrosis factor; bFGF, basic fibroblast growth factor.

$PGF_2\alpha$ associated with luteolysis. Estradiol appears to be involved in initiating this positive feedback loop by increasing the numbers of endometrial receptors for oxytocin and thus increasing its sensitivity to this hormone (Fairclough *et al.*, 1980; Flint and Sheldrick, 1983) and by stimulating oxytocin secretion from the posterior pituitary and/or corpus luteum (Sheldrick and Flint, 1985; Hixon and Flint, 1987).

Oxytocin also appears to have direct effects on the corpus luteum. Exposure of luteal cells to very high doses of oxytocin inhibited basal and gonadotropin-stimulated progesterone production (Tan *et al.*, 1982). Lower doses of oxytocin appeared to stimulate the release of progesterone by luteal cells. Subsequent studies have failed to substantiate these observations and indicate that progesterone production by luteal cells from newly formed corpora lutea (Days 4 and 5 postovulation) was inhibited by oxytocin (Schams, 1987), but no effect was observed on "midcycle" luteal cells (Rodgers *et al.*, 1985). A study by Miyamoto and Schams (1991) indicates that more consistent results with oxytocin may be obtained through the use of a microdialysis system that maintains more of the original structure of the corpus luteum. These authors observed a clear, dose-dependent stimulation of progesterone release with oxytocin that was comparable in magnitude to progesterone release in the presence of LH. Oxytocin binding sites in ovine, bovine, and human corpora lutea (Sernia *et al.*, 1989; Fuchs *et al.*, 1990) provide more evidence for an effect of oxytocin on luteal cells. Because these studies have used ovarian tissues that contain steroidogenic as well as nonsteroidogenic cells (i.e., fibroblasts, endothelial cells, smooth muscle cells), the possibility exists that these reports have described oxytocin binding sites that are separate from steroidogenic cells. Sernia *et al.* (1989) did not observe oxytocin binding sites in nonpregnant animals which should have similar nonsteroidogenic components in the corpus luteum. In a study by Fuchs *et al.* (1990), oxytocin binding was significantly different between Days 7–14 and 17–21 of the bovine estrous cycle when dramatic changes in the nonsteroidogenic populations of cells would not be expected. Thus, there is some evidence for a direct role of oxytocin in regulating luteal steroidogenesis. However, further studies are required before a clear role for oxytocin can be established.

In summary, regulation of the secretion of progesterone is an extremely complex process involving the actions of multiple hormones on at least two different steroidogenic cell types.

IV. Maternal Recognition of Pregnancy

It is now generally accepted that the conceptus in sheep and cattle secretes a trophoblastic interferon (oTP-1; bTP-1) which inhibits pulsatile secretion of $PGF_2\alpha$ from the pregnant uterus (for reviews see R. M. Roberts, 1991; Bazer *et al.*, 1992a). There is considerable homology between the structures of oTP-1 and interferon-α (Imakawa *et al.*, 1989), and both molecules have antiviral activity (Poniter *et al.*, 1988) and reduce pulsatile secretion of $PGF_2\alpha$ from the uterus (Vallet *et al.*, 1988). However, if frequent uterine-venous blood samples are collected to allow precise determination of secretory patterns, it appears that basal levels of $PGF_2\alpha$ are higher in pregnant ewes than in cycling ewes (Silvia, 1985). Thus, although pulsatile secretion of $PGF_2\alpha$ is attenuated, the total quantities of $PGF_2\alpha$ secreted are not dramatically reduced in pregnant ewes (Ellinwood *et al.*, 1979; Roberts, 1991). Secretion of PGE_2 is also increased during the time of maternal recognition of pregnancy (Ellinwood *et al.*, 1979; Marcus *et al.*, 1981). PGE_2 has been shown to stimulate production of progesterone by ovine and bovine luteal tissue (Marsh, 1975; Fitz *et al.*, 1984) and recently has been demonstrated to attenuate the increase in intracellular levels of calcium induced by $PGF_2\alpha$ (Wiepz *et al.*, 1993). It is not clear whether the increased secretion of PGE_2 is due to a direct effect of oTP-1 on the endometrium, whether it is secreted from the conceptus itself (LaCroix and Kann, 1982), or whether it is due to some other mechanism. However, it does appear unequivocal that oTP-1 is one of the major factors involved in maternal recognition of pregnancy and that its effects are mediated at the level of the uterus to alter secretion of prostaglandins.

There is also considerable evidence that factors other than oTP-1 are involved in maternal recognition of pregnancy. For example, when embryos were flushed from the uterus of ewes 14–15 days postmating, the interestrous interval was extended to at least 35 days (Moor and Rowson, 1966; Ellinwood *et al.*, 1979) while removal prior to Day 12 had no effect. However, in numerous studies, if oTP-1 (Vallet *et al.*, 1988; Ott *et al.*, 1991) or interferon-α (Nephew *et al.*, 1990; Davis *et al.*, 1992) was administered into the uterus during the critical period for maternal recognition of pregnancy (Days 12–16) the interestrous interval ranged from 17 to 28 days. This strongly suggests that some factor other than oTP-1 is required to extend the interestrous interval to >35 days. Martal *et al.* (1990) observed no effect from 80 or 170 μg of recombinant oTP-1 and found a very variable effect with 340 μg of oTP-1 on interestrous intervals. On the other hand, Bazer *et al.* (1992b) extended interestrous intervals dramatically with 200 μg of recombinant oTP-1. Dose, method of administration, purity of the preparations, or the relatively small number of animals studied may have influenced the results. It is clear from this discussion that there is still controversy regarding the effects of oTP-1 and it seems very likely that it is not the only factor involved in maternal recognition of pregnancy in sheep.

A second and perhaps more important line of evidence also suggests that factors other than oTP-1 are involved in maternal recognition of pregnancy. At least three studies from three different laboratories have demonstrated that the presence of an embryo(s) in the uterus renders corpora lutea resistant to luteolytic actions of $PGF_2\alpha$ during the time of maternal recognition on pregnancy. In the initial study, Inskeep *et al.* (1975) concluded that the presence of an embryo made the corpus luteum resistant to $PGF_2\alpha$ administered on Day 12 of pregnancy. Silvia and Niswender (1984) injected 4 mg/58 kg body weight of $PGF_2\alpha$ into cycling or pregnant ewes on Days 10 and 13 postestrus. Following injection, the decrease in serum levels of progesterone was similar in cycling ewes injected on Days 10 or 13 and in pregnant ewes injected on Day 10, and were indicative of premature luteolysis. However, $PGF_2\alpha$ injection had no effect in ewes 13 days pregnant, clearly indicating a resistance to $PGF_2\alpha$-induced luteolysis due to pregnancy on Day 13 postmating. In a subsequent study, these investigators further demonstrated that the resistance to $PGF_2\alpha$ was a function of day of pregnancy. Luteolysis was observed in ewes on Days 10, 19, 22, 26, and 30 of pregnancy but serum concentrations of progesterone did not change in pregnant ewes treated 13 or 16 days postmating (Silvia and Niswender, 1986). In fact, the level of protective action appears to be correlated with the number of embryos present (Nancarrow *et al.*, 1982; Silvia *et al.*, 1984). These data clearly indicated that a specific factor associated with pregnancy caused the corpus luteum of pregnancy to be resistant to the luteolytic actions of $PGF_2\alpha$.

Wiltbank *et al.* (1992) recently reported a 35-kDa (SP35) protein secreted from Day 15 ovine conceptuses incubated *in vitro* which inhibited both the antisteroidogenic and cytotoxic effects of $PGF_2\alpha$ on purified ovine large luteal cells. The factor (SP35) did not bind $PGF_2\alpha$ nor did it prevent $PGF_2\alpha$ binding to its receptor on luteal cells. Thus, it appears to have a direct, receptor-mediated effect on large luteal cells. A protein with biological actions similar to SP35 is secreted from the bovine conceptus on Days 17 and 18 postmating (Diskin, Wiltbank, and Niswender, unpublished data). This protein is interesting for three reasons: (1) It is a protein that is clearly secreted from the embryo; (2) it is secreted during the critical time when maternal recognition of pregnancy is taking place; and (3) it has biological effects which could explain the resistance of the corpus luteum of pregnancy to $PGF_2\alpha$. However, further research is required to establish whether it plays a biologically significant role in maternal recognition of pregnancy in ruminants.

References

Abel, J. H., Jr., Verhage, H. G., McClellan, M. C., and Niswender, G. D. (1975). Subcellular compartmentalization of the luteal cell in the ovary of the dog. *Cell Tissue Res.* 158, 461–480.

Adams, E. C., and Hertig, A. T. (1969). Studies on the human corpus luteum. I. Observations on the ultrastructure of development and regression of luteal cells during the menstrual cycle. *J. Cell Biol.* 41, 696–715.

Alila, H. W., and Dowd, J. P. (1991). The control of corpus luteum function in domestic ruminants. *Oxford Rev. Reprod. Biol.*, 13, 203–237.

Alila, H. W., and Hansel, W. (1984). Origin of different cell types in the bovine corpus luteum as characterized by specific monoclonal antibodies. *Biol. Reprod.* 31, 1015–1025.

Alila, H. W., Corradino, R. A., and Hansel, W. (1988a). A comparison of the effects of cyclooxygenase prostanoids on progesterone production by small and large bovine luteal cells. *Prostaglandins* 36, 259–270.

Alila, H. W., Dowd, J. P., Corradino, R. A., Harris, W. V., and Hansel, W. (1988b). Control of progesterone production in small and large bovine luteal cells separated by flow cytometry. *J. Reprod. Fertil.* 82, 645–655.

Alila, H. W., Corradino, R. A., and Hansel, W. (1989). Differential effects of luteinizing hormone on intracellular free Ca^{2+} in small and large bovine luteal cells. *Endocrinology* 1124, 2314–2320.

Allen, R. B., Su, H. C., Snitzer, J., and Dimino, M. J. (1988). Rapid decreases in phosphatidylinositol in isolated luteal phasma membranes after stimulation by luteinizing hormone. *Biol. Reprod.* 38, 79–83.

Arakawa, S., Isobe, S., Mori, H., Kambegawa, A., Kan, R., Okinaga, S., and Arai, K. (1990). The effect of epidermal growth factor (EGF) on progestin secretion and cyclic AMP synthesis in cultured luteal cells from preegnant rats. *Endocrinology (Japan)* 37, 479–487.

Archbald, L. F., Al-Bagdadi, F., and Godke, R. A. (1981). A light and electron microscopic study of the periparturient bovine corpus luteum. *Theriogenology* 16, 27–37.

Armstrong, D. T., and Hansel, W. (1959). Alteration in the bovine estrous cycle with oxytocin. *J. Dairy Sci.* 42, 533–542.

Auletta, F. J., Paradis, D. K., Wesley, M., and Duby, R. T. (1984). Oxytocin is luteolytic in the rhesus monkey. *J. Reprod. Fertil.* 72, 401–406.

Azmi, T. I., and Bongso, T. A. (1985). The ultrastructure of the corpus luteum of the goat. *Pertanika* 8, 215–222.

Bazer, F. W., Mirando, M. A., Ott, T. L., Harney, J. P., Dubois, D. H., Schalue, T. K., Pontzer, C. H., Hostetler, C., Johnson, H. M., and Ogle, T. (1992a). Roles of ovine trophoblast protein 1 and oestradiol/prolactin in the establishment of pregnancy in sheep and pigs. *Reprod. Fertil. Dev.* 4, 335–340.

Bazer, F. W., Olmsted, J. T., VanHeeke, G., and Ott, T. L. (1992b). Intrauterine injection of recombinant ovine trophoblast interferon, produced from a synthetic gene in *S. cerevisiae*, blocks luteolysis in sheep. *Biol. Reprod.* 45(Suppl 1), 76.

Behrman, H. R., Orczyk, G. P., MacDonald, G. J., and Greep, R. O. (1970). Prolactin induction of enzymes controlling luteal cholesterol ester turnover *Endocrinology* 87, 1251–1256.

Behrman, H. R., MacDonald, G. J., and Greep, R. O. (1971). Regulation of ovarian cholesterol esters: evidence for the enzymatic sites of prostaglandin-induced loss of corpus luteum function. *Lipids* 6, 791–796.

Behrman, H. R., Luborsky-Moore, J. L., Pang, C. Y., Wright, K., and Dorflinger, L. J. (1979). Mechanisms of $PGF_{2\alpha}$ action in functional luteolysis. *Adv. Exp. Med. Biol.* 112, 557–561.

Belapure, A. K., Caicedo, I. C., Kawada, K., Watt, D. S., Rexroad, C. E., Jr., and Fitz, T. A. (1989). Multiple classes of prostaglandin $F_2\alpha$ binding sites in subpopulations of ovine luteal cells. *Biol. Reprod.* 41, 385–392.

Belfiore, C. J. (1993). "Control of Cholesterol Side-Chain Cleavage in the Ovine Corpus Luteum." Ph.D. dissertation, Colorado State University, Fort Collins, CO.

Belfiore, C. J., Hawkins, D. E., and Niswender, G. D. (1992). Regulation of the mRNA encoding cytochrome P450 side-chain cleavage ($P450_{scc}$) in the ovine corpus luteum (CL). *Proc. West. Sec. Anim. Sci.* 43, 55–58.

Benhaim, A., Bonnamy, P. J., Mittre, H., and Leymarie, P. (1990). Involvement of the phospholipase C second messenger system in the regulation of steroidogenesis in small bovine luteal cells. *Mol. Cell. Endocrinol.* 68, 105–111.

Benyo, D. F., and Pate, J. L. (1992). Tumor necrosis factor-alpha alters bovine luteal cell synthetic capacity and viability. *Endocrinology* 130, 854–860.

Betts, J. G., Forrest, D. W., Humphrey, W. D., Randel, R. D., Harrison, L. M., and Lovering, S. L. (1985). Cloprostenol induced luteal regression in the beef cow. II. Histological evaluation of corpora lutea and correlation with luteal progesterone. *Theriogenology* 23, 523–534.

Billheimer, J. T., Strehl, L. L., Davis, G. L., Strauss, J. F., III, and Davis, L. G. (1990). Characterization of a cDNA encoding rat sterol carrier protein. *DNA Cell Biol.* 9, 159–165.

Bisgaier, C. L., Treadwell, C. R., and Vahouny, G. V. (1979). Activation of sterol ester hydrolase of bovine corpus luteum by $N^6,O^{2'}$-dibutyryl cyclic adenosine 3′:5′-phosphate. *Lipids* 14, 1–4.

Blanchette, E. F. (1966). Ovarian steroid cells. II. The lutein cell. *J. Cell Biol.* 31, 517–542.

Braden, T. D., Gamboni, F., and Niswender, G. D. (1988). Effects of prostaglandin $F_2\alpha$-induced luteolysis on the populations of cells in the ovine corpus luteum. *Biol. Reprod.* 39, 245–253.

Brown, M. S., and Goldstein, J. L. (1975). Regulation of the activity of the low density lipoprotein receptor in human fibroblasts. *Cell* 6, 307–316.

Brown, M. S., and Goldstein, J. L. (1986). A receptor-mediated pathway for cholesterol homeostasis. *Science* 232, 34–47.

Caffrey, J. L., Fletcher, P. W., Diekman, M. A., O'Callaghan, P. L., and Niswender, G. D. (1979a). The activity of ovine luteal cholesterol esterase during several experimental conditions. *Biol. Reprod.* 21, 601–608.

Caffrey, J. L., Nett, T. N., Abel, J. H., Jr., and Niswender, G. D. (1979b). Activity of 3β-hydroxy-Δ^5-steroid dehydrogenase/Δ^5-Δ^4-isomerase in the ovine corpus luteum. *Biol. Reprod.* 20, 279–287.

Carlson, J. C., Buhr, M. M., and Riley, J. C. M. (1989). Plasma membrane changes during corpus luteum regression. *Can. J. Physiol. Pharmacol.* 67, 957–961.

Cavazos, L. F., Anderson, L. L., Belt, W. D., Hendricks, D. M., Kraeling, R. R., and Melampy, R. M. (1969). Fine structure and progesterone levels in the corpus luteum of the pig during the estrous cycle. *Biol. Reprod.* 1, 83–106.

Chedrese, P. J., Schott, D., Zhang, D., and Murphy, B. D. (1991). The signal transduction system in luteotrophic stimulation of expression of the 3β-HSD gene in porcine granulosa cells in culture. *In:* "Signaling Mechanisms and Gene Expression in the Ovary" (G. Gibori, ed.) pp. 280–284. Springer-Verlag, New York.

Chegini, N., Lei, Z. M., Rao, Ch. V., and Hansel, W. (1991). Cellular distribution and cycle phase dependency of gonadotropin and eicosanoid binding sites in bovine corpora lutea. *Biol. Reprod.* 45, 506–513.

Christensen, A. K., and Gillim, S. W. (1969). The correlation of fine structure and function in steroid secreting cells with emphasis on those of the gonads. *In* "The Gonads" (K. W. McKerns, ed.) pp. 415–488. Appleton–Century–Crofts, New York.

Constantino, C. X., Keyes, P. L., and Kostyo, J. L. (1991). Insulin-like growth factor-I stimulates steroidogenesis in rabbit luteal cells. *Endocrinology* 128, 1702–1708.

Cooke, B. A., Janszen, F. H. A., Clotscher, W. F., and van der Molen, H. J. (1975). Effect of protein-synthesis inhibitors on testosterone production in rat testis interstitial tissue and Leydig-cell preparations. *Biochem. J.* 150, 413–418.

Cooke, R. G., and Knifton, A. (1981). Oxytocin-induced oestrus in the goat. *Theriogenology* 16, 95–97.

Corner, G. W. (1919). On the origin of the corpus luteum of the sow from both granulosa and theca interna. *Am. J. Anat.* 26, 117–183.

Couet, J., Martel, C., Dupont, E., Luu-The, V., Sirard, M.-A., Zhao, H.-F., Pelletier, G., and Labrie, F. (1990). Changes in 3β-hydroxysteroid dehydrogenase/Δ^5-Δ^4 isomerase messenger ribonucleic acid, activity and protein levels during the estrous cycle in the bovine ovary. *Endocrinology* 127, 2141–2148.

Crisp, T. M., Dessouky, D. A., and Denys, F. R. (1973). The fine structure of the human corpus luteum of term pregnancy. *Am. J. Obstet. Gynecol.* 115, 901–911.

Crombie, P. R., Burton, R. D., and Ackland, N. (1971). The ultrastructure of the corpus luteum of the guinea pig. *Z. Zellforsch.* 115, 473–493.

Davis, J. S., Weakland, L. L., Farese, R. V., and West, L. A. (1987). Luteinizing hormone increases inositol trisphosphate and cytosolic free Ca_{2+} in isolated bovine luteal cells. *J. Biol. Chem.* 262, 8515–8521.

Davis, J. S., Alila, H. W., West, L. A., Corradino, R. A., and Hansel, W. (1988). Acute effects of prostaglandin $F_{2\alpha}$ on inositol phospholipid hydrolysis in the large and small cells of the bovine corpus luteum. *Mol. Cell. Endocrinol.* 58, 43–50.

Davis, J. S., Alila, H. W., West, L. A., Corradino, R. A., Weakland, L. L., and Hansel, W. (1989). Second messenger systems and progesterone secretion in the small cells of the bovine corpus luteum: effects of gonadotropins and prostaglandin F2a. *J. Steroid Biochem.* 32, 643–649.

Davis, M. A., Ott, T. L., Mirando, M. A., Moser, M. T., and Bazer, F. W. (1992). Effect of recombinant alpha interferons on fertility and interestrous interval in sheep. *Theriogenology* 38, 867–875.

Davis, W. W., and Garren, L. D. (1968). On the mechanism of action of adrenocorticotropic hormone. *J. Biol. Chem.* 243, 5153–5157.

Deane, H. W., Hay, M. F., Moor, R. M., Rowson, L. E. A., and Short, R. V. (1966). The corpus luteum of the sheep: relationships between morphology and function during the oestrous cycle. *Acta Endocrinol.* 51, 245–263.

Donaldson, L., and Hansel, W. (1965). Histological study of bovine corpora lutea. *J. Dairy Sci.* 48, 905–909.

Dowd, J. P., Alila, H. W., and Hansel, W. (1990). Phorbol ester receptors in bovine luteal cells: relationship to protein kinase C. *Mol. Cell. Endocrinol.* 69, 199–206.

Ellinwood, W. E., Nett, T. M., and Niswender, G. D. (1979). Maintenance of the corpus luteum of early pregnancy in the ewe. II. Prostaglandin secretion by the endometrium in vitro and in vivo. *Biol. Reprod.* 21, 845–856.

Enders, A. C., and Lyons, W. R. (1964). Observations of the fine structure of lutein cells. II. The effects of hypophysectomy and mammotropic hormone in the rat. *J. Cell. Biol.* 22, 128–141.

Fairchild, D. L., and Pate, J. L. (1991). Modulation of bovine luteal cell synthetic capacity by interferon-gamma. *Biol. Reprod.* 44, 357–363.

Fairclough, R. J., Moore, L. G., McGowan, L. T., Peterson, A. J., Smith, J. F., Teriut, H. R., and Watkins, W. B. (1980). Temporal relationships between plasma concentrations of 13,14,dihydro-15-keto-prostaglandin F and neurophysin I/II around luteolysis in sheep. *Prostaglandins* 20, 199–208.

Farin, C. E., Moeller, C. L., Sawyer, H. R., Gamboni, F., and Niswwender, G. D. (1986). Morphometric analysis of cell types in the ovine corpus luteum throughout the estrous cycle. *Biol. Reprod.* 35, 1299–1308.

Farin, C. E., Moeller, C. L., Mayan, H., Gamboni, F., Sawyer, H. R., and Niswender, G. D. (1988). Effect of luteinizing hormone and human chorionic gonadotropin on cell populations in the ovine corpus luteum. *Biol. Reprod.* 38, 413–421.

Farin, C. E., Sawyer, H. R., and Niswender, G. D. (1989). Analysis of cell types in the corpus luteum of the sheep. *J. Reprod. Fertil. Suppl.* 37, 181–187.

Ferreri, K., and Menon, K. M. J. (1990). Detection of a 58-kilodalton high density lipoprotein binding protein in the membrane fraction of luteinized rat ovaries. *Endocrinology* 126, 2137–2144.

Fields, P. A. (1984). Ultrastructural localization of relaxin in corpora lutea of pregnant, pseudopregnant and cycling gilts using porcine relaxin antiserum and goal anti-rabbit IgG-colloidal gold. *Biol. Reprod.* 30 (Suppl 1), 116A.

Fitz, T. A., Mayan, M. H., Sawyer, H. R., and Niswender, G. D. (1982). Characterization of two steroidogenic cell types in the ovine corpus luteum. *Biol. Reprod.* 27, 703–711.

Fitz, T. A., Mock, E. J., Mayan, M. H., and Niswender, G. D. (1984). Interaction of prostaglandins with subpopulations of ovine luteal cells. II. Inhibitory effects of $PGF_2\alpha$ and protection by PGE_2. *Prostaglandins* 28, 127–138.

Fletcher, P. W., and Niswender, G. D. (1982). Effect of $PGF_{2\alpha}$ on progesterone secretion and adenylate cyclase activity in ovine luteal tissue. *Prostaglandins* 23, 803–818.

Flint, A. P. F., and Armstrong, D. T. (1973). Activities of enzymes responsible for steroid biosynthesis and cholesterol ester metabolism in rabbit ovarian interstitial tissue and corpora lutea. *Biochem. J.* 132, 301–311.

Flint, A. P. F., and Sheldrick, E. L. (1983). Evidence for a systemic role for ovarian oxytocin in luteal regression in sheep. *J. Reprod. Fertil.* 67, 215–225.

Flint, A. P. F., Sheldrick, E. L., McCann, T. J., and Jones, D. S. C. (1990). Luteal oxytocin: characteristics and control of synchronous episodes of oxytocin and $PGF_2\alpha$ secretion at luteolysis in ruminants. *Dom. Anim. Endocrinol.* 7, 111–124.

Fuchs, A.-R., Behrens, O., Helmer, H., Vangsted, A., Ivanisevic, M., Grifo, J., Barros, C., and Fields, M. (1990). Oxytocin and vasopressin binding sites in human and bovine ovaries. *Am. J. Obstet. Gyencol.* 163, 1961–1967.

Gamboni, F., Fitz, T. A., Hoyer, P. B., Wise, M. E., Mayan, M. H., and Niswender, G. D. (1984). Effect of human chorionic gonadotropin on induced ovine corpora lutea during the anestous season. *Dom. Anim. Endocrinol.* 1, 79–88.

Gemmell, R. T., Stacy, B. D., and Nancarrow, C. D. (1977). Secretion of granules by the luteal cells of the sheep and goat during the estrous cycle and pregnancy. *Anat. Rec.* 189, 161–168.

Gemmell, R. T., and Stacy, B. D. (1979). Effect of cycloheximide on the ovine corpus luteum: the role of granules in the secretion of progesterone. *J. Reprod. Fertil.* 57, 87–89.

Ghosh, D. K., Dunham, W. R., Sands, R. H., and Menon, K. M. J. (1987). Regulation of cholesterol side-chain cleavage enzyme activity by gonadotropin in rat corpus luteum. *Endocrinology* 121, 21–27.

Gibori, G., Khan, I., Warshaw, M. L., McLean, M. P., Puryear, T. K., Nelson, S., Durkee, T. J., Azhar, S., Steinschneider, A., and Rao, M. C. (1988). Placental-derived regulators and the complex control of luteal cell function. *Recent Prog. Horm. Res.* 44, 377–425.

Glass, J. D., Fitz, T. A., and Niswender, G. D. (1984). Cytosolic receptor for estradiol in the corpus luteum of the ewe: variation throughout the estrous cycle and distribution between large and small steroidogenic cell types. *Biol. Reprod.* 31, 967–974.

Golos, T. G., and Strauss, J. F. III (1987). Regulation of low density lipoprotein receptor gene expression in cultured human granulosa cells: roles of human chorionic gonadotroopin, 8-bromo-3′,5′-cyclic adenosine monophosphate, and protein synthesis. *Mol. Endocrinol.* 1, 321–326.

Golos, T. G., and Strauss, J. F. III (1988). 8-bromoadenosine cyclic 3′,5′-phosphate rapidly increases 3-hydroxy-3-methylglutaryl coenzyme A reductase mRNA in human granulosa cells: role of cellular sterol balance in controlling the response to tropic stimulation. *Biochemistry* 27, 3503–3506.

Golos, T. G., Miller, W. L., and Strauss, J. F. III (1987). Human chorionic gonadotropin and 8-bromo cyclic adenosine monophosphate promote an acute increase in cytochrome P450scc and adrenodoxin messenger RNAs in cultured human granulosa cells by a cycloheximide-insensitive mechanism. *J. Clin. Invest.* 80, 896–899.

Gospodarowicz, D., and Gospodarowicz, F. (1975). The morphological transformation and inhibition of growth of bovine luteal cells in tissue culture induced by luteinizing hormone and dibutyryl cyclic AMP. *Endocrinology* 96, 458–467.

Grusenmeyer, D. P., and Pate, J. L. (1992). Localization of prostaglandin $F_2\alpha$ inhibition of lipoprotein use by bovine luteal cells. *J. Reprod. Fertil.* 94, 311–318.

Gulyas, B. J., Stouffer, R. L., and Hodgen, G. D. (1979). Progesterone synthesis and fine structure of dissociated monkey (*Macaca mulatta*) luteal cells maintained in culture. *Biol. Reprod.* 20, 779–792.

Gwynne, A., and Condon, W. A. (1982). Effects of cytochalasin B, colchicine, and vinblastine on progesterone synthesis and secretion by bovine luteal tissue *in vitro*. *J. Reprod. Fert.* 65, 151–156.

Gwynne, J. T., and Strauss, J. F. III (1982). The role of lipoproteins in steroidogenesis and cholesterol metabolism in steroidogenic glands. *Endocr. Rev.* 3, 299–329.

Halkerston, I. D. K., Eichorn, J., and Hechter, O. (1961). A requirement for reduced triphosphopyridine nucleotide for cholesterol side-chain cleavage by mitochondrial fractions of bovine adrenal cortex. *J. Biol. Chem.* 236, 374–380.

Hansel, W., Concannon, P. W., and Lukaszewska, J. H. (1973). Corpora lutea of large domestic animals. *Biol. Reprod.* 8, 222–245.

Hansel, W., Alila, H. W., Dowd, J. P., and Yang, X. Z. (1987). Control of steroidogenesis in small and large bovine luteal cells. *Aust. J. Biol. Sci.* 40, 331–347.

Hansel, W., Alila, H. W., Dowd, J. P., and Milvae, R. A. (1991). Differential origin and control mechanisms in small and large bovine luteal cells. *J. Reprod. Fertil. Suppl.* 43, 77–89.

Harrison, L. M., Kenny, N., and Niswender, G. D. (1987). Progesterone production, LH-receptors, and oxytocin secretion by ovine luteal cell types on days 6, 10 and 15 of the oestrous cycle and day 25 of pregnancy. *J. Reprod. Fertil.* 79, 539–548.

Hawkins, D. E., Belfiore, C. J., Kile, J. P., and Niswender, G. D. (1993). Regulation of mRNA encoding 3β-hydroxysteroid dehydrogenase/Δ^5-Δ^4-isomerase (3β-HSD) in the ovine corpus luteum. *Biol. Reprod.* 48, 1185–1190.

Hedin, L., Rodgers, R. J., Simpson, E. R., and Richards, J. S. (1987). Changes in content of cytochrome $P450_{17\alpha}$, cytochrome $P450_{scc}$, and 3-hydroxy-3-methylglutaryl CoA reductase in developing rat ovarian follicles and corpora lutea: correlation with theca cell steroidogenesis. *Biol. Reprod.* 37, 211–223.

Hichens, M., Grinwich, D. L., and Behrman, H. R. (1974). $PGF_{2\alpha}$-induced loss of corpus luteum gonadotrophin receptors. *Prostaglandins* 7, 449–458.

Hixon, J. E., and Flint, A. P. F. (1987). Effect of a luteolytic dose of oestradiol benzoate on uterine oxytocin receptor concentration, phosphoinositide turnover and prostaglandin $F_2\alpha$ secretion in sheep. *J. Reprod. Fertil.* 79, 457–467.

Hoyer, P. B., and Marion, S. L. (1989). Influence of agents that affect intracellular calcium regulation on progesterone secretion in small and large luteal cells of the sheep. *J. Reprod. Fertil.* 86, 445–455.

Hoyer, P. B., and Kong, W. (1989). Protein kinase A and C activities and endogenous substrates in ovine small and large luteal cells. *Mol. Cell. Endocrinol.* 62, 203–215.

Hoyer, P. B., and Niswender, G. D. (1986). Adenosine 3′,5′-monophosphate-binding capacity in small and large ovine luteal cells. *Endocrinology* 119, 1822–1829.

Hoyer, P. B., Keyes, P. L., and Niswender, G. D. (1986). Size distribution and hormonal responsiveness of dispersed rabbit luteal cells during pseudopregnancy. *Biol. Reprod.* 34, 905–910.

Hunzicker-Dunn, M. L. B. (1976). Adenylyl cyclase activities in ovarian tissues. IV. Gonadotropin induced desensitization of the luteal adenylyl cyclase throughout pregnancy and pseudopregnancy. *Endocrinology* 99, 211–222.

Imakawa, K., Hansen, T. R., Malathy, P.-V., Anthony, R. V., Polites, H. G., Maroth, K. R., and Roberts, R. M. (1989). Molecular cloning and characterization of complementary deoxyribonucleic acids corresponding to bovine trophoblast protein-1: A comparison with ovine trophoblast protein-1 and bovine interferon-αII. *Mol. Endocrinol.* 3, 127–139.

Ingebritsen, T. S., Geelen, M. J. H., Parker, R. A., Evenson, K. J., and Gibson, D. M. (1979). Modulation of hydroxymethylglutaryl-CoA reductase activity, and cholesterol synthesis in rat hepatocytes in response to insulin and glucagon. *J. Biol. Chem.* 254, 9986–9989.

Inskeep, E. K., and Murdoch, W. J. (1980). Relation of ovarian functions to uterine and ovarian secretion of prostaglandins during the estrous cycle and early pregnancy in the ewe and cow. *In* "International Review of Physiology: Reproductive Physiology" (R. O. Greep, ed.) 22nd ed., Vol. 3, pp. 325–356. University Park Press, Baltimore, MD.

Inskeep, E. K., Smutny, W. J., Butcher, R. L., and Pexton, J. E. (1975). Effects of intrafollicular injections of prostaglandins in non-pregnant and pregnant ewes. *J. Anim. Sci.* 41, 1098–1104.

Jacobs, A. L., Homanics, G. E., and Silvia, W. J. (1987). Effects of prostaglandin (PG) $F_2\alpha$, PGE_2 and luteinizing hormone (LH) on activity of phospholipase C in ovine luteal tissue. *J. Anim. Sci.* 65(S1), 365–366.

Jefcoate, C. R., Simpson, E. R., and Boyd, G. S. (1974). Spectral properties of rat adrenal-mitochondrial cytochrome P-450. *Eur. J. Biochem.* 42, 539–551.

Juengel, J. L., Garverick, H. A., Johnson, A. L., Youngquist, R. S., and Smith, M. F. (1993). Apoptosis during luteal regression in cattle. *Endocrinology* 132, 249–254.

Keyes, P. L., and Armstrong, D. L. (1968). Endocrine role of follicles in the regulation of corpus luteum function in the rabbit. *Endocrinology* 83, 509–515.

Keyes, P. L., and Nalbandov, A. V. (1967). Maintenance and function of corpora lutea in rabbits depend on estrogen. *Endocrinology* 80, 938–946.

Kimura, T. (1981). ACTH stimulation on cholesterol side chain cleavage activity of adrenocortical mitochondria. *Mol. Cell. Biochem.* 36, 105–122.

Koering, M. J., and Kirton, K. T. (1973). The effects of prostaglandin $F_2\alpha$ on the structure and function of the rabbit ovary. *Biol. Reprod.* 9, 226–245.

Koering, M. J., and Thor, M. J. (1978). Structural changes in the regressing corpus luteum of the rabbit. *Biol. Reprod.* 17, 719–733.

Koo, Y. B., Ji, I., Slaughter, R. G., and Ji, T. H. (1991). Structure of the luteinizing hormone receptor gene and multiple exons of the coding sequence. *Endocrinology* 128, 2297–2308.

Koos, R. D., and Hansel, W. (1981). The large and small cells of the bovine corpus luteum: ultrastructural and functional differences. *In* "Dynamics of Ovarian Function" (N. B. Schwartz and M. Hunzicker-Dunn, eds.) pp. 197–203. Raven Press, New York.

LaCroix, M. C., and Kann, G. (1982). Comparative studies of prostaglandin $F_2\alpha$ and E_2 in late cyclic and early pregnant sheep: *in vitro* synthesis by endometrium and conceptus. Effects of *in vivo* indomethacin treatment on establishment of pregnancy. *Prostaglandins* 23, 507–526.

Lambeth, J. D., Seybert, D. W., Lancaster, J. R., Jr., Salerno, J. C., and Kamin, H. (1982). Steroidogenic electron transport in adrenal cortex mitochondria. *Mol. Cell. Biochem.* 45, 13–31.

LaPolt, P. S., Oikawa, M., Jai, X.-C., Dargan, C., and Hseuh, A. J. W. (1990). Gonadotropin-induced up- and down-regulation of rat ovarian LH receptor message levels during follicular growth, ovulation and luteinization. *Endocrinology* 126, 3277–3279.

Lauber, M. E., Waterman, M. R., and Simpson, E. R. (1991). Expression of genes encoding steroidogenic enzymes in the bovine corpus luteum. *J. Reprod. Fertil.* Suppl. 43, 57–64.

Leavitt, W. W., Basom, C. R., Bagwell, J. N., and Blaha, G. C. (1973). Structure and function of the hamster corpus luteum during the estrous cycle. *Am. J. Anat.* 136, 235–249.

Lemon, M., and Loir, M. (1977). Steroid release *in vitro* by two luteal cell types in the corpus luteum of the pregnant sow. *J. Endocrinol.* 72, 351–359.

Loosfelt, H., Misrahi, M., Atger, M., Salesse, R., Thi, M. T. V. H.-L., Jolivet, A., Guiochon-Mantel, A., Sar, S., Jallal, B., Garnier, J., and Milgrom, E. (1989). Cloning and sequencing of porcine LH-hCG receptor cDNA: variants lacking the transmembrane domain. *Science* 245, 525–528.

Luskey, K. L. (1988). Regulation of cholesterol synthesis: mechanism for control of HMG CoA reductase. *Rec. Prog. Horm. Res.* 44, 35–51.

Manns, J. G., Niswender, G. D., and Braden, T. (1984). FSH receptors in the bovine corpus luteum. *Theriogenology* 22, 321–328.

Marcus, G. J. (1981). Prostaglandin formation by the sheep embryo and endometrium as an indication of maternal recognition of pregnancy. *Biol. Reprod.* 25, 56–64.

Marsh, J. M. (1975). The role of cyclic AMP in gonadal function. *In* "Advances in Cyclic

Nucleotide Research" (P. Greengard and G. A. Robison, eds.), Vol. 6, pp. 137–200. Raven Press, New York.

Martal, J., Degryse, E., Charpigny, G., Assal, N., Reinaud, P., Charlier, M., Gaye, P., and Lecocq, J. D. (1990). Evidence for extended maintenance of the corpus luteum by uterine infusion of a recombinant trophoblast α-interferon (trophoblastin) in sheep. *J. Endocrinol.* 127, R5–R8.

McArdie, C. A., and Holtorf, A. P. (1989). Oxytocin and progesterone release from bovine corpus luteal cells in culture: effects of insulin-like growth factor I, insulin and prostaglandins. *Endocrinology* 124, 1278–1286.

McClellan, M. C., Diekman, M. A., Abel, J. H., and Niswender, G. D. (1975). Luteinizing hormone, progesterone and the morphological development of normal and superovulated corpora lutea in sheep. *Cell Tissue Res.* 164, 291–307.

McCracken, J. A., Schramm, W., Barcikowski, B., and Wilson, L. (1981). The identification of prostaglandin $F_2\alpha$ as a uterine luteolytic hormone and the hormonal control of its synthesis. *Acta Vet. Scand.* (Suppl 77), 71–81.

McFarland, K. C., Sprengel, R., Phillips, H. S., Kohler, M., Rosemblit, N., Nikolics, K., Segaloff, D. L., and Seeburg, P. H. (1989). Lutropin-choriogonadotropin receptor: an unusual member of the G protein-coupled receptor family. *Science* 245, 494–499.

McGuire, W. J., Hawkins, D. E., and Niswender, G. D. (1992). Activation of protein kinase (PK)-C inhibits progesterone productin *in vivo*. *Biol. Reprod.* 46 (Suppl 1), 84.

McNatty, K. P., Gibb, M., Dobson, C., Ball, K., Coster, J., Heath, D., and Thurley, D. C. (1982). Preovulatory follicular development in sheep treated with PMSG and/or prostaglandin. *J. Reprod. Fertil.* 65, 111–123.

Milvae, R. A., Alila, H. W., and Hansel, W. (1986). Involvement of lipoxygenase products of arachidonic acid metabolism in bovine luteal function. *Biol. Reprod.* 35, 1210–1215.

Miyamoto, A., and Schams, D. (1991). Oxytocin stimulates progesterone release from microdialyzed bovine corpus luteum *in vitro*. *Biol. Reprod.* 44, 1163–1170.

Miyauchi, F., and Midgley, A. R., Jr. (1990). Morphologically and functionally distinct subpopulations of steroidogenic cells in corpora lutea during pregnancy in rats. *Endocrinol. Jpn.* 37, 649–663.

Moor, R. M., and Rowson, L. E. A. (1966). The corpus luteum of the sheep: effect of the removal of embryos on luteal function. *J. Endocrinol.* 34, 497–502.

Nancarrow, C. D., Evison, B. M., and Connell, P. J. (1982). Effect of embryo on luteolysis and termination of early pregnancy in sheep with cloprostenol. *Biol. Reprod.* 26, 263–269.

Nakamura, K., Minegishi, T., Takakura, Y., Miyamoto, K., Hasegawa, Y., Ibuki, Y., and Igarashi, M. (1991). Hormonal regulation of gonadotropin receptor mRNA in rat ovary during follicular growth and luteinization. *Mol. Cell. Endocrinol.* 82, 259–263.

Nelson, S. E., McLean, M. P., Jayatilak, P. G., and Gibori, G. (1992). Isolation, characterization, and culture of cell subpopulations forming the pregnant rat corpus luteum. *Endocrinology* 130, 954–966.

Nephew, K. P., McLure, K. E., Day, M. L., Xie, S., Roberts, R. M., and Pope, W. F. (1990). Effects of intramuscular administration of recombinant bovine interferon-α1 during the period of maternal recognition of pregnancy. *J. Anim. Sci.* 68, 2766–2770.

Nett, T. M., McClellan, M. C., and Niswender, G. D. (1976). Effects of prostaglandins on the ovine corpus luteum: blood flow, secretion of progesterone and morphology. *Biol. Reprod.* 15, 66–78.

Niswender, G. D., and Nett, T. M. (1988). The corpus luteum and its control. *In* "The Physiology of Reproduction" (E. Knobil and J. Neill, eds.) pp. 489–525. Raven Press, New York.

Niswender, G. D., Reimers, T. J., Diekman, M. A., and Nett, T. M. (1976). Blood flow: a mediator of ovarian function. *Biol. Reprod.* 14, 64–81.

Niswender, G. D., Sawyer, H. R., Chen, T. T., and Endres, D. B. (1980). Action of luteinizing hormone at the luteal cell level. *In* "Advances in Sex Hormone Research" (J. A. Thomas and R. L. Singhal, eds.), Vol. 4, pp. 153–185. Urban and Schwarzenberg, Baltimore and Munich.

Niswender, G. D., Schwall, R. H., Fitz, T. A., Farin, C. E., and Sawyer, H. R. (1985). Regulation of luteal function in domestic ruminants: new concepts. *Recent Prog. Horm. Res.* 41, 101–151.

Noland, B. J., Arebalo, R. E., Hansbury, E., and Scallen, T. J. (1980). Purification and properties of sterol carrier protein$_2$. *J. Biol. Chem.* 255, 4282–4289.

Nothnick, W. B., and Pate, J. L. (1990). Interleukin-1β is a potent stimulator of prostaglandin synthesis in bovine luteal cells. *Biol. Reprod.* 43, 898–903.

Ohara, A., Mori, T., Taii, S., Ban, C., and Narimoto, K. (1987). Functional differentiation in steroidogenesis of two types of luteal cells isolated from mature human corpora lutea of menstrual cycle. *J. Clin. Endocrinol. Metab.* 65, 1192–1200.

O'Shea, J. D., Nightingale, M. G., and Chamley, W. A. (1977). Changes in the small blood vessels during cyclical luteal regression in sheep. *Biol. Reprod.* 17, 162–177.

O'Shea, J. D., Cran, D. G., and Hay, M. F. (1980). Fate of the theca interna following ovulation in the ewe. *Cell Tissue Res.* 210, 305–319.

O'Shea, J. D., Rodgers, R. J., and Wright, P. J. (1986). Cellular composition of the sheep corpus luteum in the mid- and late luteal phases of the oestrous cycle. *J. Reprod. Fertil.* 76, 685–691.

O'Shea, J. D., Rodgers, R. J., and D'Occhio, M. J. (1989). Cellular composition of the cyclic corpus luteum of the cow. *J. Reprod. Fertil.* 85, 483–487.

Oonk, R. B., Krasnow, J. S., Beattie, W. G., and Richards, J. S. (1989). Cyclic AMP-dependent and -independent regulation of cholesterol side chain cleavage cytochrome P-450 (P-450$_{scc}$) in rat ovarian granulosa cells and corpora lutea. *J. Biol. Chem.* 264, 21934–21942.

Ott, T. L., VanHeeke, G., Fliss, F. M. V., Johnson, H. M., and Bazer, F. W. (1991). Antiluteolytic and antiviral activities of recombinant ovine trophoblast protein-1 (roTP-1) over produced from a synthetic gene in *S. cerevisiae*. *Biol. Reprod.* 45 (Suppl 1), 89.

Oxberry, B. S., and Greenwald, G. W. (1982). An autoradiographic study of the binding of ^{125}I-labeled follicle stimulating hormone, human chorionic gonadotropin and prolactin to the hamster ovary throughout the estrous cycle. *Biol. Reprod.* 27, 505–516.

Paavola, L. G., and Strauss, J. F., III (1983). Uptake of lipoproteins by in situ perfused rat ovaries: identification of binding sites for high density lipoproteins. *J. Cell Biol.* 97, 593–606.

Paavola, L. G., Strauss, J. F., III, Boyd, C. O., and Nestler, J. E. (1985). Uptake of gold- and [^{3}H]cholesteryl linoleate-labeled human low density lipoprotein by cultured rat granulosa cells: cellular mechanisms involved in lipoprotein metabolism and their importance to steroidogenesis. *J. Cell Biol.* 100, 1235–1247.

Pate, J. L., and Condon, W. A. (1989). Regulation of steroidogenesis and cholesterol synthesis by prostaglandin $F_2\alpha$ and lipoproteins in bovine luteal cells. *J. Reprod. Fertil.* 87, 439–446.

Phariss, B. B., and Wyngarden, L. (1969). The effect of prostaglandin $F_2\alpha$ on the progestogen content of the ovaries of pseudopregnant rats. *Proc. Soc. Exp. Biol. Med.* 130, 92–94.

Pitzel, L., Jarry, H., and Wuttke, W. (1991). Effects of substance P and neuropeptide-Y on in vitro steroid release by porcine granulosa and luteal cells. *Endocrinology* 129, 1059–1065.

Poniter, C. H., Torres, B. A., Vallet, J. L., Bazer, F. W., and Johnson, H. M. (1988). Antiviral activity of the pregnancy recognition hormone ovine trophoblast protein-1. *Biochem. Biophys. Res. Commun.* 152, 801–807.

Pon, L. A., and Orme-Johnson, N. R. (1986). Acute stimulation of steroidogenesis in corpus luteum and adrenal cortex by peptide hormones. *J. Biol. Chem.* 261, 6594–6599.

Rajkumar, K., Couture, R. L., and Murphy, B. D. (1985). High density lipoprotein uptake and utilization by rat corpora lutea: the effects of prolactin and prostaglandin $F_2\alpha$. *In* "Lipoprotein and Cholesterol Metabolism in Steroidogenic Tissues" (J. F. Strauss, III and K. M. J. Menon, eds.), pp. 147–154. George F. Strickley, Philadelphia.

Richards, J. S., and Williams, J. J. (1976). Luteal cell receptor content for prolactin (PRL) and luteinizing hormone (LH): regulation by LH and PRL. *Endocrinology* 99, 1571–1581.

Richards, J. S., Hedin, L., and Caston, L. (1986). Differentiation of rat ovarian thecal cells: evidence for functional luteinization. *Endocrinology* 118, 1660–1668.

Roberts, J. S., McCracken, J. A., Gavagan, J. R., and Soloff, M. S. (1976). Oxytocin-stimulated release of prostaglandin $F_2\alpha$ from ovine endometrium in vitro: correlation with estrous cycle and oxytocin-receptor binding. *Endocrinology* 99, 1107.

Roberts, R. M. (1991). A role for interferons in early pregnancy. *Bioessays* 13, 121–126.

Rodgers, R. J., and O'Shea, J. D. (1982). Purification, morphology, and progesterone production and content of three cell types isolated from the corpus luteum of the sheep. *Aust. J. Biol. Sci.* 35, 441–445.

Rodgers, R. J., O'Shea, J. D., and Bruce, N. W. (1984). Morphometric analysis of the cellular composition of the ovine corpus luteum. *J. Anat.* 138, 757–769.

Rodgers, R. J., O'Shea, J. D., and Findlay, J. K. (1985). Do small and large cells of sheep interact in the production of progesterone? *J. Reprod. Fertil.* 75, 85–94.

Rodgers, R. J., Waterman, M. R., and Simpson, E. R. (1986). Cytochromes P-450_{scc}, P450-$_{scc17\alpha}$, adrenoxin, and reduced nicotinamide adenine dinucleotide phosphate-cytochrome P-450 reductase in bovine follicles and corpora lutea. *Endocrinology* 118, 1366–1374.

Rodgers, R. J., Waterman, M. R., and Simpson, E. R. (1987). Levels of messenger ribonucleic acid encoding cholesterol side-chain cleavage cytochrome P-450, 17α-hydroxylase cytochrome P-450, adrenoxin, and low density lipoprotein receptor in bovine follicles and corpora lutea throughout the ovarian cycle. *Mol. Endocrinol.* 1, 274–279.

Rodriguez, M. C., Xie, Y.-B., Wang, H., Collison, K., and Segaloff, D. L. (1992). Effects of truncations of the cytoplasmic tail of the luteinizing hormone/chorionic gonadotropin receptor on receptor-mediated hormone internalization. *Mol. Endocrinol.* 6, 327–336.

Rodriguez, M. C., and Segaloff, D. L. (1990). The orientation of the LH/CG receptor in rat luteal cells as revealed by site-specific antibodies. *Endocrinology* 127, 674–681.

Rousseau-Merk, M. F., Misrahi, M., Atger, M., Loosfelt, H., Milgrom, E., and Berger, R. (1990). Localization of the human luteinizing hormone/choriogonadotropin receptor gene (LHCGR) to chromosome 2q21. *Cytogenet. Cell Genet.* 54, 77–79.

Samuels, L. T., Helmreich, M. L., Lasater, M. B., and Reich, H. (1951). An enzyme in endocrine tissues which oxidizes Δ^5-3-hydroxy steroids to α,β unsaturated ketones. *Science* 113, 490–491.

Sawyer, H. R., Abel, J. H., Jr., McClellan, M. C., Schmitz, M., and Niswender, G. D. (1979). Secretory granules and progesterone secretion by ovine corpora lutea *in vitro*. *Endocrinology* 104, 476–486.

Sawyer, H. R., Moeller, C. L., and Kozlowski, G. P. (1986). Immunocytochemical localization of neurophysin and oxytocin in ovine corpora lutea. *Biol. Reprod.* 34, 543–548.

Sawyer, H. R., Niswender, K. D., Braden, T. D., and Niswender, G. D. (1990). Nuclear changes in ovine luteal cells in response to $PGF_2\alpha$. *Dom. Anim. Reprod.* 7, 229–238.

Schams, D. (1987). Ovarian peptides in the cow and sheep. *J. Reprod. Fertil.* Suppl. 34, 225–231.

Schuler, L. A., Flickinger, G. L., Strauss, III, J. F. (1978). Effects of luteinizing hormone on the lipid composition of rat ovaries. *J. Endocrinol.* 78, 233–238.

Schuler, L. A., Scavo, L., Kirsch, T. M., Flickinger, G. L., and Strauss, J. F., III (1979). Regulation of de novo biosynthesis of cholesterol and progestins, and formation of cholesteryl ester in rat corpus luteum by exogenous sterol. *J. Biol. Chem.* 254, 8662–8668.

Schuler, L. A., Langenberg, K. K., Gwynne, J. T., and Strauss, J. F., III (1981). High density lipoprotein utilization by dispersed rat luteal cells. *Biochim. Biophys. Acta* 664, 583–601.

Schwall, R. H., Gamboni, F., Mayan, M. H., and Niswender, G. D. (1986). Changes in the distribution of sizes of ovine luteal cells during the estrous cycle. *Biol. Reprod.* 34, 911–918.

Segaloff, D. L., Wang, H., and Richards, J. S. (1990). Hormonal regulation of luteinizing hormone/chorionic gonadotropin receptor mRNA in rat ovarian cells during follicular development and luteinization. *Mol. Endocrinol.* 4, 1856–1865.

Sernia, C., Gemmell, R. T., and Thomas, W. G. (1989). Oxytocin receptors in the ovine corpus luteum. *J. Endocrinol.* 121, 117–123.

Sheldrick, E. L., and Flint, A. P. F. (1985). Endocrine control of uterine oxytocin receptors in the ewe. *J. Endocrinol.* 106, 249–258.

Shikita, M., and Hall, P. F. (1974). The stoichiometry of the conversion of cholesterol and hydroxycholesterols to pregnenolone (3β-hydroxypregn-5-en-20-one) catalyzed by adrenal cytochrome P-450. *Proc. Natl. Acad. Sci. USA* 71, 1441–1445.

Silavin, S. L., Moss, G. E., and Niswender, G. D. (1980). Regulation of steroidogenesis in the ovine corpus luteum. *Steroids* 36, 229–241.

Silvia, W. J. (1985). "Maintenance of the Corpus Luteum of Early Pregnancy in the Ewe." Ph.D. dissertation, Colorado State University, Fort Collins, CO.

Silvia, W. J., and Niswender, G. D. (1984). Maintenance of the corpus luteum of early pregnancy in the ewe. III. Differences between pregnant and nonpregnant ewes in luteal responsiveness to prostaglandin $F_2\alpha$. *J. Anim. Sci.* 59, 746–753.

Silvia, W. J., and Niswender, G. D. (1986). Maintenance of the corpus luteum of early pregnancy in the ewe. IV. Changes in luteal sensitivity to prostaglandin $F_2\alpha$ throughout early pregnancy. *J. Anim. Sci.* 63, 1201–1207.

Silvia, W. J., Fitz, T. A., Mayan, M. H., and Niswender, G. D. (1984). Cellular and molecular mechanisms involved in luteolysis and maternal recognition of pregnancy in the ewe. *Anim. Reprod. Sci.* 7, 57–74.

Soto, E. A., Kliman, H. J., Strauss, III, J. F., and Paavola, L. G. (1986). Gonadotropins and cyclic adenosine 3′,5′-monophosphate (cAMP) alter the morphology of cultured human granulosa cells. *Biol. Reprod.* 34, 559–569.

Spies, H. G., Hilliard, J., and Sawyer, C. H. (1968). Maintenance of corpora lutea and pregnancy in hypophysectomized rabbits. *Endocrinology* 83, 354–367.

Stacy, B. D., Gemmell, R. T., and Thorburn, G. D. (1976). Morphology of the corpus luteum in the sheep during regression induced by prostaglandin $F_2\alpha$. *Biol. Reprod.* 14, 280–291.

Steinschneider, A., McLean, M. P., Billheimer, J. T., Azhar, S., and Gibori, G. (1989). Protein kinase C catalyzed phosphorylation of sterol carrier protein 2. *Endocrinology* 125, 569–571.

Strauss, J. F., III, MacGregor, L.C., and Gwynne, J. T. (1982). Uptake of high density lipoproteins by rat ovaries *in vivo* and dispersed ovarian cells *in vitro:* direct correlation of high density lipoprotein uptake with steroidogenic activity. *J. Steroid. Biochem.* 16, 525–531.

Suter, D. E., Fletcher, P. W., Sluss, P. M., Reichert, Jr., L. E., and Niswender, G. D. (1980). Alterations in the number of ovine luteal receptors for LH and progesterone secretion induced by homologous hormone. *Biol. Reprod.* 22, 205–210.

Talavera, F., and Menon, K. M. J. (1989). Regulation of rat luteal cell high density lipoprotein receptors: up-regulation in response to changes in intracellular cholesterol concentration. *Endocrinology* 125, 2015–2021.

Tan, G. J. S., Tweedale, R., and Biggs, J. S. G. (1982). Effects of oxytocin on the bovine corpus luteum of early pregnancy. *J. Reprod. Fertil.* 66, 75–78.

Taylor, M. J., and Clark, C. L. (1992). Basic fibroblast growth factor inhibits basal and stimulated relaxin secretion by cultured porcine luteal cells: analysis by reverse hemolytic plaque assay. *Endocrinology* 130, 1951–1956.

Trzeciak, W. H., Waterman, M. R., and Simpson, E. R. (1986). Synthesis of the cholesterol side-chain cleavage enzymes in cultured rat ovarian granulosa cells: induction by follicle-stimulating hormone and dibutyryl adenosine 3′,5′-monophosphate. *Endocrinology* 119, 323–330.

Umo, I. (1975). Effect of prostaglandin $F_2\alpha$ on the ultrastructure and function of sheep corpora lutea. *J. Reprod. Fertil.* 43, 287–292.
Ursely, J., and Leymarie, P. (1979). Varying response to luteinizing hormone of two luteal cell types isolated from bovine corpus luteum. *J. Endocrinol.* 83, 303–310.
Vallet, J. L., Bazer, F. W., Fliss, M. F., and Thatcher, W. W. (1988). Effect of ovine conceptus secretory proteins and purified ovine trophoblast protein-1 on interoestrous interval and plasma concentrations of prostaglandins $F_2\alpha$ and E and of 13,14-dihydro-15-keto-prostaglandin $F_2\alpha$ in cyclic ewes. *J. Reprod. Fertil.* 84, 493–504.
Van Lenny, E. W., and Madden, L. M. (1965). Electron microscope observations on the involution of the human corpus luteum of menstruation. *Z. Zellforsch.* 66, 365–380.
van Noort, M., Rommerts, F. F. G., van Amerongen, A., and Wirtz, K. W. A. (1986). Localization and hormonal regulation of the non-specific lipid transfer protein (sterol carrier $protein_2$) in the rat testis. *J. Endocrinol.* 109, R13–R15.
Vollmar, A. M., Mytzka, C., Arendt, R. M., and Schultz, R. (1988). Atrial natriuretic peptide in bovine corpus luteum. *Endocrinology* 123, 762–767.
Warbritton, V. (1934). The cytology of the corpora lutea of the ewe. *J. Morphol.* 56, 181–202.
Wegner, J. A., Martinez-Zaguilan, R., Wise, M. E., Gillies, R. J., and Hoyer, P. B. (1990). Prostaglandin $F_2\alpha$-induced calcium transient in ovine large luteal cells. I. Alterations in cytosolic-free calcium levels and calcium flux. *Endocrinology* 127, 3029–3037.
Wiepz, G. J., Wiltbank, M. C., Kater, S. B., Niswender, G. D., and Sawyer, H. R. (1993). PGE_2 attenuates $PGF_2\alpha$-induced increases in free intracellular calcium in ovine large luteal cells. *Prostaglandins* 45, 167–176.
Wilkinson, R. F., Anderson, E., and Aalberg, J. (1976). Cytological observations of dissociated rat corpus luteum. *J. Ultrastruct. Res.* 57, 168–184.
Wiltbank, M. C., Guthrie, P. B., Mattson, M. P., Kater, S. P., and Niswender, G. D. (1989a). Hormonal regulation of free intracellular calcium concentrations in small and large ovine luteal cells. *Biol. Reprod.* 41, 771–778.
Wiltbank, M. C., Knickerbocker, J. J., and Niswender, G. D. (1989b). Regulation of the corpus luteum by protein kinase C. I. Phosphorylation activity and steroidogenic action in large and small ovine luteal cells. *Biol. Reprod.* 40, 1194–1200.
Wiltbank, M. C., Diskin, M. G., Flores, J. A., and Niswender, G. D. (1990). Regulation of the corpus luteum by protein kinase C. II. Inhibition of lipoprotein-stimulated steroidogenesis by prostaglandin $F_{2\alpha}$. *Biol. Reprod.* 42, 239–245.
Wiltbank, M. C., Diskin, M. G., and Niswender, G. D. (1991). Differential actions of second messenger systems in the corpus luteum. *J. Reprod. Fertil. Suppl.* 43, 65–75.
Wiltbank, M. C., Wiepz, G. J., Knickerbocker, J. J., Belfiore, C., and Niswender, G. D. (1992). Proteins secreted from the early ovine conceptus block the action of prostaglandin $F_2\alpha$ on large luteal cells. *Biol. Reprod.* 46, 475–482.
Wiltbank, M. C., Belfiore, C. J., and Niswender, G. D. (1993). Steroidogenic enzyme activity after acute activation of protein kinase (PK) A and PKC in ovine small and large luteal cells. Submitted for publication.
Yanagibashi, K., Ohno, Y., Kawamura, M., and Hall, P. F. (1988). The regulation of intracellular transport of cholesterol in bovine adrenal cells: purification of a novel protein. *Endocrinology* 123, 2075–2082.
Yerle, M., Galman, O., Lahbib-Mansais, Y., and Gellin, J. (1992). Localization of the pig luteinizing hormone/choriogonadotropin receptor gene (LHCGR) by radioactive and nonradioactive in situ hybridization. *Cytogenet. Cell Genet.* 59, 48–51.
Zammit, V. A., and Easom, R. A. (1987). Regulation of hepatic HMG-CoA reductase in vivo by reversible phosphorylation. *Biochim. Biophys. Acta* 927, 223–228.
Zhang, R., Tsai-Morris, C. H., Kitamura, M., Buczko, E., and Dufau, M. L. (1991). Changes in binding activity of luteinizing hormone receptors by site directed mutagenesis of potential glycosylation sites. *Biochem. Biophys. Res. Comm.* 181, 804–808.

10

Cell Biology of the Oviduct and Endometrium

LOIS A. SALAMONSEN & COLIN D. NANCARROW

I. Introduction

The oviduct, or fallopian tube, forms the final component of the tubiform link between the external environment of a female, from whence the male gametes originate, and the internal structures, which provide, among other things, the female gametes. Thus, it functions as a conduit and storage organ for spermatozoa, a collecting vessel for ova, a site for fertilization and the first few embryonic cell divisions, and a means of transporting the embryo to the uterus. Since pregnancies have been obtained for several species using only *in vitro* techniques, from oocyte maturation through to transfer of blastocysts to the uterus, it is a moot point whether the oviduct provides more than physical support to the gametes. However, the cells of the oviduct secrete a range of compounds into the lumen. The function of many of these compounds is unknown, but we now know that some of them influence the behavior of the gametes and embryo. This evidence leads to the conclusion that the oviduct is an active organ that responds to endocrine changes in a vital manner. Despite the oviduct's relatively simple anatomical and tissue structure, little is known about the mechanisms that control oviduct secretions. This chapter discusses some of the issues involved.

The primary function of the uterus, and in particular the endometrial lining, is to provide an environment in which the conceptus can be nurtured from the preimplantation stage until birth. Recent advances in the techniques of cell biology and their application to reproductive tract biology

Molecular Biology of the Female Reproductive System

are leading to a greater understanding of the cellular and molecular aspects of the processes of menstruation, nidation, and pregnancy. Such cellular and intercellular mechanisms are the major focus of this chapter.

It is logical, initially, to consider the endometrium, for which a considerable amount of data are available, and then to use this material to draw comparisons with the cellular mechanisms known in the oviduct. Where possible, recent reviews rather than individual papers are cited to provide as broad a coverage of the literature as possible.

II. Morphology of the Endometrium

A. Cellular Components

The major cellular components of the endometrium are the epithelial cells, the cells of the stroma, and those associated with blood vessels (Findlay *et al.*, 1990). Epithelial cells are classified into luminal and glandular epithelium, the luminal cells forming a continuum with the epithelial cells of the glands. Within these compartments there is considerable heterogeneity; immunohistochemistry with a range of antisera suggests that inner glandular cells (close to the myometrial junction) are functionally quite different from those of the outer glands. Even within the luminal epithelium, positive immunostaining for secretory products is often confined to only a proportion of the cells (e.g., endothelin; Salamonsen and Marsh, unpublished observations). There is evidence, at least in the rabbit (Gerschenson *et al.*, 1979), that the inner glandular cells are the daughter cells that give rise to both the outer glands and the luminal epithelium, but this requires verification in other species. In the human, where the two outer layers of endometrium (the zona compacta and zona spongiosa) are lost at menstruation, progenitors of both glands and luminal epithelium must be present in the remaining basalis endometrial layer. Uterine epithelial cells, unlike epithelial cells in most other organs, are of mesenchymal origin (Ekblom, 1989) and this may explain some of their functions that are not typical of epithelial cells.

Endometrial stroma is composed of a heterogeneous population of cells. Stromal fibroblasts are the major cell type but both regional and cyclical differences can be demonstrated by reaction with a range of monoclonal antibodies (Tabibzadeh, 1991a), underscoring a heterogeneity not exemplified by their morphology. In humans, such regional differences possibly reflect predecidual changes which characteristically occur around blood vessels and subsequently around glands (Blaustein, 1982). It is not known whether such heterogeneity also exists in species in which decidualization is not observed. In addition to fibroblasts, a wide variety of lymphoid cells are found within endometrial stroma and these can be scattered as single cells, within glands, or as aggregates primarily adjacent to glands in the basalis region (Tabibzadeh *et al.*, 1986). Most are T lymphocytes and monocytes, but monocytes/macrophages (Hunt, 1989; De and Wood, 1990), mast cells (Finn, 1986), B cells, dendritic cells, and natural killer cells (NK) are also present (Clark and Muzikova, Chapter 11 this volume; Tabibzadeh, 1990).

The endometrium, particularly its inner layer, is richly endowed with blood vessels, and the cells associated with the endothelium and surrounding muscular layers are an important part of endometrial tissue. Angiogenesis occurs in association with other cyclical changes, and major changes in blood vessel permeability are seen at implantation sites (Finn, 1986).

B. The Extracellular Matrix

1. Composition

There are two types of extracellular matrix (ECM)—the interstitial matrix, which is the collagen-rich ground substance in which stromal fibroblasts are located, and the basement membrane, the specialized matrix underlying epithelial and endothelial cell layers from which its components are deposited in a polarized fashion. Basement membranes also completely surround such cells as muscle, fat, and nerve cells, and in the case of the human uterus, the decidual cells. In general, interstitial matrix contains some or all of collagen types I, II, III, V, and XI; glycosaminoglycans; proteoglycans; and fibronectin, whereas basement membranes are composed of collagen IV, the glycoproteins laminin and entactin, and heparan sulfate proteoglycans (HSPGs). These latter matrices, which have specialized support, sieving, and cell regulatory functions, are created through specific binding interactions between the unique glycoprotein and proteoglycan protomers, and the covalently stabilized polyclonal framework of collagen type IV chains (Timpl and Dziadek, 1986; Yurchenco and Schnittny, 1990).

Components of the ECM have been identified in the endometrium of humans (Aplin, 1989; Aplin *et al.*, 1988; Damjanov and Wewer, 1991; Faber *et al.*, 1986), rats (Karkavelas *et al.*, 1988; Grinnell *et al.*, 1982; Glasser *et al.*, 1987; Welsh and Enders, 1991), mice (Wewer *et al.*, 1986), and cows (Abedin *et al.*, 1982; Ayad *et al.*, 1985). The composition and organization of the ECM in the human endometrium has been described in detail (Aplin *et al.*, 1988) (Table 1) and changes in the endometrium have been shown to accompany the various stages of the menstrual cycle. During the preovulatory phase and in association with cell division, the production of ECM components leads to recovery of functional thickness after menstruation, whereas during the secretory phase the tissue becomes more hydrated

TABLE 1 Composition of the Extracellular Matrix in the Human Endometrium

Proliferative endometrium		Secretory endometrium	
Collagen type I		Collagen type I	
Collagen type III		Collagen type III	
Collagen type V		Collagen type V	
Collagen type VI		Collagen type VI (decreasing with decidualization)	
Fibronectin			
Entactin, Laminin, HSPG[a], Collagen Type IV	Confined to basement membrane and blood vessel walls	Entactin, Laminin, HSPG[a], Collagen Type IV	Begin to appear in stroma around decidual cells

Adapted from Aplin (1989) and Aplin *et al.* (1988).
[a]Heparan sulfate proteoglycan.

(suggesting the presence of glycosaminoglycans and proteoglycans as hydrated space-filling molecules), the stromal collagen becomes more loosely organized, production of collagen type I and fibronectin is slowed, and collagen type VI is extensively degraded and eventually disappears (Aplin *et al.*, 1988). Collagen types III and VI have been postulated to act as attachment proteins, integrating different components of the ECM (Karkavelas *et al.*, 1988), and their disappearance as the matrix loosens would support this hypothesis.

2. Synthesis and degradation

Stromal fibroblasts are probably responsible for production of the interstitial matrix that accompanies tissue thickening. While it is generally accepted that the epithelial cells provide the components of their own basal lamina, it cannot be assumed that they do not also contribute to the ECM of stroma, although fibronectin is produced by human (Siegfried *et al.*, 1984) and ovine (Cherny and Findlay, 1990) endometrial stromal but not epithelial cells in culture. Vascular smooth muscle and endothelial cells are likely also to deposit their own characteristics matrices as they do in other tissues. Stromal cells (human, mouse, rat) which have undergone decidualization synthesize basement membrane products (Damjanov and Wewer, 1991; Glasser *et al.*, 1987; Karkavelas *et al.*, 1988) and become surrounded by the resultant membrane.

The matrix metalloproteinases (MMPs) are a group of enzymes involved in degradation of matrix and remodeling of tissue. Along with tissue inhibitors of MMPs (TIMPs), these enzymes can be synthesized and secreted locally by the resident cells of a tissue or by recruited inflammatory cells. They have been studied extensively in inflammatory connective tissue disorders (Emonard and Grimaud, 1990) but their potential role in remodeling endometrial tissue or in implantation has been explored only recently.

MMPs are neutral zinc proteinases, showing optimal activity at the pH of the extracellular space (pH~7). They are secreted from cells which synthesize them, but such synthesis is negligible in normal connective tissue cells. The MMPs can be classified into major subfamilies on the basis of preferential activity toward specific ECM components; together they are capable of degrading the major macromolecular components of ECM (Matrisian, 1990; Nagase *et al.*, 1992) (Table 2). Unique features of MMPs that distinguish them from other metalloendopeptidases are that each is secreted as a latent zymogen which can be activated by treatment with mercurial compounds or proteinases *in vitro*, and all are inhibited by the formation of 1:1 complexes with TIMPs (Nagase *et al.*, 1992). They are also inhibited by the serum protein α2-macroglobulin. The mechanism of their activation *in vivo* is not known but is thought to be through the action of certain proteinases which can activate them *in vitro*. Induction of genes for MMPs and their activators by factors such as tumor necrosis factor-α (TNFα), epidermal growth factor (EGF), basic fibroblast growth factor

TABLE 2 Substrate Specificities of Matrix Metalloproteinases Described in the Endometrium

MMP No.	Alternative names	Extracellular matrix substrates
MMP-1	Tissue collagenase	Collagens I, II, III, VII, X Gelatin, proteoglycans
MMP-2	Gelatinase A 72 kDa gelatinase	Gelatin Collagens IV, V, VII, X Laminin, elastin, fibronectin Proteoglycan
MMP-3	Stromelysin 1	Proteoglycan Collagens IVm V, IX, X Laminin, fibronectin Procollagen peptides Activates proMMP-1 and proMMP-9
MMP-7	Matrilysin Putative uterine metalloproteinase	Fibronectin, laminin, gelatin Collagen IV Proteoglycan ProMMP-1
MMP-9	Gelatinase B 92-kDa gelatinase	Gelatin Collagens IV, V Laminin, elastin

Adapted from Nagase *et al.* (1992).

(βFGF), interleukin-1 (IL-1) and phorbol myristate acetate (PMA), and of TIMP and matrix proteins by transforming growth factor-β (TGFβ), platelet-derived growth factor (PDGF), EGF, and PMA, have been described in a range of tissues and cell cultures (Matrisian, 1990). Recent studies have shown that the genes encoding a number of MMPs may be independently regulated in cells (Overall *et al.*, 1989; MacNaul *et al.*, 1990), including those of the endometrium (Rawdanowicz *et al.*, 1992). A number of the MMP family members have now been cloned and sequenced (Nagase *et al.*, 1992).

Within the uterus there are four major functions in which MMPs are likely to play a role. In the first two—angiogenesis and nidation—the invading cells must breach barriers to their migration, including cell–cell junctions, basement membranes, and the stromal matrix. Such actions require specific proteolytic enzymes capable of degrading each component of the complex ECM. Degradative enzymes must also have a role in endometrial remodeling and menstruation, and expression of the mRNA for MMP-1 and MMP-3 is seen immediately prior to and during menstruation (Salamonsen, unpublished observations). The fourth likely function of MMPs is the activation of growth factors such as fibroblast growth factor (FGF) and the colony-stimulating factors (CSFs) by releasing them from the ECM (see below).

There is now evidence of MMP production by sheep and human endometrium. MMPs 1, 2, and 3 are produced by sheep endometrial stromal but not epithelial cells in culture, and release of MMPs 1 and 3, but not MMP-2, can be stimulated by PMA (Salamonsen *et al.*, 1991a, 1993b). In the human, endometrial stromal cells in culture secrete at least four MMPs, and these are differently regulated by IL-1 and TNF (Rawdanowicz *et al.*, 1992). Matrilysin (MMP-7) mRNA and protein have been detected in the endometrial epithelium but not the stroma and are abundant in the proliferative and the late secretory phases (Rodgers *et al.*, 1993). The release of MMPs 2 and 9 and their activation are inhibited by progesterone in endometrial explants (Marbaix *et al.*, 1992). Therefore MMPs clearly have a role in the endometrium. Their production is probably restrained by progesterone and modulated by other locally produced factors, allowing for increased production when the enzymes are required to break down tissue matrices (e.g., menstruation in the human). It will be of particular interest to establish whether similar regulation by progesterone occurs in species that do not menstruate. Although there is as yet no information on TIMP production within the endometrium, the serum inhibitor of MMPs, α2-macroglobulin, is synthesized by human endometrial tissue (Fay and Grudzinskas, 1991). Given that the MMP system is regulated at several levels, including protein production, activation, and formation of complexes with inhibitors, it will be some time before its role in regulating degradation of endometrial matrices is fully understood.

Human embryos also produce collagen-degrading activity (Puistola *et al.*, 1989) and expression of DNA for collagenase, stromelysin, and TIMP in mouse embryos increases at the blastocyst stage (Brenner *et al.*, 1989). Likewise in the sheep, preimplantation blastocysts produce MMPs 2 and 3, which have been identified in uterine flushings (Salamonsen *et al.*, 1993a). This suggests a nidatory role for MMPs of embryonic origin. It remains to be established whether MMPs of endometrial origin are also important for implantation.

III. Interactions between Structural Entities within the Endometrium

A. Cell Adhesion Molecules and Receptors

Usually, cells in the living animal must abut in order for development and expression of their differentiated properties to take place. Since cells that usually remain attached *in vivo* must also be capable of diminishing adhesive forces in order to round and divide, interactions of cells with their environment are dynamic. Most components of the ECM have sequences which render them adhesive. The major adhesive molecules for epithelial and mesenchymal cells include fibronectin, vitronectin, laminin, entactin, the fibrillar collagen types I, II, III, V, and XI, and the type IV collagen in basement membrane (Watt, 1986; Levesque *et al.*, 1991). These ECM molecules interact with a variety of cell-surface receptors or cellular adhesion molecules (CAMs). Four major classes of such receptors have been identified (Hynes, 1992; Hynes and Lander, 1992).

1. The cadherins mediate homologous adhesion between cells in the presence of calcium. Cells expressing different cadherins separate from each other to form separate aggregates.
2. The immunoglobulin superfamily members are mostly found on leukocytes but they are also located on other cells. Some mediate homologous and some heterologous adhesion, but binding can also occur with other adhesion molecules such as HSPG and integrins.
3. Selectins (LECAMS) are expressed on various blood cells and/or endothelial cells and mediate heterologous interactions during leukocyte homing and adhesion. Both selectins and the immunoglobulin superfamily are likely to be important for leukocyte movement into and within the endometrium, and for formation of leukocyte aggregates (Tabibzadeh and Poubouridis, 1990).
4. Integrins bind primarily to various ECM components and mediate cell–matrix interactions, which are important during cell migration and in attaching cells to basement membranes.

Most cells can express several integrins and can therefore bind to several adhesive ECM molecules. Furthermore, some of the integrins can recognize a given matrix molecule at several sites. The selectivity of cells for different ECM proteins is therefore controlled at least in part by the pattern in which they express integrins, which can be selectively varied. Adhesive specificity and affinity of integrins can also be regulated post-translationally. Integrins have a globular extracellular domain made up of α and β subunits which binds the ECM ligand or the immunoglobulin counter-receptor in the presence of divalent ions. Each subunit also has a transmembrane and a cytoplasmic domain. Some of the integrin recognition sites on ECM ligands have been identified. The RGDS sequence on fibronectin, vitronectin, and a variety of other adhesive molecules has been well characterized (Yamada, 1991) and is recognized by a number of integrins. Other recognition sequences include the KOAGDV sequence in fibronectin (Hynes, 1992) and the DGEA sequence in type 1 collagen (Staatz *et al.*, 1991). Laminin has multiple functional sites that mediate its interactions with integrins and other components of the ECM (Mecham, 1991).

There is substantial evidence that integrins mediate information transfer into cells, and it has been postulated that cell adhesion signals transmitted through integrins interact with those triggered by soluble growth factors and oncogenes (Hynes, 1992). Such synergy would explain the dependence of growth on anchorage. Moreover, the cytoplasmic domain is often associated with the cytoskeletal elements of the cell (Albelda and Buck, 1990; Humphries, 1990) and thus binding of ligand to receptor is likely to be one means by which information is transmitted into the cell. Many, if not all, integrins undergo activation, that is stimulated by such soluble mediators as cytokines and growth factors, and solid phase reactants such as ECM or other cells. The inactivation of the integrins is equally important but little is known about the mechanism that prevents the cell from attaching to the wrong structure at the wrong time. Thus, cell adhesion is a result of a versatile and complex array of interactions, modulations, and signaling events in which integrins play a central role.

1. Adhesion molecules and endometrial physiology

a. Integrins and other adhesion molecules in normal and abnormal endometrium

Both regional and menstrual cycle-related changes in the endometrial expression of a range of integrin molecules have been demonstrated (Tabibzadeh, 1992; Lessey *et al.*, 1992). The greatest number of integrin molecules are exhibited on endometrial epithelial and endothelial cells, followed by stromal and then lymphoid cells, though this expression can differ between the basalis and the functionalis regions. Furthermore, collagen and laminin receptors on endometrial epithelial cells are expressed only basolaterally, and are probably involved in anchoring the epithelium to the basement membrane. Fibronectin receptors are present in the stromal compartment, which is the fibronectin-rich region of the endometrium. In some but not all cases, integrin expression is related to the time of the menstrual cycle. For example, intercellular adhesion molecule-1 (ICAM-1) expression in vascular endothelial cells in the endometrium (which is confined to only a very small proportion of vessels) increases at menstruation, and the associated massive influx of leukocytes suggests a role for this integrin in the endothelial migration of leukocytes (Tawia *et al.*, 1993). In contrast, immunostaining for platelet endothelial cell adhesion molecule (PECAM) in endothelial cells is strong at all stages of the cycle. Endometrial epithelial cells also express ICAM-1, whereas endometrial T-cells express its reciprocal molecule, LFA-1, so that it is conceivable that T-cell binding to epithelial cells is based on the LFA-1:ICAM-1 interaction (Tabibzadeh and Poubouridis, 1990; Malegapuru *et al.*, 1988). Integrin expression has also been examined in endometrial carcinomas (Lessey *et al.*, 1993) and in the neoplastic phenotype shown to be associated with alterations in integrin expression, correlates better with grade than with stage of the carcinoma.

b. Integrins and implantation

Interactions among embryonic and uterine cells during implantation are examples of highly regulated cell adhesion events. The patterns of expression for cell surface and extracellular matrix glycoproteins change markedly in both embryos and uterine cells during the peri-implantation stage of development; in the embryos these changes appear to be under developmental control. Mouse trophoblasts can differentiate, adhere, and migrate on surfaces coated with a range of ECM components (Armant *et al.*, 1986a; Yagel *et al.*, 1988). Trophoblast invasiveness corresponds with the cellular expression of integrins, including recognition molecules for the RGD sequence in fibronectin and an alternative site in laminin (Armant, 1991; Armant *et al.*, 1986b; Sutherland *et al.*, 1988; Carson *et al.*, 1988; Farach *et al.*, 1987). Other glycoproteins that promote cell adhesion [uvomorulin (E-cadherin), and cell CAM 105] are expressed by blastocysts at the peri-implantation stage (Svalander *et al.*, 1987; Vestweber *et al.*, 1987) and collagens may be present on the external surface of the trophectoderm (Leivo *et al.*, 1980; Sherman *et al.*, 1980). On the maternal side, changes in glycoprotein expression seem to be related to the hormonal milieu and the sexual maturity of the animal but these responses are poorly understood. Fibronectin and laminin are synthesized by decidual cells and deposited in the endometrial ECM at the time of trophoblast invasion (Wewer *et al.*, 1985; Kisalus *et al.*, 1987). Glycoconjugate expression on the apical cell surface of the uterine epithelium also changes markedly as the uterus develops to a receptive state (Anderson *et al.*, 1986; Enders and Schlafke, 1974). It is therefore likely that integrins and other attachment molecules mediate multiple aspects of embryo attachment and implantation.

B. Cell–Cell Interactions

There are several mechanisms by which interconnecting cells in various tissue can communicate in the absence of gap junctions (Singer, 1992).

1. Direct binding of a cell surface ligand on one cell to its specific transmembrane receptor on the other cells. For example, membrane-bound precursors of growth factors, such as the cell surface glycoprotein, pro-TGFα, may accumulate on membranes and be available to form a complex with receptors on adjacent cells (Anklesaria *et al.*, 1990). Thus the pair can function as mediators of intercellular adhesion and in addition this interaction may promote a mitogenic response. Such direct interactions between cells are known as "juxtacrine."

2. Indirect communication can occur when a cell is stimulated to secrete a product into the narrow intercellular gap where another cell is in contact and the soluble products are bound to specific receptors on the adjacent cell. This mechanism may require stable intercellular adhesion and a finite time to induce secretion of the stimulated product. This is the common mechanism of cell stimulation by exogenous soluble growth factors, but is strictly limited to communication between two cells.

3. Transcellular signaling may involve binding small ligands contained in the ECM to cell surface receptors; in this case adhesion between the two cells will be indirect through the ECM. Alternatively, communication may be inhibited by the interposition of an ECM between the cells (Baker, 1990).

To date, the nature of the cell–cell interactions within the endometrium has not been established but all of these events probably occur in particular cases and should be considered in future research.

C. Cell–Matrix Interactions

The ECM can alter the hormonal responsiveness, cell shape, and gene expression of cells in a specific manner (Enat *et al.*, 1984; Reid, 1990) and conversely, the cell can alter the synthesis and degradation of the ECM (Getzenberg *et al.*, 1990). Our understanding of cell–matrix interactions is rapidly evolving as new binding proteins are identified and new sequences with biological activity are found in matrix macromolecules (Levesque *et al.*, 1991). A number of ECM proteins have mitogenic properties. For example, the central part of the laminin molecule contains domains with mitogenic properties which are quite separate from their adhesion properties (Panayotou *et al.*, 1989).Such proteins synergize with interleukins and CSFs in their mitogenic influences on the immune and hemopoietic systems. They are of low affinity but their high concentration and long half-life compensate for this. They are always present in excess, implying that their mitogenic effect is controlled at the level of the receptor. ECM proteins also contain some sequences that are partly homologous with growth factors

(Weis *et al.*, 1984; Melchers *et al.*, 1985; Bazan, 1990), and other ECM components are able to sequester extracellular growth factors and modulate their activity. For example, when PDGF is bound to osteonectin, it is unable to bind to its cognate receptor (Lane *et al.*, 1991) and HSPGs bind FGFs (Burgess and Maciag, 1989), thus protecting the FGF from degradation and providing a matrix- or cell surface-bound reservoir of FGFs. The HSPG may also change the conformation of the FGF so that it acquires an ability to bind to its receptor complex (Klagsbrun and Baird, 1991). Other growth factors using a similar mechanism are GM-CSF, TGFβ, and IL-3. Thus it appears that many growth factors and cytokines act on their target cells over short ranges only; their immobilization at the cell surfaces and in ECM accomplishes this goal.

This picture is not simple, and new complexities accrue over time (see, e.g., Nathan and Sporn, 1991). Adherence of monocytes and macrophages to fibronectin results in transcription of GM-CSF and CSF-1 genes, whereas adherence to collage induces the production of IL-1 and TNF. Within the ECM, TNF stimulates the secretion of other cytokines only when neutrophils are bound to matrix proteins (Nathan *et al.*, 1989). Therefore the ability of the ECM to present cytokines in either an active or inactive form may be a further way in which the tissue can exert control over cytokine responses of the cells contained within it. The combined result of these known mechanisms is tight spatial control of growth factor and cytokine activities (Ruosiahti and Yamaguchi, 1991). Such control is clearly important in the endometrium, in which a wide range of growth factors and cytokines have now been identified (see review by Murphy and Ballejo, Chapter 12, this volume).

Interaction between a basement membrane and the epithelial cell surface is essential for the formation of epithelial cell sheets. Epithelial cells have a polar morphology, with basal, lateral, and apical sides, the latter containing properties specific for each cell type. The formation of uterine epithelial cells from mesenchymal cells during embryonic development is associated with a gradual increase in intercellular adhesion, with mesenchymal adhesion molecules (N-CAM, interstitial collagens and fibronectin) being replaced by epithelial adhesion molecules (laminin and uvomorulin) (Ekblom, 1989). Laminin, a component of the basement membrane, appears to regulate development of epithelial cell polarity whereas uvomorulin is probably involved in lateral cell–cell interaction.

IV. *In Vitro* Approaches to the Study of Endometrial Function

Cell culture techniques have contributed greatly to our understanding of the interactions between the cell and its microenvironment and to the evaluation of cell–cell communication. Findlay *et al.* (1990) have discussed the topic in considerable detail as it applies to endometrial function, but with the rapid developments in cell biology outlined above, new approaches to cell culture are constantly evolving.

The major goals of cell culture are to:

1. isolate a group of cells that retain the specific characteristics observed *in vitro;*
2. characterize and determine the mechanisms of action of the genetic and environmental factors that influence cell behavior;
3. reconstruct the cell and its environment into a defined and experimentally controlled system to study the determinants of properties and responses of cells in a dynamic context.

A. Isolation of Purified Cell Types

Although cultures of mixed cells derived from a tissue have provided and continue to provide much useful information on regulation of endometrial cell function, pure cell cultures are essential for investigating paracrine and autocrine regulation. Several methods have been described for dissociation of uterine tissue cells which have then been separated into individual types (Findlay *et al.*, 1990). Identification of cell types by a combination of methods is obligatory; the methods include phase-contrast and transmission electron microscopy, histochemistry to detect specific enzyme markers, and immunocytochemical staining using a variety of antisera against specific antigens. Antisera against the intermediate filament proteins (keratin, vimentin, desmin) have been most widely used and will differentiate between epithelial, stromal, and decidual cells respectively (Findlay *et al.*, 1990). In addition, endothelial cells have been identified by their expression of factor VIII, stromal cells by fibronectin, and lymphocytes by the lymphocyte common antigen. Individual immunocytes may now be identified by a range of monoclonal antibodies directed against specific antigenic determinants for each class of immunocyte. Evidence is accumulating to suggest that there is heterogeneity of both stromal and epithelial cells not exemplified by their morphology (Gerschenson *et al.*, 1979; Tabibzadeh, 1991b). The separation and culture of such subsets, and the transformation of one subset to another to modify tissue responses, have yet to be achieved.

B. Culture in Defined Medium

The current challenge for cell culture laboratories is to create the conditions needed to sustain specific cell populations in the absence of serum, which contains a wide range of cell regulatory factors. For instance, the serum protein α2-macroglobulin can reversibly bind a number of growth factors (nerve growth factor, PDGF, TGFβ, βFGF), cytokines (IL-1β and IL-6) (James, 1990), and MMPs (see Section II,B,2). There are other serum proteins which bind a significant fraction of steroids added to cultures (Pavlik and Katzenellenbogen, 1978). Moreover, fetal calf serum itself contains near-physiological levels of prostaglandins and steroid hormones which can be removed by charcoal stripping of the serum. Such treatment appears not to remove attachment or spreading factors and cells plated in as little as 5% stripped fetal calf serum do not behave differently on plastic in terms of total protein synthesis than cells in complete serum (Salamonsen, 1986). One must always be aware of additives such as the phenol red indicator, which has demonstrable estrogenic activity (Berthois *et al.*, 1986).

In practice, most cell cultures are first established in serum (5–20%) and, once attached and spreading, cells are maintained in some formulated serum-free mixture. What critical components of serum must be replaced in these serum-free media? So far, only vitronectin has been established as an adhesion/spreading factor for cell culture (Underwood and Bennett, 1989) although this role was previously ascribed to fibronectin. For epithelial endometrial cell culture, it is generally accepted that a mixture of insulin, transferrin, selenite, hydrocortisone, and a low concentration of some protein (bovine serum albumin or lactoglobulin) is required. In our laboratory, such a recipe has been compatible with higher rates of endothelin production from cultured human endometrial epithelial cells than if medium with no additives was used (Hampton *et al.*, 1992). A wide range of commercial media supplements are now available and must be carefully assessed for individual needs.

C. Culture on Basement Membrane Materials

The stroma may dictate the rate of proliferation and cytodifferentiation of its epithelial counterpart by controlling the secretion of the anchoring basement membrane or it may induce the epithelium to secrete its own matrix (Gospodarowicz *et al.*, 1979, 1982). For culture of highly purified cell populations, it has become important that such cells be plated on appropriate basement membrane material. No ideal formulation has been achieved. The most commonly used is Matrigel, a naturally produced basement membrane material obtained from cultured Engelbreth–Holm–Swann sarcoma cells, the main components of which are type IV collagen and laminin with minor amounts of entactin and HSPG (Norman *et al.*, 1989). The composition of ECM proteins in Matrigel is very similar to that in the secretory phase of the human endometrium (Glasser *et al.*, 1988). While this material has the advantage of being a naturally produced basement membrane, it is unlikely that when it is applied to culture dishes it takes up the natural three-dimensional matrix form that it has *in vivo*. Significant levels of growth factors and protease activity are found within its components (McGuire and Seeds, 1989). ECMs have also been prepared from stromal cells homologous with the desired epithelial cell (Reid and Rojkind, 1979) but this has not proven useful in the case of ovine endometrial cells (Salamonsen, unpublished observations), presumably because the basement membrane underlying epithelial cells *in vivo* is of epithelial and not stromal origin. The other commonly used biomatrix is collagen type I derived from rat tails and formed in a gel on the plastic, but again, while some laboratories have found this useful, it does not approximate the natural basement membrane, which is primarily collagen type IV.

Studies on growth of endometrial cells on different extracellular matrices (Matrigel, collagen types I and IV, laminin, fibronectin) have used cells from humans (Casselen and Harper, 1991; Schatz *et al.*, 1991), guinea pigs (Mahfoudi *et al*, 1991), rats (Glasser *et al.*, 1988; Welsh and Enders, 1989; Branham *et al.*, 1991), sheep (Cherny, 1991), and cows (Helmer *et al.*, 1993). While they demonstrated the importance of the source of ECM on which the cells (particularly epithelial) are grown, these studies failed to provide a consensus on the most appropriate ECM for culturing endometrial epithelial cells. Since the functional criteria of the cells differed widely in these studies, it may be that different matrices favor different cellular functions.

D. Polarized Epithelial Cells

In vivo, endometrial epithelial cells form highly organized cell sheets lining the uterine lumen and these are contiguous with the glandular epithelial cells. As with all epithelia, they have vectorial transport and barrier functions, and are highly secretory. These functions are dependent on the polarized distribution of proteins and lipids between the outward-facing apical domain and the inward-facing basolateral domain (Simons and Fuller, 1985). Apical and basolateral components are prevented from intermixing by the formation of a barrier near the apical surface consisting of tight junctions between the lateral plasma membranes of adjacent cells (Gumbiner, 1987). Establishment and maintenance of this cell–surface polarity require proper cell–substrate attachment, formation of intercellular contacts and junctions, polarized cytoskeletal pathways, and specialized membrane transport pathways (Simons and Wandinger-Ness, 1990).

The culture of endometrial epithelial cells on matrix-coated permeable surfaces encourages the formation of these tight junctions between the cells which isolate the apical and basal compartments (rat, Glasser *et al.*, 1988; sheep, Cherny and Findlay, 1990; human, Schatz *et al.*, 1990; rabbit, Mani *et al.*, 1991). The formation of tight junctions results in the development of electrical resistance across the cell layer and this resistance can be used to test the integrity of the barrier. Such cells demonstrate both morphological and functional polarity, secreting proteins (Glasser *et al.*, 1988; Salamonsen *et al.*, 1992), prostaglandins (Cherny and Findlay, 1990; Jacobs *et al.*, 1990; Schatz *et al.*, 1991), and proteoglycans (Carson *et al.*, 1991) in a polarized manner. They also express cell surface molecules appropriately. For example, uvomorulin is expressed on the basolateral domain of all rat endometrial epithelial cells whereas the adhesion molecule CAM105 is expressed only on the apical surface of some cells (Julian *et al.*, 1992a). Development of polarity and of these secretory characteristics can be attributed not only to the growth of the cells on ECM but also to the porous membrane which allows exposure of the basal cell surfaces to culture medium. By careful selection of culture conditions, Julian *et al.* (1992b) have established cultures which exhibit responses *in vitro* similar to those expressed by estrogen-stimulated uterine epithelial cells *in vivo* under serum-free conditions. One function of estradiol in these cultures is to maintain the relationship of the epithelial cell with its basement membrane. In addition, these cells provide an apical surface that allows an embryo to become attached.

Such engineered cell systems will continue to provide important models for identifying the actions of steroid hormones and other regulators on epithelial cell function, examining the interactions of epithelial cells with stromal and trophoblast cells, and identifying those epithelial components obligatory for blastocyst attachment.

E. Decidualization *in Vitro*

Human endometrial stromal cells in culture respond to ovarian steroids with increased proliferation and production of markers characteristic of decidualization, such as desmin and prolactin (Maslar and Riddick, 1979; Faber *et al.*, 1986; Irwin *et al.*, 1989), and provide a useful model for decidualization *in vivo*. Moreover, these cells can be maintained in a defined medium. Under such conditions, PDGF, TGFα, and βFGF stimulate stromal cell growth in a progesterone-dependent manner while EGF and βFGF produce differential effects on growth and differentiation that are modulated by progesterone (Irwin *et al.*, 1991). Such studies support the concept that autocrine and paracrine regulatory loops involving growth factors and steroid hormones are important in the endometrium.

F. Cell–Cell Interactions *in Vitro*

1. Epithelial–stromal interactions

Both estrogen and progesterone receptors have been demonstrated within the endometrium in a number of species, including primates (human and nonhuman), rodents, rabbits, and sheep (for example, Bergman *et al.*, 1992; Ohta *et al.*, 1993; Cherny *et al.*, 1991; Brenner *et al.*, 1990;

Sheridan and Weaker, 1992). Broadly, the pattern of expression observed is one of abundance on the days prior to and immediately following ovulation, and minimal expression during the progestogenic phase. Both overall receptor protein concentration and cellular localization vary markedly with the stage of the reproductive cycle. Receptor protein (estrogen and progesterone) and mRNA (estrogen) (Koji and Brenner, 1993) have both been detected in the nuclei of epithelial, stromal, vascular, and myometrial cells. The marked regional variability seen during the cycle is clearly linked to varying and possibly also local concentrations of estrogen and progesterone.

It is now apparent that cell–cell interactions are necessary to mediate many steroid-regulated functions. For example, in the rhesus monkey, there are many instances where both estrogen and progesterone receptors are present only in stromal cells while the adjacent mucosal epithelium undergoes dramatic estrogen- or progestin-dependent changes. In particular, in the endometrium during the luteal-follicular transition, stromal but not glandular cells develop both receptors yet glandular epithelial cells undergo estrogen-dependent DNA synthesis and mitosis that can be inhibited by progesterone treatment. Such observations point strongly toward the presence of soluble mediators that transmit information between stromal and epithelial cells. The challenge to the cell biologist is to establish the identity of the mediators and the mechanisms of their action.

There is ultrastructural evidence of interactions between the glandular epithelium and the superficial stromal cells in the endometrium. In the human endometrium, the basal lamina develops breaks under estrogen domination during the proliferative phase and epithelial cell projections extend through this opening (Roberts *et al.*, 1986). However, the basal lamina of the rat remains intact under the influence of estrogen (Lavia and Larson, 1991), suggesting that in these two species stromal–epithelial interactions may be achieved by different means. Useful models exist for the examination of these interactions. White *et al.* (1990) described a system for culture of human endometrial cells that supports formation of a simple columnar epithelium, a stromal component, and epithelial glands. Polarized epithelial cells are generally grown on ECM-coated Millicell inserts which are easily added to and removed from culture wells containing attached stromal cells. However, such a system is only useful for examining soluble mediators and what is needed is a system in which two cell types can be grown on opposite sides of an extracellular matrix. Schatz *et al.* (1990, 1991) seeded human proliferative phase glands and stroma on Matrigel in such a manner that the stromal cells formed clusters within the Matrigel whereas the glands formed polarized cuboidal-columnar epithelial cell monolayers on the Matrigel surface. This configuration promotes epithelial–stromal interactions through cell-derived diffusible paracrine effectors such as growth factors and prostanoids, and through insoluble ECM proteins.

2. Stromal–immunocyte interactions

The interaction of immunocytes with other cells is fundamental to the expression of many normal lymphocyte functions and conversely, many lymphokines are essential for normal function of other cells. Lymphocytes adhere to and explore the surface of fibroblasts in tissue culture as a result of interactions between specific adhesion molecules (Abraham *et al.*, 1989); lymphocyte activity can be accompanied by the direct transfer of certain lysosomal enzymes and other potential intercellular modulators from the lymphocytes (Olsen *et al.*, 1983). Lymphocytes are present in large numbers both within the stroma and in association with epithelial cells in the endometrium (see Section II,A), and soluble lymphocyte products such as cytokines have regulatory effects in the endometrium (Tabibzadeh, 1991a). For example, when human endometrial T cells were added to epithelial cell cultures derived from autologous glands, morphological changes as well as HLA-DR molecules were induced in the epithelial cells. The colonies of T cells formed *in vitro* were similar to the aggregates of lymphoid cells that are characteristically localized in the basalis stroma *in vivo* (Tabibzadeh, 1991a). Further coculture studies on the interaction between immunocytes and endometrial epithelial and stromal cells seem timely and appropriate.

V. Function of the Oviduct

In most vertebrate species, the oviduct is essential for the collection of ova and provision of a suitable physical and chemical environment for the final stages of sperm capacitation and acrosome reaction, fertilization, and the early stages of embryo development. However, the fact that fertilization and embryo development for human, bovine, and ovine species, at least, may be effected entirely in culture (Fukui *et al.*, 1988; Eyestone *et al.*, 1991) argues against the need for specific products from the oviduct, apart from a suitable buffered medium containing amino acids, simple carbohydrates, and plasma proteins, with appropriate partial pressures of oxygen and carbon dioxide. The problem is that *in vitro* production of preimplantation embryos of all species is grossly inefficient, suggesting that culture media lack some unidentified facilitative factors.

For several years, the culture of ruminant embryos was inhibited around the third and fourth cell divisions, the stage at which the embryonic genome is activated (Gandolfi and Moor, 1987; Barnes and First, 1991). This block in embryo development was initially overcome by coculture with oviduct epithelial cells, particularly those taken just after estrus, following the estrogen surge (Gandolfi and Moor, 1987; Rexroad and Powell, 1988; Eyestone and First, 1989). Other sources of cells or proteins, such as heat-inactivated human serum and conditioned medium, have also been used with limited success (Walker *et al.*, 1988; Eyestone *et al.*, 1991), suggesting that oviduct secretion of nonspecific cellular embryotropic products is necessary. Walker *et al.* (1989, 1992) have found that while embryos developing in this medium showed advanced morphological development, they did so at lower mitotic rates, and subsequent viability on transfer to uteri of surrogate ewes remained low. Unaccountable effects on fetal growth appear to occur because pregnancies obtained from these cultured embryos often terminated in the difficult birth of large lambs (Walker *et al.*, 1992; Nancarrow and Marshall, unpublished observations). In a preliminary experiment to test the embryotropic nature of ovine oviduct secretions, we found that including a fraction enriched with the estrus-associated glycoprotein (EGP) (Sutton *et al.*, 1984a) in these culture media increased the cell number of blastocysts to that developed by blastocysts *in vivo* (Hill *et al.*, 1992). These observations, taken as a whole, support the concept (see review by Bavister, 1988) that the oviduct secretes substances that regulate the development and enhance the viability of embryos.

VI. Morphology of the Oviduct

A. Cellular Components

The three sections of the oviduct—the infundibulum or fimbria, the ampulla, and the isthmus, which lead to the uterus—are composed of different proportions of similar cell types, although this by no means implies that the secretions are necessarily the same (see Section VII,A). The external tissues of the oviduct consist of the longitudinal and circular musculature. The obvious anatomical difference between the ampulla and isthmus relates to the greater degree of invagination of the epithelium of the former (Figure 1A) and the greatly increased smooth muscle component of the latter (Figure 1B). A far greater secretory surface area is available to the oocyte or embryo in the ampulla, and this is reflected in the fact that the fluid secretion rate is also far greater. When oviducts of anesthetized ewes were locally perfused with fixative *in situ* (Hollis *et al.*, 1984), numerous

FIGURE 1 Transverse section through the sheep oviduct fixed *in vitro* (A) Ampulla (Day 1) and (B) isthmus (Days 10–12) showing the highly invaginated epithelium and little musculature of the ampulla, extensive musculature of the isthmus, and the relationship of the vasculature to the stroma and epithelium. E, epithelium; S, stroma; B, blood vessel; L, longitudinal smooth muscle; C, circular smooth muscle. Light micrograph, magnification: A, ×54; B, ×109. From Hollis and Nancarrow, unpublished results.

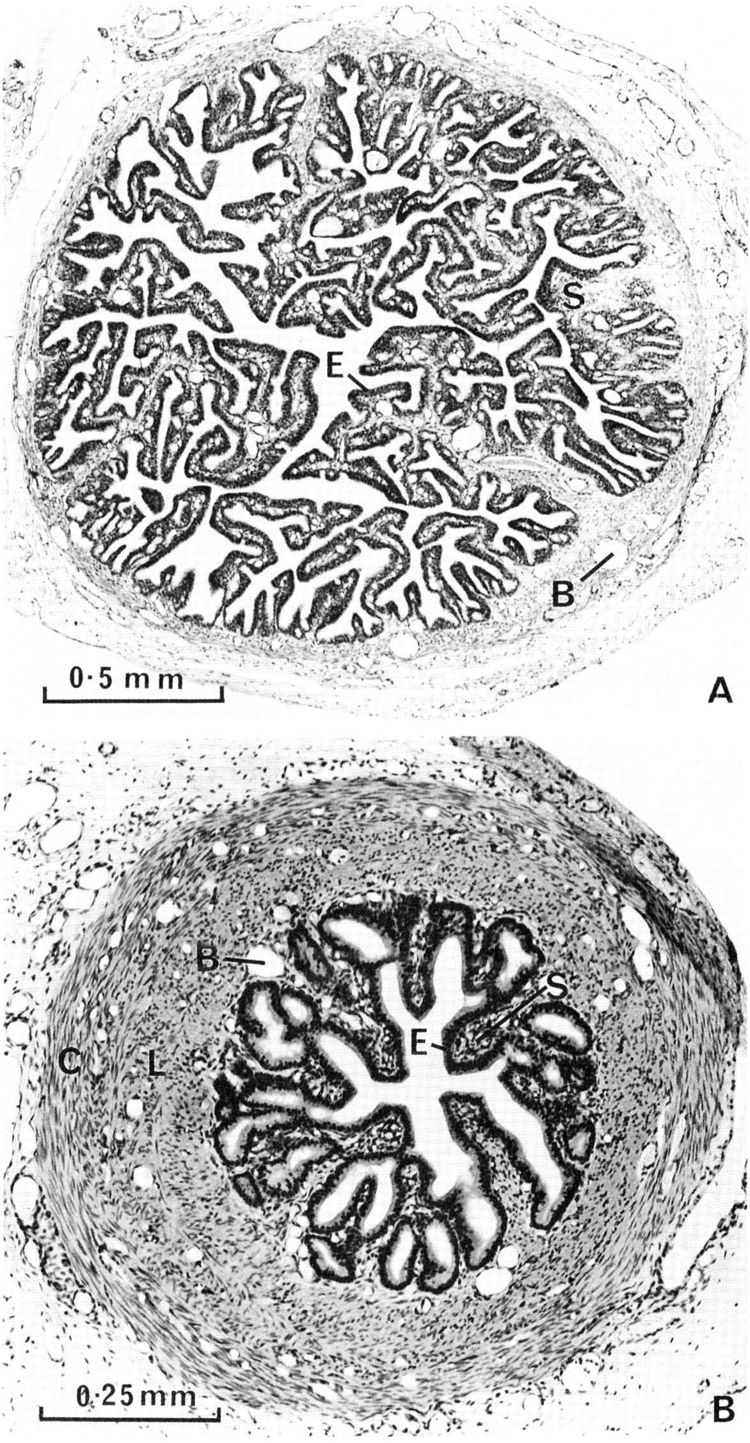
S
E
B
0·5 mm
A
B
S
E
C
L
0·25mm
B

blood capillaries were observed threading through the lamina propria, bringing any circulating plasma factors into close proximity to the cells of the lamina, the epithelium, and any embryos contained within the lumen (Figure 2). It may be expected then, that nutrients, growth factors, gases, and other communicative molecules can be readily exchanged within this environment.

The epithelial cells consist mainly of columnar ciliated cells and secretory cells with microvilli, while sparse lymphoid or basal cells are associated with the basement membrane (Figure 2). Odor (1974) questioned whether these basal cells were precursors of epithelial or stromal cells. They were found also within the connective tissue in the rabbit and were thought to be indifferent, reserve or clear cells. They are small and oval,

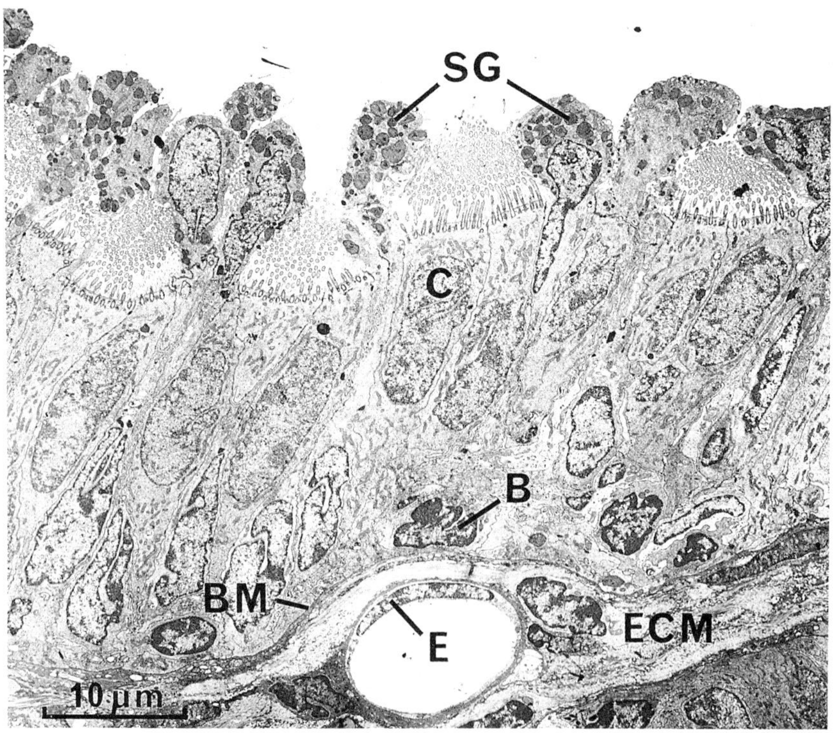

FIGURE 2 Transverse section through the ampullary epithelium of sheep oviduct fixed by *in situ* perfusion via the tubal branch of the ovarian artery on Day 10 of the estrus cycle. C, ciliated cell; SG, secretory granules in secretory cell; E, endothelial cell; B, basal cell; BM, basement membrane; ECM, extracellular matrix. Transmission electron micrograph, magnification ×2000. Reproduced with permission from Hollis *et al.* (1984).

with hyperchromatic nuclei and prominent nucleoli, and contain a scant halo of cytoplasm. Because they had moderately large pseudopodia, they were thought to be motile. Odor (1974) concluded that they were not fibroblasts, macrophages, plasma cells, or monocytes of connective tissue, but lymphocytes, a decision disputed by van Bogaert *et al.* (1978) for similar cells in human oviducts. However, these cells possess quite different characteristics from the dendritic basal cells that make up to 10% of the epithelial population in sheep (Figure 2), and Hollis *et al.* (1984) classify them as lymphocytic or phagocytic.

Underlying the basal lamina, the cells associated with the subepithelial stroma, or lamina propria, are sparse and ill-defined and surround the endothelial cells of the vasculature. They consist of large connective tissue (stromal) cells interspersed with layers of collagen fibrils and surrounded by varying proportions of inner and outer smooth muscle. Elongated smooth muscle cells are arranged in loose sheets and contain cytoplasm with interspersed bundles of myofilaments. Other cell types infrequently described in different species have been mast cells (McDaniel *et al.*, 1968; Hollis *et al.*, 1984), fibroblasts and lymphocytic round cells (Hadek, 1955; Nellor, 1965), plasma cells (Kutteh *et al.*, 1988), and scattered lymphocytes and occasional polymorphonuclear leukocytes, any of which may infiltrate the epithelium, for example, on Day 1 of the human menstrual cycle (Clyman, 1966) or during the progestational diestrous stage in sheep (Abdalla, 1968). There have been infrequent examples of mitosis in oviduct tissue, but cell death and apoptosis have been noted in monkeys (Brenner *et al.*, 1983; Verhage *et al.*, 1990).

It is clear that after many years of descriptive research, the nature of the interaction among the constituent cells of the oviduct remains mysterious. The detailed information concerning the function of equivalent uterine cells discussed earlier suggests that appropriate methods are now available to address this problem. However, the variable morphology attributed to the oviduct is recognized and we attempt to relate these observations to the secretion and function of the estrus-associated group of glycoproteins detailed in Table 3.

1. Ciliated cells

Columnar ciliated cells are attached to the basal lamina and have microvilli present on their luminal surface; their nuclei are elongated and medially or basally located. There is little or no change in the degree of ciliation occurring during the estrous cycle of sheep (Willemse, 1975a; Nayak, 1977; Hollis *et al.*, 1984), in contrast to that of primates (Brenner, 1969; Verhage *et al.*, 1979; Odor *et al.*, 1980), which have a prolonged follicular phase. Differentiation and ciliation are induced by estradiol, but this function is attenuated or blocked by progesterone in human (Verhage *et al.*, 1979; Jansen,

TABLE 3 Characteristics of Major Secretory Glycoproteins of the Oviduct in Some Species

Species	$M_r \times 10^{-3}$	p*I*	Method, properties, and source	Reference
Sheep	70–90	4.7	Oviduct Secretion Days −1 to 6 of cycle; induced by estrogen *in vivo*; terminal glycosides are fucose, galactose, and *N*-acetylgalactosamine	Sutton *et al.*, (1984a, 1985, 1986)
Sheep	92	5.0	Epithelial cell culture, secreted for 5 days after estrus, 4–5 days in culture, bound to zona and blastomeres, ampullary source	Gandolfi *et al.* (1989b, 1991)
Sheep	90–92	7.5 8.5	Explant culture after *in vivo* estrogen induction, two proteins, ampullary source	Buhi *et al.* (1991)
Sheep	90 92		Flushings and explant culture after *in vivo* estrogen induction, progesterone modulated, two proteins, ampullary source	Murray (1992)
Cow	85–97	5.0	Explant culture, greatest at Day 0, estrus induced	Malayer *et al.* (1988)
Cow	97	5.5–8.1	Explant culture and flushings, greatest at estrus	Boice *et al.* (1990a)
Cow	80–95		Oviduct secretion throughout cycle, highest around estrus.	Gerena and Killian (1990)
Cow	97 (2)	4.7	Explant culture, greatest at estrus, terminal glycosides, *N*-acetylgalactosamine *N*-acetylglucosamine, ampullary source.	Wegner and Killian (1992)
Pig	85, 115		Explant culture, associated with estrus, ampullary source.	Buhi *et al.* (1989)
Pig	75 and 85 100 and 100	<4 4.5, >8	Fucose incorporated (75/85), all ampullary source	Buhi *et al.* (1990)
Pig	80 100	<4	Explant culture following *in vivo* treatment with estrogen	Buhi *et al.* (1992)
Baboon	130	5, 6, >8	Flushings and minced tissue culture, mid-to late follicular stage	Fazleabas and Verhage (1986)

(*continues*)

TABLE 3 (*Continued*)

Species	$M_r \times 10^{-3}$	p*I*	Method, properties, and source	Reference
Baboon	100–120	4.5	Explant culture following	Verhage and Fazleabas
	120–130	8	estrogen/progesterone treatment,	(1988)
	130	5	Ampullary, isthmic, entire source, resp.	
Human	120–130	Acidic	Explant culture	Verhage *et al.* (1988)
		Basic	Midcycle	

1984), baboon (Verhage *et al.*, 1990), and cynomolgus macaques (Brenner *et al.*, 1983). Apical protrusions containing vesicles with filamentous material in isthmic ciliated cells on the day before estrus, which occasionally appeared to have ruptured, provided evidence of exocrine secretion into the lumen by this segment (Hollis *et al.*, 1984). This is in accord with the observation that Schiff-positive secretory material fills the isthmic lumen in sheep (Hadek, 1955; Abdalla, 1968). Isthmic secretion may bind to sperm (Sutton *et al.*, 1984b; Lippes and Wagh, 1989; McNutt *et al.*, 1992), affecting either their motility or function. There is little evidence for ciliated cells of the ampulla acting in a similar secretory manner (Hollis *et al.*, 1984).

2. Secretory cells

The internal structures of the nonciliated columnar secretory cells in all species examined are typical of cells actively synthesizing protein, with well-established rough endoplasmic reticulum, supranuclear Golgi zones, and various secretory granules (SG) of differing structure and electron density. Nuclei are located in the mid- to basal areas, but under progesterone domination they are usually found apically in sheep (Hadek, 1955; Abdalla, 1968; Willemse, 1975b; Nayak *et al.*, 1976) and cattle (McDaniel *et al.*, 1968; Nayak and Ellington, 1977). However, Hollis *et al.* (1984) also found nuclei in apical protrusions during the ovine estrogenic phase (Day −1). This was unusual because there have been several demonstrations of the rapid basal response of nuclei to estradiol (Willemse, 1975b; Nayak *et al.*, 1976). Possibly these examples were obtained just prior to complete progesterone withdrawal. The numerical density of SG was far greater in ampullary than in isthmic cells at all stages. Apical protrusions were evident in isthmic cells prior to estrus, but no evidence of secretion could be found (Hollis *et al.*, 1984), although Jansen (1984) quotes examples of apocrine secretion in the isthmus pre- and postovulation in humans.

Unlike the ampulla, SG of the isthmus were nonlamellated, suggesting that the contents and secretory mechanisms are dissimilar.

B. Steroid Receptors

The cells of the oviduct contain cytosolic and nuclear receptors to estrogen and progesterone. They accumulate these steroid hormones in amounts up to 25 times plasma concentrations (see reviews, Jansen, 1984; Verhage and Jaffe, 1986) and consequently respond with morphological changes associated either with the estrous cycle or with seasonal changes. In the absence of estradiol, or alternatively under progesterone domination, the columnar epithelium dedifferentiates; deciliation may occur and in general the secretory cells lose their synthetic structures—rough endoplasmic reticulum, Golgi apparatus, and secretory granules (Jansen, 1984). Cytosolic and nuclear estrogen and progesterone receptors vary throughout the cycle in the human but are generally in greatest concentration in the ampulla during the proliferative phase (Pollow *et al.*, 1981, 1982), the decrease being correlated with progesterone increase (Verhage *et al.*, 1980; Pino *et al.*, 1982). West and Brenner (1983) maintain that, in cynomolgus macaques, it is the progesterone decrease rather than the estrogen increase that precedes the rise in estrogen receptor concentration. This response to release from progesterone influence does not appear to occur in sheep because production of EGP by treatment with estradiol *in vivo* was unaffected even during the progesterone-dominated luteal phase or in the presence of progesterone-releasing implants (Sutton *et al.*, 1986).

Preliminary evidence of the distribution of estrogen receptors in the tissues of the sheep oviduct has been obtained using immunohistochemical localization (Hill and Nancarrow, unpublished observations). Receptors were found in the cells of both the lamina propria and the epithelial mucosa around estrus. Estrogen and progesterone receptors were often found only in the stromal cells of monkey uterine endometrium during responses to steroids (Brenner *et al.*, 1990), invoking a role for soluble mediators of endocrine function (Hunt, 1989; Salamonsen, 1992). Using a macaque oviductal model, Brenner *et al.* (1990) demonstrated both estrogen and progesterone receptors in secretory and stromal cells, but not in ciliated cells during the follicular phase. Under progesterone domination, these receptors were absent in both epithelial cell types, and stromal cells lost the estrogen receptors. These authors suggested then, that either stromal or secretory cells mediate estrogen action on differentiation of ciliated cells and that the suppressive effects of progesterone are mediated by factors emanating from the stroma. Further examination of the nature and distribution of steroid receptors in the oviduct is necessary to understand the changes in morphology and secretion that occur during the periods of sperm maturation, fertilization, and the early stages of embryo development.

VII. Secretions of the Oviduct

A. Synthesis of Estrus-Associated Glycoproteins

Several proteins of unspecified function are synthesized and secreted by the oviducts of all species examined. Some of these appear responsive to cyclical hormonal influence, while others are synthesized or secreted at constant rates. There may also be a gradient in secretion of some proteins from the fimbria through the ampulla and into the isthmus, or even an absolute distinction. However, a major group of oviducal glycoproteins exhibiting an electrophoretic molecular weight (M_r) in the range of 70–130 kDa after sodium dodecyl sulfate treatment appears common to many species, including sheep, cattle, swine, baboons, and humans. These proteins, with few exceptions, exhibit similar characteristics among these species and are major contenders for an embryotropic role.

Specific protein secretion by oviducts has been examined either by staining electrophoretic gels of oviduct fluid obtained *in vivo*, or, more specifically, by incubating epithelial cells or tissue explants with radiolabeled substrates and examining the conditioned medium so obtained by electrophoretic and autoradiographic techniques. Although several proteins of varying M_r have been described, the most noticeable feature of this work has been the appearance of several related proteins or isoforms, with a monomeric M_r of about 90–100 kDa. These data are summarized in Table 3 for the five species. The most common feature is that they are induced by estrogen; hence they may be associated with the final stages of folliculogenesis, estrus, or ovulation, and are coincidentally present in the oviduct during sperm transport, fertilization, and the early stages of embryonic development.

1. Sheep

Sutton *et al.* (1984a, 1985), in their pioneering work with Merino sheep, described a diffuse estrus-associated acidic glycoprotein (EGP) with a broad M_r range of 70–90 kDa, which was secreted over a period of up to 8 days. In its native form, the M_r of the glycoprotein appeared to be $>10^6$, which led to the suggestion that EGP might be secreted as a polymer or even an aggregate of several proteins. Specific lectin binding techniques showed that the terminal glycosyl residues included galactose, *N*-acetylgalactosamine, and fucose (Sutton *et al.*, 1985), and it was proposed that EGPs diffuse form might be due to variation in the nature or degree of glycosylation. Indwelling catheters were used to collect ovine oviduct fluid over sequential estrous cycles and during steroid treatment of the ewes. The EGP was invariably secreted within 3 days of a single treatment with estrogen, whatever the physiological state of the ewe, even in the presence of progesterone (Sutton *et al.*, 1986). Subsequently, Gandolfi *et al.* (1989b, 1991) prepared a monoclonal antibody (MAC 264) to a 92-kDa protein (SOP 92) secreted in culture by ovine oviducal epithelial cells obtained from Welsh mountain ewes. This material was associated with the zonae pellucidae of oviduct-recovered eggs and plasma membranes of embryos, but not with the cytoplasm or nuclei of the blastomeres. From these data, it was suggested that the protein might be involved in fertilization and embryo development. We now have obtained preliminary evidence of a role for EGP in positively regulating embryonic mitosis in culture (Hill *et al.*, 1992).

There is reasonable agreement about the characteristics of ovine EGP in that it is secreted in response to treatment with estrogen (Sutton *et al.*, 1986; Buhi *et al.*, 1991; Murray, 1992), is synthesized by the ampulla (Gandolfi *et al.*, 1989b, 1991; Buhi *et al.*, 1991; Murray, 1992; Hill *et al.*, 1994), has an isoelectric point (p*I*) in the range of 4 to 6.5 (Sutton *et al.*, 1985; Gandolfi *et al.*, 1989b), and exists *in vivo* as a large polymer or aggregate with an M_r of $>10^6$ (Sutton *et al.*, 1984a, 1985). Following solubilization in sodium dodecyl sulfate, this aggregate dissociates into at least two discrete proteins (Buhi *et al.*, 1991; Murray, 1992; Hill and Nancarrow, unpublished observations). The disturbing feature of these observations is that the proteins described by Buhi *et al.* (1991) had a p*I* in the basic range, 7.5–8.5. These authors suggest several possible reasons for this discrepancy, the major one being that the method of culturing the explant tissue disturbed the pattern of glycosylation. There must be scope for size and p*I* variability in this particular protein group, but a strong indication of the similar nature of proteins produced by different breeds and described by different authors is that the monoclonal antibody to SOP92 from Welsh mountain sheep, MAC 264 (Gandolfi *et al.*, 1989b), cross-reacts with the 70- to 90-kDa EGP collected through indwelling catheters or from conditioned medium of ampullary explants from Australian Merino ewes (Hill *et al.*, 1994).

2. Cattle

The secretion of a major specific protein by cow oviducts was similar to that of sheep, in that it was associated with estrus and the ampulla (Table 3). The characteristics of proteins examined following incubation of explants with radiolabeled substrates (Malayer *et al.*, 1988; Boice *et al.*, 1990a) and collections from flushings (Boice *et al.*, 1990a) or catheters (Gerena and Killian, 1990) were all in general agreement with an M_r ranging between 80 and 97 kDa. However, at least two forms of the protein were described, one with a p*I* of 4.7–5.5 (Malayer *et al.*, 1988; Boice *et al.*, 1990a; Wegner and Killian, 1992) and the other with p*I* up to 8.1 (Malayer *et al.*, 1988; Boice *et* *al.*, 1990a). In one-dimensional SDS–PAGE gels, two bands in this M_r range were clearly identified (Boice *et al.*, 1990a; Gerena and Killian, 1990; Wegner and Killian, 1992). The relationship between these forms and those described in sheep remains to be determined. It is not clear whether there are differences in the degree or nature of glycosylation, or whether similar but distinct proteins are involved. One difference that should be examined is that the terminal carbohydrates described by Wegner and Killian (1992) as *N*-acetylgalactosamine and *N*-acetylglucosamine partly differed from those detected on sheep EGP (Sutton *et al.*, 1985).

3. Pigs and Primates

In these species, the glycoproteins secreted in association with estrus or the late follicular phase, or induced *in vivo* by estrogen, were slightly larger than those of the sheep and cow (Table 3). There appear to be three acidic proteins in swine with an M_r range of 75–100 kDa and one basic protein of 100 kDa M_r (Buhi *et al.*, 1989, 1990, 1992). Radiolabeled fucose was incorporated into the 75- and 85-kDa proteins, suggesting similarity to EGP on the basis of several criteria (Sutton *et al.*, 1985). Oviducts from the primate species secreted both acidic and basic forms of proteins with an M_r between 100 and 130 kDa. These were secreted before ovulation in the mid- to late follicular stage of both baboons (Fazleabas and Verhage, 1986) and humans (Verhage *et al.*, 1988), and were shown to be estrogen-induced in the baboon (Verhage and Fazleabas, 1988; Verhage *et al.*, 1990) and pig (Buhi *et al.*, 1992).

B. Function of Estrus-Associated Glycoproteins

The evidence is compelling that the ampullary oviduct synthesizes and secretes one or a group of estrogen-dependent glycoproteins that, while varying both within and between species, may all be intimately involved in the early development of embryos. The use of both monoclonal and polyclonal antibodies has clearly demonstrated that oviducal glycoproteins, most likely EGP at least in the sheep, can be found associated with the zona and blastomere membranes but not yet the cytosol or nucleus of oocytes and embryos of sheep (Gandolfi *et al.*, 1991; Hill and Nancarrow, unpublished observations), cow (Wegner and Killian, 1991), and baboon (Boice *et al.*, 1990b). The binding of various oviduct proteins to sperm suggests that they may also be involved in sperm maturation or fertilization (Sutton *et al.*, 1984b; Lippes and Wagh, 1989; McNutt *et al.*, 1992) but there has been no clear distinction between ampullary and isthmic proteins. The issue is further complicated because two bovine proteins of 85–95 kDa, equivalent to ovine EGP, were found to bind cattle sperm (McNutt *et al.*, 1992), whereas specific binding of EGP to sperm could not be detected using MAC 264 following incubation in oviducal fluid (Hill and Nancarrow, 1993).

C. Formation and Release of Secretory Granules

The mechanism of estrogen induction of these proteins has not been elucidated, given that we were unable to prolong the synthesis of EGP in tissue explants (Hill and Nancarrow, unpublished observations) with estradiol, yet a single treatment *in vivo* invariably resulted in prolonged EGP secretion (Sutton *et al.*, 1986). We propose that unidentified factors associated with other structures and cell types within the ampullary tissues (Section VI,A) may be involved in this mechanism. The recent cloning of a cDNA sequence specifying the 120-kDa glycoprotein from baboon (Donnelly *et al.*, 1991) is a most important advance, and further investigation using those techniques may lead to answers to these questions of identity, conservation of genetic information, and endocrine control of secretion.

Hollis *et al.* (1984) described five stages in the formation of lamellated ampullary secretory granules (SGs); these were similar to the observations of Willemse and Van Vorstenbosch (1975a), Nayak *et al.*, (1976), and Russe and Liebich (1979) in sheep; Bjorkman and Fredricsson (1961) and Nayak and Ellington (1972) in cattle; and Odor *et al.* (1983) in monkeys. The SG were closely associated with the Golgi membranes only during Days −1 to 3 of the estrous cycle, precisely the period when *de novo* synthetic activity has been found in sheep, cattle, and pigs, and around ovulation in primates (Table 3). The final stages were associated with lamellar structures which Hollis *et al.* (1984) concluded must break down prior to exocytosis of the contents at midcycle, under progesterone influence. Reports about the ultrastructural mechanism of secretion of ovine ampullary SGs do not give a clear picture, but evidence for exocytotic (Bjorkman and Fredricsson, 1961; Brower and Anderson, 1969; Russe and Liebich, 1979; Odor *et al.*, 1983; Murray, 1992), apocrine (Nayak, 1977; Odor *et al.*, 1983; Brenner *et al.*, 1983), and holocrine (Clyman, 1966) secretion has been reported in various species. Synthesis and secretion of EGP, however, clearly occur from Day −1 through Day 6 (Table 3) since material staining with PAS (Schiff positive) (Hadek, 1955; Abdalla, 1968) and reactive to MAC 264 (Hill *et al.*, 1994) has been located on the epithelial glycocalyx and as an amorphous material in the ampullary lumen over this period. Further evidence supporting this latter observation was obtained by immunogold localization of MAC 264-reactant material associated with these granules in the metestrus and diestrus stage of sheep (Gandolfi *et al.*, 1991) and by polyclonal immunocytochemical detection of the 100- to 130-kDa glycoprotein in apical granules of the baboon during the follicular phase (Verhage *et al.*, 1989).

However, subsequent immunogold labeling identified every type of SG in both the isthmus and ampulla (Verhage *et al.*, 1990), leaving us unable to draw precise conclusions about the nature of synthesis and secretion of unique proteins.

There is general agreement from workers studying cyclic sheep (Willemse, 1975a; Willemse and Van Vorstenbosch, 1975a; Russe and Liebich, 1979), or estradiol-treated ovariectomized sheep (Willemse, 1975b; Willemse and Van Vorstenbosch, 1975b) that the synthesis and formation of SGs is controlled by estrogen. With this in mind, we suggest that our previous conclusion (Hollis *et al*, 1984), that SGs are released in midcycle and are out of phase with EGP production, is incorrect. However, there are at least two granule types containing dissimilar material that are secreted differently with respect to time and mechanism. Certainly, the sheep ampulla is capable of *de novo* synthesis of non-EGP proteins during this period (Gandolfi *et al.*, 1989b; Buhi *et al.*, 1991; Murray, 1992); some of these proteins may be induced or increased with progesterone. Both apical protrusions and SG were at a maximum from midcycle to the day before estrus, and at a minimum on Day 1, when EGP synthesis (Gandolfi *et al.*, 1989b) and secretion (Sutton *et al.*, 1986; Hill *et al.*, 1994) is at a maximum. While the primary role of estrogen in inducing EGP production appears clear, that of progesterone remains enigmatic, since it has been found to inhibit the action of estradiol, particularly in primates. However, in studies with sheep, SGs clearly accumulate during the early luteal phase (Willemse, 1975a; Hollis *et al.*, 1984), and EGP can be secreted for extended periods *in vivo* with a single treatment of estradiol during progestational dominance (Sutton *et al.*, 1986). Therefore we suggest that EGP in sheep can be synthesized for up to 7 days in response to estrogen, and EGP-containing granules are released rapidly from Days −1 to 4 by a mechanism not readily observed, but further release is inhibited by progesterone in the absence of estrogen. This results in the observed accumulation of SGs over the diestrus period in cycling sheep (Hollis *et al.*, 1984) and implies that EGP is secreted only under estrogen domination.

D. Cell–Cell Interactions

We have some confidence in concluding from this evidence that estradiol induces EGP synthesis and secretion and SG formation in the oviduct. This action probably occurs by means of the cytosolic and nuclear estradiol receptors. The mechanism by which SG content is released is still conjectural and there may be some influence of progesterone that varies among species. Our recent findings that EGP secretion cannot be maintained *in vitro* (Hill and Nancarrow, unpublished observations), whereas it is always induced by *in vivo* treatment (Sutton *et al.*, 1986), leads us to the conclusion that discrete cell–cell interactions between components of the stroma and epithelial mucosa may be involved. In this context, it is interesting to note that proliferation of epithelial cells from human oviducts in a monolayer culture could not be stimulated with estradiol, but could be stimulated with EGF (Takeuchi *et al.*, 1991). The implication of this finding is that estrogen may act through receptors within the stroma, causing mobilization of growth factors from the ECM, or other cells, which diffuses across the basement membrane to elicit a paracrine response on the epithelial cells.

Precedents have been established for these mechanisms within the uterus and the oviduct (see Section III and reviews by Hunt, 1989; Brenner *et al.*, 1990; and Salamonsen, 1992). Since these organs are of the same embryonic origin, it is reasonable to expect similar mechanisms to operate in the oviduct. A variety of cells occupy the lamina propria, and cells of probable lymphocytic origin, including T cells (Segerson *et al.*, 1991), can be found in all tissues (Section VI,A). There are estrogen receptors in cells of both the lamina and the epithelial mucosa in macaques (Brenner *et al.*, 1990) and sheep (Hill and Nancarrow, unpublished observations). Growth factor activity has been found in medium conditioned by oviduct cells (Gandolfi *et al.*, 1989a). The proximity of the endothelial cells to stromal cells (which appear to synthesize protein in response to estrogen; Brinsfield *et al.*, 1974) and epithelial cells would permit ready access to cytokine communication through cell–cell or cell–matrix interactions (Sections III,B and C). In Section IV we examined how proteins of the ECM and support of the basal membrane are of integral importance to cell function. We propose that these cell isolation and cell culture techniques will be fundamental to the elucidation of physiological mechanisms in the oviduct.

VIII. Conclusions

Advances in cell biology have been accompanied by increases in the depth of our understanding of endometrial physiology. While we know a considerable amount concerning the nature of the proteins secreted by the oviduct, our understanding of oviducal physiology lags far behind that of the endometrium. The contribution of cell culture models has been modest in view of their enormous potential. With the present availability of a wide range of cell regulators, we are now in a position to pursue more actively the factors controlling endometrial and oviducal cell biology and the interactions among these cells, steroid hormones, and the ECM. The elucidation of these details will aid us greatly in further understanding female reproductive tract functions in both humans and animals.

References

Abdalla, O. (1968). Observations on the morphology and histochemistry of the oviducts of the sheep. *J. Anat.* 102, 333–344.

Abedin, M. Z., Ayad, S., and Weiss, J. B. (1982). Isolation and native characterization of cysteine-rich collagens from bovine placental tissues and uterus and their relationship to types IV and V collagens. *Biosci. Rep.* 2, 493–502.

Abraham, D., Bou-Gharios, G., Muir, H., and Olsen, I. (1989). Adhesion of lymphoid cells to fibroblasts in tissue culture. *Cell. Immunol.* 122, 33–47.

Albelda, S. M., and Buck, C. A. (1990). Integrins and other cell adhesion molecules. *FASEB J.* 4, 2868–2880.

Anderson, T. L., Olson, G. E., Hoffman, L. (1986). Stage-specific alterations in the apical membrane glycoproteins of endometrial epithelial cells related to implantation in rabbits. *Biol. Reprod.* 34, 701–720.

Anklesaria, P., Teixido, J., Laiho, M., Pierce, J. H., Greenberger, J. S., and Massague, J. (1990). Cell-cell adhesion mediated by binding of membrane-anchored transforming growth factor α to epidermal growth factor receptors promotes cell proliferation. *Proc. Natl. Acad. Sci. USA* 87, 3289–3293.

Aplin, J. D. (1989). Cellular biochemistry of the endometrium. *In* "Biology of the Uterus" (R. M. Wynn and W. P. Jollie, eds.), pp. 89–129. Plenum Medical Book Co., New York.

Aplin, J. D., Charlton, A. K., and Ayad, S. (1988). An immunohistochemical study of human endometrial extracellular matrix during the menstrual cycle and first trimester of pregnancy. *Cell Tissue Res.* 253, 231–240.

Armant, D. R. (1991). Cell interactions with laminin and its proteolytic fragments during outgrowth of mouse primary trophoblast cells. *Biol. Reprod.* 45, 664–672.

Armant, D. R., Kaplan, H. A., and Lanmarz, W. J. (1986a). Fibronectin and laminin promote in vitro attachment and outgrowth of mouse blastocysts. *Dev. Biol.* 116, 519–523.

Armant, D. R., Kaplan, H. A., Mover, H., and Lennarz, W. J. (1986b). The effect of hexapeptides on attachment and outgrowth of mouse blastocysts cultured *in vitro*: evidence for the involvement of the cell recognition tripeptide arg-gly-asp. *Proc. Natl. Acad. Sci. USA* 83, 6751–6755.

Ayad, S., Chamgers, C. A., Shuttleworth, C. A., and Grant, M. E. (1985). Isolation from bovine elastic tissues of collagen type VI and characterization of its form *in vivo*. *Biochem. J.* 230, 465–474.

Baker, N. W., Mlodzik, M., Rubin, G. M. (1990). Spacing differentiation in the developing *Drosophila* eye: a fibrinogen-related lateral inhibitor encoded by scabrous. *Science* 250, 1370–1377.

Barnes, F. L., and First, N. L. (1991). Embryonic transcription in in vitro cultured bovine embryos. *Mol. Reprod. Dev.* 29, 117–123.

Barrett, A. J., and Starkey, P. M. (1973). The interaction of a α-macrogloblin with proteinases. *Biochem. J.* 133, 709–724.

Bavister, B. D. (1988). Role of oviductal secretions in embryonic growth *in vivo* and *in vitro*. *Theriogenology* 29, 143–153.

Bazan, J. F. (1990). Structural design and molecular evolution of a cytokine receptor superfamily. *Proc. Natl. Acad. Sci. USA* 87, 6934–6938.

Bergman, M. D., Schachter, B. S., Karelus, K., Combatsiaris, E. P., Garcia, T., and Nelson, J. F. (1992). Up-regulation of the uterine estrogen receptor and its messenger ribonucleic acid during the mouse estrous cycle: the role of estradiol. *Endocrinology* 130, 1923–1930.

Berthois, Y., Katzenellenbogen, J. A., and Katzenellenbogen, B. S. (1986). Phenol red in tissue culture media is a weak estrogen: implications concerning the study of estrogen-responsive cells in culture. *Proc. Natl. Acad. Sci. USA* 83, 2496–2500.

Bjorkman, N., and Fredricsson, B. (1961). The bovine oviduct epithelium and its secretory process as studied with the electron microscope and histochemical tests. *Z. Zellforsch. Mikrosk. Anat.* 55, 500–513.

Blaustein, A. (ed.) (1982). "Pathology of the Female Genital Tract." 2nd ed. pp. 235–278. Springer-Verlag, New York.

Boice, M. L., Geisert, R. D., Blair, R. M., and Verhage, H. G. (1990a). Identification and characterization of bovine oviductal glycoproteins synthesized at estrus. *Biol. Reprod.* 43, 457–465.

Boice, M. L., McCarthy, T. J., Mavrogianis, P. A., Fazleabas, A. T., and Verhage, H. G. (1990b). Localization of oviductal glycoproteins within the zona pellucida and perivitelline space of ovulated ova and early embryos in baboons (*Papio anubis*). *Biol. Reprod.* 43, 340–346.

Branham, W. S., Lyn-Cook, B. D., Andrews, A., McDaniel, M., and Sheehan, D. M. (1991). Growth of neonatal rat uterine luminal epithelium on extracellular matrix. *In Vitro Cell. Dev. Biol.* 27A, 442–446.

Brenner, C. A., Adler, R. R., Rappolee, D. A., Pedersen, R. A., and Werb, Z. (1989). Genes for extracellular matrix-degrading metalloproteinases and their inhibitor, TIMP, are expressed during early mammalian development. *Genes. Dev.* 3, 848–859.

Brenner, R. M. (1969). Renewal of oviduct cilia during the menstrual cycle of the rhesus monkey. *Fertil. Steril.* 20, 599–611.

Brenner, R. M., Carlisle, K. S., Hess, D. L., Sandow, B. A., and West, N. B. (1983). Morphology of the oviducts and endometria of cynomolgus macaques during the menstrual cycle. *Biol. Reprod.* 29, 1289–1302.

Brenner, R. M., West, N. B., and McClellan, M. C. (1990). Estrogen and progestin receptors in the reproductive tract of male and female primates. *Biol. Reprod.* 42, 11–19.

Brinsfield, T. H., Fisher, M., and Hawk, H. W. (1974). Ultrastructure of endometrial stromal cells in the ewe during the estrous cycle and early pregnancy. *J. Anim. Sci.* 39, 565–574.

Brower, L. K., and Anderson, E. (1969). Cytological events associated with the secretory process in the rabbit oviduct. *Biol. Reprod.* 1, 130–148.

Buhi, W. C., Vallet, J. L., and Bazer, F. W. (1989). De novo synthesis and release of polypeptides from cyclic and early pregnant porcine oviductal tissue in explant culture. *J. Exp. Zool.* 252, 79–88.

Buhi, W. C., Alvarez, I. M., Sudhipong, V., and Dones-Smith, M. M. (1990). Identification and characterization of de novo-synthesized porcine oviductal secretory proteins. *Biol. Reprod.* 43, 929–938.

Buhi, W. C., Bazer, F. W., Alvarez, I. M., and Mirando, M. A. (1991). *In vitro* synthesis of oviductal proteins associated with estrus and 17β-estradiol-treated ovariectomized ewes. *Endocrinology* 128, 3086–3095.

Buhi, W. C., Ashworth, C. J., Bazer, F. W., and Alvarez, I. M. (1992). *In vitro* synthesis of oviductal secretory proteins by estrogen-treated ovariectomized gilts. *J. Exp. Zool.* 262, 426–435.

Burgess, W. H., and Macaig, T. (1989). The heparin-binding (fibroblast) growth factor family of proteins. *Annu. Rev. Biochem.* 58, 575–606.

Carson, D. D., Tang, J. P., and Gay, S. (1988). Collagens support embryo attachment and outgrowth in vitro: effects of the arg-gly-asp sequence. *Dev. Biol.* 127, 368–375.

Carson, D. D., Raboudi, N., and Jacobs, A. L. (1991). Glycoprotein expression and function in embryo-uterine interactions. *In* "Cellular Signals Controlling Uterine Function" (L. A. Lavia, ed.), pp. 107–117. Plenum, New York.

Casselen, B. G., and Harper, M. J. K. (1991). Human endometrial epithelial cells grown on collagen in serum-free medium: estrogen responsiveness and morphology. *Acta Endocrin. (Copenh.)* 125, 101–108.

Cherny, R. A. (1991). "Paracrine Regulation of Endometrial Function in the Ewe." Ph.D. thesis, Monash University, Melbourne, Australia.

Cherny, R. A., and Findlay, J. K. (1990). Separation and culture of ovine endometrial epithelial and stromal cells: evidence of morphological and functional polarity. *Biol. Reprod.* 43, 241–250.

Cherny, R. A., Salamonsen, L. A., and Findlay, J. K. (1991). Immunocytochemical localization of oestrogen receptors in the endometrium of the ewe. *Reprod. Fertil. Dev.* 3, 321–331.

Clyman, M. J. (1966). Electron microscopy of the human fallopian tube. *Fertil. Steril.* 17, 281–301.

Damjanov, I., and Wewer, U. M. (1991). Uterine extracellular matrix remodelling in pregnancy. *In* "Cellular Signals Controlling Uterine Function" (L. A. Lavia, ed.), pp. 99–106. Plenum, New York.

De, M., and Wood, G. W. (1990). Influence of oestrogen and progesterone on macrophage distribution in the mouse uterus. *J. Endocrinol.* 126, 417–424.

Donnelly, K. M., Fazleabas, A. T., Verhage, H. G., Mavrogianis, P. A., and Jaffe, R. C. (1991). Cloning of a recombinant complementary DNA to a baboon (*Papio anubis*) estradiol-dependent oviduct-specific glycoprotein. *Mol. Endocrinol.* 5, 356–364.

Drudy, L., Sheppard, B., and Bonnar, J. (1991). Mast cells in the normal uterus, and in dysfunctional uterine bleeding. *Eur. J. Obs. Gyn. Reprod. Biol.* 39, 193–201.

Ekblom, P. (1989). Developmentally regulated conversion of mesenchyme to epithelium. *FASEB J.* 3. 2141–2150.

Emonard, H., and Grimaud, J.-A. (1990). Matrix metalloproteinases: a review. *Cell. Mol. Biol.* 36, 131–153.

Enat, R., Jefferson, D. M., Ruiz-Poazo, N., Gatmaitin, Z., Leinwand, L. A., Reid, L. M. (1984). Hepatocyte proliferation in vitro: its dependence on the use of serum-free hormonally defined medium and substrate of extracellular matrix. *Proc. Natl. Acad. Sci. USA* 81, 1411–1415.

Enders, A. C., and Schlafke, S. (1974). Surface coats of the mouse blastocyst and uterus during the preimplantation period. *Anat. Rec.* 180, 31–45.

Epelu-Opio, J. (1978). Cytological changes associated with the functions of the oviduct epithelium during proestrus and estrus in impala, *Aepyceros melampus* (Lichtenstein, 1812). *Anat. Anz.* 144, 36–47.

Eyestone, W. H., and First, N. L. (1989). Co-culture of early cattle embryos to the blastocyst stage with oviducal tissue or in conditioned medium. *J. Reprod. Fertil.* 85, 715–720.

Eyestone, W. H., Jones, J. M., and First, N. L. (1991). Some factors affecting the efficiency of oviduct tissue-conditioned medium for the culture of early bovine embryos. *J. Reprod. Fertil.* 92, 59–64.

Faber, M., Wewer, U. M., Berthelsen, J. G., Liotta, L. A., and Albrechsten, R. (1986). Laminin production by human endometrial stromal cells relates to the cyclic and pathologic state of the endometrium. *Am. J. Pathol.* 124, 384–398.

Farach, M. C., Tang, J. P., Decker, G. L., and Carson, D. D. (1987). Heparan/heparan sulfate is involved in attachment and spreading of mouse embryos *in vitro*. *Dev. Biol.* 123, 401–410.

Fay, T. N., and Grudzinskas, J. G. (1991). Human endometrial peptides: a review of their potential role in implantation and placentation. *Hum. Reprod.* 6, 1311–1326.

Fazleabas, A. T., and Verhage, H. G. (1986). The detection of oviduct-specific proteins in the baboon (*Papio anubis*). *Biol. Reprod.* 35, 455–462.

Findlay, J. K., Salamonsen, L. A., and Cherny, R. A. (1990). Endometrial function: studies using isolated cells *in vitro*. *In* "Oxford Reviews of Reproductive Biology" (S. Milligan, ed.), Vol. 12, pp. 182–223. Oxford University Press, New York.

Finn, C. A. (1986). Implantation, menstruation and inflammation. *Biol. Rev.* 61, 313–328.

Fukui, Y., Glew, A. M., Gandolfi, F., and Moor, R. M. (1988). *In vitro* culture of sheep oocytes matured and fertilized *in vitro*. *Theriogenology* 29, 883–891.

Gandolfi, F., and Moor, R. M. (1987). Stimulation of early embryonic-development in the sheep by co-culture with oviduct epithelial cells. *J. Reprod. Fertil.* 81, 23–28.

Gandolfi, F., Brevini, T. A. L., and Moor, R. M. (1989a). Effect of oviduct environment on embryonic development. *J. Reprod. Fertil.* Suppl. 38, 107–115.

Gandolfi, F., Brevini, T. A. L., Richardson, L., Brown, C. R., and Moor, R. M. (1989b). Characterization of proteins secreted by sheep oviduct epithelial cells and their function in embryonic development. *Development* 106, 303–312.

Gandolfi, F., Modina, S., Brevini, T. A. L., Galli, C., Moor, R. M., and Lauria, A. (1991). Oviduct ampullary epithelium contributes a glycoprotein to the zona pellucida, perivitelline space and blastomeres membrane of sheep embryos. *Eur. J. Bas. Appl. Histochem.* 35, 383–392.

Gerena, R. L., and Killian, G. J. (1990). Electrophoretic characterization of proteins in oviduct fluid of cows during the estrous cycle. *J. Exp. Zool.* 256, 113–120.

Gerschenson, L. E., Conner, E. A., Yang, J., and Anderson, M. (1979). Hormonal regulation of proliferation in two populations of endometrial cells in culture. *Life Sci.* 24, 1337–1344.

Getzenberg, R. H., Pienta, K. J., and Coffey, D. S. (1990). The tissue matrix: cell dynamics and hormone action. *Endocrinol. Rev.* 11, 399–417.

Glasser, S. R., Lampelo, S., Munir, M. I., and Julian, J. (1987). Expression of desmin, laminin and fibronectin during *in situ* differentiation (decidualization) of rat uterine stromal cells. *Differentiation* 35, 132–142.

Glasser, S. R., Julian, J., Decker, G. L., Tang, J.-P., and Carson, D. D. (1988). Development of morphological and functional polarity in primary cultures of immature rat uterine epithelial cells. *J. Cell. Biol.* 107, 2409–2423.

Gospodarowicz, D., Vlodavsky, I., Greenburg, G., and Johnson, L. K. (1979). Cellular shape is determined by the extracellular matrix and is responsible for the control of cellular growth and function. *Cold Spring Harbor Conference on Cell Proliferation* 6, 561–592.

Gospodarowicz, D., Cohen, D., and Fujii, D. K. (1982). Regulation of cell growth by the basal lamina and plasma factors: relevance to embryonic control of cell proliferation and differentiation. *Cold Spring Harbor Conference on Cell Proliferation* 9, 95–124.

Grinnell, F., Head, J. R., and Hoffpauir, J. (1982). Fibronectin and cell shape *in vivo*: studies on the endometrium during pregnancy. *J. Cell Biol.* 94, 597–606.

Gumbiner, B. (1987). Structure, biochemistry, and assembly of epithelial tight junctions. *Am. J. Physiol.* 253, 749–758.

Hadek, R. (1955). The secretory process in the sheep's oviduct. *Anat. Rec.* 121, 187–205.

Hampton, A. L., Findlay, J. K., and Salamonsen, L. A. (1992). Endothelin production by human endometrial cells in primary culture: regulation of secretion by serum. *Proc. Aust. Soc. Reprod. Biol.* 24, 124.

Helmer, S. D., Zavy, M. T., and Geisert, R. D. (1993). Isolation of bovine endometrial epithelial and stroma cells. I. *In vitro* growth of highly enriched bovine uterine surface epithelium, glandular epithelium and stromal cells grown on different extracellular matrices. *Biol. Reprod.* (in press).

Hill, J. L., and Nancarrow, C. D. (1993). Absence of binding of an ovine oestrus-associated oviducal glycoprotein to ram spermatozoa. *Proc. Aust. Soc. Reprod. Biol.* 25, 73.

Hill, J. L., Marshall, J. T., and Nancarrow, C. D. (1992). Effect of an oestrus-associated oviducal glycoprotein on early ovine embryo development. *Proc. Aust. Soc. Reprod. Biol.* 24, 104.

Hill, J. L., Gandolfi, F., and Nancarrow, C. D. (1994). Immunological analysis of the presence and secretion of an oestrus-associated glycoprotein in the oviduct of cyclic sheep. *J. Reprod. Fert.*, submitted.

Hollis, D. E., Frith, P. A., Vaughan, J. D., Chapman, R. E., and Nancarrow, C. D. (1984). Ultrastructural changes in the oviductal epithelium of Merino ewes during the estrous cycle. *Am. J. Anat.* 171, 441–456.

Humphries, M. J. (1990). The molecular basis and specificity of integrin-ligand interactions. *J. Cell. Sci.* 97, 585–592.

Hunt, J. S. (1989). Cytokine networks in the uteroplacental unit: macrophages as pivotal regulatory cells. *J. Reprod. Immunol.* 16, 1–17.

Hynes, R. O. (1992). Integrins: versatility, modulation, and signalling in cell adhesion. *Cell* 69, 11–25.

Hynes, R. O., and Lander, A. D. (1992). Contact and adhesive specificities in the associations, migrations, and targetting of cells and axons. *Cell* 68, 303–322.

Irwin, J. C., Kirk, D., King, R. J. B., Quigley, M. M., and Gwatkin, R. B. L. (1989). Hormonal regulation of human endometrial stromal cells in culture: an *in vitro* model for decidualization. *Fertil. Steril.* 52, 761–768.

Irwin, J. C., Utian, W. H., and Eckert, R. L. (1991). Sex steroids and growth factors differentially regulate the growth and differentation of cultured human endometrial stromal cells. *Endocrinology* 129, 2385–2392.

Jacobs, A. L., Decker, G. L., Glasser, S. R., Julian, J., and Carson, D. D. (1990). Vectorial secretion of prostaglandins by polarized rodent uterine epithelial cells. *Endocrinology* 126, 2125–2136.

James, K. (1990). Interactions between cytokines and α-2 macroglobulin. *Immunol. Today* 11, 163–166.

Jansen, R. P. (1984). Endocrine response in the fallopian tube. *Endocr. Rev.* 5, 525–551.

Joshi, M. S. (1988). Isolation, cell culture and immunocytochemical characterization of oviduct epithelial cells of the cow. *J. Reprod. Fertil.* 83, 249–261.

Julian, J., Carson, D. D., and Glasser, S. R. (1992a). Polarized rat uterine epithelium *in vitro*: Responses to estrogen in defined medium. *Endocrinology* 130, 68–78.

Julian, J., Carson, D. D., and Glasser, S. R. (1992b). Polarized rat uterine epithelium *in vitro*: Constitutive expression of estrogen-induced proteins. *Endocrinology* 130, 79–87.

Karkavelas, G., Kefalides, N. A., Amenta, P. S., and Martinez-Hernandez, A. (1988). Comparative ultrastructural localization of collagen types III, IV, VI and laminin in rat uterus and kidney. *J. Ultrastruct. Mol. Struct. Res.* 100, 137–155.

Kisalus, L. L., Herr, J. C., Little, C. D. (1987). Immunolocalization of extracellular matrix proteins and collagen synthesis in first trimester human decidua. *Anat. Rec.* 218, 402–415.

Klagsbrun, M., and Baird, A. (1991). A dual receptor system is required for basic fibroblast growth factor activity. *Cell* 67, 229–231.

Koji, T., and Brenner, R. M. (1993). Localization of estrogen receptor messenger ribonucleic acid in rhesus monkey uterus by nonradioactive *in situ* hybridization with digoxigenin-labeled oligodeoxynucleotides. *Endocrinology* 132, 382–392.

Kutteh, W., Hatch, K., Blackwell, R., and Mestecky, J. (1988). Secretory immune system of the female reproductive tract: I. Immunoglobulin and secretory component containing cells. *Obstet. Gynecol.* 71, 56–60.

Lane, T. F., Iruela-Arispe, L., Ross, R., Sage, H., and Raines, E. W. (1991). SPARC binds PDGF AB and PDGF BB and inhibits the binding of PDGF to its receptor. *J. Cell. Biochem.* 15F, 182.

Lavia, L. A., and Larson, B. A. (1991). Rat endometrial stromal-epithelial response to estrogen infusion. *Steroids* 56, 123–130.

Leese, H. J. (1988). The formation and function of oviduct fluid. *J. Reprod. Fertil.* 82, 843–856.

Leivo, I., Vaheri, A., Timpl, R., and Wartiovaara, J. (1980). Appearance and distribution of collagens and laminins in the early mouse embryo. *Dev. Biol.* 76, 100–114.

Lessey, B. A., Damjanovich, L., Coutifaris, C., Castelbaum, A., Albelda, S. M., and Buck, C. A. (1992). Integrin molecules in the human endometrium: correlation with the normal and abnormal menstrual cycle. *J. Clin. Invest.* 90, 188–195.

Lessey, B. A., Berchuck, A., Kohler, M., Yeh, I., Albelda, S. M., and Buck, C. A. (1993). Distribution of integrin adhesion molecules in normal and neoplastic human endometrium. *Conn. Tissue Res.* 28(4) (in press).

Levesque, J.-P., Hatzfeld, A., and Hatzfeld, J. (1991). Mitogenic properties of major extracellular proteins. *Immunol. Today* 12, 258–262.

Lippes, J., and Wagh, P. V. (1989). Human oviductal fluid (hOF) proteins. IV. Evidence for hOF proteins binding to human sperm. *Fertil. Steril.* 51, 81–88.

MacNaul, K. L., Chartrain, N., Lark, M., Tocci, M. J., Hutchinson, N. I. (1990). Discoordinate expression of stromelysin, collagenase and tissue inhibitor of metalloproteinases-I in rheumatoid synovial fibroblasts. *J. Biol. Chem.* 265, 17238–17245.

Mahfoudi, A., Nicollier, M., Propper, A. Y., Coumes-Marquet, S., and Adessi, G. L. (1991). Establishment of endometrial glandular epithelial cell subculture in serum-free, hormonally defined medium, on a basement membrane matrix. *Biol. Cell* 71, 255–265.

Malayer, J. R., Hansen, P. J., and Buhi, W. C. (1988). Secretion of proteins by cultured bovine oviducts collected from estrus through early diestrus. *J. Exp. Zool.* 248, 345–353.

Malegapuru, W. M., Saunders, M. E., Ginther Luce, G. E., Dustin, M. L., Springer, T. A., Clark, E. A., Mannoni, P., and Shaw, S. (1988). ICAM-1 a ligand for LFA-1-dependent adhesion of B, T and myeloid cells. *Nature* 331, 86–88.

Mani, S. K., Decker, G. L., and Glasser, S. R. (1991). Hormonal responsiveness by immature rabbit uterine epithelial cells polarized *in vitro*. *Endocrinology* 128, 1563–1573.

Marbaix, E., Donnez, J., Courtoy, P. J., and Eeckhout, Y. (1992). Progesterone regulates the activity of collagenase and related gelatinases A & B in human endometrial explants. *Proc. Natl. Acad. Sci. USA* 89, 11789–11793.

Maslar, I. A., and Riddick, D. H. (1979). Prolactin production by human endometrium during the normal menstrual cycle. *Am. J. Obstet. Gynecol.* 135, 751–754.

Matrisian, L. M. (1990). Metalloproteinases and their inhibitors in matrix remodelling. *Trend. Genet.* 6, 121–125.

McDaniel, J. W., Scalzi, H., and Black, D. L. (1968). Influence of ovarian hormones on histology and histochemistry of the bovine oviduct. *J. Dairy Sci.* 51, 754–761.

McGuire, P. G., and Seeds, N. W. (1989). The interaction of plasminogen activator with a reconstituted basement membrane matrix and extracellular macromolecules produced by cultured epithelial cells. *J. Cell. Biochem.* 40, 215–217.

McNutt, T., Rogowski, L, Vasilatos-Younken, R., and Killian, G. (1992). Adsorption of oviductal fluid proteins by the bovine sperm membrane during *in vitro* capacitation. *Mol. Reprod. Dev.* 33, 313–323.

Mecham, R. P. (1991). Receptors for laminin on mammalian cells. *FASEB J.* 5, 2538–2546.

Melchers, F., Erdei, A., Schulz, T., and Dierich, M. P. (1985). Growth control of activated, synchronized murine B cells by the C3d fragment of human complement. *Nature* 317, 264–267.

Murray, M. K. (1992). Biosynthesis and immunocytochemical localization of an estrogen-dependent glycoprotein and associated morphological alterations in the sheep ampulla oviduct. *Biol. Reprod.* 47, 889–902.

Nagase, H., Barrett, A. J., and Woessner, J. F., Jr. (1992). Nomenclature and glossary of the matrix metalloproteinases. *Matrix.* Suppl I, 421–424.

Nathan, C., and Sporn, M. (1991). Cytokines in context. *J. Cell. Biol.* 113, 981–986.

Nathan, C., Srimal, S., Farber, C., Sanchez, E., Kabbash, L., Asch, A., Gailit, J., and Wright, S. D. (1989). Cytokine-induced respiratory burst of human neutrophils: dependence on extracellular matrix proteins and CD11/CD18 integrins. *J. Cell. Biol.* 109, 1341–1349.

Nayak, R. K. (1977). Scanning electron microscopy of the camel uterine tube (oviduct). *Am. J. Vet. Res.* 38, 1049–1054.

Nayak, R. K., and Ellington, E. F. (1977). Ultrastructural and ultracytochemical cyclic changes in the bovine uterine tube (oviduct) epithelium. *Am. J. Vet. Res.* 38, 157–168.

Nayak, R. K., Albert, E. N., and Kassira, W. N. (1976). Cyclic ultrastructural changes in ewe uterine tube (oviduct) infundibular epithelium. *Am. J. Vet. Res.* 74, 923–933.

Nellor, J. E. (1965). The leukocyte-like cells of the oviducts during the normal estrous cycle and their modification by progestin and estrogen treatment. *Anat. Rec.* 1551, 171–181.

Norman, J. T., Kujubu, D., and Fine, L. G. (1989). Regulation of epithelial cell growth in vitro. *In* "Functional Epithelial Cells in Culture" (K. S. Matlin, J. D. Valentich, eds.), pp. 165–191. A. R. Liss, New York.

Odor, D. L. (1974). The question of "basal" cells in oviductal and endocervical epithelium. *Fertil. Steril.* 25, 1047–1062.

Odor, D. L., Gaddum-Rosse, P., Rumery, R. E., and Blandeau, R. J. (1980). Cyclic variations in the oviductal ciliated cells during the menstrual cycle and after estrogen treatment in the pig-tailed monkey, *Macaca nemestrina*. *Anat. Rec.* 198, 35–57.

Odor, D. L., Gaddum-Rosse, P., and Rumery, R. E. (1983). Secretory cells of the oviduct of the pig-tailed monkey, *Macaca nemestrina*, during the menstrual cycle and after estrogen treatment. *Am. J. Anat.* 166, 149–172.

Ohta, Y., Sato, T., and Iguchi, T. (1993). Immunocytochemical localization of progesterone receptor in the reproductive tract of adult female rats. *Biol. Reprod.* 48, 205–213.

Olsen, I., Muir, H., Smith, R., Fensom, A., and Watt, D. J. (1983). Direct enzyme transfer from lymphocytes is specific. *Nature* 306, 75–77.

Overall, C. M., Wrana, J. L., and Sodek, J. (1989). Independent regulation of collagenase, 72kDa progelatinase and metalloproteinase inhibitor expression in human fibroblasts by transforming growth factor-β. *J. Biol. Chem.* 264, 1860–1889.

Panayotou, G., End, P., Aumailley, M., Timpil, R., and Engel, J. (1989). Domains of laminin with growth-factor activity. *Cell* 56, 93–101.

Pavlik, E. J., and Katzenellenbogen, B. S. (1978). Human endometrial cells in primary tissue culture: estrogen interaction and modulation of cell proliferation. *J. Cell. Endocrinol. Metab.* 47, 333–344.

Pino, A. M., Devoto, L., Soyo, E., Castro, O., and Sierralta, W. (1982). Changes in cytosolic and nuclear estradiol receptors of normal fallopian tube throughout the menstrual cycle. *J. Steroid Biochem.* 16, 193–197.

Pollow, K., Inthraphuvasak, J., Manz, B., Grill, H.-J., and Pollow, B. (1981). A comparison of cytoplasmic and nuclear estradiol and progesterone receptors in human fallopian tube and endometrial tissue. *Fertil. Steril.* 36, 615–622.

Pollow, K., Inthraphuvasak, J., Grill, H.-J., and Manz, B. (1982). Estradiol and progesterone binding components in the cytosol of normal human fallopian tubes. *J. Steroid Biochem.* 16, 429–435.

Puistola, U., Ronnberg, L., Martikainen, H., and Turpeanniemi-Hujanen, T. (1989). The human embryo produces basement membrane (type IV collagen)-degrading protease activity. *Hum. Reprod.* 4, 309–311.

Rawdanowicz, T., Hampton, A. L., Nagase, H., Woolley, D. E., and Salamonsen, L. A. (1992). Human endometrial stromal cells secrete matrix metalloproteinases in culture: identification of MMP-1, MMP-2 and MMP-3. *Proc. Aust. Soc. Med. Res.* P18.

Reid, L. M. (1990). Stem cell biology, hormone/matrix synergies and liver differentation. *Curr. Op. Cell Biol.* 2, 121–130.

Reid, L. M., and Rojkind, M. (1979). New techniques for culturing differentiated cells: reconstituted basement membranes rafts. *In* "Methods in Enzymology" (W. B. Jakoby and I. Pastan, eds.), Vol. 58, pp. 263–278. Academic Press, San Diego, CA.

Rexroad, C. E., and Powell, A. M. (1988). Co-culture of ovine ova with oviductal cells in medium 199. *J. Anim. Sci.* 66, 947–953.

Roberts, D. K., Walker, N. J., and Lavia, L. A. (1986). Ultrastructural evidence of stromal/epithelial interactions in the human endometrial cycle. *Am. J. Obstet. Gynecol.* 158, 854–861.

Rodgers, W. H., Osteen, K. G., Matrisian, L. M., Navre, M., Guidice, L. C., and Gorstein, F. (1993). Expression and localization of matrilysin, a matrix metalloproteinase, in human endometrium during the reproductive cycle. *Am. J. Obstet. Gynecol.* 168, 253–260.

Ruosiahti, E., and Yamaguchi, Y. (1991). Proteoglycans as modulators of growth factor activities. *Cell* 64, 867–869.

Russe, I., and Liebich, H. G. (1979). Maturation of secretory granules in the endosalpinx one to four days post-coitum in sheep. *Cell Tissue Res.* 201, 145–158.

Salamonsen, L. A. (1992). Local regulators and the establishment of pregnancy: a review. *Reprod. Fertil. Dev.* 4, 125–134.

Salamonsen, L. A., Doughton, B. W., and Findlay, J. K. (1986). The effects of the preimplantation blastocyst *in vivo* and *in vitro* on protein synthesis and secretion by cultured epithelial cells from sheep endometrium. *Endocrinology* 119, 622–628.

Salamonsen, L. A., Nagase, H., and Woolley, D. W. (1991a). Production of matrix metalloproteinase 3 (stromelysin) by cultured ovine endometrial cells. *J. Cell. Sci.* 100, 351–385.

Salamonsen, L. A., Cherny, R. A., and Findlay, J. K. (1991b). *In vitro* studies of the effects of interferons on endometrial metabolism in sheep. *J. Reprod. Fert.* Suppl. 43, 27–38.

Salamonsen, L. A., Nagase, H., and Woolley, D. E. (1993a). Matrix metalloproteinases are expressed by cultured ovine endometrial cells and are present in uterine flushings from pregnant and non-pregnant ewes. *Conn. Tissue Res.* 28(4) (in press).

Salamonsen, L. A., Suzuki, R., Nagase, H., and Woolley, D. E. (1993b). Production of matrix metalloproteinase 1 (interstitial collagenase) and matrix metalloproteinase 2 (gelatinase A: 72kDa gelatinase) by ovine endometrial cells *in vitro*: different regulation and preferential expression by stromal fibroblasts. *J. Reprod. Fertil.* 98(2) (in press).

Schatz, F., Gordon, R. E., Laufer, N., and Gurpide, E. (1990). Culture of human endometrial cells under polarizing conditions. *Differentiation* 42, 184–190.

Schatz, F., Hausknecht, V., Gordon, R. E., Heller, D., Markiewicz, L., Deligdisch, L., and Gurpide, E. (1991). Studies on human endometrial cells in primary culture. *Ann. N.Y. Acad. Sci.* 622, 80–88.

Segerson, E. C., Matterson, P. M., and Gunsett, F. C. (1991). Endometrial T-lymphocyte subset infiltration during the ovine-estrous cycle and early pregnancy. *J. Reprod. Immunol.* 20, 221–236.

Sheridan, P. J., and Weaker, F. J. (1992). Cellular distribution of nuclear steroid receptors in endometrial blood vessels. *In* "Steroid Hormones and Uterine Bleeding" (N. J. Alexander and C. d'Arcangues, eds.), pp. 179–184. American Assocation for the Advancement of Science.

Sherman, M. I., Gay, R., Gay, S., and Miller, E. J. (1980). Association of collagen with preimplantation and peri-implantation mouse embryos. *Dev. Biol.* 74, 470–478.

Siegfried, J. M., Nelson, K. G., Martin, J. L., Kaufman, D. G. (1984). Histochemical identification of cultured cells from human endometrium. *In Vitro Cell. Dev. Biol.* 20, 25–32.

Simons, K., and Fuller, S. D. (1985). Cell surface polarity in epithelia. *Annu. Rev. Cell Biol.* 1, 243–288.

Simons, K., and Wandinger-Ness, A. (1990). Polarized sorting in epithelia. *Cell* 62, 207–210.

Singer, S. J. (1992). Intercellular communication and cell-cell adhesion. *Science* 255, 1671–1677.

Staatz, W. D., Fok, K. F., Zutler, M. M., Adams, S. P., Rodriguez, B. A., and Santoro, S. A. (1991). Identification of a tetrapeptide recognition sequence for the alpha 2 beta 1 integrin in collagen. *J. Biol. Chem.* 266, 7363–7367.

Stern, J. E., Gardner, S., Quirk, D., and Wira, C. R. (1992). Secretory immune system of the male reproductive tract: effects of dihydrotesterone and estradiol in IgA and secretory component levels. *J. Reprod. Immunol.* 22, 73–85.

Sutherland, A. E., Calarco, P. G., and Damsky, C. H. (1988). Expression and function of cell surface extracellular matrix receptors in mouse blastocyst attachment and outgrowth. *J. Cell Biol.* 106, 1331–1348.

Sutton, R., Nancarrow, C. D., Wallace, A. L. C., and Rigby, N. W. (1984a). Identification of an oestrus-associated glycoprotein in oviducal fluid of the sheep. *J. Reprod. Fertil.* 72, 415–442.

Sutton, R., Wallace, A. L. C., Engel, H., and Nancarrow, C. D. (1984b). Binding of sheep oviducal fluid proteins to spermatozoa. *In* "Reproduction in Sheep" (D. R. Lindsay and D. T. Pearce, eds.), pp. 140–143. Australian Academy of Science, Canberra.

Sutton, R., Wallace, A. L. C., and Nancarrow, C. D. (1985). Characterisation of glycoprotein in oviductal fluid by two-dimensional electrophoresis and lectin binding to protein gel blots. *Electrophoresis* 6, 516–520.

Sutton, R., Nancarrow, C. D., and Wallace, A. L. C. (1986). Oestrogen and seasonal effects on the production of an oestrus-associated glycoprotein in oviducal fluid of sheep. *J. Reprod. Fertil.* 77, 645–653.

Svalander, P. C., Odin, P., Nilsson, B. O., and Obrink, B. (1987). Trophectoderm surface expression of the cell adhesion molecule cell-CAM 105 on rat blastocysts. *Development* 100, 653–660.

Tabibzadeh, S. (1990). Proliferative activity of lymphoid cells in human endometrium throughout the menstrual cycle. *J. Clin. Endocrinol. Metab.* 70, 437–443.

Tabibzadeh, S. (1991a). Human endometrium: an active site of cytokine production and action. *Endocrine Rev.* 12, 272–290.

Tabibzadeh, S. (1991b). Distinct subsets of stromal cells confined to unique microenvironments in human endometrium throughout the menstrual cycle. *Am. J. Reprod. Immunol.* 26, 5–10.

Tabibzadeh, S. (1992). Patterns of expression of integrin molecules in human endometrium throughout the menstrual cycle. *Hum. Reprod.* 7, 876–882.

Tabibzadeh, S., and Poubouridis, D. (1990). Expression of leukocyte adhesion molecules in human endometrium. *Am. J. Clin. Pathol.* 93, 183–189.

Tabibzadeh, S., Bettica, A., and Gerber, M. A. (1986). Variable expression of Ia antigens in human endometrium and in chronic endometriosis. *Am. J. Clin. Pathol.* 86, 153.

Takeuchi, K., Maruyama, I., Yamamoto, S., Oki, T., and Nagata, Y. (1991). Isolation and monolayer culture of human fallopian tube epithelial cells. *In Vitro Cell Dev. Biol.* 27A, 720–724.

Tawia, S. A., Beaton, L. A., and Rogers, P. A. W. (1993). Immunolocalization of the cellular adhesion molecules, intercellular adhesion molecule (ICAM-1) and platelet endothelial cell adhesion molecule (PECAM) in human endometrium throughout the menstrual cycle. *Hum. Reprod.* 8, 175–181.

Timpl, R., and Dziadek, M. (1986). Structure, development, and molecular pathology of basemenet membranes. *In* "International Review of Experimental Pathology" Vol. 29, pp. 1–112. Academic Press, Orlando, Fla.

Underwood, P. A., and Bennett, F. A. (1989). A comparison of the biological activities of the cell-adhesive proteins vitronectin and fibronectin. *J. Cell Sci.* 93, 641–649.

van Bogaert, L.-J., Maldague, P., and Abarca, J. (1978). Granulocytic-like cells in the human fallopian tube. *J. Reprod. Med.* 20, 297–300.

Verhage, H. G., and Fazleabas, A. T. (1988). The *in vitro* synthesis of estrogen-dependent proteins by the baboon (*Papio anubis*) oviduct. *Endocrinology* 123, 552–558.

Verhage, H. G., and Jaffe, R. C. (1986). Hormonal control of the mammalian oviduct: morphological features and the steroid receptor systems. *In* "The Fallopian Tube: Basic Studies and Clinical Contributions" (S. M. Siegler, ed.), pp. 107–117. Futura, New York.

Verhage, H. G., Bareither, M. L., Jaffe, R. C., and Akbar, M. (1979). Cyclic changes in ciliation, secretion and cell height of the oviductal epithelium in women. *Am. J. Anat.* 156, 505–522.

Verhage, H. G., Akbar, M., and Jaffe, R. C. (1980). Cyclic changes in cytosol progesterone receptor of human fallopian tube. *J. Clin. Endocrinol. Metab.* 51, 776–780.

Verhage, H. G., Fazleabas, A. T., and Donnelly, K. (1988). The *in vitro* synthesis and release of proteins by the human oviduct. *Endocrinology* 122, 1639–164.

Verhage, H. G., Boice, M. L., Mavarogianis, P. A., Donnelly, K., and Fazleabas, A. T. (1989). Immunological characterization and immunocytochemical localization of oviduct-glycoproteins in the baboon (*Papio anubis*). *Endocrinology* 124, 2464–2472.

Verhage, H. G., Mavrogianis, P. A., Boice, M. L., Li, W., and Fazleabas, A. T. (1990). Oviductal epithelium of the baboon: Hormonal control and the immuno-gold localization of oviduct-specific glycoproteins. *Am. J. Anat.* 187, 81–90.

Vestweber, D., Gossler, A., Boller, K., and Kemler, R. (1987). Expression and distribution of cell adhesion molecule uvomorulin in preimplantation embryos. *Dev. Biol.* 124, 451–456.

Walker, S. K., Seamark, R. F., Quinn, P., Warnes, G. M., Ashman, R. J., Smith, D. H., and Ancell, P. (1988). Culture of pronuclear embryos of sheep in a simple medium. *Proc. 11th Int. Cong. of Anim. Reprod. A.I.* 4, 483–485.

Walker, S. K., Lampe, R. J., and Seamark, R. F. (1989). Culture of sheep zygotes in synthetic oviduct medium with different concentrations of sodium bicarbonate and HEPES. *Theriogenology* 32, 797–804.

Walker, S. K., Heard, T. M., and Seamark, R. F. (1992). In vitro culture of sheep embryos without co-culture: successes and perspectives. *Theriogenology* 37, 111–126.

Watt, F. M. (1986). The extracellular matrix and cell shape. *TIBS* 11, 482–485.

Wegner, C. C., and Killian, G. J. (1991). *In vitro* and *in vivo* association of an oviduct estrus-associated protein with bovine zona pellucida. *Mol. Reprod. Dev.* 29, 77–84.

Wegner, C. C., and Killian, G. J. (1992). Origin of oestrus-associated glycoproteins in bovine oviductal fluid. *J. Reprod. Fertil.* 95, 841–854.

Weiss, J. J., Tedder, T. F., and Fearon, D. T. (1984). Identification of a 145,000 Mr membrane protein as the C3d receptor (CR2) of human B lymphocytes. *Proc. Natl. Acad. Sci. USA* 81, 881–885.

Welsh, A. O., and Enders, A. C. (1989). Comparisons of the ability of cells from rat and mouse blastocysts and rat uterus to alter complex extracellular matrix in vitro. *In* "Blastocyst Implantation" (K. Yoshinaga, ed.), pp. 55–74. Adams Publishing, Boston.

Welsh, A. O., and Enders, A. C. (1991). Chorioallantoic placenta formation in the rat: I. Luminal epithelial cell death and extracellular matrix modifications in the mesometrial region of implantation chambers. *Am. J. Anat.* 192, 215–231.

West, N. B., and Brenner, R. M. (1983). Estrogen receptor levels in the oviducts and endometria of cynomolgus macaques during the menstrual cycle. *Biol. Reprod.* 29, 1303–1312.

Wewer, U. M., Faber, M., Liotta, L. A., and Albrechsten, R. (1985). Immunochemical and ultrastructural assessment of the nature of the pericellular basement membrane of human decidual cells. *Lab. Invest.* 53, 624–633.

Wewer, U. M., Damjanov, A., Weiss, J., Liotta, L. A., and Damjanov, I. (1986). Mouse endometrial stromal cells produce basement-membrane components. *Differentiation* 32, 49–59.

White, T. E. K., di sant'Agnese, P. A., and Miller, R. K. (1990). Human endometrial cells grown on an extracellular matrix form simple columnar epithelia and glands. *In Vitro Cell. Dev. Biol.* 26, 636–642.

Willemse, A. H. (1975a). The secretory activity of the epithelium of the ampulla tubae in cyclic ewes: a light microscopical study. *Tijdschr. Diergeneesk.* 100, 84–94.

Willemse, A. H. (1975b). The influence of estradiol benzoate and progesterone on the secretory activity of the epithelium of the ampulla tubae in ovariectomized ewes: a light microscopical study. *Tijdschr. Diergeneesk.* 100, 436–444.

Willemse, A. H., and Van Vorstenbosch, C. J. A. H. V. (1975a). The secretory activity of the epithelium of the ampulla tubae in cyclic ewes: an electron microscopical study. *Tijdschr. Diergeneesk.* 100, 95–105.

Willemse, A. H., and Van Vorstenbosch, C. J. A. H. V. (1975b). The influence of oestradiol-benzoate and progesterone on the secretory activity of the epithelium of the ampulla tubae in ovariectomized ewes: an electron microscopical study. *Tijdschr. Diergeneesk.* 100, 763–774.

Yagel, S., Parhar, R. S., Jeffery, J., and Lala, P. (1988). Normal nonmetastatic human trophoblast cells share *in vitro* invasive properties of malignant cells. *J. Cell Physiol.* 136, 455–462.

Yamada, K. M. (1991). Adhesive recognition sequences. *J. Biol. Chem.* 266, 12809–12812.

Yurchenco, P. D., and Schittny, J. C. (1990). Molecular architecture of basement membranes. *FASEB J.* 4, 1577–1590.

11

Cell Biology of Migratory Cells in the Endometrium

DAVID A. CLARK & EWA MUZIKOVA

I. Introduction

Endometrium is a unique tissue in that it undergoes cyclic breakdown and regeneration in preparation for pregnancy. Individual components of endometrium include the epithelium (surface and glandular), the stroma, and blood vessels. Endometrial surface epithelium is involved in attachment of the blastocyst to the wall of the uterus, and for this reason it is considered to have a special role in many species. The endometrial stroma is a complex structure which consists of various mesenchymal cells, some of which proliferate in response to the implanting blastocyst and transform into decidual cells. Transformation of endometrium into decidua occurs only in those species with hemochorial placentation.

An important component of endometrium that is present in both the epithelium and stroma are lymphomyeloid cells (leukocytes) which migrate to endometrium through the blood from bone marrow and lymphoid tissue. In the human, stromal leukocytes may represent 8% of the cells in preimplantation endometrium, 23% in peri- and early postimplantation endometrium, and up to 32% of the cells in early pregnancy decidua. There are also lymphomyeloid cells within epithelium. Table 1 lists the main migratory cell types.

This chapter reviews current information about lymphomyeloid cell types in endometrium, concentrating on the main types—macrophages, B cells, T cells, and granulated non-T, non-B cells. The discussion of their identification, function, origin, and fate is based on work done with tissue

Molecular Biology of the Female Reproductive System

TABLE 1 Phenotypes of Migratory Cells in Human Endometrium

Cell type	Phenotypic features	Stroma	Epithelium	Reference
Macrophage	CD14+ HLA-DR/DP/DQ$^+$ CD4$^{+/-a}$ CD11b (OKM-I)$^-$ CD15 (Leu-M$_1$)$^-$	+	+	Bulmer *et al.* (1987 a,b,c, 1988, 1991)
B cells	Surface Ig of one isotype	+	−	Kamat and Isaacson (1987); Morris *et al.* (1985); Parr and Parr (1985a,b, 1992)
T cells	CD2$^+$/3$^+$/5$^+$	+	+	Morris *et al.* (1985); Christmas *et al.* (1990); Dietl *et al.* (1990); Tabibzadeh (1991b); Yeh *et al.* (1990); Kamat and Isaacson (1987); Nandi and Allison, (1991); Bulmer *et al.* (1987abc, 1988, 1991)
Non-T, non-B granulated lymphoid cells	CD2$^{+/-}$ CD8$^{\text{bright-dim} +/-}$ CD56 (NKH-1)$^+$ CD16$^+$ (Leu-11) or 16$^-$ CD11a$^+$/− CD11b (OKM-1)−	+	+	Bulmer *et al.* (1987abc, 1988); Slukvin *et al.* (1992); Starkey *et al.* (1988); Kurai *et al.* (1992)

Note: Macrophages in endometrium are CD4$^-$ but become CD4$^+$ with decidualization. Anti-CD4 cannot be used to discriminate between T cells and macrophages in decidua.

from humans and rodents since these tissues have supplied most of the information. [Kearns and Lala (1982) have proposed that decidual cells are also derived from bone marrow, but this thesis has received little experimental support from other workers (Fowlis and Ansel, 1985; Rossant and Croy, 1985).]

II. Macrophages

Macrophages in human endometrium can be found scattered in the stratum functionalis and in aggregates adjacent to endometrial gland epithelium in the basal zone. Maternal macrophages can be detected throughout pregnancy in decidualized endometrium and are often closely associated with extravillous fetal trophoblast (Bulmer and Johnson, 1985; Kabawat *et al.*, 1985). In the rat, macrophages migrate away from the uterine surface epithelium prior to implantation. Shortly after implantation in the rat, macrophages were present near the implantation site but none were found within the decidua (Tachi and Tachi, 1989). Macrophages are excluded from the primary decidual reaction around the conceptus in studies of the laboratory mouse (Head, 1987). Data on macrophages at the time of human implantation are not available at present.

Macrophages at nonuterine sites may perform a variety of functions, including phagocytosis and degradation of nonviable or foreign material, presentation of degradation products as antigen to the immune system, production of enzymes and free radicals, and release of a wide spectrum of cytokines and other small informational molecules. Cytokines are peptides with a molecular weight usually <100 kDa, which exert autocrine, paracrine, and occasionally endocrine effects. The differences between cytokines and classic hormones have been reviewed elsewhere (Clark, 1992). Macrophage-derived cytokines may include interleukin-1 (IL-1), tumor necrosis factor-α (TNFα), interleukin-6 (IL-6), colony-stimulating factors for macrophages (M-CSF or CSF-1) and granulocyte–macrophage growth and differentiation (GM–CSF), other growth factors such as epidermal growth factor (EGF) and transforming growth factor-β (TGFβ), and inflammatory peptides of the IL-8 family (Clark, 1992, 1993a). Other mediators produced by macrophages include procoagulants, plasminogen activator inhibitors, and prostaglandins-leukotrienes.

It is unclear which cytokines and mediator may be produced by macrophages during different phases of the menstrual cycle, but a prominent role has been proposed for macrophage-derived cytokines in endometrial sloughing and repair (Clark and Daya, 1990; Clark, 1992). Some recent studies of TNFα mRNA expression in decidua have been done in the mouse and human (Yelavarthi *et al.*, 1991; Chen *et al.*, 1992) and surprisingly, TNFα messenger ribonucleic acid (mRNA) was associated primarily with epithelium and nonmacrophage cells. Tabibzadeh (1991a), using immunohistochemistry, found TNFα to be widespread throughout the stroma of human endometrium and present also in the epithelium of secretory endometrium; TNFα was also present in decidua.

Macrophage-like cells (M cells) may be present within the epithelium lining nonuterine mucosal tissue such as the gut and respiratory tract, and take up antigen from the lumen. In the uterus, however, such cells are absent, so that antigens that enter the uterus would not normally be ingested and processed for immune stimulation (Parr and Parr, 1992). It is likely, however, that pathogenic bacteria or viruses transgressing the epithelium would be dealt with appropriately. The lymphatic drainage from superficial endometrium is minimal (Head, 1987), but studies have not been done on inflamed endometrium.

Macrophages may be present in uterine luminal fluid; they may possibly migrate from adjacent endometrium but more likely come from peritoneal fluid which enters the fallopian tube (reviewed in Clark, 1991).

These macrophages may increase in number and secrete IL-1α and TNFα in conditions such as endometriosis. These cytokines may interfere with fertilization and development of the early preimplantation embryo (Hill *et al.*, 1987; Clark, 1991).

Tachi and Tachi (1989), looking at the function of macrophages in the endometrium at the time of implantation, suggested that the blastocyst altered their morphology and function. *In vitro* studies showed a "rounding up" and loss of ability to secrete leukotriene C4 (Tachi and Tachi, 1989). Hunt (1989) indicated that this type of macrophage *in vivo* is most likely inactive. Redline and colleagues (1990) described active paralysis of macrophages in murine decidua; no conceptus was required for this effect. McKay *et al.* (1992) have shown that the matrix of decidua is most likely responsible for paralysis. Active paralysis of certain macrophage functions is consistent with the intimate and benign association of macrophages with trophoblast cells during pregnancy (Bulmer *et al.*, 1987a,b,c). It is therefore not clear why it should be necessary to exclude macrophages from the vicinity of the implanting blastocyst (in rodents).

A surprising feature of macrophages in endometrium and in decidua is expression of Class II major histocompatibility complex (MHC) (HLA DR/DP/DQ) antigens. Expression of class II MHC is usually viewed as a sign of macrophage "activation." Macrophages of first trimester human decidua present antigen to T cells when they are removed from their milieu and prepared in the form of a single cell suspension (Oksenberg *et al.*, 1987, 1988); similar data have been provided using mouse tissues (Searle *et al.*, 1987). It is *not* clear that macrophages *in situ* in decidua can present antigen.

Macrophages prepared by disaggregation of human or mouse decidua may also release immunosuppressive quantities of prostaglandin-E2 (PGE_2) *in vitro* (Wood *et al.*, 1988; Matthews and Searle, 1987; Lala *et al.*, 1988; Parhar *et al.*, 1989; Scodras *et al.*, 1990). Indeed, it has been proposed that release of PGE_2 *in situ* is responsible for suppressing the rejection of the allogeneic embryo (Lala *et al.*, 1990). Recent data indicate this is unlikely to be correct (Smith and Kelly, 1988; Clark *et al.*, 1993a). Release of high levels of PGE_2 is associated with tissue damage, caused either by disaggregation and removal of progesterone or by the death and spontaneous abortion of the embryo (Clark, 1991; Clark *et al.*, 1993a; Branch *et al.*, 1992). Indeed, prior to the onset of abortion in DBA/J mice, the levels of PGE_2 in fetoplacental units destined to abort are slightly increased (Gendron *et al.*, 1992).

CSF-1 and TGFβ1 merit discussion as macrophage cytokines that are detectable *in situ* in the uterus. TGFβ1 is present in the murine endometrium in the pre- and peri-implantation period, and may synergize with EGF to enhance embryo cell growth (Paria and Dey, 1990; Tamada *et al.*, 1990). However, to affect growth of the preimplantation embryo, TGFβ1 would have to be produced and secreted by epithelial cells, and this appears to be true (Tamada *et al.*, 1990); specific localization to macrophages has not been shown. With respect to CSF-1, the situation is now clear as a result of the availability of mutant mice of genotype op/op, which lack this cytokine. In op/op mice, macrophages and osteoclasts are deficient, the latter leading to osteoporosis. Female op/op mice mated to op/op males fertilize but do not implant (Pollard *et al.*, 1991). By treating the mice from birth with CSF-1, uterine macrophages are restored, as is implantation; short-term CSF-1 given from the day of mating (too short a treatment period to restore tissue macrophage numbers) is ineffective, an observation suggesting an important role for macrophages in implantation—notwithstanding their exclusion from the primary decidual zone in rodents (Pollard, 1992).

However, to restore implantation, it is also important to treat the male with CSF-1, for reasons as yet unclear. In the decidua in humans, macrophages may be in intimate contact with interstitial trophoblast, and macrophages may secrete CSF-1 and GM-CSF, which may affect trophoblastic cells. Macrophages may therefore promote trophoblast growth and function in order to enhance the success of the embryo. Conversely, administration of macrophage activators such as lipopolysaccharide can stimulate TNFα production and cause abortions. Although there is more TNFα or RNA per milligram of protein in the uterus than in other organs after intraperitoneal liposaccharide (LPS) challenge of mice (Giroir *et al.*, 1992), it is not clear that the TNFα involved in initiation of abortion is being locally produced by decidual macrophages because the levels produced locally may be very low (Gendron *et al.*, 1990; Clark *et al.*, 1991a). The embryolethal effects of LPS may be mediated by a thrombogeneic action on the vasculature, and this may involve systemic activation of natural killer (NK) cells (Clark *et al.*, 1993a). Low doses of recombinant CSF-1 given to normal mice in early pregnancy enhanced postimplantation failure (i.e., resorption) (Tartakovsky *et al.*, 1991). Taken together, existing data support the thesis that lymphomyeloid, stromal, and epithelial cells may all produce cytokines that promote critical processes in pregnancies or if present in too high a concentration at the wrong time, may be lethal.

The fate of uterine macrophages is unclear. Since the cells enter from bone marrow, there must be loss through emigration or local cell death. Kinetic studies defining the fate of most of the migratory cell subsets in endometrium and decidua are not available. In the human, loss with menstrual tissue would be predicted.

III. B Cells

Lymphocytes (surface immunoglobulin (Ig) positive) are present in human endometrial stroma but are a minor population. In rodents, B cells (bone-marrow-derived lymphocytes), particularly surface-IgA^+ cells, localize in endometrium, which is consistent with the idea that the genital tract is part of a common mucosal immune system (Parr and Parr, 1985a,b). In the rodent, sperm are ejaculated into the uterine lumen together with bacteria, and immunoglobulin and complement may be found in the uterine lumen (Parr and Parr, 1988). There is also activation of inflammation-related cytokines (McMaster *et al.*, 1992; Sanford *et al.*, 1992). Nevertheless, surface-IgA$^+$ B cells also migrate to the rodent cervix (McDermott *et al.*, 1982); IgA is also produced in the human cervix, where the exogenous bacterial challenge to the endometrium following ejaculation is minimal. IgA, IgG, and IgM are also found at the implantation site in the mouse, and IgG is present on the first trimester human trophoblast (Parr and Parr, 1985a,b; Dresser *et al.*, 1989; J. F. Mowbray, personal communication). A local origin for these molecules is uncertain. When there is endometritis (infection), follicles of B cells can develop in endometrium. B cells can therefore be functionally activated and active in endometrium, but their role in normal endometrial functions and in implantation is a matter of speculation. The fate of B cells in endometrium is unclear, but in the human, loss with menstrual tissue would be anticipated.

IV. T Cells

The second major population of migratory cells in endometrium, thymus-derived (T) cells, may be divided into two compartments. T cells are present within the epithelium, with CD8$^+$ T cells being four times more frequent than CD4$^+$ cells. In endometrium, a slightly greater proportion of T cells are CD8$^+$, and most of the cells are in the stratum functionalis (Morris *et al.*, 1985; Bulmer, 1987a,b,c; Bulmer *et al.*, 1988).

In nonuterine sites, the function of T cells is to recognize and react to antigen by proliferation and differentiation into effector cells. CD8$^+$ cells "see" antigen associated with class I MHC, and CD4$^+$ cells "see" antigen associated with class II MHC; T-cell receptor (TCR) binds to antigen, and the CD4 or 8 molecule on the T cell to the MHC on the antigen-presenting cell. Effector functions include suppression (mostly CD8$^+$), cytotoxicity (mostly CD8$^+$), and "help" (mostly CD4$^+$). "Help" is largely mediated through production of cytokines such as IL-2, IL-3, IL-4, IL-5, IFN-γ, and GM-CSF. The function of T cells in endometrium and in pregnancy decidua is controversial. There is an increase in the number of T cells in the mid-secretory phase of the human endometrium, but in the late secretory phase and in areas of predecidual change, T cells are scant (Bulmer *et al.*, 1987a,b,c). In the decidua of early pregnancy, the level of TCR on T cells is subnormal, and the cells respond poorly to conventional T-cell stimulants (Dietl *et al.*, 1990; reviewed in Clark, 1993b). Endometrial T-cell TCR levels and T-cell function in human endometrium has not been studied. T-cell receptor on peripheral T cells is usually composed of two chains, α and β. TCRαβ$^+$ cells predominate, but some T cells bear a different type of TCR, γδ (reviewed in Clark, 1993b). TCRγδ$^+$ cells are also present in vaginal epithelium (Nandi and Allison, 1991) and in uterine pregnancy epithelium (Yeh *et al.*, 1990). TCR γδ$^+$ cells are of interest because they can recognize "super antigens" without the need for CD4/CD8–MHC interactions. T-like cells in epithelium may manifest NK (Tagliabue *et al.*, 1982). T cells have been cloned from "decidua" using the mitogen (PHA) and IL-2 (± a feeder cell layer), and various functions described, but it is uncertain if the precursors of the clones were cells *in* the decidual stroma or were from contaminating blood lymphocytes (Christmas *et al.*, 1990).

There is some information to suggest unique functional activities for T cells in endometrium and decidua. In the human, allogeneic activation of CD8$^+$ T cells upregulates progesterone receptors, and progesterone, if present, stimulates release of a 34-kDa molecule that inhibits the activity of natural killer cells (Szekeres-Bartho *et al.*, 1990). The progesterone-receptor blocking drug RU486 causes abortion in humans and mice, but in mice, abortion by RU486 can be prevented by coadministration of the 34-kDa factor (Szekeres-Bartho *et al.*, 1990). Abortion in rodents is mediated by activated NK-lineage cells, and the need for CD8$^+$ cells for a successful pregnancy seems to vary among strains and animal colonies (reviewed in Clark, 1991). For example, T- and B-cell-deficient SCID mice and T-cell-deficient nude mice held under specific pathogen-free (defined flora) conditions reproduce successfully, and partial depletion of CD8$^+$ cells by *in vivo* injection of anti-CD8 has minimal effects in female Lewis rats and in female C57/Bl and DBA/1 mice (Sulila *et al.*, 1988; Mattsson and Holmdahl, 1987). However, in DBA/2-mated CBA/J female mice, which are particularly sensitive to the abortion-inducing and NK-activating effects of the synthetic nucleotide poly(I:C) and lipopolysaccharide, anti-CD8 antibody dramatically increases the spontaneous abortion rate (Clark *et al.*, 1989). Large CD8$^+$ suppressor cells appear in murine endometrium in response to the hormonal changes of early pregnancy and suppress generation of cytotoxic T lymphocytes (CTL) *in vitro* (the 34-kDa factors may also be inhibitory) (Clark *et al.*, 1991a). Similar large-sized cells which block the T-cell proliferative response to concanavalin A also develop in human endometrium in the midsecretory (luteal) phase (Daya *et al.*, 1985). Furthermore, after fertilization, the embryo itself appears to produce a low-molecular-weight factor that stimulates the ovary and oviduct to form a molecular complex called early pregnancy factor (EPF) which then activates the suppressor function of CD8$^+$ T cells (Rolfe *et al.*, 1989).

By using a monoclonal antibody (No. 14-30) to a CD4$^+$-cell-derived factor that activates CD8$^+$ suppressor cells, Beaman and Hoversland (1988), were able to completely block implantation in mice. They used this monoclonal antibody to measure tissue levels and found that the uterus contained more factor than other tissues. This is surprising since T cells are scant, so cells other than CD4$^+$ T cells may produce the factor reacting with 14-30 (Hoversland and Beaman, 1990). Interestingly, antibody to EPF given prior to implantation also causes loss of embryonic viability in animals (Athanasas-Platsis *et al.*, 1989). The ability of anti-CD8 antibody to block implantation in mice cannot be assessed at present because this antibody has always been given *after* implantation to study effects on abortion; however, successful implantation does occur in nude and SCID mice deficient in T cells (Clark *et al.*, 1990a; Croy *et al.*, 1991b), so classic T cells are not necessarily required for implantation. These animals may have an intact intraepithelial lymphocyte population, however, which may include T-like cells ($\gamma\delta$?).

Taken together, the preceding data suggest that local endometrial CD8$^+$ cells play an important role in early pregnancy under certain situations, but exact mechanisms remain to be elucidated. There are four caveats which must be kept in mind when drawing inferences from experimental data. First, antibody injection may "activate" cells rather than eliminate them. Second, systemic effects of antibody in T cells may be important. Third, it is uncertain if the immunosuppressive CD8$^+$ cells isolated from endometrium are derived from the stroma, epithelium, or contaminating blood present in endometrial blood vessels. There has been one suggestion that T cells in pregnancy decidua may be activated, however, and that is based in the presence of an "activation marker"; its presence on stromal T cells *in situ* remains to be shown (reviewed in Clark, 1993b). Finally, while there are nonclassical pathways which could activate T cells in endometrium and decidua (as described earlier and elsewhere, Clark, 1993), actual production of suppressor molecules (or any other cytokine) by stromal T cells (i.e., activation of function) has not yet been demonstrated. The fate of T cells in the uterus has not been defined.

V. Granulated Non-T and Non-B Cells (Endometrial/Decidual Granulated Lymphocytes)

The third major cell population in endometrium and decidua, and perhaps the most interesting, consists of the granulated non-T, non-B cells (endometrial granulated lymphocytes, eGL, and decidual granulated lymphocytes, dGL). These cells are most readily detected by the presence of CD56 (NKH-1), a ligand for the adhesion molecule NCAM-1. A proportion express CD2 (a ligand for LFA-3) and CD11a (a ligand for LFA-1). Slukvin *et al.* (1992) suggested that a subpopulation might also express CD8$^+$, but the levels are very low in comparison with CD8$^+$ (bright) T cells. CD56$^+$ cells are present in peripheral blood, where they act as NK cells. Endometrial and decidual CD56$^+$ cells appear to differ, however. Blood CD56$^+$ NK cells express CD16 (Fc receptor for IgG) but eGL and dGL are $CD16^-$ and are deficient in natural killer activity in decidua, even though they may contain the molecule perforin, which appears to be involved in lysis of target cells (Haynes and Smith, 1992; Haller *et al.*, 1992; Starkey, 1992). Some dGL do express the high-affinity component of the IL-2 receptor and can be readily activated by IL-2 into killer cells that lyse NK-resistant target cells, such as human and murine trophoblasts (Starkey, 1992; King and Loke, 1990; Clark, 1991). In this respect, it is particularly interesting that eGL accumulate (possibly by local division in the stratum functionalis) in the second half of the secretory phase endometrium, and persist throughout menstruation, suggesting that the accumulation is mediated by adhesion receptors rather than by hormone, and that these cells could provide a continuous source of perforin leading to cell lysis at the time of menstruation. In pregnancy decidua, dGL are the dominant cell type and can often be seen near interstitial and endovascular trophoblast. These data taken together suggest that eGL could be involved in destruction of tissue (menstruation) and/or embryos (occult or as a miscarriage). Indeed, dGL are absent at sites of tubal ectopic implantation (Bulmer *et al.*, 1987a) and this could explain resistance of these implants to elimination by RU486 treatment (i.e., there may be no effector cells present) (Avrech *et al.*, 1991). The cytotoxic $CD56^+$ cell population isolated from decidua is $CD16^+$ and this subcomponent may be derived either from the $CD16^+$ subset which represents 1–5% of stromal $CD56^+$ cells, or through contamination with blood containing the cytotoxic $CD56^+$ $CD16^+$ cells.

There are currently no data concerning the function of eGL of the $CD56^+$ 16^- type. However, there are some useful data concerning dGL which may provide insight into the possible function of eGL cases. In the mouse, granulated cells of NK lineage accumulate after implantation (as assessed by the asialo GM1 and 3C2 marker), both at the implant site and 2–4 days after implantation in the area where the decidua basalis forms (reviewed in Clark, 1991; Linnemeyer and Pollack, 1992). The accumulation of these cells appears to correlate with a brief burst of conventional NK activity, which depends on the presence of an embryo (Croy *et al.*, 1985; Gambel *et al.*, 1985; Kiso *et al.*, 1992). Four days after implantation, a number of the NK cells appear to transform into large granule-containing cells called metrial gland (MG) cells, which express the 4H12 marker (in preference to 3C2) (Linnemeyer and Pollack, 1992). These cells may secrete cytokines such as IL-1 and CSF-1, and may be directly or indirectly cytotoxic to the trophoblast (Croy and Kassouf, 1989; Croy *et al.*, 1991).

In the DBA/2-mated CBA/J mouse, which has a high rate of recurrent spontaneous resorption (abortion), an increased accumulation of asiolo $GM1^+$ cells has been associated with embryos that subsequently abort (reviewed in Clark, 1991). However, 4 days after implantation in the mouse, a second "granulated" cell population that is smaller and phenotypically distinct from asiolo GM1$^+$ cells may be detected through its immunosuppressive function (Clark *et al.*, 1991a). This function is mediated by the secretion of a novel family of TGFβ-2-like molecules (Clark *et al.*, 1990a, 1991b, 1992, 1993b; Lea *et al.*, 1992). At implantation sites where TGFβ-2 production is deficient, the probability of resorption is increased (Clark *et al.*, 1990b, 1991a). Although there is as yet no CD surface marker for these cells in the mouse, a homologous population of suppressor cells producing the TGFβ2-related factor has been demonstrated in the human, and these cells are CD56$^+$ CD16$^-$ (Clark *et al.*, 1993b). (Fc receptor may be present on the murine suppressor cell but although this is not consistent with human data, CD32 and CD64-type Fc receptors have not been tested on human CD56$^+$16$^-$ cells). The activation of a suppressor function in mouse decidua requires signals from trophoblast cells, and in the human, the trophoblast also appears able to activate suppressor function (Slapsys *et al.*, 1988; Daya *et al.*, 1989; Clark, 1991; Clark *et al.*, 1991a; Petrovic *et al.*, 1992). Some other cytokines have been detected in dGL by polymerase chain reaction (PCR), including TNFα,γ IFN, GM-CSF, CSF-1, and leukemia inhibitory factor (LIF) (Kurai *et al.*, 1992); eGL may contain TNFα, IL-1, and CSF-1 (Bulmer *et al.*, 1991), although the extent to which these molecules are produced by eGL is uncertain (Tabibzadeh, 1991a). Based on these data, it would be reasonable to propose two lineages for eGL.

1. CD56$^+$ CD16$^+$ cells with cytotoxic NK activity; their roles may include antibacterial action (Garcia-Penarrubia *et al.*, 1989), elimination of the embryo, or participation in tissue destruction with menstruation.
2. CD56$^+$ CD16$^-$ cells which prevent cytotoxic rejection of the embryo through release of TGFβ-2-like factor, and which may play a role in tissue regeneration after the onset of menses (Clark and Daya, 1990; Clark, 1992b).

It seems clear that in addition to more detailed phenotypic characterization of eGL, the information needed to understand the role of CD56$^+$ subsets in the endometrium will require testing for production of functional products at different phases of the menstrual cycle. For example, changes in function of eGL during menses, particularly with respect to cytokine and perforin secretion, have not been studied. It would then be appropriate to look for cells with receptors for these products, and for alterations in function associated with pathology, such as infertility, endometriosis, menorrhagia, and intermenstrual bleeding.

Acknowledgment

This work was supported by awards from MRC Canada. We owe special thanks to Ann Lawton for her secretarial assistance.

References

Athanasas-Platsis, S., Quinn, K. A., Wong, T.-Y., Rolfe, B. E., Cavanagh, A. C., and Morton, H. (1989). Passive immunization of pregnant mice against early pregnancy factor causes loss of embryonic viability. *J. Reprod. Fertil.* 87, 495–502.

Avrech, O. M., Yolan, A., Heinraub, Z., Bukovski, I., and Caspi, E. (1991). Mephisterone (RU486) alone or in combination with a prostaglandin analogue for termination of early pregnancy: a review. *Fertil. Steril.* 56, 385–393.

Beaman, K. D., and Hoversland, R. C. (1988). Induction of abortion in mice with a monoclonal antibody specific for suppressor T-lymphocyte molecules. *J. Reprod. Fertil.* 82, 691–696.

Branch, D. W., Dudley, D. J., and Mitchell, M. D. (1992). Prostaglandin E2 (PGE2) production by murine decidua is increased in response to abortifacient doses of lipopolysaccharide (LPS). *Am. J. Reprod. Immunol.* 27, 30.

Bulmer, J. N., and Johnson, P. M. (1985). Immunohistological characterization of the decidual leukocyte infiltrate related to endometrial gland epithelium in early human pregnancy. *Immunology* 55, 35–44.

Bulmer, J. N., and Ritson, A. (1988). The decidua in early pregnancy. *In* "Early Pregnancy Loss: Mechanisms and Treatment" (R. W. Beard and F. Sharp, eds.), pp. 171–180. Peacock Press, Ashton-Under-Lyne, England.

Bulmer, J. N., Ritson, A., Earl, V., and Hollings, D. (1987a). Immunocompetent cells in human decidua. *In* "Reproductive Immunology. Materno-fetal Relationship" (G. Chaouat, ed.), Vol. 154, pp. 89–100. John Libbey Colloque INSERM, Paris, France.

Bulmer, J. N., Johnson, P. M., and Bulmer, D. (1987b). Leukocyte populations in human decidua and endometrium. *In* "Immunoregulation and Fetal Survival" (T. J. Gill, III, and T. G. Wegmann, eds.), pp. 111–134. Oxford University Press, New York.

Bulmer, J. N., Hollings, D., and Ritson, A. (1987c). Immunocytochemical evidence that endometrial stromal granulocytes are granulated lymphocytes. *J. Pathol.* 153, 281–288.

Bulmer, J. N., Lunny, P., and Hagin, S. V. (1988). Immunohistochemical characterization of stromal leukocytes in non-pregnant human endometrium. *Am. J. Reprod. Immunol. Microbiol.* 17, 83–90.

Bulmer, J. N., Morrison, L., Longfellow, M., and Ritson, A. (1991). Leucocytes in human decidua: investigation of surface markers and function. *In* "Cellular and Molecular Biology of the Materno-fetal Relationship" (G. Chaouat and J. Mowbray, eds.), Vol. 212, pp. 189–196. John Libbey Colloque INSERM, Paris, France.

Chang, Y., Pollard, J. W., Khalili, H., Goyert, J., and Diamond, D. (1992). The antigen presenting activity of placental macrophage cell lines. *Am. J. Reprod. Immunol.* 27, 28.

Chen, H. L., Yang, Y., Xiao-Ling, H., Yelavarthi, K. K., Fishback, J. L., and Hunt, J. J. (1992). Tumor necrosis factor α mRNA and protein are present in human placental and uterine cells at early and late stages of gestation. *Am. J. Pathol.* 139, 327–335.

Christmas, S. E., Bulmer, J. N., Meager, A., and Johnson, P. M. (1990). Phenotypic and functional analysis of human $CD3^-$ decidual leucocyte clones. *Immunology* 71, 182–189.

Clark, D. A. (1991). Controversies in reproductive immunology. *Crit. Rev. Immunol.* 11, 215–247.

Clark, D. A. (1993a). Cytokines, decidua and early pregnancy. *In* "Oxford Reviews of Reproductive Biology" (S. Milligan, ed.), Vol. 15, pp. 83–111. Oxford University Press, New York.

Clark, D. A. (1992). Cytokines and uterine bleeding. *In* "Steroid Hormones and Uterine Bleeding" (N. Alexander and C. d'Arcanques, eds.), pp. 263–275. AAAS Press, Washington, D.C.

Clark, D. A. (1993b). Paraimmunology and receptors on decidual T-cells: conflict or confirmation? *Am. J. Reprod. Immunol.* 29, 35–38.

Clark, D. A., and Daya, S. (1990). Macrophages and their migratory cells in endometrium: relevance to endometrial bleeding. *In* "Contraception and Mechanisms of Uterine

Bleeding" (C. D'Arcangues, I. S. Fraser, J. R. Newton, and V. Odlind, eds.), pp. 363–382. Cambridge University Press, Cambridge.

Clark, D. A., Brierley, J., Banwatt, D., and Chaouat, G. (1989). Hormone-induced preimplantation Lyt2^{+} murine uterine suppressor cells persist after implantation and may reduce the spontaneous abortion rate in CBA/J mice. *Cell. Immunol.* 123, 334–343.

Clark, D. A., Flanders, K. C., Banwatt, D., Millar-Book, W., Manuel, J., Stedronska-Clark, J., and Rowley, B. (1990a). Murine pregnancy decidua produces a unique immunosuppressive molecule related to transforming growth factor β-2. J. *Immunology* 144, 3008–3014.

Clark, D. A., Drake, B., Head, J. R., Stedronska-Clark, J., and Banwatt, D. (1990b). Decidua associated suppressor activity and viability of individual implantation sites of allopregnant C3H mice. *J. Reprod. Immunol.* 17, 253–264.

Clark, D. A., Lea, R. G., Podor, T., Daya, S., Banwatt, D., and Harley, C. (1991a). Cytokines determining the success or failure of pregnancy. *Ann. N.Y. Acad. Sci.* 626, 524–536.

Clark, D. A., Lea, R. G., Rowley, B., Denburg, J., Banwatt, D., Manuel, J., Damji, N., Underwood, J., Michel, M., Mowbray, J., Daya, S., and Chaouat, G. (1991b). Transforming growth factor-β related suppressor factor in mammalian pregnancy decidua: homologies between the mouse and human in successful pregnancy and in recurrent unexplained abortion. *In* "Cellular and Molecular Biology of the Materno-fetal Relationship" (G. Chaouat and J. Mowbray, eds.), Vol. 212, pp. 171–179. John Libbey Colloque INSERM, Paris, France.

Clark, D. A., Lea, R. G., Flanders, K. C., Banwatt, D., and Chaouat, G. (1992). Role of a unique species of TGF-β in preventing rejection of the conceptus during pregnancy. *In* "Progress in Immunology VIII" (J. Gergely, M. Benczúr, A. Erdei, A. Falus, G. Füst, G. Medgyesi, G. Petrányi, and E. Rajnavölgyi, eds.), pp. 841–847. Springer-Verlag, Heidelberg.

Clark, D. A., Banwatt, D., and Chaouat, G. (1993a). Effect of prostaglandin synthesis inhibitors on spontaneous and endotoxin-induced abortion in mice. *J. Reprod. Immunol.* 24, 29–44.

Clark, D. A., Flanders, K. C., Starkey, P., Vince, G., and Hirte, H. (1993b). Murine and human decidua-associated natural suppressor cells release a novel type of TGF-β2. *J. Immunol.* 150, 4A.

Croy, B. A., and Kassouf, S. A. (1989). Evaluation of the murine metrial gland for immunological function. *J. Reprod. Immunol.* 15, 51–69.

Croy, B. A., Gambel, P., Rossant, J., and Wegmann, T. G. (1985). Characterization of murine decidual natural killer (NK) cells and their relevance to the success of pregnancy. *Cell. Immunol.* 93, 315–326.

Croy, B. A., Guilbert, L. J., Browne, M. A., Gough, N. M., Stinchcomb, D. T., Reed, N., and Wegmann, T. G. (1991). Characterization of cytokine production by the metrial gland and granulated metrial gland cells. *J. Reprod. Immunol.* 19, 149–166.

Croy, B. A., Chapeau, C., Reed, N., Stewart, I. J., and Peel, S. (1991b). Is there an essential requirement for bone-marrow-derived cells at the fetomaternal interface during successful pregnancy? A study of pregnancies in immunodeficient mice. *In* "Molecular and Cellular Immunobiology of the Maternal Fetal Interface" (T. G. Wegmann, T. J. Gill, III, and E. Nisbet-Brown, eds.), pp. 168–188. Oxford University Press, New York.

Daya, S., Clark, D. A., Devlin, C., Jarrell, J., and Chaput, A. (1985). Preliminary characterization of two types of suppressor cells in the human uterus. *Fertil. Steril.* 44, 778–785.

Daya, S., Johnson, P. M., and Clark, D. A. (1989). Trophoblast induction of suppressor type cell activity in human endometrial tissue. *Am. J. Reprod. Immunol.* 19, 65–72.

Dietl, J., Horny, H.-P., Ruck, P., Marzusch, K., Kaiserling, E., Griesser, H., and Kabelitz, D. (1990). Intradecidual T lymphocytes lack immunohistochemically detectable T cell receptors. *Am. J. Reprod. Immunol.* 24, 33–36.

Dresser, D. W., Popham, A. M., Herrera, N. E., and Carter, J. (1989). Immunoglobulin-secreting cells in close proximity to the mid-gestation mouse embryo. *J. Reprod. Immunol.* 16, 55–70.

Fowlis, D. J., and Ansell, J. D. (1985). Evidence that decidual cells are not derived from bone marrow. *Transplantation.* 39, 445–446.

Gambel, P., Croy, B. A., Moore, W. D., Hunziker, R. D., Wegmann, T. G., and Rossant, J. (1985). Characterization of immune effector cells present in early murine decidua. *Cell. Immunol.* 93, 303–314.

Garcia-Penarrubia, P., Koster, K. T., Kelley, R. O., McDowell, T. D., and Bankhurst, A. D. (1989). Antibacterial activity of human natural killer cells. *J. Exp. Med.* 169, 99–113.

Gendron, R. L., Nestel, F. P., Lapp, W. S., and Baines, M. G. (1990). Lipopolysaccharide-induced fetal resorption in mice is associated with the intrauterine production of tumour necrosis factor-α. *J. Reprod. Fertil.* 90, 395–402.

Gendron, R. L., Farookhi, R., and Baines, M. G. (1992). Murine pregnancies predisposed to spontaneous resorption show alterations in the concentrations of leukotriene B4 and prostaglandin E2. *Biol. Reprod.* 47, 72–75.

Giroir, B. P., Johnson, J. H., Brown, T., Allen, G. L., and Beutler, B. (1992). The tissue distribution of tumor necrosis factor biosynthesis during endotoxemia. *J. Clin. Invest.* 90, 693–698.

Haller, H., Sinjicki-Bezic, N., Gudelj, L., Rukavina, D., and Podack, E. R. (1992). An immunohistochemical study of perforin positive cells in early pregnancy decidua. *Riv. Immunol. Immunofarm.* 12, 99.

Haynes, M. K., and Smith, J. B. (1992). Phenotypic and functional characterization of cultured human decidual lymphoid cells. *Am. J. Reprod. Immunol.* 27:27–28.

Head, J. R. (1987). Lymphoid components in the rodent uterus. *In* "Immunoregulation and Fetal Survival" (T. J. Gill, III, T. G. Wegmann, and E. Nisbet-Brown, eds.), pp. 46–59. Oxford University Press, Oxford.

Hill, J. A., Haimovici, F., and Anderson, D. J. (1987). Products of activated lymphocytes and macrophages inhibit mouse embryo development *in vitro*. *J. Immunol.* 139, 2250–2254.

Hoversland, R. C., and Beaman, K. D. (1990). Embryo implantation associated with increase in T-cell suppressor factor in the uterus and spleen of mice. *J. Reprod. Fertil.* 88, 135–139.

Hunt, J. S. (1989). Cytokine networks in the uteroplacental unit: macrophage as a pivotal regulatory cell. *J. Reprod. Immunol.* 16, 1–17.

Kabawat, S., Mostoufi-Zadeh, M., Driscoll, S., and Bhan, A. K. (1985). Implantation site in normal pregnancy: a study with monoclonal antibodies. *Am. J. Pathol.* 1, 76–84.

Kamat, B. R., and Isaacson, P. G. (1987). The immunocytochemical distribution of leukocyte subpopulations in human endometrium. *Am. J. Pathol.* 127, 66–73.

Kearns, H., and Lala, P. K. (1982). Bone marrow origin of decidual cell precursors in the pseudopregnant mouse uterus. *J. Exp. Med.* 155, 1537–1554.

King, A., and Loke, Y. W. (1990). Human trophoblast and JEG choriocarcinoma cells are sensitive to lysis by IL-2 stimulated decidual NK cells. *Cell. Immunol.* 129, 435–448.

Kiso, Y., McBey, B. A., Mason, L., and Croy, A. (1992). Histological assessment of the mouse uterus from birth to puberty for the appearance of LGL-1^+ natural killer cells. *Biol. Reprod.* 47, 227–232.

Kurai, N., Santo, S., Nishikawa, K., Morii, T., Enomoto, M., Narctaiv, I., and Kihijo, M. (1992). Detection of cytokine mRNA in decidual mononuclear cells and $CD16^-$ $CD56^{bright}$ cells by using RT-PCR method. *Riv. Immunol. Immunofarm.* 12, 111.

Lala, P. K., Kennedy, T. G., and Parhar, R. S. (1988). Suppression of lymphocyte alloreactivity by early gestational human decidua. II. Characterization of the suppressor mechanisms. *Cell. Immunol.* 116, 411–422.

Lala, P. K., Scodras, J. M., Graham, C., Lysiak, J. J., and Parhar, R. S. (1990). Activation of maternal killer cells in the pregnant uterus with chronic indomethacin therapy, IL-2 therapy or a combination therapy is associated with embryonic demise. *Cell. Immunol.* 127, 368–381.

Lea, R. G., Flanders, K. C., Harley, C. B., Manuel, J., Banwatt, D., and Clark, D. A. (1992). Release of a TGF-β2-related suppressor factor from postimplantation murine decidual tissue can be correlated with the detection of a subpopulation of cells containing RNA for TGF-β2. *J. Immunol.* 148, 778–787.

Linnemeyer, P. A., and Pollack, S. B. (1992). Phenotypic changes of NK cells at uterine implantation sites. *Riv. Immun. Immunofarm.* 12, 114.

Mathews, C. J., and Searle, R. F. (1987). The role of prostaglandin in the immunosuppressive effects of supernatants from adherent cells of murine decidual tissue. *J. Reprod. Immunol.* 12, 109–124.

Mattson, R., and Holmdahl, R. (1987). Maintained allogeneic pregnancy in rats depleted of T cytotoxic/suppressor cells by OX8 monoclonal antibody treatment. *Am. J. Reprod. Immunol.* 12, 23–34.

McDermott, M. R., Clark, D. A., and Bienenstock, J. (1982). Selective localization of mesenteric T lymphoblasts in intestinal respiratory and reproductive tissues. *Fed. Proc.* 41, 607.

McKay, D. B., Vazquez, M. A., Redline, R. W., and Lu, C. Y. (1992). Macrophage functions are regulated by murine decidual and tumour extracellular matrices. *J. Clin. Invest.* 89, 134–142.

McMaster, M. T., Newton, R. C., Dey, S. K., and Andrews, G. K. (1992). Activation and distribution of inflammatory cells in the mouse uterus during the pre-implantation period. *J. Immunol.* 148, 1699–1705.

Morris, H., Edwards, J., Tiltman, A., and Emms, M. (1985). Endometrial lymphoid tissue: an immunohistological study. *J. Clin. Pathol.* 38, 644–652.

Nandi, D., and Allison, J. P. (1991). Phenotypic analysis and γδ-T cell receptor repertoire of murine T cells associated with the vaginal epithelium. *J. Immunol.* 147, 1773–1778.

Oksenberg, J. R., Mor-Josef, S., Ezra, Y., and Brauthbar, C. (1987). Antigen presenting cells in human decidual tissue. II. Accessory cells for the development of antitrinitrophenyl cytotoxic T lymphocytes. *J. Reprod. Immunol.* 10, 309–318.

Oksenberg, J. R., Mor-Josef, S., Ezra, Y., Brauthbar, C. (1988). Antigen presenting cells in human decidual tissue. III. Role of accessory cells in the activation of suppressor cells. *Am. J. Reprod. Immunol. Microbiol.* 16, 151–158.

Pace, D., Longfellow, M., and Bulmer, J. N. (1991). Characterization of intra-epithelial lymphocytes in human endometrium. *J. Reprod. Fertil.* 91, 165–174.

Parhar, R. S., Yagel, S., and Lala, P. K. (1989). PGE-2 mediated immunosuppression by first trimester human decidual cells blocks activation of maternal leukocytes in the decidua with potential anti-trophoblastic activity. *Cell. Immunol.* 120, 61–74.

Paria, B. C., and Dey, S. K. (1990). Preimplantation embryo development in vitro. Cooperative interactions among embryos and role of growth factors. *Dev. Biol.* 87, 4756–4760.

Parr, E. L., and Parr, M. B. (1985a). Localization of immunoglobulins in the mouse uterus, embryo, and placenta during the second half of pregnancy. *J. Reprod. Immunol.* 8, 153–171.

Parr, E. L., and Parr, M. B. (1985b). Immunohistochemical localization of immunoglobulins A, G and M in the mouse female genital tract. *J. Reprod. Fertil.* 74, 361–370.

Parr, E. L., and Parr, M. P. (1988). Deposition of C3 on bacteria in the mouse uterus after mating. *J. Reprod. Immunol.* 12, 315–319.

Parr, E. L., and Parr, M. P. (1992). Secretory immune responses in the female genital tract. *Am. J. Reprod. Immunol.* 27:44.

Petrovic, O., Haller, H., Gudelj, L., Rubesa, G., and Rukavina, D. (1992). Immunoregulatory potential of decidual leukocytes in pathological pregnancies. *Riv. Immunol. Immunofarm.* 12, 129.

Pollard, J. W. (1992). Role of CSF in pregnancy. *In* "Proc. 5th Int. Congress Reprod. Immunol.," Serono Symposia, Rome, Italy (in press).

Pollard, J. W., Hunt, J. S., Wiktor-Jedrzejczak, W., and Stanley, R. (1991). A pregnancy defect in the osteopetrotic (*op/op*) mouse demonstrates the requirement for CSF-1 female fertility. *Dev. Biol.* 148, 273–283.

Redline, R. W., McKay, D., Vazgez, A., Papaidannou, V. E., and Lu, C. H. Y. (1990). Macrophage functions are regulated by the substratum of murine decidual stromal cells. *J. Clin. Invest.* 85, 1951–1958.

Rolfe, B., Quinn, K., Athanasas, S., Cavanagh, A., and Morton, H. (1989). Genetically-restricted effector molecules released by human lymphocytes in response to early pregnancy factor. *Immunol. Cell. Biol.* 67, 205–208.

Rossant, J., and Croy, B. A. (1985). Genetic identification of tissue of origin of cellular populations with the mouse placenta. *J. Embryol. Exp. Morphol.* 86, 177–189.

Sanford, T. R., De, M., Wood, G. W. (1992). Expression of colony stimulating factors and inflammatory cytokines in the uterus of CD1 mice during days 1 to 3 of pregnancy. *J. Reprod. Fertil.* 94, 213–220.

Scodras, J. M., Parhar, R. S., Kennedy, T. G., and Lala, P. K. (1990). Prostaglandin-mediated inactivation of natural killer cells in the murine decidua. *Cell. Immunol.* 127, 352–367.

Searle, R. F., Elcock, J. M., and Matthews, C. J. (1987). Antigen presentation by mouse decidua and placenta. *In* "Immunoregulation and Fetal Survival" (T. Gill, III, T. G. Wegmann, and E. Nisbet-Brown, eds.), pp. 96–110. Oxford University Press, New York.

Slapsys, R. M., Younglai, E. G., and Clark, D. A. (1988). Active suppression of host-versus-graft reaction in pregnancy mice. A novel suppressor cell is recruited to decidua by fetal trophoblast cells. *Reg. Immunol.* 1, 182–189.

Sluvkin, I. I., Bondarenko, G. I., and Chernishov, V. P. (1992). Flow cytometric characterization of human decidual $CD8^+$ lymphocytes. *Am. J. Reprod. Immunol.* 27, 28.

Smith, S. K., and Kelly, R. W. (1988). The release of PGF2α and PGE2 from separated cells of human endometrium and decidua. *Prost. Leuk. Essen. Fatty Acids* 33, 91–96.

Sulila, P., Holmdahl, R., Hansson, I., Bernodotte, F., Mattson, A., and Mattson, R. (1988). An investigation of allogeneic pregnancy in multiparous mice subjected to *in vivo* depletion of CD8(Ly2)-positive lymphocytes by monoclonal antibody treatment. *J. Reprod. Immunol.* 14, 235–246.

Starkey, P. M. (1992). Natural killer cells/large granular lymphocytes in pregnancy. *In* "The Natural Immune System: The Natural Killer Cell" (C. E. Lewis and J. O. D. McGee, eds.), pp. 206–240, IRL Press, Cambridge (U.K.)

Starkey, P. M., Sargent, I. L., and Redman, C. W. G. (1988). Cell populations in human early decidua: characterization and isolation of large granular lymphocytes by flow cytometry. *Immunology* 65, 129–134.

Szerekes-Bartho, J., Kilar, F., Falkay, G., Csernus, V., Torok, A., and Pacsa, A. S. (1985). The mechanism of the inhibitory effect of progesterone on lymphocyte cytotoxicity. I. Progesterone treated lymphocytes release a substance inhibiting cytotoxicity and prostaglandin synthesis. *Am. J. Reprod. Immunol. Microbiol.* 9, 15–18.

Szerekes-Bartho, J., Varga, P., Kinsky, R., and Chaouat, G. (1990). Progesterone-mediated immunosuppression and the maintenance of pregnancy. FORUM 1990. *Res. Immunol. (Institut Pasteur)* 141, 175–181.

Tabibzadeh, S. (1991a). Ubiquitous expression of TNF-α/cachectins immunoreactivity in human endometrium. *Am. J. Reprod. Immunol.* 26, 1–4.

Tabibzadeh, S. (1991b). Distinct subsets of stromal cells confined to unique microenvironments in human endometrium throughout the menstrual cycle. *Am. J. Reprod. Immunol.* 26, 5–10.

Tachi, C., and Tachi, S. (1989). Role of macrophages in the maternal recognition of pregnancy. *J. Reprod. Fertil.* 37, 63–68.

Tagliabue, A., Befus, A. D., Clark, D. A., and Bienenstock, J. (1982). Characterization of natural killer cells in the murine intestinal epithelium and lamina propria. *J. Exp. Med.* 155, 1785–1796.

Tamada, M., McMaster, M. T., Flanders, K., Andrews, G. K., and Dey, S. K. (1990). Cell type specific expression of TGF-β1 in the mouse uterus during the peri-implantation period. *Mol. Endocrinol.* 4, 965–972.

Tarkatovsky, B., Goldstein, O., and Ben-Yair, E. (1991). *In vivo* modulation of pre-implantation embryonic development by cytokines. *In* "Cellular and Molecular Biology of the Materno-Fetal Relationship" (G. Chaouat and J. Mowbray, eds.), pp. 239–244. John Libbey Eurotext, Colloque INSERM, Paris, France.

Wood, G. W., Kamel, S., and Smith, K. (1988). Immunoregulation and prostaglandin production by mechanically-derived and enzyme-derived murine decidual cells. *J. Reprod. Immunol.* 13, 235–248.

Yeh, C. H. J., Bulmer, J. N., Hui, B. L., Tian, W. T., and Rittershaus, C. H. (1990). Monoclonal antibodies to T cell receptor γ/δ complex react with human endometrial glandular epithelium. *Placenta* 11, 253–261.

Yelavarathi, K. K., Chen, H. L., Yang, Y., Gowley, J. V., Fishback, B. D., and Hunt, J. S. (1991). Tumour necrosis factor α mRNA and protein in rat uterine and placental cells. *J. Immunol.* 146, 3840–3848.

12

Growth Factor and Cytokine Expression in the Endometrium

LIAM J. MURPHY & GUSTAVO BALLEJO

I. Introduction

Human endometrium is composed of glandular structures embedded in stromal tissue. Both components are responsive to steroid hormones and appear to be interdependent in terms of their function and their responsiveness to steroid hormone. During the normal menstrual cycle there is loss of the outer layer of the endometrium, with preservation of the basal layer containing the deep glandular epithelium. Stem cells located in this layer are able to be recruited and committed to differentiate into the various endometrial cell types in the presence of the appropriate hormonal stimulation so that the entire endometrium is able to regenerate (Padykula, 1991). In addition to proliferation and differentiation, there is neovascularization and invasion of the endometrium with a variety of blood-derived cells. Thus the endometrium is composed of a variety of cell types in addition to the glandular epithelial cells and stromal cells. These include vascular endothelial and smooth muscle cells, fibroblasts, a variety of lymphoid cells, polymorphonuclear leukocytes, mast cells, and macrophages (Bulmer *et al.*, 1991). Each of these cell types may synthesize growth factor and cytokines. In addition to cellular proliferation and differentiation, there is extensive remodeling of the extracellular matrix and neovascularization in the endometrium during the menstrual cycle. As will be discussed below, the changes in extracellular matrix can alter cellular responsiveness to growth factors and cytokines; in addition, extracellular matrix can affect cell phenotype by regulating gene expression.

Estrogen appears to be the most important hormone in terms of endometrial proliferation. Its effects are predominantly seen in the outer layer of the endometrium during the proliferative phase of the menstrual cycle. In the primate endometrium, DNA synthesis is most marked in the luminal and glandular epithelial cells but can also be demonstrated in the stromal cells (Padykula, 1991). Following ovulation, with rising concentrations of progesterone, DNA synthesis diminishes in the epithelial cells but persists in the stromal cells. In contrast, in the basal layers of the endometrium, the DNA labeling index in both epithelial and stromal cells increases following ovulation (Padykula, 1991). The latter observation suggests that the cells in the basal layer may escape progesterone inhibition and/or that under appropriate circumstances progesterone may be able to stimulate proliferation of epithelial cells. The effects of steroids on the proliferation of other cell types are not known. Estrogen receptors have been localized in uterine arteries, but there are no reports to date that this hormone can stimulate proliferation of vascular cells (Perrot-Applanat *et al.*, 1988). The effects of estrogen on the blood-derived cells and their functional role in the endometrium remain to be determined. These cells may produce cytokines and growth factors necessary in normal endometrial physiology. For example, conditioned medium from macrophages enhances endometrial stromal cell proliferation (Olive *et al.*, 1991).

The ovarian steroids also play an essential role in preparing the endometrium for blastocyst implantation. The response of endometrial glandular and stromal cells to steroids is difficult to demonstrate *in vitro* although some workers have developed isolation and culture techniques which preserve the steroid hormone responsiveness of endometrial cells (Surrey and Halme, 1991; Kirk *et al.*, 1978). However, these responses are modest compared with the effects seen *in vivo*. Thus, the proliferative responses of endometrial cells to steroid hormones are likely to involve complex interactions of cells with each other and with noncellular components. It has become increasingly apparent that at least some of the changes induced by steroid hormones are mediated by local factors such as cytokines and growth factors acting in a paracrine, juxtacrine, or even autocrine fashion (Murphy and Ghahary, 1990; Tabibzadeh, 1991). The list of potential growth factors, cytokines, and chemotactic agents identified in uterine tissue continues to grow and now includes a variety of cytokines in addition to the better-characterized growth factors (Table 1).

In this chapter we examine the potential mechanisms involved in cell–cell and cell–matrix interactions in the endometrium, and review the expression of cytokines and growth factors, as well as their receptors, which underlies these interactions. Also reviewed is the evidence indicating secretion of these cytokines and growth factors by endometrial cells *in vitro*. Finally, the potential biological roles of locally produced cytokines and growth factors in endometrium biology are discussed.

TABLE 1 Growth Factor and Cytokine Expression in the Endometrium

	Regulation by steroids	Cell type
Growth factor[a]		
IGF-I	+	Stromal cells macrophages
IGF-II	+	?
EGF	+	Epithelial cells
TGFα	+	Stromal/epithelial cells
TGFβ	+	Stromal/epithelial cells
bFGF	+	?
PDGF	?	Macrophages + others?
Cytokines		
CSF-1	+	Epithelial cells
TNFα	?	Macrophages + others?
INFγ	?	Lymphocytes + others?
INGα	?	?

[a]IGF-I, insulin-like growth factor-I; IGF-II, insulin-like growth factor-II; EGF, epidermal growth factor; TGFα, transforming growth factor-α; TGFβ, transforming growth factor-β; bFGF, basic fibroblast growth factor; PDGF, platelet-derived growth factor; CSF-1, colony-stimulating factor-1, TNFα, tumor necrosis factor-α; INF-γ, interferon-γ; INFα, interferon α.

II. Mechanisms of Cellular Interactions in the Endometrium

A. Autocrine, Paracrine, and Juxtacrine Growth Factors and Cytokines

Cell–cell and cell–extracellular matrix interactions appear to be very important both for proliferation and for the maintenance of differentiated function in a variety of tissues, including the endometrium. Cells may interact with other cells and with the noncellular components of their microenvironment through a variety of mechanisms which are listed in Table 2. The best-described mechanism involves diffusible growth regulators which interact with surface receptors in adjacent cells. Polypeptide growth factors which appear to function in this paracrine fashion have received considerable attention; however, other growth regulators such as the prostaglandins may function in a paracrine fashion. Data to support a paracrine role for insulin-like growth factors (IGFs), epidermal growth factor (EGF), transforming growth factor-α (TGFα) and colony-stimulating factor (CSF-1) in the uterus have been reported (Bartocci *et al.*, 1986; DiAugustine *et al.*, 1988; Han *et al.*, 1987; Murphy and Ghahary, 1990). Stromal–epithelial interaction may be also mediated through molecules anchored on the cell membrane. These include adhesion molecules, growth factors, and growth

TABLE 2 Mechanisms Involved in Cellular Interactions

Cell–Cell interactions
- Autocrine interactions
- Paracrine growth modulators
  - Peptide growth factors
  - Secreted proteases
  - Prostaglandins and other nonpeptide modulators
- Juxacrine interactions
  - Peptide growth factors
  - Proteases
  - Integrins and adhesion factors
  - Gap junctions
- Electrotonic coupling

Cell–matrix interactions
- Matrix-sequestrated growth modulators
  - Juxtacrine action
  - Release of sequestrated growth modulators
  - Secreted and membrane-bound proteases
  - Low pH in cellular microenvironment
- Matrix-sequestrated proteases
  - Activation of membrane-bound growth modulators
  - Activation of secreted growth factor precursors
- Effects of matrix on gene expression

factor receptors. The term "juxtacrine" has been used to describe this latter type of cell–cell communication (Massague, 1991). In addition to their other modes of action, both TGFα and EGF have been clearly demonstrated to have a juxtacrine function in some systems (Massague, 1991). Membrane-bound proteases may also activate growth modulators secreted by, or anchored to adjacent cells or extracellular matrix. Such a mechanism may be involved in processing EGF from the large-molecular-weight precursor present in the uterus (DiAugustine *et al.*, 1988). Additional mechanisms of cell–cell interaction involve electrotonic coupling in which ions and small molecules of less than 1000 Da are transferred between cells. This type of cell–cell interaction has been clearly documented in such organs as the ovary, adrenal, testes, and uterus (Garfield, *et al.*, 1977).

B. Extracellular Matrix as a Source of Growth Factors and Cytokines

Growth factors, proteases, and adhesion molecules may also be sequestrated in extracellular matrix. Both TGFβ and the heparin-binding growth factors have been shown to be closely associated with components of extracellular matrix (Massague, 1990; Baird and Ling, 1987). Matrix-bound growth modulators may interact directly with surface-bound growth factors or may be released by changes in pH in the cell microenvironment or by membrane-bound or secreted proteases. Some of these processes may be regulated by steroid hormones. Thus, extracellular matrix not only serves a mechanical support role but may also provide appropriate signals to allow for differentiation and/or proliferation.

The importance of cell–cell interaction for estrogen action has been clearly demonstrated in rodent tissue both in the uterus and in other hormone-responsive tissues. In the neonatal mouse luminal epithelium demonstrated estrogen responsiveness in the absence of demonstrable estrogen binding (Bigsby and Cunha, 1986). From these kinds of data, Bigsby and Cunha have proposed that the proliferative response of the epithelium is mediated through tropic factors produced by adjacent mesenchymal cells. A similar stromal–epithelial cell interaction has been observed in the mammary gland. Cultured normal rodent mammary epithelial cells are not estrogen responsive but when epithelial and stromal cells were cocultured, the epithelial cells proliferated in response to estrogen (Bigsby and Cunha, 1986). This appears to be dependent on cell–cell interaction since uterine epithelial cells from mice grown on complex matrix, such as basement membrane extract derived from Engelbreth-Holm-Swarm sarcoma, did not proliferate in response to estrogen (Ghosh *et al.*, 1991). Epithelial–stromal cell interaction represents only one of the many potential types of cell–cell interactions which can occur in the uterus. Interactions involving the same cell types are difficult to study but they are likely to be important. The interaction of stromal and epithelial cells in the uterus with blood-derived cells is also likely to be important in endometrial physiology.

In addition to the obvious morphological changes which occur in cells when grown *in vitro*, molecular tools demonstrate that the matrix on which cells are grown can dramatically affect the expression of a variety of genes. For example, both casein and transferrin receptor gene expression is extinguished when mammary epithelial cells are grown on plastic (Lee *et al.*, 1985). Furthermore, IGF-I gene expression may be extinguished in endometrial stromal cells cultured on plastic, as appears to be the case in hepatocytes (Johnson *et al.*, 1989; Murphy *et al.*, 1991).

III. Insulin-like Growth Factors

A. Expression in Uterine Tissue

Insulin-like growth factors, IGF-I and -II, were first identified as circulating mediators of growth hormone action. To date, two IGFs have been identified, IGF-I and II, along with their variants. As the name implies, both IGF-I and IGF-II are structurally and functionally similar to insulin (Rinderknecht and Humbel, 1978). However, unlike insulin, the IGFs are expressed at least at some stages of development, in virtually all tissues (Lund *et al.*, 1986; Murphy *et al.*, 1987a). With the availability of sensitive molecular probes, it was discovered that these growth factors were highly expressed in reproductive organs such as the ovary, testis, and uterus.

In most species so far examined, immunoreactive IGF-I and -II can be demonstrated in uterine luminal fluid or tissue extracts. The mRNAs for both IGF-I and -II have also been detected in uterine tissue of the rat, mouse, pig, cow, and in the human uterus (Ogasawara *et al.*, 1989; Geisert *et al.*, 1991; Hoppener *et al.*, 1988). A truncated form of IGF-I with enhanced bioactivity was originally isolated from porcine uterine tissue as uterine-derived growth factors (Ogasawara *et al.*, 1989). This IGF-I is missing the N-terminal tripeptide, glycine–proline–glutamine, and has enhanced bioactivity *in vitro* because of its lack of affinity for the insulin-like growth factor binding proteins (see Section III,D). The mechanisms involved and the physiological relevance of cleavage of IGF-I to the more active truncated form are not clear. Whether this process occurs *in vivo* is not known. If this process does occur *in vivo*, it may represent yet another step by which bioactivation of IGF-I can be regulated. A variant form of IGF-II has also been identified. This variant results from alternative splicing of the IGF-II mRNA and can be detected in the human reproductive tract (Yeh *et al.*, 1991).

While IGF-I is able to mimic many of the effects of growth hormone in a growth hormone-deficient animal model, it is now realized that this growth factor has many other functions in addition to its growth-promoting activities. IGF-II has no growth-promoting effect in the hypophysectomized rate model. In the rodent and a number of other species, but not man, circulating IGF-II levels are high in prenatal life and low in the postnatal period (Brown *et al.*, 1986). It has been suggested that IGF-II functions as a fetal growth factor in the rodent (Schoenle *et al.*, 1985). Supporting this hypothesis is the recent observation that IGF-II gene deletion using homologous recombination in mice results in retardation of intrauterine and perinatal growth (DeChiara *et al.*, 1990). The role of IGF-II in man remains controversial.

B. Steroid Hormone Regulation of Endometrial IGF-I Expression

In the rodent uterus, both IGF-I and IGF-II mRNAs increase following estrogen administration (Murphy *et al.*, 1988. The IGF-I response is more marked and more consistently observed, possibly because in the rodent expression of IGF-II declines rapidly after birth. Immunoreactive IGF-I can be extracted from rat uterine tissue, and the concentration of IGF-I increases after estrogen is administered (Murphy *et al.*, 1987b). This increase presumably reflects both increased blood-derived IGF-I and locally synthesized IGF-I. This estrogen action on uterine IGF-I is direct and does not require continuing protein synthesis (Murphy and Luo, 1989). Furthermore, it is pituitary-independent (Murphy and Luo, 1989) and appears to be restricted to a few select tissues. To date, estrogen-induced IGF-I expression has been demonstrated only in the ovary, bone, and uterus (Hernandez *et al.*, 1989; Ernst and Rodan, 1991). In most other tissues, the major regulator of IGF-I expression is growth hormone and in some tissues such as the liver, estrogen can antagonize induction of IGF-I expression by growth hormone (Murphy and Friesen, 1988). Interestingly, growth hormone can antagonize the estrogen response in the uterus and ovary (Murphy and Friesen, 1988; Hernandez *et al.*, 1989). Estrogen induction of IGF-I expression has also been demonstrated in the pig uterus (Simmen *et al.*, 1990). IGF-I expression may be estrogen dependent in human endometrium since secretion of IGF-I and IGF-II by fibroid explant cultures is reduced in women in whom ovulation has been suppressed by treatment with a gonadotropin-releasing hormone agonist (Rein *et al.*, 1990).

While the changes in uterine IGF-I expression following estrogen administration to ovariectomized or immature animals are easily demonstrated, the response to the physiological changes in circulating estradiol during the reproductive cycle are less dramatic (Murphy *et al.*, 1988). There are at least two possible explanations for this. First, in the rat, repeated injections of estrogen result in a gradual decrease in the maximal IGF-I mRNA response and a gradual rise in the basal levels of IGF-I mRNA detected immediately prior to each injection (Murphy and Friesen, 1989). Second, in the rat, as in the human, most of the IGF-I mRNA is located in the myometrium (Ghahary and Murphy, 1990), a layer of the uterus which shows little proliferative response to the cyclic variations in estrogen levels in the mature animal. In the bovine uterus, modest increases in both IGF-I and IGF-II concentrations were found in luminal washings from proestrous animals (Geisert *et al.*, 1991). This increase was not associated with any increase in IGF-I or IGF-II mRNA levels. These data suggest that much of the increase in luminal IGF levels may represent enhanced transport from the plasma. There is very little change in IGF-I levels in uterine luminal fluid from ewes throughout the reproductive cycle (Ko *et al.*, 1991). In endometrial biopsies from women who were having elective tubal ligation, there was no difference in amounts of IGF-I or IGF-II in samples collected in the luteal and the proliferative phase of the menstrual cycle (Murphy *et al.*, 1991). However, Boehm *et al* (1990) have reported that IGF-I transcripts are most abundant in human endometrium from the late proliferative stage of the menstrual cycle.

The effect of other steroid hormones on uterine IGF expression has not received as much attention as the estrogen-induced IGF-I response. Norstedt *et al.* (1989) have reported that progesterone can enhance the estrogen-induced increase in uterine IGF-I mRNA levels in the immature rat. Another group has reported no effect of progesterone on uterine IGF-I expression in hypophysectomized rats (Carlsson and Billig, 1991). During induction of pseudodecidualization in the rat, progesterone inhibits the effects of estrogen on uterine IGF-I expression (Croze *et al.*, 1990). Similarly in the pig, progesterone treatment of estrogen-primed immature gilts reduced uterine IGF-I mRNA levels while an increase was seen in mature ovariectomized gilts given progesterone within 4 days of castration (Simmen *et al.*, 1990). As discussed earlier, IGF-I mRNA levels may decline in human endometrium following ovulation, suggesting that progesterone may attenuate IGF-I expression in the human endometrium also (Boehm *et al.*, 1990).

C. IGF-II Expression in the Endometrium

The amount of uterine IGF-II mRNA also appears to be regulated by estrogen; however, the response to estrogen has not been observed by all groups (Murphy *et al.*, 1987b). Norstedt *et al.* (1989) failed to find an estrogen effect on IGF-II expression in the rat, and Simmen *et al.* (1990) reported that estrogen decreased IGF-II mRNA levels in porcine endometrium. IGF-II transcripts are most abundant in human endometrium obtained from the early proliferative phase of the menstrual cycle (Boehm *et al.*, 1990). Furthermore, IGF-II mRNA secretion by leiomyomata is lower in tissue obtained from women treated with gonadotropin-releasing hormone agonist than in fibroids obtained from untreated women (Rein *et al.*, 1990).

In both human and rodent uterus the greater part of the IGF-I and IGF-II mRNA is localized to the myometrial smooth muscle cell layer (Hoppener *et al.*, 1988; Murphy *et al*, 1988). In the endometrium, IGF-I mRNA is localized to the stromal cells, with little expression in the glandular or luminal epithelium (Ghahary and Murphy, 1990). In the rat it appears to be particularly abundant in the deeper layer or perimyometrial stromal layer, and in the periglandular stromal cells (Ghahary and Murphy, 1990). The localization of IGF-II mRNA in rodent uterine tissue has not as yet been reported; however, high levels of IGF-II transcripts are detectable in human leiomyomata and in normal human myometrial tissue (Hoppener *et al.*, 1988). As might be anticipated from the cellular localization of IGF-I expression in the endometrium, human endometrial adenocarcinoma cell lines do not appear to express IGF-I (Murphy *et al.*, 1991). In contrast, IGF-II expression is detectable in at least some human endometrial adenocarcinoma cell lines (Murphy *et al.*, 1991).

D. IGF Receptors and Binding Proteins in the Endometrium

Specific receptors for IGF-I and II have been demonstrated in uterine tissue. The affinity, capacity, and molecular mass of the IGF-I receptor in uterine tissue is in agreement with IGF receptors in other tissues (Ghahary and Murphy, 1989). Administration of estradiol increases the number of uterine IGF-I receptors in both immature and ovariectomized rats, with no significant effect on receptor affinity (Ghahary and Murphy, 1989). In addition, IGF-I receptors show a variation throughout the estrous cycle that is consistent with the changes in circulating estradiol. Autoradiographic studies in the rat demonstrate that uterine IGF-I receptors are most abundant in the myometrium; however, IGF-I receptors are also demonstrable in stromal and epithelial cells from the rat uterus (Ghahary and Murphy, 1989). IGF-I receptors are present in both human endometrium and myometrium (Reynolds *et al.*, 1990). Changes in endometrial IGF-I receptors throughout the menstrual cycle have yet to be documented but there are some data on the effects of steroid hormones on these receptors in primary cell cultures (Reynolds *et al.*, 1990). Estrogen has no effect on IGF-I receptor number or affinity in primary cultures of normal endometrial, stromal, and glandular tissue. In contrast, progesterone upregulated IGF-I receptors in endometrial glandular tissue (Reynolds *et al.*, 1990). Specific receptors for IGF-II are also present in rat uterine tissue but vary throughout the estrous cycle (Chandrasekhar *et al.*, 1991). There are as yet no reports on IGF-II receptors in human endometrial tissue.

An additional complexity of the IGF system is the presence of high-affinity binding proteins which can compete with the receptor for binding. They can both enhance and inhibit the action of IGF-I in various *in vitro* assays systems. So far six binding proteins have been identified and characterized. At least four distinct binding proteins (IGFBPs) are present in uterine extracts and rat luminal fluid. No consistent changes were observed in the relative abundance of IGFBPs during the estrous cycle when rat uterine extracts and luminal fluid were analyzed by ligand blotting (Murphy, 1991).

In terms of the uterus, the most intensely studied of the binding proteins is IGFBP-1. This binding protein was originally purified from amniotic fluid and placenta (Drop *et al.*, 1979; Koistinen *et al.*, 1986). In the rat, immunohistochemistry and *in situ* hybridization localize IGFBP-1 immunoreactivity and IGFBP-1 mRNA to the uterine luminal epithelium and the stromal glandular tissue (Croze *et al.*, 1990). In the primate endometrium, IGFBP-1 is predominantly localized to the stromal cells (Fazleabas *et al.*, 1989). In the immature rat, estrogen administration results in a marked decrease in amounts of uterine IGFBP-1 mRNA (Murphy, 1991). The changes in IGFBP-1 levels during the estrous cycle are the inverse of the changes seen in uterine IGF-I receptors. The levels of IGFBP-1 mRNA are highest in diestrus and lowest in proestrus. If IGFBP-1 actually inhibits IGF-I action in the uterine tissue, then the attenuation of expression of this binding protein may increase the bioavailability and functional activity of uterine IGF-I.

Progesterone also appears to be important in induction of IGFBP-1

TABLE 3 Potential Sites of Regulation of the IGF Action in the Endometrium

1. Expression and synthesis of the IGFs
2. Proteolytic activation of IGFs to decrease binding protein affinity
3. Expression and synthesis of the IGF receptors
4. Expression and synthesis of the IGFBPs
5. Proteolytic degradation of the IGFBPs
6. Storage and release of IGFs from extracellular matrix
7. Postreceptor interactions through activation of *transacting* factors

expression in the rat uterus since there is a dramatic increase in IGFBP-1 mRNA when decidualization is induced with a combination of estrogen and progesterone (Croze *et al.*, 1990). In primates, progesterone appears to be the major regulator of IGFBP-1 expression. In human endometrial tissue, IGFBP-1 levels are highest in the follicular phase of the menstrual cycle (Waites *et al.*, 1988).

Plasmin is able to dissociate IGFs from their binding proteins (Campbell *et al.*, 1992). As discussed in the next section, both plasminogen activators and inhibitors are present in uterine tissue, and the activity of these competing systems is modulated by TGFβ and by steroid hormones. Since some of the IGFBPs have arginine–glycine–aspartic acid recognition sequence for binding to extracellular matrix proteins, it has been postulated that extracellular matrix may also serve as a storage site for the IGFs (Murphy *et al.*, 1990a; Linkhart and Mohan, 1989). It is of interest that a major portion of the IGF-I extracted from uterine tissue is truncated (Ogasawara *et al.*, 1989). It is possible that activation of the proteolytic enzyme or enzymes responsible for this process is also regulated by steroid hormones and this is yet another potential site where steroid hormones could regulate the uterine IGF system (Table 3).

IV. The Epidermal Growth Factor Family

A. Epidermal Growth Factor

Although EGF was first identified in mouse submaxillary gland extract as an agent which was able to induce premature opening of the eyelids in neonatal mice (Carpenter, 1981), it has many diverse functions and is expressed in a number of tissues. Immunoreactive EGF is present in mouse uterine luminal fluid and the concentration increased following administration of estrogen to immature and ovariectomized mice (DiAugustine *et al.*, 1988). This effect takes a number of days to become fully manifest, is of relatively small magnitude (two- to fivefold), and is accompanied by an increase in EGF mRNA.

EGF is synthesized as a large precursor which has a sequence homology with a diverse variety of membrane-anchored molecules. Interestingly, the membrane-anchored pro-EGF is biologically active and thus can function in a juxtacrine fashion in some systems. Mature EGF is derived from the precursor by proteolytic cleavage, which appears to be regulated by estrogens in the murine uterus (DiAugustine *et al.*, 1988).

Both EGF and its mRNA have been detected in human endometrium, and the levels of EGF mRNA appear to be similar in secretory and proliferative phase endometrial biopsies (Murphy *et al.*, 1991; Haining *et al.*, 1991). A recent immunohistochemical study showed that the EGF protein is predominantly localized in the stromal cells and that its levels increase after ovulation (Hoffmann *et al.*, 1991). These data suggest that progesterone rather than estrogen may have a role in regulating expression of EGF in human endometrium, which is consistent with the regulation of EGF expression in human breast cancer tissue (Murphy *et al.*, 1990b). A functional role for EGF in estrogen-induced uterine proliferation has been clearly demonstrated in ovariectomized immature mice. In this animal model, EGF implants alone are sufficient to cause histological changes in the uterus that are indistinguishable from those induced by estrogen itself (Nelson *et al.*, 1991).

Epidermal growth factor is also expressed by some human endometrial adenocarcinoma biopsy samples and cell lines (Anzai *et al.*, 1992; Leake *et al.*, 1991). Its regulation in endometrial adenocarcinoma cells has not yet been reported. EGF is able to stimulate proliferation of epithelial cells from both murine and normal human uterine tissue (Leake *et al.*, 1991; Tomooka *et al.*, 1986), and human endometrial cancer cells (Anzai *et al.*, 1992). Of more interest is the observation that EGF has other nonmitogenic effects in uterine tissue. It is able to suppress production of prolactin from human decidual cells (Yamamoto *et al.*, 1989) and to stimulate uterine smooth muscle contraction (Gardner *et al.*, 1987). EGF is angiogenic in the hamster cheek pouch assay system (Schreiber *et al.*, 1986), and in human endometrium is highly expressed in stromal cells surrounding the spiral arterioles (Hoffmann *et al.*, 1991). These observations suggest that it may play a role in angiogenic processes in the endometrium.

B. Epidermal Growth Factor Receptors

The receptor for EGF is a 175-kDa transmembrane protein which is the homologue of the *v erb B* oncogene (Downward *et al.*, 1984). Classical radioligand binding studies have demonstrated the presence of the EGF receptor in rat and human endometrium (Mukku and Stancel, 1985; Sheets *et al.*, 1985). Both glandular epithelium and stromal cells show intense staining with monoclonal antibodies to the EGF receptor (Berchuck *et al.*, 1989a). Binding of radiolabeled EGF to uterine membranes is upregulated some two- to threefold following administration of estradiol, and the changes in uterine EGF receptors parallel the changes in estradiol during the estrous cycle in the rat (Mukku and Stancel, 1985; Gardner *et al.*, 1989). The effects of estrogen on EGF receptors in human endometrium are controversial. Two groups have reported higher EGF receptor numbers during the proliferative phase than in the secretory phase of the human menstrual cycle (Taketani and Mizuno, 1989; Troche *et al.*, 1991). However, others have reported no differences in EGF binding (Sheets *et al.*, 1985) or in staining when immunohistochemistry was used to examine EGF receptor content of endometrial biopsies from different phases of the menstrual cycle (Berchuck *et al.*, 1989a). *In vitro* estradiol decreased EGF binding in human endometrial glandular cells but not in stromal cells (Reynolds *et al.*, 1990). In the Ishikawa endometrial adenocarcinoma cell line, estradiol decreased binding, whereas tamoxifen enhanced EGF binding (Gong *et al.*, 1992). In human endometrial stromal cells, progesterone rather than estrogen enhanced EGF receptor expression (Reynolds *et al.*, 1990; Taketani and Mizuno, 1989). It should be noted that in mammary adenocarcinoma cells, progesterone rather than estrogen is the major regulator of EGF receptors (Murphy *et al.*, 1986). Approximately 30% of endometrial adenocarcinomas express high levels of EGF receptors; however, there is no correlation between EGF receptor expression and histological grade, depth of invasion, extrauterine metastasis, or recurrent disease (Berchuck *et al.*, 1989b).

C. Transforming Growth Factor-α

Transforming growth factor-α was first identified as one of the factors present in conditioned media from cancer cells which had the ability to confer several biological properties indicative of malignant transformation onto nontransformed fibroblasts. These properties included anchorage-independent growth in soft agar, a reduced serum requirement, and a loss of density-dependent growth inhibition. TGFα is a 50-amino-acid, single-chain polypeptide that is structurally very similar to epidermal growth factor. TGFα interacts with the EGF receptor and has actions that are qualitatively similar to EGF in most tissues.

Since it now appears that TGFα interacts with a slightly different region of the EGF receptor than EGF itself, it is not surprising that in some cell types the EGF receptor has a higher affinity for TGFα than EGF, and that TGFα is even more potent than EGF in stimulating cell proliferation (Lax *et al.*, 1988; Winkler *et al.*, 1989). TGFα is synthesized as a larger precursor and the TGFα precursor shares sequence similarities with a family of integral membrane proteins with very diverse functions (Derynck, 1988; Appella *et al.*, 1988). The precursor can be converted to the mature 5.6-kDa TGFα by proteolytic digestion. The mature TGFα sequence is flanked by

Ala–Val dipeptides that are thought to serve as a proteolytic cleavage site for an elastase-like protease (Salomon *et al.*, 1990). Cleavage of this membrane precursor is regulated and can be stimulated by agents which activate protein kinase C (Pandiella and Massague, 1991). It is not clear whether this process is also regulated by steroid hormones. Furthermore, the tissue distribution and cellular and subcellular locations of the proteolytic activating enzyme for TGFα are not known. Since prepro-TGFα is present on the cell surface, this enzyme could be membrane bound or secreted. A membrane-bound activating enzyme located on cell types different from the TGFα precursor could mediate cell–cell interactions. In addition, TGFα precursors associated with the cell membrane are able to interact with the EGF receptors on adjacent cells in a juxtacrine fashion. Under most circumstances TGFα stimulates cellular proliferation, but like EGF, under certain circumstances high concentrations of TGFα can inhibit growth in certain cell lines (Salomon *et al.*, 1990).

The physiological role of TGFα in the uterus has not been determined. It is expressed in normal human endometrium and there appears to be no marked difference in the abundance of its mRNA in the different stages of the menstrual cycle (Murphy *et al.*, 1991). However, these data are derived from rather insensitive techniques and would not exclude changes in TGFα mRNA levels in some cell populations at different stages of the cycle. Furthermore, there are a number of post-translational steps in the synthesis of bioactive TGFα that are potential sites of regulation by steroid hormones. Preliminary immunohistochemical data indicate that TGFα is localized in stromal cells during the proliferative phase, and in both surface and glandular epithelial cells during the secretory phase of the menstrual cycle (Horowitz *et al.*, 1992). As discussed in Section IV,B, EGF receptors are present on both epithelial and stromal cells in human endometrium. Endometrial adenocarcinoma cells express TGFα, which stimulates growth in these cell lines (Gong *et al.*, 1992). In Ishikawa cells, TGFα expression is upregulated by estrogen (Gong *et al.*, 1992). Furthermore, antiserum to the EGF receptor is able to block proliferation of Ishikawa cells, which suggests that in this cell line either TGFα, EGF, or some other member of the EGF family functions in an autocrine manner (Murphy *et al.*, 1992).

D. Other Members of the Epidermal Growth Factor Family

Recently a heparin binding growth factor secreted by macrophages has been shown to share sequence similarity with members of the EGF family (Higashiyama *et al.*, 1991). This new HB-EGF also binds to the EGF receptor. It appears to have a greater affinity for EGF receptors on smooth muscle cells than other cell types, and is a potent mitogen for these cells (Higashiyama *et al.*, 1991). The expression of this new growth factor in the endometrium has not been reported. It remains to be determined whether HB-EGF is produced by endometrial macrophages and acts in a paracrine fashion on myometrial or vascular smooth muscle cells.

V. The Transforming Growth Factor-β Gene Family

A. Biosynthesis of the Transforming Growth Factor-β

In addition to TGFα, a second transforming activity can be identified in the conditioned medium of a variety of cancer cell lines. The protein responsible for this second type of bioactivity can be clearly distinguished from TGFα since it has little transforming activity itself but rather potentiates the anchorage-independent, growth-promoting activity of TGFα and does not depend on interaction with the EGF receptor (Wakefield *et al.*, 1990). The TGFβs, a group of homodimeric proteins of approximately 25,000 Da, are responsible for this second type of transforming activity. To date, five different structurally related TGFβ subtypes have been identified. TGFβ subtypes 1, 2, and 3 are expressed in mammalian tissues, while subtypes 4 and 5 have been identified in other species (Wakefield *et al.*, 1990). The TGFβs are part of a much larger supergene family which has diverse effects on morphogenesis, proliferation, and differentiation. Other members of this gene family include Müllerian-inhibiting substance, inhibin, activin, and the pentaplegic complex-C gene of *Drosophila* (Roberts and Sporn, 1988; Wakefield *et al.*, 1990). Like other members of this gene family, the functionally active protein involves dimerization of two monomers of TGFβ. The TGFβs expressed in mammalian tissues share between 70 and 80% amino acid identity, with the N-terminal region being the most divergent (Derynck *et al.*, 1988). Although the TGFβs have both stimulatory and inhibitory actions, depending on the cell line and culture conditions used, the different TGFβs appear to have qualitatively similar activities. The different TGFβs can form functionally active heterodimers and can compete for the same receptor binding sites (Graycar *et al.*, 1989). The TGFβs do not interact with the EGF receptor but rather with specific membrane receptors. A series of TGFβ binding proteins have been identified by affinity cross-linking studies (see Section V,B).

TGFβ_1, the most ubiquitious TGFβ, is expressed in normal human endometrium and endometrial cancer cells (Murphy *et al.*, 1991; Gong *et al.*, 1992). It is derived from a larger precursor molecule by proteolytic digestion. This precursor complex, which is biologically inactive, consists of the homodimeric mature TGFβ_1 cleaved from propeptide but still noncovalently attached to the propeptide. In addition, the propeptide portion is joined by disulfide bonds to a structurally unrelated 125- to 160-kDa glycoprotein (Lawrence *et al.*, 1984; Wakefield, *et al.*, 1988). Interestingly, this glycoprotein contains multiple EGF-like repeats (Kanzaki *et al.*, 1989). It does not appear to bind activated $TGF\beta_1$ nor does it inhibit binding of activated $TGF\beta_1$ to its receptor (Massague, 1990). $TGF\beta_1$ can be activated by low pH and a variety of nonphysiological means.

Plasmin is able to activate $TGF\beta_1$ *in vitro,* and plasmin or an enzyme with similar specificity may be important in the activation of this growth factor *in vivo.* In both the rodent and human endometrium, plasminogen activator is upregulated by estradiol (Peltz *et al.*, 1983; Casslen *et al.*, 1986). Both urokinase (UPA) and tissue plasminogen activator (TPA) are present in the human endometrium. Endometrial and luminal fluid PA activity increases during the proliferative phase, reaching a maximum at midcycle and subsequently decreasing during the luteal phase (Casslen and Ohlsson, 1981). Inhibitors of UPA and TPA can also be extracted from human endometrium, although the exact nature and regulation of these inhibitors remains controversial (Littlefield, 1991). Interestingly, the expression of type 1 plasminogen activator inhibitor (PAI-1) is enhanced by TGFβ in both cell cultures and *in vivo* in a variety of different cell types and tissues (Sawdey and Loskutoff, 1991). A variety of other proteins expressed in the endometrium can also activate latent TGFβ. These include cathepsin D, alkaline phosphatase, and interferon-γ (Miyazono *et al.*, 1990). Thus, there are many potential levels of regulation of TGFβ in the uterus.

$TGF\beta_1$ is expressed in both epithelial and stromal cells in the human endometrium. There appears to be little variation in the amounts of its mRNA throughout the menstrual cycle (Murphy *et al.*, 1991), although a more detailed analysis with *in situ* hybridization is required to determine whether TGFβ expression in certain cell types varies throughout the menstrual cycle. Rossi *et al.* (1992) have presented preliminary findings from a study in which they used immunohistochemistry to demonstrate the presence of $TGF\beta_1$ and $TGF\beta_2$ protein in luminal and glandular epithelial cells of the human endometrium. Graham and colleagues (1992) also used immunohistochemistry to examine the localization of $TGF\beta_2$ in human decidua. Intense labeling was observed in the extracellular matrix of first trimester decidua and in the cytoplasm of decidual cells in biopsies from the third trimester. Unfortunately the immunohistochemical approach cannot distinguish between active and latent forms of TGFβ, and the decidual extracellular matrix may serve as a storage site for inactive TGFβ precursor. $TGF\beta_1$ mRNA has also been demonstrated in murine endometrial epithelium during the preimplantation period (Akhurst *et al.*, 1990) and some 1–4 days later the mature TGFβ protein can be detected in the adjacent stroma (Tamada *et al.*, 1990). A recent report suggests that $TGF\beta_2$ is present in mouse amniotic fluid and may function as a local immunosuppressant, preventing rejection of maternal–fetal allografts (Altman *et al.*, 1990).

B. Transforming Growth Factor-β Receptors

The TGFβs do not interact with the EGF receptor but with specific membrane receptors. At least three groups of high-affinity cell membrane proteins have been identified by affinity cross-linking studies employing iodinated TGFβ as a ligand. These proteins, termed type I, II, and III TGFβ receptors, appear to have distinct biological functions. The type III receptor, also known as betaglycan, is present in cells which do not appear to respond to TGFβ in any way (Massague, 1990). Thus, it has been proposed that the type III receptor is not involved directly in signal transduction but may concentrate TGFβ and facilitate its interaction with type I and II receptors. Both the type III and type II receptors have recently been cloned (Lopez-Casillas *et al.*, 1991; Wang *et al.*, 1991; Lin *et al.*, 1992a). The type III receptor gene encodes a transmembrane protein with a short cytoplasmic domain which has no similarity to any other receptor cytoplasmic domain (Wang *et al.*, 1991). The exact role of the type III receptor is unclear. In contrast, the cytoplasmic domain of the type II receptor has sequence similarity with serine/theonine kinases, suggesting that this activity may be involved in TGFβ signal transduction. The receptor for activin, a closely related protein, has also been cloned and it too appears to be a transmembrane serine kinase (Mathews and Vale, 1991). As yet there have been no reports of TGFβ receptors in human endometrium; however, receptors for this growth factor appear to be ubiquitous and the demonstration that TGFβ can exert biological actions in endometrial cells suggests that receptors will be found in this tissue.

C. Biological Actions of the Transforming Growth Factor-βs

The TGFβs are usually inhibitory to normal epithelial cells while malignant transformation of epithelial cells may be associated with a reduced sensitivity to the inhibitory effects of this agent (Hoosein *et al.*, 1989; Hubbs *et al.*, 1989). Furthermore, resistance to the growth inhibitory effects of TGFβ_1 appears to increase with malignant progression (Hubbs *et al.*, 1989). In addition, in a small number of cancer cell lines, TGFβ_1 actually stimulates cell proliferation (Massague, 1990). For example, the growth of the Ishikawa human endometrial adenocarcinoma cell line, which is only moderately tumorigenic in nude mice, is inhibited by TGFβ, whereas the growth of the HEC-50 endometrial adenocarcinoma cell line, which is highly tumorigenic, is stimulated by TGFβ (Anzai *et al.*, 1992). Each of these two endometrial cell lines expresses TGFβ and it is possible to demonstrate autocrine feedback loops in each cell type using antibodies to TGFβ and antisense oligonucleotides (Murphy *et al.*, 1992). Although expression of TGFβ in normal human endometrium appears to be similar in the follicular and the luteal phase of the menstrual cycle, it is possible to demonstrate regulation by steroid hormones in a variety of cell lines, including human endometrial adenocarcinoma cells (Gong *et al.*, 1991). Estradiol reduced TGFβ_1 expression in some but not all endometrial cancer cell lines, whereas progestins generally enhanced TGFβ_1 expression in these lines (Gong *et al.*, 1991).

The inhibitory effect of TGFβ on cell proliferation appears to involve more complex mechanisms than simple inhibition of mitogenesis induced by other growth factors. In human hepatoma cells, TGFβ_1 induces apoptosis. In this cell line, the cytocidal effects of TGFβ_1 involve activation of a preformed endogenous nuclease, which results in fragmentation of the cellular DNA (Lin *et al.*, 1992b). A similar mechanism may be involved in the shedding of the outer layers of the endometrium, which occurs following involution of the corpus luteum each menstrual cycle.

TGFβ has a variety of other actions which may be relevant to the physiological changes that occur in the endometrium throughout the menstrual cycle and during implantation and decidualization. It has a variety of effects on extracellular matrix. It enhances expression of fibronectin, collagen, proteoglycans, tenascin, osteonectin, osteopontin, and thrombospondin, thus favoring the production of extracellular matrix (Massague, 1990). In addition, TGF-β inhibits the degradation of extracellular matrix through upregulation of pAI-1 and tissue metalloproteinase inhibitors (TIMPs), and downregulation of collagenase, transin/stromelysin, and plasminogen activator. The effects of TGFβ on PAI-1, a plasmin inhibitor and plasminogen activator, are particularly important since plasmin appears to be able to activate TGFβ. Thus the expression of TGFβ may be regulated by feedback loops. As discussed above, both plasminogen activators and their inhibitors are expressed in the human endometrium, and expression is regulated either directly or indirectly by steroid hormones (Casslen and Ohlsson, 1981). Steroid hormone could potentially regulate TGFβ biosynthesis and action at a number of levels.

TGFβ upregulates a variety of receptors in some cell lines which are involved in cell–matrix and cell–cell adhesion. These include integrin receptors for collagen, fibronectin, laminin, and fibronectin (Ignotz and Massague, 1987). These receptors are important in mediating cell–cell and cell–extracellular matrix interactions. Clearly, our understanding of the physiological functions of the TGFβs in the endometrium is far from complete.

VI. Platelet-Derived Growth Factors

A. Structure

Platelet-derived growth factors (PDGFs) are potent mitogens for mesenchymal cells and were first isolated from blood platelets (Raines *et al.*,

1990). They represent a family of dimeric proteins composed of varying combinations of two polypeptides held together by disulfide bonds. The A and B polypeptide chains are encoded by different genes located on human chromosomes 7 and 22 and having similar exon–intron organization (Westermark and Heldin, 1991). Three dimer forms—AA, AB, and BB—have been identified and purified from platelets and transformed cells.

The PDGF-A chain gene is transcribed into three mRNA molecules (1.9, 2.3, and 2.9 kb). The PDGF-B chain gene is the cellular homologue of the *v-sis* oncogene and is transcribed predominantly into a 3.8-kb mRNA molecule and into a less abundant 1-kb mRNA transcript. The primary translation products of the A and B chain mRNAs are 23- and 28-kDa proteins, respectively, which after dimer formation are processed at the amino and carboxy terminal ends to give rise to the mature secreted 30-kDa PDGFs (Westermark and Heldin, 1991).

Boehm and colleagues (1990) have demonstrated the presence of PDGF-B chain mRNA transcripts in human endometrial tissue from the late proliferative stage of the menstrual cycle. The relative amounts of PDGF-B chain transcript were similar in endometrium and myometrium RNA samples. In this study, PDGF-A chain mRNA was barely detectable in uterine tissue or not at all (Boehm *et al.*, 1990). However, another group employing a more sensitive nuclease protection assay did demonstrate both PDGF-A chain and PDGF-B chain transcripts in RNA prepared from human myometrium obtained from nonpregnant and pregnant women. The levels of PDGF-A chain transcripts increased toward the end of pregnancy while no changes were observed in the relative amounts of the PDGF-B chain transcript (Boehm *et al.*, 1990). It is still not clear which cell types are responsible for PDGF expression in the endometrium. However, PDGF is secreted by macrophages (Shimokado *et al.*, 1985). Immunocytochemical analysis of pregnant and nonpregnant uterine samples with anti-PDGF antiserum revealed intense staining only in the myometrial cells from the pregnant uterus (Mendoza *et al.*, 1990). In addition, there is an increase in the levels of PDGF-A chain transcripts observed as pregnancy progresses, suggesting that PDGF may be involved in the production of myometrium hyperplasia and the hypertrophy occurring during pregnancy (Mendoza *et al.*, 1990).

B. Receptors in the Endometrium

PDGFs interact with two different cell membrane receptors named α and β receptors. Both receptors are transmembrane proteins with intrinsic tyrosine kinase activity (Raines *et al.*, 1990). Interestingly, although all three dimeric isoforms of PDGF are biologically active, only PDGF BB can bind with high affinity to the PDGFβ receptor. There are as yet little data on PDGF receptors in the endometrium. However, human endometrial stromal cells are responsive to PDGF *in vitro*. Surrey and Halme (1991) and

Mendoza *et al.* (1990) have demonstrated the presence of PDGF-β type receptor transcripts in myometrial samples. Thus, the potential exists for PDGF paracrine interactions within the endometrium between macrophages and stromal cells, and between endometrial and myometrial cells.

VII. The Fibroblast Growth Factor Family

A. Structure and Receptors

The fibroblast growth factor (FGF) family consists of at least seven related gene products which influence the growth and function of a variety of cell types (Goldfarb, 1990). Acidic and basic FGF, the prototypes of the family, are monomeric proteins which have been isolated from various tissues. The FGF-related proteins int-2, hst/k, and FGF-5 were first identified by sequence analysis of genes originally isolated as oncogenes. FGF-6 cDNA was found by low stringency hybridization of a cDNA library with hst probes. Another member [FGF 7, keratinocyte growth factor (KGF)] of this family was identified after purification from conditioned medium of human embryonic fibroblasts and is a potent mitogen for epithelial cells. Interestingly, acidic and basic FGF are synthesized without the hydrophobic amino-terminal signal sequence that characterizes secretory proteins. Variant forms of bFGF have been described which differ in amino-terminal sequence as a result of differences in translational start sites (Finch *et al.*, 1989).

FGFs can act as chemoattractants, mitogens, angiogenic factors, and inducer molecules in the embryo. The diverse biological effects of the FGFs may be explained in part by the complexity of the receptors for these growth factors. High-affinity cellular receptors for FGFs belong to the transmembrane tyrosine-kinase-like receptor family as do the receptors for CSF-1, PDGF, and vascular permeability factor (Hou *et al.*, 1991). Four human genes have been identified that encode distinct high-affinity receptors. Each of these encodes multiple proteins derived from alternative mRNA splicing. For example, analysis of the nucleotide sequence of different cDNAs isolated from hepatoma cells encoding for FGF receptors indicates that three amino terminal domain motifs, two juxtamembrane motifs and two intracellular C-terminal domains combine to form a minimum of 6 and potentially 12 iso receptors (Hou *et al.*, 1991). Specific iso receptors may mediate a subset of the biological effects elicited by FGFs. In addition they may differ in their affinity for the various members of the FGF family.

B. Expression in the Endometrium

Human, porcine, and bovine uterine extracts contain acidic and basic FGF-like proteins (Brigstock *et al.*, 1989; Milner *et al.*, 1989). Human endometrial carcinoma cell lines HEC-1A, HEC-1B, and AN3CA appear to produce bFGF, as indicated by Western blot and immunoprecipitation studies (Presta, 1988). In all three cell lines estradiol-17β stimulated bFGF synthesis. Progesterone is able to inhibit the effect of estrogen on bFGF production by endometrial adenocarcinoma cells (Presta, 1988).

One aspect of FGF's action which is important when considering its potential effects in the endometrium is that cell surface heparan sulfate is needed for FGF to activate the FGF receptor (Rapraeger *et al.*, 1991). FGF binds with low affinity to cell surface and extracellular heparan sulfate proteoglycans. It has been shown that the endometrium basal lamina contains high amounts of proteoglycans (Aplin *et al.*, 1988). Whether endometrial proteoglycans contribute to FGF stabilization and storage also remains to be explored. However, in other systems, bFGF has been found in association with extracellular matrix (Baird and Ling, 1987). Heparanase and other extracellular matrix-degrading enzymes may modulate the bioavailability of stored FGFs and thus play an important physiological role in cellular invasion and vascularization (Tryggvason *et al.*, 1987). As discussed in Section V, these extracellular matrix-degrading enzymes are present in the endometrium and are modulated either directly or indirectly by steroid hormones and TGFβ. Clearly, these enzymes have important roles in decidualization, neovascularization, and embryo implantation, and in addition to remodeling extracellular matrix, they may liberate growth factors such as the FGFs and the TGFβs.

VIII. Hepatocyte Growth Factor

Hepatocyte growth factor (HGF) was originally isolated as the factor responsible for proliferation of hepatocytes after partial hepatectomy. It was initially thought to be liver specific but has now been shown to be a potent mitogen for a variety of epithelial and endothelial cells, and to be expressed in other tissues such as the kidney, in addition to the liver. HGF is the most potent mitogen for mature parechymal hepatocytes in primary culture so far described. It induces tyrosine phosphorylation of a 145-kDa membrane protein that has recently been shown to correspond to the β-subunit of the c-met proto-oncogene product (Battaro *et al.*, 1991). The presence of immunoreactive c-met protein has been demonstrated in superficial and deep glandular epithelium from the normal human endometrium as well as in endometrial carcinomas (Prat *et al.*, 1991).

IX. Angiogenic Factors

Angiogenesis is an important component of the changes which occur in the endometrium each menstrual cycle, and is an important part of decidualization and implantation. It is a complex, multistep process involving both migration and proliferation. As discussed earlier, EGF, TGFs, and the FGFs have angiogenic activity in various bioassays. More recently, a variety of other growth factors and cytokines have been shown to have very specific effects on vascular tissue, although their role in the endometrium still requires investigation.

A. Endothelin-1

Endothelin-1 (ET-1) is a growth factor for both fibroblasts and vascular smooth muscle cells (Takuwa *et al.*, 1989; Komuro *et al.*, 1988). PreproET-1 mRNA and endothelin immunoreactivity have been detected in human endometrium (Economos e*et al.*, 1992a; Salamonsen *et al.*, 1992). ET-1 is secreted by human endometrial stromal cells and glandular epithelium. In HEC-1A cells, expression is upregulated by TGFβ (Economos *et al.*, 1992b). A variety of tumor cells produce ET-1 but appear to lack receptors for this growth factor. The secretion of ET-1 by tumor cells lacking ET-1 receptors is thought to represent a paracrine mechanism by which the tumor promotes growth of contiguous mesenchymal cells and thus stimulates vascularization.

B. Vascular Permeability Factor

Vascular permeability factor (VPF) is a 40- to 45-kDa disulfide-linked, homodimeric, heparin-binding glycoprotein that promotes endothelial cell growth; it was originally isolated from conditioned medium of a variety of cell lines (Keck *et al.*, 1989). More recently, a cDNA encoding a closely related protein has been isolated from a human placental library (Maglione *et al.*, 1991). The VPFs both enhance vascular permeability and stimulate angiogenesis. They share homology with regions of the PDGF molecules. In addition, as the name suggests, VPF enhances blood vessel permeability (Keck *et al.*, 1989; Leung *et al.*, 1989). Four molecular species of VPF have been described which are generated from a single gene by alternative mRNA splicing (Houck *et al.*, 1991). Each molecular form differs in secretion pattern and biological activity. De Vrieo and colleagues (1992) have reported that VPF receptor is a transmembrane tyrosine kinase named *Flt* that is related to the *c-Fms* proto-oncogene product.

The expression of VPF and its receptor has not been examined in the endometrium. Phillips and colleagues (1990) have reported that VPF is expressed in the rat ovary and that expression is under steroid hormone control. This information suggests that similar studies should be undertaken in the endometrium. Considering the biological actions of VPF, it appears to be a good candidate for mediating the increase in vascular permeability and blood vessel growth induced by estrogens in the uterus.

X. Cytokine Expression in the Endometrium

One of the most intriguing phenomena which may be relevant to cell–cell interaction is the infiltration of the uterine stromal tissue with blood-derived cells following estrogen administration. It has been known for more than three decades that one of the markers of estrogen action in the rodent uterus is the infiltration of the stroma with eosinophils (Tchernitchin, 1979). This response is seen within a few hours after a single injection of estrogen in an immature rat. Estrogen appears to enhance uterine expression of a specific eosinophil chemotactic factor (Lee *et al.*, 1989). The exact nature of this agent and the mechanisms by which estrogen induces this protein are not clear. The phenomenon is less well marked in other species, but infiltration of uterine tissue with a variety of other blood-derived cells, such as leukocytes, mast cells, and macrophages, has been described in many species, including man. Since these blood-derived cells produce a variety of growth factors and cytokines, including PDGFs, IGFs, FGFs, TGFs, and colony-stimulating factors, an appealing hypothesis is that these cells are in some way intimately involved in the normal response of the uterus to steroid hormones. Glucocorticoids which block uterine eosinophil infiltration do not inhibit estrogen induced mitogenesis in this tissue; however, appropriate experiments to address the role of other blood-derived cells in the estrogen-induced uterine proliferative response have yet to be reported (Tchernitchin *et al.*, 1975). Macrophages produce a factor or factors which enhance stromal cell proliferation (Olive *et al.*, 1991). A variety of cytokines originally isolated because of their effects on hemopoietic and lymphoid cells have now been shown to be expressed in the uterus. The role of these cytokines has been extensively reviewed by Tabibzadeh (1991) and is discussed only briefly in the next section.

A. Colony-Stimulating Factor-1

Colony-stimulating factor-1 is a homodimeric glycoprotein of 90 kDa. It is also known as macrophage colony-stimulating factor and is expressed in both the rodent and human uterus. CSF-1 is localized predominantly to the glandular epithelium in both the mouse and human endometrium (Pollard *et al.*, 1987; Daiter *et al.*, 1992). It interacts with a specific 165-kDa receptor which has tyrosine kinase functions and is identical to the c-fms protooncogene product (Sherr *et al.*, 1985). CSF-1 expression is regulated by both estradiol and progesterone (Pollard *et al.*, 1987). The levels of CSF-1 increase in the uterus and in a variety of other tissues throughout pregnancy (Sherr, 1988). The receptor for CSF-1 is also expressed in the pregnant uterus and shows a complex pattern of expression (Pollard *et al.*, 1991). CSF-1 receptor expression can be demonstrated in the preimplantation embryo, and after implantation, the receptor is also found in the maternal decidual cells and in the trophectoderm. Later in pregnancy the receptor can be detected in the giant trophoblasts of the developing placenta (Pollard *et al.*, 1991). A mutation in the CSF-1 gene has been demonstrated in the toothless osteopetrotic (op/op) mouse (Pollard *et al.*, 1991). Female mice carrying this mutation are subfertile, although the male mice of this strain have normal fertility. Fetal and/or paternal seminal sources of CSF-1 appear to be able to substitute for maternal CSF-1 since matings between females of the mutant stain and normal male mice result in normal placental and fetal development. The exact role of CSF-1 remains unclear, but it may be involved in placental development and in modulation of the maternal immune response to the allogenic fetus (Pollard *et al.*, 1991). Human endometrial adenocarcinoma cells express both CSF-1 receptor and CSF-1 mRNA (Croxtall *et al.*, 1992). In addition, CSF-1 appears to act in an autocrine fashion to induce proliferation of adenocarcinoma cells. It also enhances expression of lipocortin II by these cells and thus may indirectly regulate phospholipase A_2 activity and prostaglandin availability (Huang *et al.*, 1986; Crompton *et al.*, 1988; Croxtall *et al.*, 1992).

B. Interleukins

Interleukin-1 is cytokine synthesized by activated macrophages and monocytes. Like basic FGF, IL-1 has no signal peptide and thus is not secreted in the conventional manner. It appears to be released as a result of cell lysis. It has a variety of biological activities relevant to the uterus. It is able to stimulate smooth muscle cell proliferation (Ross, 1991) and collagen synthesis by fibroblasts (Schmidt *et al.*, 1984). In addition, IL-1 stimulates the production of prostaglandin-estradiol inhibitor (PGE_2) and IL-6 by human endometrium (Tabibzadeh *et al.*, 1989, 1990) and inhibits decidualization of endometrial stromal cells (Kariya *et al.*, 1991).

Interleukin-2 is expressed in the syncytiotrophoblast of the human placenta (Boehm *et al.*, 1989). It is able to transform decidual lymphocytes into killer cells which are able to lyse trophoblasts and other cell types (Tatsumi *et al.*, 1987). Interleukin-2 receptor has also been demonstrated in human and bovine uterine fluid (Tatsumi *et al.*, 1987; Segerson and Libby, 1990). The role of interleukin-2 in the nongestational endometrium has not been established.

Interleukin-6 is also known as B-cell growth factor-2, plasmacytoma growth factor, and interferon β_2. It is a 184-amino acid protein which is modified post-translationally by both glycosylation and phosphorylation (Tabibzadeh, 1991). It is produced by endometrial stromal cells (Tabibzadeh *et al.*, 1989).

C. Tumor Necrosis Factor-α

Cachetin, or TNFα, is produced by activated macrophages and is an important mediator of endotoxic shock. It is synthesized as a larger precursor and is subject to proteolytic cleavage. It exerts a variety of effects which are part of the inflammatory reaction. TNFα mRNA has been demonstrated in term decidual as well as endometrial stromal cells from nonpregnant individuals (Jattela *et al.*, 1988; Casey *et al.*, 1989). However, the mRNA for this cytokine is constitutively expressed and regulation of expression appears to be at the translational and post-translational steps (Tracy *et al.*, 1989). It has been proposed that TNFα may have a function as an immune modulator in the decidualized endometrium. It is not clear what role, if any, it has in the endometrium in the nonpregnant state.

D. Interferons

The interferons were first isolated as proteins with antiviral activity. Interferon-γ (IFNγ) is a homodimer of 166 amino acid residues which is modified by N-linked glycosylation. IFNγ is released by activated T lymphocytes in the endometrium and has profound effects on adjacent epithelial cells. In these cells it enhances expression of HLA-DR molecules and inhibits proliferation (Tabibzadeh *et al.*, 1986, 1988).

Ovine trophoblast protein-1 is an interferon-α (IFNα) variant that is secreted by the sheep conceptus (Godkin *et al.*, 1982). It has a variety of actions in the uterus, including inhibition of prostaglandin $F_{2\alpha}$ production by the uterus. IFNα also has potent antiproliferative effects on a variety of cells in culture (Pontzer *et al.*, 1991). The interferons and ovine trophoblast protein-1 are discussed in more detail in Chapter 14.

XI. Conclusion

We are only now starting to identify the growth factors and cytokines which are important in endometrial physiology. The observations made so far are largely empirical, and the mechanistic interactions between the various factors so far identified have not as yet been explored. Further studies in this area will not only enhance our understanding of normal endometrial physiology but will provide insights into the pathophysiology of endometrial disease.

References

Akhurst, R. J. Lehnert, S. A., Gatherer, D., and Duffie, E. (1990). The role of TGF-β in mouse development. *Ann. N.Y. Acad. Sci.* 593, 259–266.

Altman, D. J., Schnieder, S. L., Thompson, D. A., Cheng, H.-L., and Tomasi, T. B. (1990). A transforming growth factor β2 (TGF-β2)-like immunosuppressive factor in amniotic fluid and localization of TGF-β2 in the pregnant uterus. *J. Exp. Med.* 172, 1391–1401.

Anzai, Y., Gong, Y., Holinka, C. F., Murphy, L. J., Murphy, L. C., Kuramoto, H., and

Gurpide, E. (1992). Effects of transforming growth factors and regulation of their mRNA levels in two human adenocarcinoma cell lines. *J. Steriod. Biochem. Mol. Biol.* 42, 449–457.

Aplin, J. D., Charlton, A. K., and Ayad, S. (1988). An immunohistochemical study of human endometrial extracellular matrix during the menstrual cycle and first trimester of pregnancy. *Cell Tissue Res.* 253, 231–240.

Appella, E., Weber, I. T., and Blasi, I. (1988). Structure and function of epidermal growth factor-like regions in proteins. *FEBS Lett.* 231, 1–4.

Baird, A., and Ling, N. (1987). Fibroblast growth factors are present in the extracellular matrix produced by endothelial cell *in vitro*: implications for a role of heparinase-like enzymes in the neovascular response. *Biochem. Biophys. Res. Commun.* 142, 428–435.

Bartocci, A., Pollard, J. W., and Stanley, E. R. (1986). Regulation of colony-stimulating factor 1 during pregnancy. *J. Exp. Med.* 164, 956–961.

Battaro, D. P., Rubin, J. S., Faletto, D. L., Chan, A. M. L., Kmilcik, T. E., Vande, Woude, G. F., and Aaronson, S. A. (1991). Identification of the hepatocyte growth factor receptor as the c-met proto-oncogene product. *Science* 251, 802–804.

Berchuck, A., Soisson, A. P., Soper, J. T., Charke-Pearson, D. L., Bast, R. C., and McCarty, K. S. (1989a). Reactivity of epidermal growth factor monoclonal antibodies with human uterine tissues. *Arch. Pathol. Lab. Med.* 113, 1155–1156.

Berchuck, A., Soisson, A. P., Olt, G. J., Spoer, J. T., Clarke-Pearson, D. L., Bast, R. C., and McCarty, K. S. (1989b). Epidermal growth factor receptor expression in normal and malignant endometrium. *Am. J. Obstet. Gynecol.* 161, 1247–1252.

Bigsby, R. M., and Cunha, G. R. (1986). Estrogen stimulation of deoxyribonucelic acid synthesis in uterine epithelial cells which lack estrogen receptors. *Endocrinology* 119, 390–396.

Boehm, K. D., Kelley, M. F., Ilan, J., and Ilan, J. (1989). The interleukin-2 gene is expressed in the syncytiotrophoblast of the human placenta. *Proc. Natl. Acad. Sci. USA* 86, 656–660.

Boehm, K. D., Daimon, M., Gorodeski, I. G., Shelan, L. A., Utian, W. H., and Ilan, J. (1990). Expression of the insulin-like and platelet-derived growth factor genes in human uterine tissues. *Mol. Rep. Dev.* 27, 93–101.

Brigstock, D. R., Heap, R. B., and Brown, K. D. (1989). Polypeptide growth factors in uterine tissues and secretions. *J. Reprod. Fertil.* 85, 747–758.

Brown, A. L., Graham, D. E., Nissley, S. P., Hill, D. J., Strain, A. J., and Rechler, M. M. (1986). Developmental regulation of insulin-like growth factor-II mRNA in different rat tissues. *J. Biol. Chem.* 261, 13144–13151.

Bulmer, J. N., Longfellow, M., and Ritson, A. (1991). Leukocytes and resident blood cells in endometrium. *Ann. N.Y. Acad. Sci.* 622, 57–68.

Campbell, P. G., Novak, J. F., Yanosick, T. B., and McMaster, J. H. (1992). Involvement of the plasmin system in dissociation of the insulin-like growth factor-binding protein complex. *Endocrinology* 130, 1401–1412.

Carlsson, B., and Billig, H. (1991). Insulin-like growth factor-I gene expression during development and estrous cycle in the rat uterus. *Mol. Cell. Endocrinol.* 77, 175–180.

Carpenter, G. (1981). Epidermal growth factor. *Handb. Exp. Pharmacol.* 57, 90–126.

Casey, M. L., Cox, S. M., Beutler, B., Milewich, L., and MacDonald, P. C. (1989). Cachectin/tumor necrosis factor alpha formation in human decidua: potential role of cytokines in infection-induced preterm labor. *J. Clin. Invest.* 83, 430–436.

Casslen, B., and Ohlsson, K. (1981). Cyclic variation of plasminogen activation in human uterine fluid and the influence of an intrauterine device. *Acta Obstet. Gynecol. Scand.* 60, 97–101.

Casslen, B., Andersson, A., Nilsson, I. N., and Astedt, B. (1986). Hormonal regulation of the release of plasminogen activators and of a specific activator inhibitor from endometrial tissue in culture. *Proc. Soc. Exp. Biol. Med.* 182, 419–424.

Chandrasekhar, Y., Narayan, S., Singh, P., and Nagamani, M. (1991). Receptors for insulin-like growth factor II in the rat uterus: characterization and variation throughout the estrous cycle. *Acta Endocrinol. (Copenhagen)* 124:434–441.

Crompton, M. R., Moss, S. E., and Crumpton, M. J. (1988). Diversity in the lipocortin/calpactin family. *Cell* 55, 1–3.

Croxtall, J. D., Pollard, J. W., Carey, F., Forder, R. A., and White, J. O. (1992). Colony stimulating factor-1 stimulates Ishikawa cell proliferation and lipocortin II synthesis. *J. Steroid Biochem. Mol. Biol.* 42, 121–129.

Croze, F., Kennedy, T. G., Schroedter, I. C., Friesen, H. G., and Murphy, L. J. (1990). Expression of insulin-like growth factor-I and insulin-like growth factor binding protein-1 in the rat uterus during decidualization. *Endocrinology* 127, 1995–2001.

Daiter, E., Pampfer, S., Yeung, Y. G., Barad, D., Stanley, E. R., and Pollard, J. W. (1992). Expression of colony-stimulating factor-1 in the human uterus and placenta. *J. Clin. Endocrinol. Metab.* 74, 850–858.

DeChiara, T., Efstratiadis, A., and Robertson, E. A. (1990). Growth-deficient phenotype in heterozygous mice carrying an insulin-like growth factor II gene disruption by targeting. *Nature* 345, 78–80.

De Vrieo, C., Escobedo, J. A., Veno, H., Houck, K., Ferrara, N., and Williams, L. T. (1992). The fms-like tyrosine kinase a receptor for vascular endothelial growth factor. *Science* 255, 989–991.

Derynck, R. (1988). Transforming growth factor-α. *Cell* 54, 593–595.

Derynck, R., Lindquist, P. B., Lee, A., Wen, D., Tamm, J., Graycar, J. L., Rhee, L., Mason, A. J., Miller, D. A., Coffey, R. J., Moses, H. L., and Chen, E. Y. (1988). A new type of transforming growth factor-beta, TGFbeta3. *EMBO J.* 7, 3737–3743.

DiAugustine, R. P., Petrusz, P., Bell, G. I., Brown, C. F., Korach, K. S., McLachlan, J. A., and Tebg, C. T. (1988). Influence of estrogens on mouse uterine epidermal growth factor precursor protein and messenger ribonucleic acid. *Endocrinology* 122, 2355–2363.

Downward, H. J., Yarden, Y., Mayes, E., Scrace, G., Totty, N., Stockwell, P., Ullrich, A., Schessinger, J., and Waterfield, M. B. (1984). Close similarity of epidermal growth factor and v-erb-B oncogene protein sequences. *Nature* 307, 521–527.

Drop, S. P. S., Valiquette, G., Guyda, H., Corvol, H. J., Corvol, M. T., and Posner, B. I. (1979). Partial purification and characterization of a binding protein for insulin-like activity (ILAs) in human amniotic fluid: a possible inhibitor of insulin-like activity. *Acta Endocrinol. (Copenh.)* 90, 505–518.

Economos, K., MacDonald, P. C., and Casey, M. L. (1992a). Endothelin-1 gene expression and protein biosynthesis in human endometrium: potential modulators of endometrial blood flow *J. Clin. Endocrinol. Metab.* 74, 14–19.

Economos, K., MacDonald, P. C., and Casey, M. L. (1992b). Endothelin-1 gene expression and biosynthesis in human endometrial HEC-1A cancer cells. *Cancer Res.* 52, 554–557.

Ernst, M., and Rodan, G. A. (1991). Estradiol regulation of insulin-like growth factor-I expression in osteoblastic cells: Evidence for transcriptional control. *Mol. Endocrinol.* 5, 1081–1089.

Fazleabas, A. T., Jaffe, R. C., Verhage, H. G., Waites, G., and Bell, S. C. (1989). An insulin-like growth factor binding protein in the baboon (*Papio anubis*) endometrium: synthesis, immunocytochemical localization and hormonal regulation. *Endocrinology* 124, 2321–2329.

Finch, P. W., Rubin, J. S., Miki, T., Ron, D., and Aaronson, S. A. (1989). Human KGF is FGF-related with properties of a paracrine effector of epithelial cell growth. *Science* 245, 752–755.

Gardner, R. M., Lingham, R. B., and Stancel, G. M. (1987). Contractions of isolated uterus stimulated by epidermal growth factor. *FASEB J.* 1, 224–228.

Gardner, R. M., Verner, G., Kirkland, J. L., and Stancel, G. M. (1989). Regulation of uterine epidermal growth factor (EGF) receptors by estrogen in the mature rat uterus during the estrous cycle. *J. Steroid Biochem.* 32, 339–343.

Garfield, R. E., Sims, S., and Daniel, E. E. (1977). Gap junctions: their presence and necessity in myometrium during parturition. *Science* 198, 958–960.

Geisert, R. D., Lee, C.-Y., Simmen, F. A., Zavy, M. T., Fliss, A. E., Bazer, F. W., and Simmen, R. C. M. (1991). Expression of messenger RNAs encoding insulin-like growth factor-I -II and insulin-like growth factor binding protein-2 in bovine endometrium during the estrous cycle and early pregnancy. *Biol. Reprod.* 45, 975–983.

Ghahary, A., and Murphy, L. J. (1989). Regulation of uterine insulin-like growth factor receptors by estrogen and variation throughout the estrous cycle. *Endocrinology* 125, 597–604.

Ghahary, A., and Murphy, L. J. (1990). In situ localization of the sites of synthesis and action of insulin-like growth factor-I in the rat uterus. *Mol. Endocrinol.* 4, 191–195.

Ghosh, D., Danielson, K. G., Alston, J. T., and Heyner, S. (1991). Functional differentiation of mouse uterine epithelial cells grown on collagen gels or reconstituted basement membranes. *In Vitro Cell Dev. Biol.* 27A, 713–719.

Godkin, J. D., Bazer, F. W., Moffatt, J., Sessions, F., and Roberts, R. M. (1982). Purification and properties of a major, low molecular weight protein released by the trophoblast of sheep blastocyst at day 13–21. *J. Reprod. Fertil.* 65, 141–150.

Goldfarb, M. (1990). The fibroblast growth factor family. *Cell Growth Diff.* 1, 439–445.

Gong, Y., Anzai, Y., Murphy, L. C., Ballejo, G., Holinka, C. F., Gurpide, E., and Murphy, L. J. (1991). Transforming growth factor gene expression in human endometrial adenocarcinoma cells: regulation by progestins. *Cancer Res.* 51, 5476–5481.

Gong, Y., Ballejo, G., Murphy, L. C., and Murphy, L. J. (1992). Differential effects of estrogen and antiestrogen on transforming growth factor gene expression in endometrial adenocarcinoma cells. *Cancer Res.* 52, 1704–1709.

Graham, C. H., Lysiak, J. J., McCrae, K. R., and Lala, P. K. (1992). Localization of transforming growth factor-β in the human fetal-maternal interface: role in trophoblast growth and differentiation. *Biol. Reprod.* 46, 568–572.

Graycar, J. L., Miller, D. A., Arrick, B. A., Lyons, R. M., Moses, H. L., and Derynck, R. (1989). Human transforming growth factor-β3: recombinant expression, purification and biological activities in comparison with transforming growth factors-β1 and β2. *Mol. Endocrinol.* 7, 1977–1986.

Han, V. K. M., Hunter, E. S., Pratt, R. M., Zendegui, J. G., and Lee, D. C. (1987). Expression of rat transforming growth factor alpha mRNA during development occurs predominantly in the maternal decidua. *Mol. Cell. Biol.* 7, 2335–2343.

Haining, R. E. B., Schofield, J. P., Jones, D. S., Rajput-Williams, J., and Smith, S. K. (1991). The identification of mRNA for EGF and TGF-alpha present in low copy number in human endometrium and decidua using RT PCR. *J. Mol. Endocrinol.* 6, 207–214.

Hernandez, E. R., Roberts, C. T., LeRoith, D., and Adashi, E. Y. (1989). Rat ovarian insulin-like growth factor-I (IGF-I) gene expression is granulosa cell selective: 5′-untranslated mRNA variants representation and hormonal regulation. *Endocrinology* 89, 572–574.

Higashiyama, S., Abraham, J. A., Miller, J., Fidds, J. C., and Klagsbrun, M. (1991). A heparin-binding growth factor secreted by macrophage-like cells that is related to EGF. *Science* 252, 936–939.

Hoffmann, G. E., Scott, R. T., and Bergh, P. A. (1991). Immunohistochemical localization of epidermal growth factor in human endometrium, decidua and placenta. *J. Clin. Endocrinol. Metab.* 73, 882–887.

Hoosein, N. M., McKnight, M. L., Levine, A. E., Mulder, K. M., Childress, K. E., Brattain, D. E., and Brattain, M. G. (1989). Differential sensitivity of subclasses of human colon carcinoma cell lines to the growth inhibitory effects of transforming growth factor β-$_1$ *Exp. Cell. Res.* 181, 442–453.

Hoppener, J. W. M., Mosselman, S., Roholl, P. J. M., Lambrechts, C., Slebos, R. J. C., de Pagter-Holthuzen, P., Lips, C. J. M., Jansz, H. S., and Sussenbach, J. S. (1988). Expression of insulin-like growth factor-I and -II genes in human smooth muscle tumours. *EMBO J.* 7, 1379–1385.

Horowitz, G. M., Scott, R. T., Drews, M. R., Navot, D., Grunfeld, L., and Hoffmann, G. E. (1992). Immunohistochemical localization of transforming growth factor-α in human endometrium decidual and trophoblast. *In* "Proc. 39th Annual Scientific Meeting for Gynecology Investigations." [Abstract 57].

Hou, J. Z., Kan, M. K., McKeehan, K., McBride, G., Adams, P., and McKeehan, W. L. (1991). Fibroblast growth factor receptors from liver vary in three structural domains. *Science* 251, 665–668.

Houck, K. A., Ferrara, N., Winer, J., Cachianes, G., Li, B., and Leung, D. W. (1991). The vascular endothelial growth factor family: identifications of a fourth molecular species and characterizations of alternative splicing of RNA. *Mol. Endocrinol.* 5, 1806–1814.

Huang, K., Wallner, B. P., Mattaliano, R. J., Tizard, R., Burne, C., Frey, A., Hession, C., McGray, P., Sinclair, L. K., Tang, J., Smart, J. E., and Pepinsky, R. B. (1986). Two human 35kd inhibitors of phospholipase A_2 are related to substrates of $pp60^{v\text{-}src}$ and of epidermal growth factor receptor/kinase. *Cell* 46, 191–199.

Hubbs, A. F., Hahn, F. F., and Thomassen, D. G. (1989). Increased resistance to transforming growth factor beta accompanies neoplastic progression of the rat tracheal epithelial cells. *Carcinogenesis* 10, 1599–1605.

Ignotz, R. A., and Massague, J. (1987). Cell adhesion protein receptors as targets for transforming growth factor-β action. *Cell* 51, 189–197.

Jattela, M., Kuusela, P., and Saksela, E. (1988). Demonstration of tumor necrosis factor in human amniotic fluid and supernatants of placental and decidual tissues. *Lab. Invest.* 58, 48–52.

Johnson, T. R., Blossey, B. K., Denko, C. W., and Ilan, J. (1989). Expression of insulin-like growth factor-I in cultured rat hepatocytes: effects of insulin and growth hormone. *Mol. Endocrinol.* 3, 580–587.

Kanzaki, T., Olofsson, A., Moren, A., Wernstedt, C., and Hellman, U. (1989). TGF-β_1 binding protein: a component of the large complex of TGF-β_1 with multiple repeat sequences. *Cell* 61, 1051–1061.

Kariya, M., Kanzaki, H., Takakura, K., Imai, K., Okamoto, N., Emi, N., Kariya, Y., and Mori, T. (1991). Interleukin-1 inhibits *in vitro* decidualization of human endometrial stromal cells. *J. Clin. Endocrinol. Metab.* 73, 1170–1174.

Keck, P. I., Hanser, S. D., Krivi, G., Sanzo, K., Warren, T., Feder, J., and Cannolly, D. T. (1989). Vascular permeability factor, and endothelial cell nitrogen related to PDGF. *Science* 246, 1309–1312.

Kirk, D., King, R. J. B., Heyes, J., Peachey, L., Hirsch, P. J., and Taylor, R. W. T. (1978). Normal human endometrium in cell culture. I. Separation and characterization of epithelial and stromal components *in vitro*. *In Vitro* 14, 651–662.

Ko, Y., Lee, C. Y., Ott, T. L., Davis, M. A., Simmen, R. C. M., Bazer, F. W., and Simmen, F. A. (1991). Insulin-like growth factors in sheep uterine fluids: concentrations and relationship to ovine trophoblast protein-1 production during early pregnancy. *Biol. Reprod.* 45, 135–142.

Koistinen, R., Kalkkinen, N., Huhtala, M.-L., Seppala, M., Bohn, H., and Rutanen, E. M. (1986). Placental protein 12 is a decidual protein that binds somatomedin and has an identical N-terminal amino acid sequence with somatomedin-binding protein from human amniotic fluid. *Endocrinology* 118, 1375–1378.

Komuro, I., Kurihara, H., Sugiyama, T., Takaku, F., and Yazaki, Y. (1988). Endothelin stimulates c-fos and c-myc expression and proliferation of vascular smooth muscle cells. *FEBS Lett.* 238, 249–252.

Lawrence, D. A., Pircher, R., Kryceve-Martinerie, C., and Jullien, P. (1984). Normal embryo fibroblasts release transforming growth factors in a latent form. *J. Cell. Physiol.* 121, 184–188.

Lax, I., Johnson, A., Howk, R., Sap, J., Bellot, F., Winkler, M., Ullrich, A., Vennstrom, B., Schlessinger, J., and Givol, D. (1988). Chicken epidermal growth factor (EGF) receptor:

cDNA cloning, expression in mouse cells, and differential binding of EGF and transforming growth factor alpha. *Mol. Cell. Biol.* 8, 1970–1978.

Leake, R., Carr, L., and Rinaldi, F. (1991). Autocrine and paracrine effects in the endometrium. *Ann. N.Y. Acad. Sci.* 622, 145–148.

Lee, EY-HP., Lee, W.-H., Kaetzel, S. C., Parry, G., and Bissell, M. J. (1985). Interaction of mouse mammary epithelial cell with collagen substrata: regulation of casein gene expression and secretion. *Proc. Natl. Acad. Sci. U.S.A.* 82, 1419–1423.

Lee, Y. H., Howe, R. S., Sha, S., Teuscher, C., Sheehan, D. M., and Lyttle, C. R. (1989). Estrogen regulation of an eosinophil chemotactic factor in the immature rat uterus. *Endocrinology* 125, 3022–3028.

Leung, D. W., Cachianes, G., Kuang, W. J., Goeddel, D. V., and Ferrara, N. (1989). Vascular endothelial growth factor is a secreted angiogenic nitrogen. *Science* 246, 1306–1308.

Lin, H. Y., Wang, X. F., Ng-Eaton, E., Weinberg, R. A., and Lodish, H. F. (1992a). Expression cloning of the TGF-β type II receptors, a functional transmembrane serine/threonine kinase. *Cell* 68, 775–785.

Lin, J.-K., and Chou, C.-K. (1992b). *In vitro* apoptosis in the human hepatoma cell line induced by transforming growth factor β_1. *Cancer Res.* 52, 385–388.

Linkhart, T. A., and Mohan, S. (1989). Parathyroid hormone stimulates release of insulin-like growth factor-I (IGF-I) and IGF-II from neonatal mouse calvaria in organ culture. *Endocrinology* 125, 1484–1491.

Littlefield, B. A. (1991). Plasminogen activators in endometrial physiology and embryo implantation: a review. *Ann. N.Y. Acad. Sci.* 622, 167–175.

Lopez-Casillas, F., Cheifetz, S., Doody, J., Andres, J. L., Lane, W. S., and Massague, J. (1991). Structures and expression of the membrane proteoglycan betaglycan, a component of the TGF-β receptor system. *Cell* 67, 785–795.

Lund, P. K., Moats-Staats, B. M., Hynes, M. A., Simmons, J. D., Jansen, M., D'Ercole, A. J., and Van Wyk, J. J. (1986). Somatomedin/insulin-like growth factor-I and insulin-like growth factor-II mRNAs in rat fetal and adult tissues. *J. Biol. Chem.* 261, 14539–14544.

Maglione, D., Guerriero, V., Viglietto, G., Delli-Bovi, P., and Persico, M. G. (1991). Isolation of a human placenta cDNA coding for a protein related to the vascular permeability factor. *Proc. Natl. Acad. Sci. USA* 88, 9267–9271.

Massague, J. (1990). The transforming growth factor-β family. *Annu. Rev. Cell Biol.* 6, 597–641.

Massague, J. (1991). Transforming growth factor-α: a model for membrane-anchored growth factors. *J. Biol. Chem.* 265, 21393–21397.

Mathews, L. S., and Vale, W. W. (1991). Expression cloning of an activin receptor, a predicted transmembrane serine kinase. *Cell* 65, 973–982.

Mendoza, A. E., Young, R., Orkin, S. H., and Collins, T. (1990). Increased platelet-derived growth factor A-chain expression in human uterine smooth muscle cells during the physiology hypertrophy of pregnancy. *Proc. Natl. Acad. Sci. USA* 87:2177–2181.

Milner, P. G., Li, Y.-S., Hoffman, R. M., Kodner, C. M., Siegel, N. R., and Deuel, T. F. (1989). A novel 17 kD heparin binding growth factor (HBGF-8) in bovine uterus: purification and N-terminal amino acid sequence. *Biochem. Biophys. Res. Commun.* 165, 1096–1103.

Miyazono, K., Yuki, K., Takaku, F., Wernstedt, C., Kanzaki, T., Olofsson, A., Hellman, U., and Hedlin, C.-H. (1990). Latent forms of TGF-β: structure and biology. *Ann. N.Y. Acad. Sci.* 593:51–58.

Mukku, V. R., and Stancel, G. M. (1985). Regulation of epidermal growth factor receptor by estrogen. *J. Biol. Chem.* 260, 9820–9824.

Murphy, L. J. (1991). The uterine insulin-like growth factor system. *In* "Modern Concepts of Insulin-like Growth Factors" (E. M. Spencer, ed.), pp. 275–285. Elsevier, New York.

Murphy, L. J., and Friesen, H. G. (1988). Differential effects of estrogen and growth hormone on uterine and hepatic insulin-like growth factor-I gene expression in the ovariectomized, hypophysectomized rat. *Endocrinology* 122, 325–332.

Murphy, L. J., and Friesen, H. G. (1989). Regulation of insulin-like growth factor-I gene expression by estrogen. *In* "Molecular and Cellular Biology of Insulin-like Growth Factors and Their Receptors" (D. LeRoith and M. K. Raizada, eds.), pp. 153–169. Plenum, New York.

Murphy, L. J., and Ghahary, A. (1990). Uterine insulin-like growth factor-I: regulation of expression and its role in estrogen-induced uterine proliferation. *Endocr. Rev.* 11, 443–453.

Murphy, L. J., and Luo, J. M. (1989). Effects of cycloheximide on hepatic and uterine insulin-like growth factor-I mRNA. *Mol. Cell. Endocrinol.* 64, 81–86.

Murphy, L. J., Sutherland, R. L., Stead, B., Murphy, L. C., and Lazarus, L. (1986). Progestin regulation of epidermal growth factor receptors in human mammary carcinoma cells. *Cancer Res.* 46, 728–734.

Murphy, L. J., Bell, G. I., and Friesen, H. G. (1987a). Tissue distribution of insulin-like growth factor I and II messenger ribonucleic acids in the adult rat. *Endocrinology* 120, 1279–1281.

Murphy, L. J., Murphy, L. C., and Friesen, H. G. (1987b). Estrogen induces insulin-like growth factor-I expression in the rat uterus. *Mol. Endocrinol.* 1, 445–450.

Murphy, L. J., Murphy, L. C., and Friesen, H. G. (1988). A role for the insulin-like growth factors as estromedins in the rat uterus. *Trans. Assoc. Am. Phys.* 99, 204–211.

Murphy, L. J., Seneviratne, C., Ballejo, G., Croze, F., and Kennedy, T. G. (1990a). Identification and characterization of a rat decidual insulin-like growth factor-binding protein, complementary DNA. *Mol. Endocrinol.* 4, 329–326.

Murphy, L. C., Dotzlaw, H., Wong, M. S. J., Miller, T., Mroczkowski, B., Gong, Y., and Murphy, L. J. (1990b). Epidermal growth factor: receptor and ligand expression in human breast cancer. *Sem. Cancer Biol.* 1, 305–315.

Murphy, L. J., Gong, Y., Murphy, L. C., and Bhavnani, B. (1991). Growth factors in normal and malignant uterine tissue. *Ann. N.Y. Acad. Sci.* 622, 383–391.

Murphy, L. J., Gong, Y., and Murphy, L. C. (1992). Regulation of transforming growth factor gene expression in human endometrial adenocarcinoma cells. *J. Steroid Biochem. Mol. Biol.* 41:309–314.

Nelson, K. G., Takahashi, T., Bossert, N. L., Walmer, D. K., and McLachlan, J. A. (1991). Epidermal growth factor replaces estrogen in the stimulation of female genital tract growth and differentiation. *Proc. Natl. Acad. Sci. USA* 88, 21–25.

Norstedt, G., Levinovitz, A., and Eriksson, H. (1989). Regulation of uterine insulin-like growth factor-I mRNA and insulin-like growth factor-II mRNA by estrogen in the rat. *Acta Endocrinol. (Copenh.)* 120, 466–472.

Ogasawara, M., Karey, K. P., Marquardt, H., and Sirbasku, D. A. (1989). Identification and purification of truncated insulin-like growth factor I from porcine tissue: evidence for high biological potency. *Biochemistry* 28, 2710–2721.

Olive, D. L., Montoya, I., Riehl, R. M., and Schenken, R. S. (1991). Macrophage-conditioned media enhances endometrial stromal cell proliferation *in vitro*. *Am. J. Obstet. Gynecol.* 164, 953–958.

Pandiella, A., and Massague, J. (1991). Cleavage of the membrane precursor for transforming growth factor-α is a regulated process. *Proc. Natl. Acad. Sci. USA* 88, 1726–1730.

Padykula, H. A. (1991). Regeneration in the primate uterus: the role of stem cells. *Ann. N.Y. Acad. Sci.* 622, 47–56.

Peltz, S. W., Katzenellenbogen, B. S., Kneifel, M. A., and Mangel, W. F. (1983). Plasminogen activators in tissues of the immature and estrogen-stimulated rat uterus and in uterine luminal fluid: characterization and properties. *Endocrinology* 112, 890–897.

Perrot-Applanat, M., Groyer-Picard, M. T., Garcia, E., Lorenzo, F., and Milgrom, E. (1988). Immunocytochemical demonstration of estrogen and progesterone receptors in muscle cells of uterine arteries in rabbits and humans. *Endocrinology* 123, 1511–1519.

Phillips, H. S., Hains, J., Leung, D. W., and Ferrara, N. (1990). Vascular endothelial growth factor is expressed in rat corpus luteum. *Endocrinology* 127, 965–967.

Pollard, J. W., Bartocci, A., Arceci, R. J., Orlofsky, A., Ladner, M. B., and Stanley, E. R. (1987). Apparent role of the macrophage growth factor, CSF-1, in placental development. *Nature* 330, 484–486.

Pollard, J. W., Hunt, J. S., Wiktor-Jedrzejczak, W., and Stanley, E. R. (1991). A pregnancy defect in the osteopetrotic (op/op) mouse demonstrates the requirement for CSF-1 in female fertility. *Dev. Biol.* 148, 273–283.

Pontzer, C. H., Bazer, F. W., and Johnson, H. M. (1991). Antiproliferative activity of a pregnancy recognition hormone, ovine trophoblast protein-1. *Cancer Res.* 51, 5304–5307.

Prat, M., Narsimhan, R. P., Crepalde, T., Nicotra, M. R., Natali, P. G., and Comoglio, P. M. (1991). The receptor encoded by the human c-met oncogene is expressed in hepatocytes, epithelial cells and solid tumors. *Int. J. Cancer* 49, 323–328.

Presta, M. (1988). Sex hormones modulate the synthesis of basic fibroblast growth factor in human endometrial adenocarcinoma cells: implications for the neovascularization of normal and neoplastic endmetrium. *J. Cell. Physiol.* 137, 513–517.

Raines, E. W., Bowen-Pope, D. F., and Ross, R. (1990). Platelet-derived growth factor. *In* "Handbook of Experimental Pharmacology: Peptide Growth Factors and Their Receptors" (M. B. Sporn and A. B. Roberts, eds.), pp. 173–262. Springer-Verlag, Heidelberg.

Raprager, A. C., Krufka, A., and Olwin, B. B. (1991). Requirement of heparan sulfate for bFGF-mediated fibroblast growth and myoblast differentiation. *Science* 252, 1705–1708.

Rein, M. S., Friedman, A. J., Pandian, M. R., and Heffner, L. J. (1990). The secretion of insulin-like growth factors I and II by explant cultures of fibroids and myometrium from women treated with a gonadotropin-releasing hormone agonist. *Obstet. Gynecol.* 76, 388–394.

Reynolds, R. K., Talavera, F., Roberts, J. A., Hopkins, M. P., and Menon, K. M. J. (1990). Regulation of epidermal growth factor and insulin-like growth factor-I receptors by estradiol and progesterone in normal and neoplastic endometrial cell cultures. *Gynecol. Oncol.* 38, 396–406.

Rinderknecht, E., and Humbel, R. E. (1978). The amino acid sequence of human insulin-like growth factor-I and its structural homology with proinsulin. *J. Biol. Chem.* 253, 2769–2776.

Roberts, A. B., and Sporn, M. B. (1988). Transforming growth factor β review. *Adv. Cancer Res.* 51, 107–145.

Ross, R. (1991). Polypeptide growth factors and atherosclerosis. *Trends Cardiovasc. Med.* 1, 277–282.

Rossi, M. J., Chegini, N., Flandus, K. C., and Ksander, G. A. (1992). Tissue localization and in vitro actions of transforming growth factor β in human uteri. *In* "Proc. 39th Ann. Scientific Meeting Society for Gynecologic Investigation." [Abs. 578]

Salamonsen, L. A., Butt, A. R., Macpherson, A. M., Rodgers, P. A. W., and Findlay, J. K. (1992). Immunolocalization of the vasoconstrictor endothelin in human endometrium during the menstrual cycle and in umbilical cord at birth. *Am. J. Obs. Gynecol.* 167, 163–167.

Salomon, D. S., Kim, N., Saeki, T., and Ciardiello, F. (1990). Transforming growth factor-α: an oncodevelopmental growth factor. *Cancer Cell* 2, 389–397.

Sawdey, M. S., and Loskutoff, D. J. (1991). Regulation of murine type 1 plasminogen activator inhibitor gene expression in vivo: tissue specificity and induction of lipopolysaccharide, tumor necrosis factor-α and transforming growth factor-β. *J. Clin. Invest.* 88, 1346–1353.

Schmidt, J. A., Oliver, C. N., Lepe-Zuniga, J. L., Green, I., and Gery, I. (1984). Silica-stimulated monocytes release fibroblast proliferation factors identical to interleukin 1. A potential role for interleukin 1 in the pathogenesis of silicosis. *J. Clin. Invest.* 73, 1462–1472.

Schoenle, E., Zapf, J., Hauri, C., Steiner, T., and Froesch, E. R. (1985). Comparison of *in vivo* effects of insulin-like growth factor-I and II and of growth hormone in hypophysectomized rats. *Acta Endocrinol. (Copenhagen)* 108, 167–174.

Schreiber, A. B., Winkler, M. E., and Derynck, R. (1986). Transforming growth factor-α: a more potent angiogenic mediator than epidermal growth factor. *Science* 232, 1250–1253.

Segerson, E. C., and Libby, D. W. (1990). Interaction of bovine luminal protein with interleukin-2 and interleukin-2 receptors of T lymphocytes. *Biol. Reprod.* 43, 619–628.

Sheets, E. E., Tsibris, J. C. M., Cook, N. I., Virgin, S. D., DeMay, R. M., and Spellacy, W. N. (1985). In vitro binding of insulin and epidermal growth factor to human endometrium and endocervix. *Am. J. Obstet. Gynecol.* 153, 60–65.

Sherr, C. J. (1988). The *fms* oncogene. *Biochim. Biophys. Acta* 948, 225–243.

Sherr, C. J., Rettenmier, C. W., Sacca, R., Roussel, M. F., Look, A. T., and Stanley, E. R. (1985). The c-fms protooncogene product is related to the receptor for the mononuclear phagocyte growth factor, CSF-1. *Cell* 41, 665–676.

Shimokado, K., Raines, E. W., Madtes, D. K., Barrett, T. B., Benditt, E. P., and Ross, R. (1985). A significant part of macrophage-derived growth factor consists of at least two forms of PDGF. *Cell* 43, 277–286.

Simmen, R. C. M., Simmen, F. A., Hogif, A., Farmer, S. J., and Bazer, F. W. (1990). Hormonal regulation of insulin-like growth factor gene expression in the pig uterus. *Endocrinology* 127, 2166–2174.

Surrey, E. S., and Halme, J. (1991). Effect of platelet-derived growth factor on endometrial stromal cell proliferation *in vitro*: a model for endometriosis. *Fertil. Steril.* 56, 672–679.

Tabibzadeh, S. S. (1991). Human endometrium: an active site of cytokine production and action. *Endocr. Rev.* 12, 272–290.

Tabibzadeh, S. S., Gerber, M. A., and Satyaswaroop, P. G. (1986). Induction of HLA-DR antigen in human endometrial epithelial cells cell in vitro by recombinant gamma-interferon. *Am. J. Pathol.* 125, 90–96.

Tabibzadeh, S. S., Satyaswaroop, P. G., and Rao, P. N. (1988). Antiproliferative effect of interferon gamma in human endometrial epithelial cells *in vitro*: potential local growth modulatory role in endometrium. *J. Clin. Endocrinol. Metab.* 67, 131–138.

Tabibzadeh, S. S., Santhanam, U., Sehgal, P. B., and May, L. (1989). Cytokine-induced production of interferon-β_2/interleukin-6 by freshly explanted human endometrial stromal cells: modulation by estradiol-17β. *J. Immunol.* 142, 3134–3139.

Tabibzadeh, S., Kaffka, K. L., Satyaswaroop, P. G., and Killian, P. L. (1990). Interleukin-1 (IL-1) regulation of human endometrial function: presence of IL-1 receptors correlates with IL-1-stimulated prostaglandin E_2 production. *J. Clin. Endocrinol. Metab.* 70, 1000–1006,

Taketani, Y., and Mizuno, M. (1989). Cyclic changes in epidermal growth factor receptor in human endometrium during the menstrual cycle. *Endocrinol. (Japan)* 35, 19–25.

Takuwa, N., Takuwa, Y., Yanagisawa, M., Yanashita, K., and Masaki, T. (1989). A novel vasoactive peptide endothelin stimulates mitogenesis through inositol lipid turnover in Swiss 3T3 fibroblasts. *J. Biol. Chem.* 264, 7856–7861.

Tamada, H., McMaster, M. T., Flanders, K. C., Andrew, G. K., and Dey, S. K. (1990). Cell type specific expression of transforming growth factor-b1 in the mouse uterus during pre-implantation period. *Mol. Endocrinol.* 4, 965–972.

Tatsumi, K., Mori, T., Mori, E., Kanzaki, H., and Mori, T. (1987). Immunoregulatory factor release from a cell line derived from human decidual tissue. *Am. J. Reprod. Immunol. Microbiol.* 13, 87–92.

Tchernitchin, A. (1979). The role of eosinophil receptors in the nongenomic response to oestrogens in the uterus. *J. Steroid Biochem.* 11, 417–424.

Tchernitchin, A., Rooryck, J., Tchernitchin, X., Vandenhende, J., and Galand, P. (1975). Effects of cortisol on uterine eosinophilia and other oestrogenic responses. *Mol. Cell. Endocrinol.* 2, 331–337.

Tomooka, T., DiAugustine, R. P., and McLachlan, J. A. (1986). Proliferation of mouse uterine epithelial cells *in vitro*. *Endocrinology* 118, 1011–1018.

Tracy, K. J., Vlassara, H., and Cerami, A. (1989). Peptide regulatory factors. Cachectin/tumour necrosis factor. *Lancet* i, 1122–1126.

Troche, V., O'Connor, D. M., and Schandies, R. P. (1991). Measurement of human epidermal growth factor receptor in the endometrium during the menstrual cycle. *Am. J. Obstet. Gynecol.* 165, 1499–1503.

Tryggvason, K., Hoyhtya, M., and Salo, T. (1987). Proteolytic degradation of extracellular matrix in tumor invasion. *Biochem. Biophys. Acta* 907, 191–217.

Waites, G. T., James, R. F. L., and Bell, S. C. (1988). Immunohistological localization of human endometrial secretory proteins "pregnancy-associated endometrial secretory $_1$-globulin" ($_1$-PEG), an insulin-like growth factor binding protein, during the menstrual cycle. *J. Clin. Endocrinol. Metab.* 67, 1100–1104.

Wakefield, L. M., Smith, D. M., Flanders, K. C., and Sporn, M. B. (1988). Latent transforming growth factor-β from human platelets. *J. Biol. Chem.* 263, 7646–7654.

Wakefield, L., Kim, S.-J., Glick, A., Winokur, T., Colletta, A., and Sporn, M. (1990). Regulation of transforming growth factor-β subtypes by members of the steroid hormone superfamily. *J. Cell. Sci. Suppl.* 13, 139–148.

Wang, X. F., Lin, H. Y., Ng-Eaton, E., Downward, J., Lodish, H. F., and Weinberg, R. A. (1991). Expression cloning and characterization of the TGF-β type III receptor. *Cell* 67, 797–805.

Westermark, B., and Heldin, C. H. (1991). Platelet derived growth factor in autocrine transformations. *Cancer Res.* 51, 5087–5092.

Winkler, M. E., O'Connor, L., Winget, M., and Fendly, B. (1989). Epidermal growth factor and transforming growth factor α bind differently to the epidermal growth factor receptor. *Biochemistry* 28, 6373–6378.

Yamamoto, T., Nishiyama, M., Yanoh, K., Naka, Y., Naka, A., and Sugiyama, Y. (1989). Regulation of human endometrial and decidual cell functions—role of epidermal growth factor in decidualization. *Prog. Clin. Biol. Res.* 294, 413–424.

Yeh, J., Danehy, F. T., Osathanondh, R., and Villa-Komaroff, L. (1991). mRNAs for insulin-like growth factor-II (IGF-II) and variant IGF-II are co-expressed in human fetal ovary and uterus. *Mol. Cell. Endocrinol.* 80, 75–82.

13

Paracrine Interactions in Endometrial Function

MARKKU SEPPÄLÄ & EEVA-MARJA RUTANEN

I. Introduction

It is becoming increasingly evident that, in addition to endocrine control by classic hormones, cell–cell interactions are important in the growth, development, and maintenance of normal tissue functions. One cell may respond to a hormonal stimulus by producing local regulatory factors that act on the very cell of synthesis (autocrine action) or on adjacent cells (paracrine action). The uterine endometrium is an example of a target organ in which hormonally regulated local factors play a role. The number of factors whose potential role in autocrine/paracrine regulation of endometrial functions is suggested has increased exponentially, and it is not meaningful to discuss all of them in this short review. Therefore we have chosen two major endometrial secretory proteins, namely insulin-like growth factor binding protein-1 (IGFBP-1) and progesterone-associated endometrial protein/placental protein 14 as examples of paracrine factors which may regulate endometrial function and endometrial–trophoblast interactions.

II. Endometrial Cell Types and Hormonal Regulation of Endometrial Function

Endometrium is a dynamic tissue which undergoes cyclic processes of cell proliferation, differentiation, and menstrual shedding in response to

Molecular Biology of the Female Reproductive System

ovarian endocrine events. The two major cell types in the endometrium are epithelial cells and stromal cells. Epithelial cells include those lining the uterine cavity and the glands invading the stroma. Endometrial stroma also contains fibroblast-like cells, vascular endothelial cells, macrophages, and lymphoid cells (Kamat and Isaacson, 1987). All different cell types are involved in cell–cell interactions, and their synchronous response to ovarian sex steroids is necessary for normal endometrial function.

During the follicular phase of the ovarian cycle, granulosa cells produce increasing amounts of estrogen, and the concomitant proliferative changes in the endometrium include a rapid increase of cellular mitoses and ribonucleic acid (RNA) synthesis. During the first week after ovulation, progesterone, secreted by the corpus luteum, induces differentiation and secretory activity in endometrial glands, whereas during the second postovulatory week, increased mitoses are seen in stromal lymphocytes (Tabibzadeh, 1990a,b), and part of the stromal cells undergo predecidual transformation. Concomitantly there are changes in surface epithelial cells, and their microvilli become modified to ensure contact between the implanting blastocyst and the endometrium. These changes are considered essential for successful implantation and embryo–endometrial interaction to take place. In a conception cycle, the hatched blastocyst implants about 1 week after ovulation. The implantation involves complex interactions between uterine and embryonic cells. In nonconception cycles, menstrual shedding is the final stage of endometrial development.

Locally produced peptide growth factors and proto-oncogenes are assumed to play a role in hormonally induced cell proliferation and differentiation in the endometrium. Some of the local factors are apparently synthesized and secreted throughout the menstrual cycle and may not be affected by estrogen or progesterone, while others may be more tightly bound to hormonal regulation or to the stage of differentiation, some factors obviously being triggered by the implanting embryo.

III. The Insulin-like Growth Factor System and IGF-Binding Protein-1

A. IGF-I and IGF-II

The IGF system includes insulin-like growth factors (IGF-I and IGF-II), their receptors, and soluble-binding proteins (IGFBPs). IGF-I and IGF-II are peptides with a structural homology to proinsulin and with important effects on cell proliferation, growth, differentiation, and metabolism (Daughaday and Rotwein, 1989). IGF-I is a single-chain polypeptide consisting of 70 amino acids, with an M_r of 7649 Da. IGF-II consists of 67 amino acids, with an M_r of 7471 Da. Most tissues synthesize both IGF-I and

IGF-II at some time during pre- and postnatal development. Postnatally, liver production of IGF-I and, to a lesser extent of IGF-II, is regulated by growth hormone (GH), and IGF-I apparently mediates the growth-promoting effects of GH in many tissues. In endocrine glands, IGF production is regulated by tropic hormones, the actions of which are amplified by IGFs (Holly and Wass, 1989). These peptides, notably IGF-II, are believed to play a key role in fetal growth and development (D'Ercole, 1991).

In rat uterus, estrogen stimulates both IGF-I and IGF-II, as well as type I IGF receptor mRNA expression, suggesting that IGFs mediate the proliferative action of estrogen in the uterus (Murphy and Ghahary, 1990). In human endometrium, IGF-I mRNA is most abundant in the late proliferative phase and IGF-II mRNA in the early proliferative phase (Boehm *et al.*, 1990) or in the secretory phase (Guidice *et al.*, 1991a,b) (Table 1). In terms of cell proliferation, IGF-I is considered a progression factor, allowing a cell to traverse the G_0/G_s cell cycle following exposure to competence factors (Stiles *et al.*, 1979). IGF-I and IGF-II mRNAs have been identified in human embryos, suggesting that IGFs are involved in early embryonic development (Bondy *et al.*, 1990). Circulating IGFs act in an endocrine fashion.

B. IGF Receptors

The biological effects of IGFs are mediated through two types of cell membrane receptors present in most tissues, including the endometrium (Rutanen *et al.*, 1988). Type I IGF receptor binds both IGF-I and IGF-II with

TABLE 1 The IGF System in Human Endometrium (Abundance of mRNA)

	Proliferative		Secretory	
Factor	Early	Late	Early	Late
IGF-I mRNA[a]	++	+++	+	+
IGF-II mRNA[a]	++	+	+	+++
Type I IGF receptor[b]	+	+	+	+++
Type II IGF receptor[b]	+	+	+	+++
IGFBP-1[c]	−	−	+	+++
IGFBP-2[d]	+	+	++	+++
IGFBP-3[d]	+	+	++	++
IGFBP-4[b]	+	+	+	+
IGFBP-5[b]	+	+	+	+

[a]From Boehm *et al.* (1990).
[b]From Rutanen *et al.* (1993).
[c]From Julkunen *et al.* (1990).
[d]From Giudice *et al.* (1991a,b).

high affinity, which suggests that this receptor mediates the biological responses of both IGF-I and IGF-II (Steele-Perkins *et al.*, 1988). Human placenta contains both IGF-I and IGF-II receptors (Marshall *et al.*, 1974). After binding to the ligand, type I IGF receptor undergoes autophosphorylation, and subsequent phosphorylation of endogenous substrates takes place within the cell. This process also mediates IGF-I-stimulated glucose uptake, glycogen synthesis, and DNA synthesis (Steele-Perkins *et al.*, 1988).

C. Binding Proteins

In serum and other extracellular fluids, IGFs are bound to specific binding proteins (IGFBPs) which are believed to play an important role in IGF action. Six distinct IGFBPs which differ in structure, tissue distribution, regulation, and function have been cloned and sequenced (Shimasaki and Ling, 1991). IGFBPs regulate the transport and presentation of IGFs to cell surface receptors, prolong their half-life, and modulate the biological actions of IGFs at the cellular level. Like IGFs, the IGFBPs are produced in a wide variety of cells, suggesting an autocrine or paracrine mode of action.

D. IGFBP-1 in Human Endometrium: mRNA Expression, Secretion, and Regulation

IGFBP-1 is the major IGFBP in human endometrium. Before the current nomenclature was established, IGFBP-1 had many names in the literature. N-terminal sequence analyses and cloning studies indicated that the first IGFBP purified from human amniotic fluid (Povoa *et al.*, 1984) is the same as placental protein 12 (PP12) (Rutanen *et al.*, 1982; Koistinen *et al.*, 1986), the low-molecular weight 34K IGF-binding protein (Koistinen *et al.*, 1987a), α1-pregnancy-associated endometrial globulin (α1-PEG) (Bell *et al.*, 1988), BP 28 (Brewer et al., 1988), and 25 K BP (Lee *et al.*, 1988). IGFBP-1 consists of 269 amino acids and has an M_r of 25.3 kDa (Julkunen *et al.*, 1988a). The binding protein binds IGF-I with high affinity, similar to that of type I IGF receptor (Koistinen *et al.*, 1987a). The IGFBP-1 gene resides in chromosome 7p12-p13, and it has DNA polymorphism (Alitalo *et al.*, 1989). IGFBP-1 contains the tripeptide arginine–glycine–aspartic acid (RGD) sequence common to extracellular matrix proteins that bind to integrins (Ruoslahti and Pierschbacher, 1986). So far, no receptor has been identified for IGFBP-1. Northern blot analysis (Julkunen *et al.*, 1988a), *in situ* hybridization histochemistry (Julkunen *et al.*, 1990), and reverse transcriptase polymerase chain reaction (Rutanen *et al.*, 1993) have been used to identify IGFBP-1 mRNA in pregnant and nonpregnant endometria. Radioimmunoassay (RIA) (Rutanen *et al.*, 1984, 1988; Pekonen *et al.*, 1988), immunofluorometric assays (Koistinen *et al.*, 1987b), immunoradiometric assay (IRMA) (Pekonen *et al.*, 1992), immunohistochemistry (Rutanen *et al.*,

1991), metabolic labeling with [^{35}S]methionine (Rutanen *et al.*, 1985, 1986a,b), and Western ligand- and immunoblot analyses (Giudice *et al.*, 1991a,b; Rutanen *et al.*, 1988, 1993; Pekonen *et al.*, 1992) have been used in studies on cellular localization, tissue content, secretion, and regulation of translated IGFBP-1 in the endometrium.

A 1.6-kb IGFBP-1 mRNA is expressed in secretory phase endometrium and pregnancy decidua, but not in proliferative phase endometrium, which suggests that progesterone regulates synthesis of this protein (Julkunen *et al.*, 1988a). Using *in situ* hybridization histochemistry, Julkunen and her colleagues (1990) localized IGFBP-1 mRNA to predecidualized stromal cells in late secretory phase endometrium and to decidual cells in pregnancy endometrium. The tissue content of IGFBP-1 was much higher in pregnancy decidua than in late secretory phase endometrium (Rutanen *et al.*, 1984, 1985, 1986a,b, 1988). *In vitro*, explants of early and late pregnancy decidua secreted significantly more IGFBP-1 than did explants of nonpregnant endometrium (Rutanen *et al.*, 1985, 1986a,b).

Early studies by Rutanen and her co-workers (1986a) have shown that progesterone stimulates secretion of IGFBP-1 from nonpregnant endometrium. In keeping with this, IGFBP-1 has not been found by immunohistochemical staining in proliferative phase endometrium, whereas it is readily detectable 1 week after ovulation (Rutanen *et al.*, 1991; Wahlström and Seppälä, 1984), and is present in abundance in the endometrium of women wearing an intrauterine contraceptive device that releases levonorgestrel (Pekonen *et al.*, 1992). In nonpregnant endometrium, IGFBP-1 is localized to predecidualized stromal cells, and transiently also to some epithelial cells (Rutanen *et al.*, 1991). The timing and cellular localization of IGFBP-1 in human endometrium suggest that this protein is associated with decidual differentiation of endometrial stromal cells. In keeping with this, Bell and his co-workers (1991) have shown that medroxyprogesterone-acetate stimulates IGFBP-1 secretion in cultured endometrial stromal cells. In combination with this progestin, relaxin stimulates stromal cell IGFBP-1 secretion.

In the endometrium of early pregnancy, IGFBP-1 is localized primarily to decidualized stromal cells. Weaker and uneven immunoreactivity can be seen in some glandular epithelial cells as well (Rutanen *et al.*, 1991). Because IGFBP-1 mRNA is found in stromal cells only, the epithelial localization of the protein must reflect a site of paracrine action.

E. IGFBP-1 as a Paracrine Regulator in Epithelial–Stromal Cell Interactions

Several aspects suggest a paracrine role for IGFBP-1 in endometrial physiology: (1) abundance of IGFBP-1 mRNA in the endometrium compared with other tissues and other IGFBPs; (2) cyclic change in mRNA expression and protein production; (3) tissue-specific regulation by progesterone and relaxin; (4) gene expression in stromal cells which, after implantation, surround the embryonic cells containing IGF receptors but do not express IGFBP-1 mRNA; (5) localization of IGFBP-1 protein and mRNA in different cell types; and (6) inhibition by IGFBP-1 of binding of IGF-I to its receptor.

IGFBP-1 inhibits binding of IGF-I to its endometrial IGF receptors (Rutanen *et al.*, 1988). Because IGFs bind to cell membrane receptors and also to soluble IGFBP-1 with similar affinity, the availability of IGFs to IGF receptors depends on the abundance of IGFBP-1 and the other IGFBPs in the environment.

IGF-I mRNA is stimulated by estrogen and IGFBP-1 by progesterone. Therefore it is likely that the IGF system is one of the mechanisms through which the biological actions of estrogen and the antiestrogenic effects of progesterone are mediated in the endometrium. Irwin and co-workers (1991) used cultured endometrial stromal cells as a model to study interactions between growth factors and sex steroids in the regulation of endometrial stromal cell growth and differentiation. In their experiments, IGF-I and IGF-II had neither mitogenic nor differentiation-stimulating effects on cultured endometrial stromal cells. These results, however, do not exclude the possibility that IGFs derived from stromal cells *in vivo* may have such effects. Should this be the case, then IGFBP-1, by inhibiting the action of IGFs, could mediate stromal influence on epithelial cells. Obviously some factor(s) of stromal cell origin has an inhibitory effect on epithelial cells, because endometrial epithelium adjacent to decidualized stromal cells becomes atrophic and gradually disappears in conditions in which decidual reaction in the endometrium is maintained for a longer period of time, i.e., during pregnancy or under sustained progestagen treatment (Dallenbach-Hellweg, 1987). In this regard it is of interest to note that an intrauterine contraceptive device that releases levonorgestrel induces strong decidual reaction with associated atrophy in the adjacent epithelium and yet there is abundant IGFBP-1 in decidualized endometrial stromal cells (Pekonen *et al.*, 1992). This suggests that IGFBP-1 indeed may act as a paracrine inhibitor of epithelial cell proliferation.

It is not impossible that alterations in expression or production of IGFs or IGFBPs, notably of IGFBP-1, are associated with the development of uterine neoplasias. It can be hypothesized that overexpression of IGFs, an increase in IGF receptors, or a decrease of IGFBP-1 and possibly of other IGFBPs, may lead to enhanced IGF action and thereby contribute to excessive stimulation of growth in endometrium. In estrogen-dependent proliferative disorders, such as cystic glandular hyperplasia, there is no detectable IGFBP-1 in the endometrium (Rutanen *et al.*, 1993). Adenomyosis is a benign uterine disease characterized by invasion of endometrial epithelial and stromal cells into the myometrium. In adenomyosis, IGFBP-1 is often absent from the endometrium as well as from adenomyotic foci despite the presence of a morphologically defined predecidual reaction in the endometrium (Rutanen *et al.*, 1993). Whether the absence of IGFBP-1 and endometrial invasion into the myometrium are interrelated remains to be studied. Several endometrial adenocarcinoma cell lines have been studied for IGFBP-1 production, and two of them, KLE and HEC 1B, have been shown to secrete IGFBP-1 (Pekonen *et al.*, 1991). The regulation of IGFBP-1 in KLE and HEC 1B cells differs from that in normal endometrium. Like the IGFBP-1 secreted by normal endometrium, IGFBP-1 secreted by endometrial cancer cells inhibits binding of [^{125}I]IGF-I to the receptor.

F. Potential Paracrine Role of IGFBP-1 in Implantation and Maintenance of Pregnancy

Decidual IGFBP-1 can prevent the receptor binding and biological action of IGF-I in transformed trophoblastic cells (JEG-3) (Ritvos *et al.*, 1988), as well as on placental membranes (Pekonen *et al.*, 1988), which suggests that IGFBP-1 inhibits IGF actions on these sites. Frost and Tseng (1991) have shown that IGFBP-1 produced by decidualized endometrial stromal cells is phosphorylated, and the time course of IGFBP-1 phosphorylation coincides with the arrest in growth and decidual differentiation of endometrial stromal cells. Clemmons and his co-workers (1991) reported that phosphorylated isoforms of IGFBP-1 have remarkably higher affinity for IGF-I than does nonphosphorylated IGFBP-1. This supports the concept that IGFBP-1 secreted by predecidualized/decidualized endometrial stromal cells has a high affinity for IGFs and therefore acts as an inhibitor of IGF action. According to studies by Koistinen and her co-workers (1993), the degree of IGFBP-1 phosphorylation increases from early to late gestation.

Implantation takes place around Day 6 after ovulation. A few days later, the embryo is surrounded by decidualized endometrial stromal cells. It is well established that progesterone-induced adaptive changes in the endometrium are essential for implantation to take place. Considering this, IGFBP-1 may be important for cell–cell interactions between embryonic and endometrial cells, beginning from the first days of the implantation process. This is indicated by expression of IGFBP-1 mRNA and protein secretion around the time of expected implantation, and by their dramatic increase during pregnancy. After implantation, decidual cells surrounding the embryo secrete increasing amounts of IGFBP-1, whereas the fetally derived trophoblasts possess IGF receptors and express IGF-II mRNA, but not IGFBP-1 mRNA (Marshall *et al.*, 1974; Ritvos *et al.*, 1988; Giudice, 1991b).

The presence of a number of IGF binders with equal affinities for IGFs in adjacent cells makes the hypothesis of paracrine regulation between trophoblastic and decidual cells attractive. Trophoblast cells have a high proliferative capacity, but some unknown mechanisms limit these cells from making too extensive an invasion into the uterine wall. If IGFs stimulate trophoblast proliferation, then IGFBP-1, by regulating the receptor binding of IGFs (Pekonen *et al.*, 1988; Ritvos *et al.*, 1988), may be one of the paracrine factors controlling proliferation and invasion of the trophoblast into maternal tissue.

G. IGFBP-2, -3, -4, and -5 in the Endometrium

Contrary to IGFBP-1 mRNA, expressed in secretory phase endometrium only, mRNAs for IGFBP-2 and IGFBP-3 (Giudice *et al.*, 1991b), and IGFBPs 4 and 5 (Rutanen *et al.*, 1993) are expressed in the endometrium throughout the menstrual cycle. Giudice and her co-workers (1991b) have shown that IGFBP-2 and IGFBP-3 mRNAs are expressed differentially in secretory and proliferative phase endometrium, and that the synthesis of both IGFBP-2 and IGFBP-3 proteins is regulated by estrogen and progesterone. As evident from Western ligand blotting and cross-linking studies, the amount of other IGFBPs in the endometrium is much lower compared with that of IGFBP-1 (Rutanen *et al.*, 1988, 1993; Pekonen *et al.*, 1992). Also, transformed endometrial cancer cells HEC 1A (Lamson *et al.*, 1989) secrete both IGFBP-2 and IGFBP-3. Despite steroid hormone dependence of endometrial IGFBP-1, -2, and -3 production, serum levels of these three binding proteins show no cyclic change during the menstrual cycle (Giudice 1991b; Rutanan *et al.*, 1992), nor is there any increase in circulating IGFBP-1 in women wearing a progestin-releasing intrauterine device despite the fact that IGFBP-1 is abundant in the endometrium (Pekonen *et al.*, 1992). This supports the concept of an autocrine/paracrine role for endometrial IGFBP-1.

IV. Placental Protein 14

A. Nomenclature

Placental protein 14 is an endometrial protein that has many names in the literature. Based on their immunological and physicochemical properties, progestagen-associated endometrial protein (PEP) (Joshi *et al.*, 1980; Julkunen *et al.*, 1986b), placental protein 14 (PP14) (Bohn *et al.*, 1982; Julkunen *et al.*, 1988b), pregnancy-associated α2-globulin (α2-PEG) (Bell, 1986), α-uterine protein (Sutcliffe *et al.*, 1982), and chorionic α2-microglobulin (Petrunin *et al.*, 1980) are probably the same protein. The primary structures of PP14 and α2-PEG are similar (Julkunen *et al.*, 1988b; Garde *et al.*, 1991), and they exhibit significant homology to β-lactoglobulins of various animal species (Huhtala *et al.*, 1987; Bell *et al.*, 1987). The Human Gene Mapping Nomenclature Committee (Winnipeg, Manitoba, Canada) has reserved PP for inorganic pyrophosphatase and PEP for peptidase genes. Therefore it has been suggested that PP14 should be renamed PAEP (progesterone-associated endometrial protein) instead of PP or PEP (Kämäräinen *et al.*, 1991). The confusion about the name will probably be solved when the function of the protein is better established. For these reasons, the name PAEP appears in this chapter synonymously with PP14, PEP, and α2-PEG, as they appear in the original papers.

B. Physicochemical Properties

PP14 is a glycoprotein containing 17% carbohydrate (Bohn *et al.*, 1982). Human midtrimester amniotic fluid contains up to 100 mg/liter PP14 and thus it is an excellent source for purification. Using monoclonal antibody columns we have purified PP14 in milligram quantities (Riittinen *et al.*, 1989, 1991). Purified PP14 has a molecular mass of 28 kDa in sodium dodecyl sulfate–polyacrylamide gel electrophoresis (SDS-PAGE) (Riittinen *et al.*, 1989) and, deducing from its primary structure, the predicted molecular weight of the mature protein is 18,787 (Julkunen *et al.*, 1988b). The 162-residue-long sequence of PP14 has 53.5% identity with the sequence of horse β-lactoglobulin (Julkunen *et al.*, 1988b). The human PP14 gene contains seven exons and six introns (Vaisse *et al.*, 1990), and it shows identical organization with the ovine β-lactoglobulin gene, confirming their common origin. The highest homology has been found in exon 2, which encodes the sequence involved in retinoic acid binding by β-lactoglobulin. There is N-terminal amino acid sequence homology between PP14 and human retinol-binding protein (23%), but PP14 does not bind retinol or retinoic acid (Seppälä *et al.*, 1988a).

Glucocorticoid/progesterone regulatory elements (PREs) have been found between the duplication and the "classic" *Alu* elements of the PP14 gene. These PREs bind immunopurified progesterone receptor *in vitro* (Vaisse *et al.*, 1990). The human PP14 gene has been localized to chromosome 9q34 (Van Cong *et al.*, 1991), and there is *Hinf*I restriction enzyme polymorphism in the PP14 gene: 1000 bp (allele A1), or 600 and 400 bp (allele A2). The frequency of allele A1 is 5% and that of allele A2 is 95% (Kämäräinen *et al.*, 1991). The polymorphism is not reflected in any detectable difference in the mobility of PP14 protein in PAGE.

C. Synthesis and Cellular Localization

Secretory and decidualized endometrium can incorporate labeled methionine into immunoreactive PP14 (Julkunen *et al.*, 1986a,b). Endometrial but not placental synthesis has been demonstrated by finding a 0.9-

kb PP14 mRNA in human secretory and decidualized endometrium, but not in postmenopausal endometrium, placenta, liver, kidney, and adrenals (Julkunen *et al.*, 1988b).

Studies employing a complementary (c) cRNA probe and *in situ* hybridization have shown PP14 messenger RNA (mRNA) in secretory endometrial glands (Julkunen *et al.*, 1990). Also, the deep basal glands during the first days of proliferative phase contain PP14 mRNA. This mRNA is also translated, as shown by detection of PP14 protein in deep basal glands in the same phases of the menstrual cycle (Seppälä *et al.*, 1988b). Based on the localization of both PP14 and its mRNA in the glands, PP14 appears to be exclusively a glandular marker (Seppälä *et al.*, 1988b; Julkunen *et al.*, 1990). PP14 and its mRNA disappear from the endometrium after Day 5 of the cycle and they remain undetectable in tissue until Day 5 after ovulation (Seppälä *et al.*, 1988b), i.e., around the time of implantation. Thereafter the concentration in tissue steeply increases, being highest in late luteal phase (Julkunen *et al.*, 1986a). All endometrial glands are strongly PP14 positive on Day 10 after ovulation (Seppälä *et al.*, 1988b). This is the time when significant amounts of PP14 are released by endometrial explants in tissue culture (Julkunen *et al.*, 1986a). However, a recent study by Li and his colleagues (1991) shows great variation in PP14 concentrations at different sites of the endometrium, with no consistent pattern, suggesting that the diagnostic value of the PP14 measurement in endometrial tissue is limited.

If the cycle is hyperstimulated, then PP14 may reappear in the endometrium as early as 28 hours after oocyte retrieval, indicating advanced endometrial maturation (Wahlström *et al.*, 1985). In keeping with the common embryonic origin of the uterus and the fallopian tubes, fallopian tubal mucosa also contains PP14 and its mRNA (Julkunen *et al.*, 1986c, 1990).

D. Biological Action

Bolton and his co-workers (1987) used mixed lymphocyte cultures to show that decidual extracts and seminal plasma exhibit immunosuppressive activity that is reduced but not completely inhibited by treatment with a monoclonal anti-PP14 antibody-based immunoadsorbent. They also showed that both decidual extract and purified PP14 dose-dependently suppress the mitogenic lymphocyte reactivity to phytohemagglutinin (Pockley *et al.*, 1988). These observations suggest that PP14 could exert an immunosuppressive effect in the endometrium. This line of research has been advanced by Okamoto and his co-workers (1991), who showed that PP14 suppresses the activity of T cells and natural killer (NK) cells. It can be argued that, because α2PEG/PP14 is excreted mainly into the glandular lumen, it would be difficult to conceive how the immunosuppressive action of PP14 is transferred into the implantation site surrounded by the stromal cells and stromal lymphocytes. Further studies are required to elucidate the possible effect of PP14 on endometrial lymphocytes.

V. Summary and Conclusions

Normal endometrial function and embryo–endometrial interactions require balanced secretion of ovarian hormones and well-orchestrated action by local factors that mediate cell–cell interactions to regulate cell growth and differentiation. A number of putative local regulators have been identified in the endometrium. Insulin-like growth factor-binding protein and progestagen-associated endometrial protein/PP14, reviewed in this chapter, are two major products of the secretory phase and decidualized endometrium, the first being produced by decidualized stromal cells and the latter by epithelial glands. Both the message and the concentration of the proteins increase markedly in endometrial tissue during the secretory phase. IGFBP-1 probably inhibits the local actions of IGFs, whereas the significance of the immunosuppressive properties of PP14 for reproduction remains a matter of further studies.

Acknowledgments

The authors' studies in this chapter were supported by grants from the Finnish Social Insurance Institution, the Sigrid Jusélius Foundation, the Academy of Finland, the Cancer Society of Finland, and the University of Helsinki.

References

Alitalo, T., Kontula, K., Koistinen, R., Julkunen, M., Jänne, O. A., Seppälä, M., and de la Chapelle, A. (1989). The gene encoding human low-molecular weight insulin-like growth factor binding protein (IGF-BP25): regional localization to 7p12-p13 and description of a DNA polymorphism. *Hum. Genet.* 83, 335–338.

Bell, S. C. (1986). Purification of human secretory pregnancy-associated endometrial α2-globulin (α2-PEG) from cytosol of first-trimester endometrium. *Hum. Reprod.* 1, 313–318.

Bell, S. C., Keyte, J. W., and Waites, G. T. (1987). Pregnancy-associated endometrial α2-globulin, the major secretory protein of the luteal phase and first trimester pregnancy endometrium is not glycosylated prolactin but related to β-lactoglobulins. *J. Clin. Endocrinol. Metab.* 65, 1067–1071.

Bell, S. C., Patel, S. R., Jackson, J. A., and Waites, G. T. (1988). Major secretory protein of human decidualized endometrium in pregnancy is an insulin-like growth factor-binding protein. *J. Endocrinol.* 118, 317–328.

Bell, S. C., Jackson, J. A., Ashmore, J., Zhu, H. H., and Tseng, L. (1991). Regulation of insulin-like growth factor-binding protein-1 synthesis and secretion by progestin and relaxin in long term cultures of human endometrial stromal cells. *J. Clin. Endocrinol. Metab.* 72, 1014–1024.

Boehm, K. D., Daimon, M., Gorodeski, I. G., Sheean, L. A., Utian, W. H., and Ilan, J. (1990). Expression of the insulin-like and platelet-derived growth factor genes in human uterine tissues. *Mol. Reprod. Dev.* 27, 93–101.

Bohn, H., Kraus, W., and Winckler, W. (1982). New soluble placental tissue proteins: their isolation, characterization, localization and quantification. *Placenta Suppl.* 4, 67–81.

Bolton, A. E., Pockley, A. G., Clough, K. J., Mowles, E. A., Stoker, R. J., Westwood, O. M. R., and Chapman, M. G. (1987). Identification of placental protein 14 as an immunosuppressive factor in human reproduction. *Lancet* 1, 593–595.

Bondy, C. A., Werner, H., Roberts, C. R., Jr., and LeRoith, D. (1990). Cellular pattern of insulin-like growth factor-I (IGF-I) and type I IGF receptor gene expression in early organogenesis: comparison with IGF-II gene expression. *Mol. Endocrinol.* 4, 1386–1398.

Brewer, M. T., Stetler, G. L., Squires, S. H., Thompson, R. C., Busby, W. H., and Clemmons, D. R. (1988). Cloning, characterization and expression of a human insulin-like growth factor binding protein. *Biochem. Biophys. Res. Commun.* 152, 1289–1297.

Clemmons, D. R., Camacho-Hubner, C., Jones, J. I., McCusker, R. H., and Busby, W. H. Jr. (1991). Insulin-like growth factor binding proteins: mechanisms of action at the cellular levels. *In* "Modern Concepts of Insulin-like Growth Factors" (E. M. Spencer, ed.), pp. 475–486. Elsevier, New York.

Dallenbach-Hellweg, G. (1987). The histopathology of the endometrium. *In* "Histopathology of the Endometrium," pp. 93–297. Springer-Verlag, Heidelberg.

D'Ercole, A. J. (1991). The insulin-like growth factors and fetal growth. *In* "Modern Concepts of Insulin-like Growth Factors" (E. M. Spencer, ed.), pp. 9–23. Elsevier, New York.

Daughaday, W. H., and Rotwein, P. (1989). Insulin-like growth factors I and II: peptide, messenger ribonucleic acid and gene structure, serum and tissue concentrations. *Endocr. Rev.* 10, 68–90.

Frost, R. A., and Tseng, L. (1991). Insulin-like growth factor-binding protein-1 is phosphorylated by cultured human endometrial stromal cells and multiple protein kinases in vitro. *J. Biol. Chem.* 266, 18082–18088.

Garde, J., Bell, S. C., and Eperon, I. C. (1991). Multiple forms of mRNA encoding human pregnancy-associated endometrial α2-globulin, a β-lactoglobulin homologue. *Proc. Natl. Acad. Sci. USA* 88, 2456–2460.

Giudice, L. C., Lamson, G., Rosenfeld, R. G., and Irwin, J. C. (1991a). Insulin-like growth factor-II (IGF-II) and IGF binding proteins in human endometrium. *Ann. N.Y. Acad. Sci.* 626, 295–307.

Giudice, L. C., Milkowski, D. A., Lamson, G., Rosenfeld, R. G., and Irwin, J. C. (1991b). Insulin-like growth factor binding proteins in human endometrium: steroid-dependent messenger ribonucleic acid expression and protein synthesis. *J. Clin. Endocrinol. Metab.* 72, 779–787.

Holly, J. M. P., and Wass, J. A. H. (1989). Insulin-like growth factors; autocrine, paracrine or endocrine? New perspectives of the somatomedin hypothesis in the light of recent developments. *J. Endocrinol.* 122:611–618.

Huhtala, M.-L., Seppälä, M., Närvänen, A., Palomäki, P., Julkunen, M., and Bohn, H. (1987). Amino acid sequence homology between human placental protein 14 and β-lactoglobulins from various species. *Endocrinology* 120, 2620–2622.

Irwin, J. C., Utian, W. H., and Eckert, R. L. (1991). Sex steroids and growth factors differentially regulate the growth and differentiation of cultured human endometrial stromal cells. *Endocrinology* 129, 2385–2392.

Joshi, S. G., Ebert, K. M., and Swartz, D. P. (1980). Detection and synthesis of a progestagen-dependent protein in human endometrium. *J. Reprod. Fertil.* 59, 273–285.

Julkunen, M. (1986). Human decidua synthesizes placental protein 14 (PP14) *in vitro*. *Acta. Endocrinol.* 112, 271–277.

Julkunen, M., Koistinen, R., Sjöberg, J., Rutanen, E.-M., Wahlström, T., and Seppälä, M. (1986a). Secretory endometrium synthesizes placental protein 14. *Endocrinology* 118, 1782–1786.

Julkunen, M., Raikar, R. S., Joshi, S. G., Bohn, H., and Seppälä, M. (1986b). Placental protein 14 and progestagen-dependent endometrial protein are immunologically indistinguishable. *Hum. Reprod.* 1, 7–8.

Julkunen, M., Wahlström, T., and Seppälä, M. (1986c). Human fallopian tube contains placental protein 14. *Am. J. Obstet. Gynecol.* 154, 1076–1079.

Julkunen, M., Koistinen, R., Aalto-Setälä, K., Seppälä, M., Jänne, O. A., and Kontula, K. (1988a). Primary structure of human insulin-like growth factor-binding protein/placental protein 12 and tissue specific expression of its mRNA. *FEBS Lett.* 236, 295–302.

Julkunen, M., Seppälä, M., and Jänne, O. (1988b). Complete amino acid sequence of human placental protein 14: a progesterone-regulated uterine protein homologous to β-lactoglobulins. *Proc. Natl. Acad. Sci. USA* 85, 8845–8849.

Julkunen, M., Koistinen, R., Suikkari, A.-M., Seppälä, M., and Jänne, O. A. (1990). Identification by hybridization histochemistry of human endometrial cells expressing mRNAs encoding a uterine β-lactoglobulin homologue and an insulin-like growth factor-binding protein-1. *Mol. Endocriol.* 4, 700–707.

Kämäräinen, M., Julkunen, M., and Seppälä, M. (1991). HinfI polymorphism in the human progesterone associated endometrial protein (PAEP). *Nucleic Acid Res.* 19, 5092.

Kamat, B. R., and Isaacson, P. G. (1987). The immunocytochemical distribution of leukocytic subpopulations in human endometrium. *Am. J. Pathol.* 127, 66–73.

Koistinen, R., Kalkkinen, N., Huhtala, M.-L., Seppälä, M., Bohn, H., and Rutanen, E.-M. (1986). Placental protein 12 is a decidual protein that binds somatomedin and has an identical N-terminal amino acid sequence with somatomedin-binding protein from human amniotic fluid. *Endocrinology* 118, 1375–1378.

Koistinen, R., Huhtala, M.-L., Stenman, U.-H., and Seppälä, M. (1987a). Purification of placental protein PP12 from human amniotic fluid and its comparison with PP12 from placenta by immunological, physicochemical and somatomedin-binding propeties. *Clin. Chim. Acta* 164, 293–303.

Koistinen, R., Stenman, U.-H., Alfthan, H., and Seppälä, M. (1987b). Time-resolved immunofluorometric assay for 34-kDa somatomedin-binding protein. *Clin. Chem.* 33, 1126–1128.

Koistinen, R., Angervo, M., Leinonen, P., Hakala, T., and Seppälä, M. (1993). Phosphorylation of insulin-like growth factor-binding protein 1 from different sources. *Growth Reg.* 3, 34–37.

Lamson, G., Oh, Y., Pham, H., Giudice, L. C., and Rosenfeld, R. G. (1989). Expression of two insulin-like growth factor-binding proteins in a human endometrial cancer cell line: structural, immunological, and genetic characterization. *J. Clin. Endocrinol. Metab.* 69, 852–859.

Lee, Y. L., Hintz, R. L., James, P. M., Lee, P. D. K., Shively, J. E., and Powell, D. R. (1988). Insulin-like growth factor (IGF) binding protein complementary deoxyribonucleic acid from human HEP G2 hepatoma cells: predicted protein sequence suggests an IGF binding domain different from those of the IGF-I and IGF-II receptors. *Mol. Endocrinol.* 2, 404–411.

Li, T. C., Pockley, G., Bolton, A. E., and Cooke, I. D. (1991). The variation of endometrial protein PP14 in different parts of the human endometrium. *Int. J. Gynecol. Obstet.* 34, 257–260.

Marshall, R. N., Underwood, L. E., Voine, S. J., Fousher, D. B., and Van Wyk, J. J. (1974). Characterization of insulin and somatomedin-C receptors in human placental cell membranes. *J. Clin. Endocrinol. Metab.* 39, 283–292.

Murphy, L. J., and Ghahary, A. (1990). Uterine insulin-like growth factor-1: Regulation of expression and its role in estrogen-induced uterine proliferation. *Endocr. Rev.* 11, 443–453.

Okamoto, N., Uchida, A., Takakura, K., Kariya, Y., Kanzaki, Y., Riittinen, L., Koistinen, R., Seppälä, M., and Mori, T. (1991). Suppression by human placental protein 14 of natural killer cell activity and T cell proliferative response. *Am. J. Reprod. Immunol.* 26, 137–142.

Pekonen, F., Suikkar, A.-M., Mäkinen, T., and Rutanen, E.-M. (1988). Different insulin-like growth factor binding species in human placenta and decidua. *J. Clin. Endocrinol. Metab.* 67, 1250–1257.

Pekonen, F., Nyman, T. and Rutanen, E.-M. (1991). Human endometrial adenocarcinoma cell lines HEC 1B and KLE secrete insulin-like growth factor binding protein-1 and contain IGF-I receptors. *Mol. Cell. Endocrinol.* 75, 81–87.

Pekonen, F., Nyman, T., Lähteenmäki, P., Haukkamaa, M., and Rutanen, E.-M. (1992). Intrauterine progestin induces continuous insulin-like growth factor-binding protein-1 production in the human endometrium. *J. Clin. Endocrinol. Metab.* 75, 660–664.

Petrunin, D. D., Kozlyaeva, G. A., Mesnyankina, N. V., and Shevchenko, O. P. (1980). Detection of chorionic alpha.2 microglobulin in the endometrium in the secretory phase of the menstrual cycle and in the male sperm. *Akush. i Ginek.* 3, 22–23. [In Russian]

Pockley, A. G., Mowles, E. A., Stoker, R. J., Westwood, O. M. R., Chapman, M. G., and Bolton, A. E. (1988). Suppression of *in vitro* lymphocyte reactivity to phytohemagglutinin by placental protein 14. *J. Reprod. Immunol.* 13, 31–39.

Povoa, G., Enberg, G., Jörnvall, H., and Hall, K. (1984). Isolation and characterization of a somatomedin-binding protein from mid-term human amniotic fluid. *Eur. J. Biochem.* 144, 199–204.

Riittinen, L., Julkunen, M., Seppälä, M., Koistinen, R., and Huhtala, M.-L. (1989). Purification and characterization of endometrial protein PP14 from mid-trimester amniotic fluid. *Clin. Chim. Acta* 184, 19–29.

Riittinen, L., Närvänen, O., Virtanen, I., and Seppälä, M. (1991). Monoclonal antibodies against endometrial protein PP14 and their use for purification and radioimmunoassay of PP14. *J. Immunol. Methods* 136, 85–90.

Ritvos, O., Ranta, T., Jalkanen, J., Suikkari, A.-M., Voutilainen R., Bohn, H., and Rutanen, E.-M. (1988). Insulin-like growth factor (IGF) binding protein from human decidua inhibits the binding and biological action of IGF-I in cultured choriocarcinoma cells. *Endocrinology* 122, 2150–2157.

Ruoslahti, E., and Pierschbacher, M. D. (1986). Arg-Gly-Asp: a versatile cell recognition signal. *Cell* 44, 517–518.

Rutanen, E.-M. (1992). Insulin-like growth factor binding protein-1. *Sem. Reprod. Endocrinol.* 10, 154–163.

Rutanen, E.-M., Bohn, H., and Seppälä, M. (1982). Radioimmunoassay of placental protein 12: levels in amniotic fluid, cord blood, and serum of healthy adults, pregnant women, and patients with trophoblastic disease. *Am. J. Obstet. Gynecol.* 144, 460–463.

Rutanen, E.-M., Koistinen, R., Wahlström, T., Sjöberg, J., Stenman, U.-H., and Seppälä, M. (1984). Placental protein 12 (PP12) in the human endometrium tissue concentration in relation to histology and serum levels of PP12, progesterone and oestradiol. *Br. J. Obstet. Gynecol.* 91, 1025–1030.

Rutanen, E.-M., Koistinen, R., Wahlström, T., Bohn, H., Ranta, T., and Seppälä, M. (1985). Synthesis of placental protein 12 by human decidua. *Endocrinology* 116, 1304–1309.

Rutanen, E.-M., Koistinen, R., Sjöberg, J., Julkunen, M., Wahlström, T., Bohn, H., and Seppälä, M. (1986a). Synthesis of placental protein 12 by human endometrium. *Endocrinology* 118, 1067–1071.

Rutanen, E.-M., Menabawey, M., Isaka, K., Bohn, H., Chard, T., and Grudzinskas, J. G. (1986b). Synthesis of placental protein 12 (PP12) by decidua from early pregnancy. *J. Clin. Endocrinol. Metab.* 63, 675–679.

Rutanen, E.-M., Pekonen, F., and Mäkinen, T. (1988). Soluble 34K binding protein inhibits the binding of insulin-like growth factor I to its cell receptors in human secretory phase endometrium: evidence for autocrine/paracrine regulation of growth factor action. *J. Clin. Endocrinol. Metab.* 66, 173–180.

Rutanen, E.-M., Gonzales, E., Said, J., and Braunstein, G. D. (1991). Immunohistochemical localization of the insulin-like growth factor binding protein-1 in female reproductive tissues by monoclonal antibodies. *Endocr. Pathol.* 2, 132–138.

Rutanen, E.-M., Pekonen, F., Nyman, T., and Wahlström, T. (1993). Insulin-like growth factors and their binding proteins in benign and malignant uterine diseases. *Growth Reg.* 3, 72–75.

Seppälä, M., Riittinen, L., Julkunen, M., Koistinen, R., Wahlström, T., Iino, K., Alfthan, H., Stenman, U.-H., and Huhtala, M.-L. (1988a). Structural studies, localization in tissue and clinical aspects of human endometrial proteins. *J. Reprod. Fertil. Suppl.* 36, 127–141.

Seppälä, M., Wahlström, T., Julkunen, M., Vartiainen, E., and Huhtala, M.-L. (1988b). Endometrial proteins as indicators of endometrial function. *In* "Placental and Endometrial Proteins: Basic and Clinical Aspects" (Y. Tomoda, S. Mizutani, O. Narita, and A. Klopper, eds.), pp. 35–42. VNU Science Press, Utrecht.

Shimasaki, S., and Ling, N. (1991). Identification and molecular characterization of insulin-like growth factor binding proteins (IGFBP-1, -2, -3, -4, -5 and -6). *Prog. Growth Factor Res.* 3, 243–266.

Steele-Perkins, G., Turner, J., Edman, J. C., Hari, J., Pierce, S. B., Stober, C., Rutter, W. J., and Roth, R. A. (1988). Expression and characterization of a functional human insulin-like growth factor I receptor. *J. Biol. Chem* 263, 11486–11492.

Stiles, C. D., Capone, G. T., Scher, C. D., Antoniades, H. N., Van Wyk, J. J., and Pledger, W. J. (1979). Dual control of cell growth by somatomedins and platelet-derived growth factor. *Proc. Natl. Acad. Sci. USA* 76, 1279–1283.

Sutcliffe, R. G., Joshi, S. G., Paterson, W. F., and Bank, J. F. (1982). Serological identity between human uterine protein and human progestagen-dependent endometrial protein. *J. Reprod. Fertil.* 65, 207–209.

Tabibzadeh, S. S. (1990a). Proliferative activity of lymphoid cells in human endometrium throughout the menstrual cycle. *J. Clin. Endocrinol. Metab.* 70, 437–443.

Tabibzadeh, S. S. (1990b). Evidence of T-cell activation and potential cytokine action in human endometrium. *J. Clin. Endocrinol. Metab.* 71, 645–649.

Vaisse, C., Atger, M., Potier, B., and Milgrom, E. (1990). Human placental protein 14 gene: sequence and characterization of a short duplication. *DNA Cell Biol.* 9, 401–413.

Van Cong, N., Vaisse, C., Gross, M.-S., Slim, R., Milgrom, E., and Bernheim, A. (1991). The human placental protein 14 (PP14) gene is localized on chromosome 9q34. *Hum. Genet* 86, 515–518.

Wahlström, T., and Seppälä, M. (1984). Placental protein 12 (PP12) is induced in the endometrium by progesterone. *Fertil. Steril.* 41, 781–784.

Wahlström, T., Koskimies, A. I., Tenhunen, A., Rutanen, E.-M., Yki-Järvinen, H., Julkunen, M., Sjöberg, J., and Seppälä, M. (1985). Pregnancy proteins in the endometrium after follicle aspiration for *in vitro* fertilizations. *Ann. N.Y. Acad. Sci.* 442, 402–407.

14

Molecular Biology of Trophectoderm and Placental Hormones

R. MICHAEL ROBERTS & RUSSELL V. ANTHONY

I. Placental Hormones

The placenta is a multifunction organ. It anchors the fetus within the uterus and, because it is the site of interchange of nutrients and waste materials between the conceptus and the mother, it forges a close spatial relationship with the blood supply of the uterus. It provides a cellular barrier to pathogens and an obstacle to the maternal immune system, yet participates extensively in the selective exchange of the biochemical information essential for a successful pregnancy. It is the likely target for a range of maternal as well as fetal hormones, and is itself an endocrine organ. However, the full range of hormones, growth factors, cytokines, and other bioactive factors produced by the placenta has not yet been defined. In all probability, the final list will be staggeringly large and will contain many endocrine factors synthesized in other complex organ systems (see Talamantes and Ogren, 1988). On the other hand, there are clearly a few that can, with little quibbling, be classified as "true" placental hormones because they are synthesized in large amount by the placenta and not in appreciable quantities at extraplacental sites, and have established or inferred functions during pregnancy.

The placental hormones that have been most intensively studied—the chorionic gonadotropins (CG), the trophoblast interferons (IFNτ), the placental lactogens (PL), and an expanding group, clearly not themselves PL, but nevertheless related to prolactin (PRL) and growth hormone (GH)—are the subject of this chapter. They are all stage-specific products of cells that are derived from the trophectoderm, a flattened extraembryonic epithelial layer that first becomes distinguishable from the embryo proper at the time of blastocoele formation. The developmental progression of this cell lineage varies in complexity among species and particularly according to the type of placenta that forms. In this chapter, we deal primarily with three quite different types of placenta—the epitheliochorial placenta of the horse (Renfree, 1982); the synepitheliochorial placenta, as exhibited by domestic ruminants such as cattle, sheep, and goats (Wooding, 1992); and the hemochorial placenta found in rodents and higher primates (Steven, 1975; Renfree, 1982).

II. The Placenta and Trophoblast Development

Placentas have generally been classified according to the number of cell layers that separate the blood supplies of the fetus and the mother, and by the manner in which the chorion is attached to the endometrium (see Renfree, 1982). For example, the pig has a diffuse, epitheliochorial placentation—diffuse because it attaches over the greater part of its surface, and epitheliochorial because the cell layers (epithelium, mesenchyme, and blood vessel endothelium) on both the maternal and fetal sides remain intact during the placentation. In this species the trophoblast cells exist as a continuous epithelium (the chorionic epithelium) which forms a close contact with the overlying uterine epithelium by intertwining microvilli on their opposing surfaces. Specialized domes of large absorptive cells (areolae) do develop opposite each uterine gland, but the uterine epithelium is never eroded and remains intact throughout pregnancy. It is unclear whether the chorionic epithelium (trophoblast) synthesizes a specific complement of polypeptide hormones, although it is almost certainly active in steroid biosynthesis. There is no strong evidence, however, for production of placental lactogens or chorionic gonadotropins during pregnancy in the pig, although just prior to attachment the expanding pig blastocyst produces two types of interferon (LaBonnardière *et al.*, 1991). However, these are formed in small quantities and appear to be structurally distinct from IFNτ characteristic of the ruminant at an equivalent stage in its development (Lefèvre *et al.*, 1990).

The horse placenta, discussed later in relation to CG synthesis, represents a variant of the porcine type of diffuse, epitheliochorial placentation. Attachment occurs late (~Day 35), well after maternal recognition of pregnancy, and is mediated by a belt of elongated trophoblast cells that rings the more or less spherical conceptus. Some of these cells also migrate into the maternal endometrium to form the so-called endometrial cups (Enders and Kim, 1991). It is these cup cells that produce equine chorionic gonadotropin. Later in pregnancy these structures are destroyed so that CG production is ephemeral.

The placentas of cattle, goats, and sheep have been classified as syndesmochorial cotyledonary (see Renfree, 1982). The former term implied that three fetal and two maternal cell layers remained intact as the placenta formed, but this assumption is probably simplistic since the maternal epithelium is not eroded, even within the placentomes (see later), but remains as an intact cell layer—part maternal, part trophoblast—as the result of fusion of migrating trophoblast binucleate cells and uterine epithelial cells. Wooding (1992), therefore, has suggested that a more appropriate term would by synepitheliochorial. Cotyledonary refers to the localized regions of protruding villous trophoblast (or cotyledons) that, beginning between the third and fourth weeks of pregnancy, intrude into the crypt-like maternal caruncles. These regions of fused cotyledon–caruncle are known as placentomes and provide the main regions for gas and nutrient exchange for the fetus. There is also extensive migration of trophoblast binucleate cells from the chorionic villi to the maternal epithelium within these structures. After fusion with the epithelium, the trinucleate cells so formed are able to release the content of their secretory granules directly into the maternal tissue below.

In sheep, continued binucleate cell fusion with the maternal epithelium provides a more or less continuous syncytial layer between the chorion and the maternal endometrium, but again the granules are discharged toward the maternal system (Wooding, 1992). The contents of these granules are various (Lee *et al.*, 1986; Xie *et al.*, 1991), but one of the defined components is PL (see Section IV). The binucleate cells constitute about 20% of the total trophoblast cells after placentation begins, but are first evident a few days earlier, just before attachment (Wooding, 1992). They probably arise from mononucleate cells after a nuclear division without cytokinesis. They then increase in size and form a full complement of granules before migrating. Although they may constitute a more differentiated cell type than the mononucleate population, they probably should not be regarded as the equivalent of the rodent giant cell (see later), which is also derived from a trophoblast mononucleate cell, but which has a single polyploid nucleus. Nevertheless, the differentiated giant cell is the major source of placental lactogen in the rodent.

Humans and most other primates, as well as rodent species, have a hemochorial discoid placenta. Here the absorptive villous surface of the placenta is in direct contact with maternal blood as the result of the partial erosion of all three maternal cell layers. The placenta itself forms a disk-like plate embedded partially or entirely within the uterine wall. A description of its full development in any of the species is beyond the scope of this chapter. In the human, as implantation proceeds, the proliferating cytotrophoblast differentiates at its leading invasive edge to form an extensive syncytium (Figure 1) (Aplin *et al.*, 1991). There is, therefore, most probably a developmental hierarchy, beginning with a population of stem cells, continuing through various stages of cytotrophoblast differentiation, and culminating in the syncytium, although there are probably numerous side branches from this path. As discussed later, the expression of CG mirrors this progression. The CG α-subunit is first synthesized by an intermediate stage cell, while β-subunit production follows later in more mature, transitional cytotrophoblast and syncytiotrophoblast. Only when both subunits are present can a functional heterodimer be formed. Placental lactogen, by contrast, is expressed only in the mature syncytium.

Although the pattern of placental growth is somewhat different in the mouse, there is an analogous path of trophoblast differentiation. In this species, the outermost layer of trophoblast in the junctional zone between the fetal placenta and maternal decidual cells is composed of giant trophoblast cells, which, as mentioned earlier, are large polyploid cells and seem to represent the culmination of trophoblast cell differentiation. They are also the major sites for production of the various placental lactogens and prolactin-like molecules in the rat and mouse (see Sections IV and V).

Therefore, in each of these species there appears to be a progression from a precursor stem cell, through intermediate mononucleate cells, to various differentiated end-stage derivatives that include the binucleate

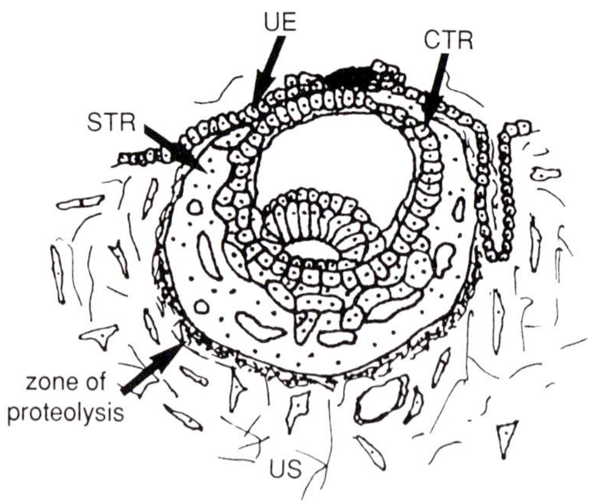

FIGURE 1 A human implantation site at approximately 9 days after conception. The outer layer of the conceptus consists of syncytiotrophoblast (STR) in which spaces (lacunae) filled with maternal blood are already visible. The syncytiotrophoblast can, therefore, release hCG directly into the maternal bloodstream. Adjacent to the outer syncytial surface at its interface with the maternal uterine stroma (US) is a zone of tissue degradation. Also shown is the uterine epithelium (UE) and cytotrophoblast layer (CTR), which gives rise to the syncytiotrophoblast. Reproduced with permission from Aplin (1991).

cells of the ruminants, the syncytiotrophoblast cells of primates, and the polyploid giant cells of rodents. Imprinted on this developmental hierarchy is the expression of various placental hormones.

III. Chorionic Gonadotropins

A. Function

Chorionic gonadotropins are considered to have their primary role in supporting the steroidogenic function of the corpus luteum (CL) during early pregnancy (see Braden and Niswender, Chapter 9, this volume). To accomplish this function, CG binds to luteal receptors for luteinizing hormone (LH), and activates the adenylate cyclase and possibly also the phosphotidylinositol second messenger systems (Davis *et al.*, 1986; Talamantes and Ogren, 1988). In this manner, CG supplements the activity of pituitary LH, thus ensuring continued production of progesterone by luteal cells. Progesterone itself is necessary for maintaining the uterus in a state that nourishes and maintains the implanting embryo and for preventing the normal cyclic turnover of the endometrium. In the absence of CG, e.g., if it is immunoneutralized, progesterone production drops, and the pregnancy fails (Hearn *et al.*, 1991). The presence of human (h) CG in maternal blood can first be detected within the first 8–10 days after conception, presumably reflecting the rapid access of the invading trophoblast to uterine capillaries (see Hearn, 1986; Hodgen and Itskovitz, 1988). These low concentrations of hCG, which are two to three orders of magnitude less than those detected 2 months later, are sufficient to induce maternal recognition of pregnancy. However, it still remains unclear whether CG alone is responsible for intervening in the normal cyclical onset of luteolysis. Conceivably other factors produced by the implanting conceptus may be involved (Hearn *et al.*, 1991). Injection of CG can extend luteal function in nonpregnant monkeys and women for a few days (Hodgen and Itskovitz, 1988) and may be of some clinical value in humans for rescuing late-developing embryos that have not yet begun to produce sufficient CG to prevent initiation of luteolysis (Nader *et al.*, 1988), but there are no reports that the CL can be maintained indefinitely by continued CG administration (Vandevoort *et al.*, 1988). Indeed, the mechanisms that trigger luteolysis during the normal menstrual cycles of primates are poorly understood (Auletta and Flint, 1988). Antiluteolytic (i.e., preventive) as well as luteotropic (i.e., stimulatory) mechanisms may be at play to maintain the CL of pregnancy for the full length of gestation.

The CL, however, although surviving until term, is only necessary for the first 4 to 8 weeks of human pregnancy (Auletta and Flint, 1988). As the conceptus increases in size, the placenta begins to assume the dominant role in steroid production, and it may be for this reason that CG production falls after the first 8 weeks (Figure 2) even though synthesis is maintained until term. Whether this continued synthesis of CG and its α-subunit, which is produced in excess and is detectable in serum in increasing amounts from the 6th to the 36th week (Ashitaka *et al.*, 1974; Reuter *et al.*, 1980), are critical for maintenance of pregnancy is unclear. Certainly the concentrations of hCG in serum at 10 weeks of pregnancy are far higher than what is needed to provide full occupation of the luteal cell receptor population. Therefore, additional roles for CG, other than in luteal support, have been proposed, including immunosuppression of maternal lymphocyte activity (see Bartocci *et al.*, 1983). Similarly, Blithe and colleagues (1991) have suggested that free α-subunit, which, because of its rapid clearance rates, never approaches the concentration in serum of the larger, intact hCG, regulates the production of decidual prolactin whose concentrations also rise markedly in the last trimester of human pregnancy (Blithe *et al.*, 1991). These aspects of CG function remain controversial and are beyond the scope of this chapter.

B. General Structure

Chorionic gonadotropin, along with lutropin (luteinizing hormone), follitropin (follicle-stimluating hormone or FSH), and thyrotropin (thyroid-stimulating hormone or TSH) are related glycoproteins, each consisting of two noncovalently associated subunits—a common α-subunit shared by all

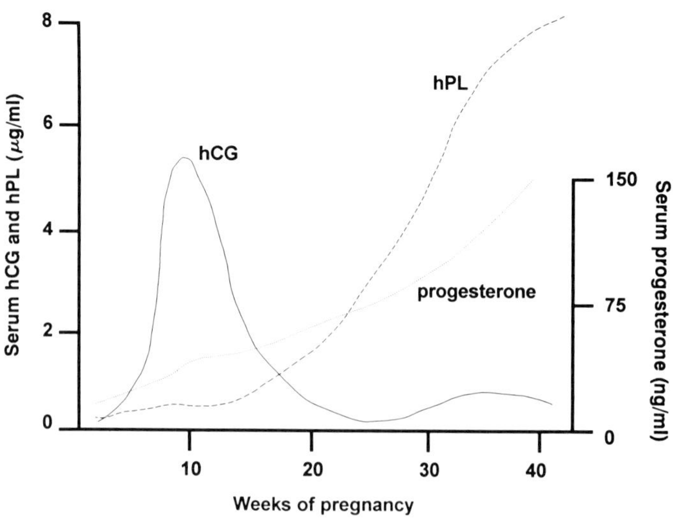

FIGURE 2 Concentrations of hCG, hPL, and progesterone over the course of pregnancy in serum of women. The figure summarizes data compiled from a variety of sources, including Tuchinsky and Hobel (1973), Ashitaka *et al.* (1974), and Braunstein *et al.* (1980).

four hormones and a β-subunit that imparts specificity (Fiddes and Goodman, 1981; Gharib *et al.*, 1990). However, it is now well established that the β-subunit itself has negligible biological activity (see Ryan *et al.*, 1987), and it is the unique conformation of the fully assembled dimer that confers receptor binding activity. Indeed, in the case of hCG, the α-subunit is probably partially sandwiched between the β-subunit and the receptor when the hormone becomes bound to its target (Ryan *et al.*, 1987; Moyle *et al.*, 1990).

All known mammals express LH, FSH, and TSH in their pituitaries (LH and FSH in the gonadotroph cells, TSH in thyrotroph cells) where each is regulated independently (Fiddes and Talmadge, 1984; Gharib *et al.*, 1990). For example, LH expression is induced by LH-releasing factor and inhibited by estrogen. In contrast, production of CG has only been unequivocally demonstrated in primates and equids and is confined to the placenta. Moreover, considerably less is known about how its expression is controlled.

The chemistry of hCG has been intensively studied because of its ready availability from placental tissue. The α-subunit, common also to the pituitary gonadotropins and highly conserved in sequence across species from mammals to fish, has a molecular mass of about 14,500 Da. Approximately 70% of the mass is contributed by the 92-residue amino acid chain (Birken *et al.*, 1980), the remainder by carbohydrate which is composed of two N-linked complex oligosaccharide chains (at positions Asn^{52} and Asn^{78}) (Ryan *et al.*, 1987; Kobata, 1988) (Figure 3). The α-subunits from most other species, including the horse, have a four-amino acid insert near the NH_2 terminus of the polypeptide, so that the carbohydrate chains are at Asn^{56} and Asn^{78} (Ryan *et al.*, 1987). All α-subunits have 10 conserved cysteines, which provide five disulfide bridges whose assignments are still not completely defined.

The β-subunit, which has a molecular mass of ~22,000 Da, consists of 145 amino acids, two complex N-linked carbohydrate chains (at Asn^{13} and Asn^{30}), and four O-linked chains linked to serine residues within 25 residues of the carboxyl terminus of the polypeptides (Figure 3). The CGβ- and LHβ-subunits, which are products of distinct genes, are very similar in sequence and presumed structure over their first 110 residues. All 12 cysteines (each in disulfide bonds) are placed identically. However, hCGβ has a 24-amino-acid extension present at its carboxyl terminus, apparently arising from two base changes in the gene that resulted in a shift in the position of the stop codon. This extension appears entirely superfluous since its absence does not influence subunit interaction or binding of hCG to the common LH/CG receptor (Talmadge *et al.*, 1984; Moyle *et al.*, 1990).

Although chemically deglycosylated (Harris *et al.*, 1989) and enzymatically desialylated (Lusthader *et al.*, 1989) hCG have been crystallized, there is at present no published three-dimensional description of the structure of either hCG or the other glycoprotein hormones. Nevertheless,

FIGURE 3 Structures of some of main types of sugar chains of hCG that have been purified from urine of pregnant women. Fuc, L-fucose; Gal, D-galactose; GlcNAc, *N*-acetyl D-glucosamine; GalNAc, *N*-acetyl D-galactosamine; Man, D-mannose; NeuAc, *N*-acetyl D-neuraminic acid. There is considerable microheterogeneity among the chains, particularly in the extent of sialic acid (NeuAc) addition. The fucosylated structures (1 and 2) and the O-linked chains (type 6) are only found on the β-subunit. Structures similar to 3 and 4 are found on both subunits. Type 5 monantennary structures are peculiar to the α-subunit. Assignments of specific N-linked chains to particular Asn residues is still incomplete. The oligosaccharides on hCG produced by choriocarcinoma cells appear to contain much less sialic acid. Used with permission from Kobata (1988). Copyright © 1988 Wiley-Liss, a division of John Wiley and Sons, Inc.

attempts have been made to model both the α- and β-subunits of hCG based on various predictive methods for secondary structure, likely placement of the disulfide bonds, data derived from cross-linking the subunits to each other, and comparative binding of monoclonal antibodies to the intact hormone and to the free subunits (Ryan *et al.*, 1987; Moyle *et al.*, 1990).

C. Expression of Chorionic Gonadotropins in Relation to Trophoblast Differentiation

Chorionic gonadotropin can be detected in maternal serum soon after the trophoblast intrudes into maternal blood vessels, i.e., after about the first week following conception in the human (Hearn *et al.*, 1991). Human blastocysts also begin to secrete CG spontaneously in culture at about the time they first form outgrowths (Fishel *et al.*, 1984). Thus, CG production is initiated at a relatively early stage of conceptus development, presumably reflecting the necessity of intervening in maternal ovarian function well before luteal regression is irreversibly launched. As the pregnancy progresses, hCG in maternal serum rises markedly to reach a maximum at about 10 weeks of pregnancy when concentrations of 5 μg/ml are typically observed (Figure 2). Levels of CG then decline, although continued placental synthesis of both subunits continues until birth. However, whereas in early pregnancy α-subunit synthesis only slightly exceeds that of β, by term the imbalance in both biosynthetic rates and mRNA concentrations is considerable (Boothby *et al.*, 1983). As a consequence, free α-subunit concentration in serum rises markedly in the last two trimesters (see Section III,A). Thus, α- and β-subunit production appear to be controlled independently, although in early pregnancy, perhaps when it is most necessary, a stoichiometry appropriate for heterodimer formation is maintained.

Synthesis of CG and placental lactogens (Section IV) has been studied intensively in cell lines that have been established from choriocarcinomas, which are highly malignant tumors of trophoblast origin. These cells, which include Jar, BeWo, and JEG 3 (Ringler and Strauss, 1990), seem to represent precursor stages on a pathway that culminates in syncytial cell formation, and can often be induced to acquire the phenotype of more advanced stages when they are treated with reagents that reduce proliferation (Otani *et al.*, 1989; Arbiser *et al.*, 1991; Sibley *et al.*, 1991). Such induced cellular differentiation markedly promotes increased hCG production. However, even within a given population of normally cultured Jar cells, there is a spectrum of cell types, with only the largest nondividing cells seemingly capable of expressing large amounts of CGβ (Hochberg et al., 1992). *In situ* hybridization (Hoshima *et al.*, 1982) and immunocytochemistry (Gaspard *et al.*, 1980; Kurman *et al.*, 1984) performed on sections of placental tissue have confirmed that expression of individual subunit genes is closely linked to stage of trophoblast differentiation. CGβ, for example, is localized only to occasional cytotrophoblast cells, but is expressed highly in syncytium. CGα, on the other hand, is more generally expressed in cytotrophoblast. Similarly, agents such as epidermal growth factor (EGF), which promote differentiation of cytotrophoblast into syncytium, induce increased CG expression (Maruo *et al.*, 1987; Morrish *et al.*, 1987), while hormones that inhibit this transition such as transforming growth factor-β (TGFβ) reduce CG synthesis (Morrish *et al.*, 1991). All these studies are consistent with the conclusion that the α-subunit gene is first transcribed at a considerably earlier stage on the pathway of trophoblast differentiation than the β-subunit, whose expression is only detectable just prior to syncytical cell formation. Transcriptional rates of both subunits become maximal in syncytium.

Although it remains unclear what controls hCG synthesis and secretion in a functioning placenta, a range of compounds, including gonadotropin-releasing hormone (GnRH), various steroids and cytokines, have been proposed (see Talamantes and Ogren, 1988; Masuhiro *et al.*, 1991). What is less controversial, however, is that hCG synthesis is strongly enhanced by increased cellular concentrations of 3′5′-cyclic adenosine monophosphate (cAMP) (see Ringler *et al.*, 1989; Talamantes and Ogren, 1988), which may be the functional second messenger system *in vivo*. The very potent effects of cAMP on isolated cytotrophoblast cells are illustrated in Figure 4. There is a relatively early increase in α-subunit mRNA expression, followed several hours later by an upsurge in β-subunit transcripts and hCG secretion. As will be discussed later, increased α gene expression can be explained on the basis of cAMP response elements (CRE) in the proximal promoter region of the gene. The enhanced expression of hCGβ, however, is more puzzling since its gene lacks a well-defined CRE and may be responding only secondarily to cAMP. Conceivably, cAMP influences cytotrophoblast cell differentiation, although there is no evidence that it promotes syncytium formation in cultured cytotrophoblast (Ringler *et al.*, 1989).

Glucocoritcoids may also influence CG synthesis, but reports have been contradictory. Ringler *et al.* (1989) reported an increase in hCG in cytotrophoblast, whereas Akerblom *et al.* (1988) showed transcription of the CG-α gene to be suppressed in JEG-3 choriocarcinoma cells.

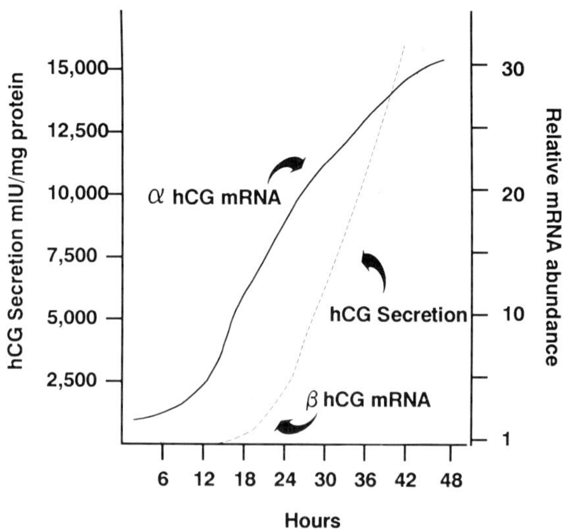

FIGURE 4 The effects of 8-bromo-cAMP on CG secretion by human cytotrophoblast and on the relative concentrations of CGα- and β-subunits. Cytotrophoblast cells were isolated from term placentas and exposed to 1.5 m*M* cAMP for periods up to 48 hours. CG was measured by immunoassay and the mRNA by Northern blot analysis. The figure is based on the data of Ringler et al. (1989). Used with permission from Ringler and Strauss (1990).

D. Gene Organization of the Chorionic Gonadotropin

1. α-Subunit gene

In humans, the CGα and β genes probably evolved from a common ancestral gene but are present on different chromosomes (6 and 19, respectively) (Naylor *et al.*, 1983). There is only a single α gene per haploid complement of DNA, and it consists of four exons separated by three introns (Figure 5), the first of which is relatively long, variable between species, and interrupts the 5′-untranslated region of the gene (reviewed by Gharib *et al.*, 1990). The α-subunit gene has been cloned from numerous species and is well conserved (Steger *et al.*, 1991; Nilson *et al.*, 1991). The transcription start site is placed normally, −23 bases behind a standard TATA box (Delegeane *et al.*, 1987), and the gene codes for a 116-amino acid polypeptide from which a 24-residue signal peptide is cleaved.

2. β-Subunit gene

In contrast, there are six structurally similar hCGβ genes clustered within a 40-kb region upstream of the single LHβ gene to which they are closely related and from which they are presumed to have evolved by an initial gene duplication event (Talmadge *et al.*, 1984). The CG and LHβ genes consist of three exons (Figure 5), and those that have been sequenced

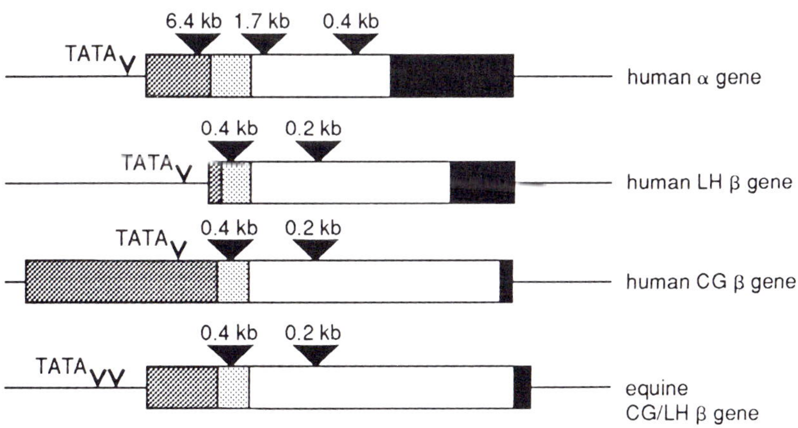

FIGURE 5 Diagram depicting the structures of the genes coding for human CG/LH α-subunit, human LHβ-subunit, human CGβ-subunit, and equine CG/LH β-subunit. Filled triangles indicate the positions of introns whose sizes are also shown. The bars represent the transcribed portions of the genes and include the untranslated regions (heavy shading), the region coding for the signal sequences (light shading), and the mature protein (unshaded). The positions of the TATA boxes are indicated. Although there are six CGβ genes, they are all structurally similar. Note that transcription of CGβ genes is not adjacent to the TATA box. This figure is based, in part, on Gharib *et al.* (1990) and on the data of Sherman *et al.* (1992). Used with permission from Gharib *et al.* (1990). Copyright © 1990 The Endocrine Society.

have well-conserved intron/exon boundaries (Policastro *et al.*, 1986; Gharib *et al.*, 1990). At least one of these six hCGβ genes (β1) undergoes some form of alternative splicing (Bo and Boime, 1992), but the nature of the protein product that ultimately results has not been characterized. Originally only three of six genes (1, 3, and 5) were considered to have functional promoters and to be expressed, but more recent experiments employing the sensitive technique of reverse transcription polymerase chain reaction (RT-PCR) to amplify rare transcripts from first trimester placental tissue and Jar choriocarcinoma cells have demonstrated that all the β genes are transcribed in trophoblast tissue, and that their mRNAs are most probably translated (Bo and Boime, 1992). The most actively transcribed are β5, 3, and 8, in that order.

Although the LHβ genes employ a TATAA sequence as a promoter, the hCG genes, although possessing an apparently conventional TATA box, are transcribed from a region lacking such a promoter ~350 nucleotides further upstream (Jameson *et al.*, 1986b) (Figure 5). The horse, which has only a single LH/CGβ gene, is anomalous and transcribes its CG/LHβ subunit mRNA conventionally in both placenta and pituitary, although two adjacent TATA elements are present (Sherman *et al.*, 1992). The presence of only a single CG/LHβ gene in horses was not unexpected since both placenta and pituitary β-subunit polypeptides had been shown to possess identical amino acid sequences (Sugino *et al.*, 1987). Interestingly, at both locations the polypeptide carries an extended carboxyl terminal analogous to but structurally dissimilar to the one on hCGβ (Sherman *et al.*, 1992).

E. Transcriptional Control of Gene Expression

1. α-Subunit gene

As discussed earlier, the α gene is expressed in certain pituitary cells, where it is itself under distinctive endocrine control, and in the placentas of primates and horses. It is not expressed in the placenta of any other known species, including rodents or domestic ruminants such as sheep and cattle. Its expression in the trophoblast depends on the stage of differentiation of the cells, and is responsive to certain hormones. In particular, it is markedly upregulated by elevations in intracellular cAMP (Jameson *et al.*, 1986a). The mechanisms that are involved in controlling the expression of the single α-subunit gene have, as a consequence, been intensively studied.

The placenta-specific expression and cAMP responsiveness have been shown by gene transfer experiments to reside in the first 180 nucleotides upstream of the transcription start site of the gene (Delegeane *et al.*, 1987). Basal expression is mediated by typical TATA and CCAAT promoter elements. However, the enhancer that has a major role in placental expression consists of two adjacent regions of DNA. The first is a trophoblast-specific element (TSE), often also referred to as the upstream regulatory element (or URE), which is located approximately between positions −180 and −146 of the gene, and which binds at least two distinct polypeptides in nuclear extracts prepared from choriocarcinoma cells (Delegeane *et al.*, 1987; Nilson *et al.*, 1991). Immediately downstream (−146 to −110) to the TSE is the second region of the enhancer that contains two tandem repeats of a well-defined cAMP response element, each of which is centered around an 8-bp palindromic core sequence (TGACGTCA) (Figure 6). The TSE confers no transcriptional activity when placed ahead of a basal promoter, but in the presence of one or both CRE increases activity in the presence of cAMP severalfold more in choriocarcinoma cells than in nontrophoblast cells.

The CRE bind an ubiquitous transcription factor known as CRE-binding protein (or CREB), which is found in most known cell types and which is activated by phosphorylation as cAMP concentrations rise (Hoeffler *et al.*, 1989). As a consequence, the CRE, unlike the TSE, are able to confer cAMP responsiveness on heterologous reporter genes in a wide variety of cells. Thus, the TSE and CRE act synergistically to provide for tissue-specific, cAMP-dependent expression of the α-subunit gene. Together, they have all the properties of a classical enhancer in that the entire unit

```
             TSE    CRE-2               CRE-1
         -149                                                -103
Human      atc aaattgacgtcatggtaa aaattgacgtcatggtaa ttacacca
Gorilla    atc aaattgacgtcatggtaa aaattgacgtcatggtaa ttacacca
Rhesus     atc ------------------ aaattgacgtcatggtaa ttacacca
Horse      atc ------------------ aaattgaTgtcatA-taa ttacaTGa
Cattle     atA ------------------ aaattgaTgtcatggtaa ttaTGTGa
Sheep      aGA ------------------ aaattgaTgtcatggtaa ttaTGTGa
Pig        atc ------------------ aaattgaTgtcatggtaa ttacacAa
Rabbit     atc ------------------ aaattgaTgtcatggtaa ttacacca
Mouse      atc ------------------ -aattgaTgtcatggtaa ttaTacca
```

FIGURE 6 The nucleotide sequences of the cAMP response elements (CRE) within the CGα genes of a variety of species. These CRE are immediately adjacent to a trophoblast-specific element (TSE, or upstream response element, URE), which together control placental expression in primate species. Note the presence of only a single functional CRE in rhesus and the presence of base changes (underlined capital letters) in the region homologous to the CRE in the genes of the other species. These base changes render the CRE incapable of binding CREβ with high affinity. The figure is based on that of Nilson *et al.* (1991). Reproduced with permission from Nilson *et al.* (1991).

can operate in either orientation and in a distant-independent manner when placed ahead of a minimal promoter (see Nilson *et al.*, 1991).

There is now good evidence that the lack of expression of an α-subunit in species other than primates can be explained solely on the basis of the sequences within this enhancer (Nilson *et al.*, 1991). The presence of the tandemly arranged CRE is certainly not necessary, since Old World monkeys—baboons and rhesus species—have only a single CRE yet produce a placental CG. Moreover, only one CRE seems required to fully activate an adjacent TSE. However, close examination of the single CRE of species known not to express CG, e.g., various Bovidae, rodents, pigs, and rabbits, has revealed the presence of one or more changes likely to render the CRE incapable of binding the transcriptional regulator CREB (Figure 6). The genes, therefore, lack cyclic AMP responsiveness and, because of the dependence of the TSE on a functional CRE element, have no placental-specific expression.

Interestingly, the TSE of many of these species, which generally appears to be well conserved (Figure 7), seems not to be defective. For example, bovine TSE binds the same trophoblast nuclear proteins as the human TSE and is active when placed adjacent to a fully functional CRE, under which circumstances it can impart trophoblast-specific expression to either transfected choriocarcinoma cells or transgenic mice (see Nilson *et al.*, 1991). The latter observation is of particular interest since it implies that all the necessary transcription factors are present in the murine trophoblast that would allow the murine α-subunit gene to be expressed if only its CRE did not contain a single T→C base change that renders it unresponsive to cAMP.

The horse, as ever, is a puzzle. Its single CRE has three base changes likely to render it inactive (Steger *et al.*, 1991), yet placental expression in

```
        -201                                                  -147
human   cacctgaaaatggctcCaaac-aaaaatgacctaagggttgaaacaagataagatc
equine  cGcctgaaaatggctcaaaacaaaaaatgatctaagAgttgaaacaagataagatc
bovine  caTctgaaaatggctcaaaacaaaaaatAatctaagggCtgaaacaagataagatA
```

FIGURE 7 The nucleotide sequences of the region of the CGα gene encompassing tissue-specific elements (TSE/URE). Sequences are shown for the human, equine, and bovine genes and clearly show the high degree of conservation across species. The bovine TSE is known to confer placental expression if placed adjacent to a functional CRE. Reproduced with permission from Nilson *et al.* (1991).

chorionic girdle cells is high. It remains unclear whether the horse TSE, in absence of a functional CRE, is still part of an enhancer that can confer tissue-specific expression (see Steger *et al.*, 1991), or whether control lies elsewhere, possibly in a more distal region of the gene.

2. β-Subunit gene

It was pointed out earlier that in primates (but not in horses) the CGβ subunit, though resembling LHβ, is coded by a set of genes that is not expressed strongly in pituitary cells. These genes also seem to be first transcribed at a slightly later stage of trophoblast cell differentiation than the α gene, although transcription of both α and β genes must be at least partially coordinated in the syncytial cell to provide for stoichometrically equivalent amounts of the two subunit polypeptides. Amounts of CGβ mRNA increase dramatically in response to cAMP in cytotrophoblast, but transcription is delayed relative to CGα mRNA (Figure 4). Despite these superficial similarities in control, there are no obvious regions of sequence identity within the promoter regions of the β- and α-subunit genes. In addition, the six β genes themselves diverge significantly in the regions upstream of the transcription start sites, and they are expressed at widely different rates in placenta. Despite these results, our understanding of the transcriptional control operating over CGβ genes remains poor, although a series of elements within the upstream promoter region have been implicated, including one that is cAMP-responsive yet does not conform to a classical CRE (Albanese *et al.*, 1991).

The proximal promoter regions from all six β genes, when placed ahead of a reporter gene and transfected into choriocarcinoma cells, are actively transcribed in the same order as the intact genes in placenta, i.e., β5 > β3 = β8 > β7, β1, β2 whereas the LHβ-subunit promoter remains silent (Hollenberg *et al.*, 1993). The entire gene cluster has also been microinjected as a unit into pronuclei of fertilized mouse oocytes, and several transgenic strains generated (Strauss *et al.*, 1993). Expression occurred in mouse placenta, with CGβ5 accounting for 68% of the transcripts and β3 and β8 accounting for the rest. The transgenic mouse, therefore, is likely to be of value for functionally dissecting the hCGβ promoter. Interestingly, however, expression was not completely placenta specific because some transcripts were detected in pituitary and brain.

F. Effects of Carbohydrate on hCG

At least three effects of carbohydrate chains on glycoproteins are illustrated by a study of hCG. First, the circulating half-life of hCG, which is greater than 24 hours (Nisula *et al.*, 1989), can be markedly shortened by removing terminal sialic acid residues from the N-linked oligosaccharide chains (Rosa *et al.*, 1984). The exposed galactosyl residues can be recognized by the asialoglycoprotein receptor of hepatocytes and the hormone cleared within minutes from the blood. Thus, carbohydrate structure influences the effective biological half-life of the hormone.

A second effect has become evident by using site-directed mutagenesis to allow selective removal of different carbohydrate chains on the α-subunit (Bielinska *et al.*, 1989). Both rates of secretion and the ease of dimer assembly are strongly influenced by the presence of carbohydrate chains at particular positions. Presumably both tertiary and quaternary interactions are influenced by carbohydrate placement and processing.

A third effect is the activation of the adenylate cyclase system (Ryan *et al.*, 1987; Sairam, 1989). Chemically deglycosylated hCG, although capable of binding to the receptor with high affinity, does not generate a cellular response. Similarly, the selective removal of any or all of the N-linked oligosaccharide on either subunit by site-directed mutagenesis has little effect on receptor binding (Matzuk *et al.*, 1989). On the other hand, receptors containing an unglycosylated β-subunit and an α-subunit lacking Asn^{52}, and therefore lacking an oligosaccharide at that position, fail to elevate cAMP concentrations or stimulate steroid production. The role of carbohydrate chains in this context is still not clear, though it is possible that they recruit accessory proteins into a fully functional receptor complex (Ryan *et al.*, 1987).

IV. Placental Lactogens

A. Function

Trophectoderm from a wide variety of species produces proteins that are related structurally to either pituitary prolactin or growth hormone. The placental lactogens (also known as chorionic somatomammotropin, CS), so named for their activity in lactogenic bioassays and radioreceptor assays, have been the most extensively studied of these placental proteins. More recently, through the use of newer technology available in protein biochemistry and molecular biology, the existence of a family of placental prolactin-related proteins, distinct from the PL, has been demonstrated (Section V). In addition, placental expression of the human growth hormone variant gene (hGH-V) has been reported (Section VI). Each of these subclasses of the prolactin/growth hormone family is discussed separately, although there is overlap in their respective structures and most probably in their function.

1. Human

The existence of PL has been demonstrated in a variety of species, including primates, rodents, and ruminants, but not in others such as the domestic cat, dog, horse, and pig (Forsyth, 1986). Also, as detailed later, the PL characterized structurally to date are either similar to pituitary-derived prolactin (PRL) or to growth hormone. As a consequence, it has been difficult to determine precisely what function(s) PL has during pregnancy. In the case of all the PLs, including hPL, it has been suggested that the hormone may be involved in regulating maternal metabolism, fetal growth and metabolism, and/or the preparation of the mammary gland for lactation. Much of the research on the biological actions of hPL has been recently reviewed (Handwerger, 1991), and is beyond the scope of this chapter. One aspect of hPL physiology that has yet to be resolved is whether its actions are being mediated through unique receptors that are truly specific for hPL. Although hPL and hGH share an 85% primary amino acid sequence identity, hPL binds the hGH receptor poorly (Lowman *et al.*, 1991). Affinity for the hPRL receptor is also low. Nevertheless, it is still possible that some of the biological actions attributed to hPL, especially within the maternal system, may be mediated through these receptors. However, recent results suggest that there are specific hPL receptors in fetal liver and skeletal muscle (Hill *et al.*, 1988), but until these are better characterized, our understanding of hPL function is unlikely to progress.

2. Rodents

In rodents, two PLs with different gestational profiles have been detected in maternal serum (see Section IV,C). Both are structurally more similar to pituitary PRL than to GH, but again the role(s) of these PLs is not understood (see Ogren and Talamantes, 1988; Soares *et al.*, 1991), nor is it clear whether there are specific PL receptors in rodents.

3. Ruminants

Prolactin-like PLs have been identified and purified from the placentas of cattle (Murthy *et al.*, 1982; Arima and Bremel, 1983), goats (Currie *et al.*, 1990), and sheep (Martal and Djiane, 1975; Chan *et al.*, 1976; Hurley *et al.*, 1977). A great deal of research has been conducted on their biological activities (see Handwerger, 1991; Byatt *et al.*, 1992). However, their functions still remain elusive. Although bovine PL (bPL) and ovine PL (oPL) bind to both GH and PRL receptors, specific binding sites have also been identified. In cattle, a bPL specific binding site has been identified in the pregnant uterine endometrium (Galosy *et al.*, 1991; Kessler *et al.*, 1991), but it has not been characterized biochemically, nor has its cDNA been cloned. The oPL binding sites present in maternal and fetal sheep liver have been the most extensively examined (Freemark *et al.*, 1986, 1987, 1988; Freemark and Comer, 1989), primarily because there is no specific GH binding to the fetal liver until term (Gluckman *et al.*, 1983). Properties of the oPL receptor reported so far indicate that it is a high-affinity binding site (K_d~0.27 nM; Freemark *et al.*, 1986, 1987) with an apparent M_r~38–44,000 (Freemark *et al.*, 1987, 1988; Freemark and Comer, 1989). Freemark and Comer (1989) reported the purification of the oPL receptor from maternal and fetal liver, but the yield was too small to allow a detailed study of its biochemical properties. Despite the PRL-like sequence of ruminant PL, recent results obtained with immunoblotting (Freemark, 1991) and immunoneutralization of binding sites (Byatt *et al.*, 1992) suggest that the ruminant PL receptors may be variant forms of the GH receptor. The purification and molecular cloning of one of the ruminant PL receptors could rapidly advance our understanding of their biological actions.

B. General Structure

1. Primate

Although the GH-like PL have been identified, partially characterized, or purified from the chimpanzee and orangutan (Forsyth, 1986), baboon (Josimovich *et al.*, 1973), and rhesus monkey (Shome and Friesen, 1971; Vinik *et al.*, 1973), the human protein has been the most extensively studied primate PL. It is a 191-amino-acid, single-chain polypeptide (M_r~22,000) that lacks carbohydrate and has two disulfide bonds. A total of 167 out of 191 amino acids are found at identical positions in human GH (hGH) (Bewley *et al.*, 1972; Li *et al.*, 1973). Of the PL whose primary amino acid sequences are known, hPL retains the least amino acid identity (44 out of 191) with human PRL (hPRL) (Shome and Parlow, 1977), although their homology to each other is increased considerably if conservative amino acid substitutions are considered (Cooke *et al.*, 1981).

2. Rodents

The rodent PLs resemble PRL more than they do pituitary GH. Figure 8 summarizes the sequence identities of rat PLs with rat PRL and GH. In each of the species there are two major types (PL-I and II) with distinctive structures and dissimilar patterns of production during pregnancy. The rat possesses a third family member, the so-called PL-I variant (Deb *et al.*, 1991a; Robertson *et al.*, 1991). Murine PL-I is a glycoprotein that separates into broad bands (M_r 29,000–32,000 and 36,500–42,000) upon electrophoresis in SDS–polyacrylamide gels. Rat PL-I is somewhat larger (M_r 40,000–50,000) (Colosi *et al.*, 1987a; Robertson *et al.*, 1982). The heterodispersity in size probably results from variability in carbohydrate composition between isoforms. Mouse and rat PL-I are both generated by removal of a 30-amino acid signal peptide from 224 and 230 amino acid precursor polypeptides respectively (Colosi *et al.*, 1987b; Robertson *et al.*, 1990). Both contain two sites for N-linked glycosylation and appear to lack the large disulfide-linked loop between Cys^{51} and Cys^{169} that is characteristic of all other members of the GH/PRL family. Instead, Cys^{169} is probably linked to closely adjacent Cys at position 167 (Colosi *et al.*, 1987b; Robertson *et al.*, 1990). All the rodent PL-IIs so far investigated are 191-amino acid polypep-

	PL-I	PL-Iv	PL-II	PLP-A	PLP-B	PLP-C	PRL	GH
PL-I		79	39	26	24	24	28	21
PL-Iv	85		43	34	26	26	32	24
PL-II	47	52		29	30	27	36	27
PLP-A	40	45	41		32	25	30	22
PLP-B	33	38	41	41		27	31	21
PLP-C	38	35	36	31	36		25	19
PRL	38	43	47	39	39	34		22
GH	31	33	33	29	30	24	32	

Sequence identity (upper, unshaded) Sequence similarity (lower, shaded)

FIGURE 8 The primary amino acid sequence identities and similarities among members of the rat placental PRL-GH family. Similar amino acids are: K and R; D and E; S and T; and V, L and I. See text for appropriate references.

tides that lack carbohydrate (Jackson *et al.*, 1986; Duckworth *et al.*, 1986a; Southard *et al.*, 1989). The rPL-Iv has an apparent M_r of 33,000 and, like rPL-I, contains two N-linked glycosylation sites and the additional Cys residue (Deb *et al.*, 1991a; Robertson *et al.*, 1991).

3. Ruminants

The primary amino acid sequences have been determined for bovine PL (Schuler *et al.*, 1988; Yamakawa *et al.*, 1990) and ovine PL (Colosi *et al.*, 1989; Warren *et al.*, 1990b). Like the rodent PL, bPL, and oPL are structurally more similar to PRL (48–50%) than they are to GH (22–24%), but there exists only one known PL within each ruminant species. Although bPL and oPL are 67% identical in amino acid sequences (Warren *et al.*, 1990b), the latter is a single nonglycosylated polypeptide with an apparent M_r of 22,000 (Warren *et al.*, 1990a), whereas bPL consists of multiple isoforms (Byatt *et al.*, 1986) with an apparent M_r on electrophoretic gels of 32–34,000 (Murthy *et al.*, 1982; Arima and Bremel, 1983). Bovine PL is glycosylated, containing both N-linked and O-linked oligosaccharides (Shimomura and Bremel, 1988; Byatt *et al.*, 1990), although enzymatic deglycosylation does not adversely effect its activity in somatotropin radioreceptor assay (Byatt *et al.*, 1990).

C. Expression of Placental Lactogens in Relation to Trophoblast Differentiation

1. Human

Human PL appears in the maternal circulation at about 21 days post-conception (Spellacy *et al.*, 1966; Beck, 1970). There is a steady rise in maternal hPL concentrations until about Weeks 34 to 35, after which time levels remain relatively constant (~10 μg/ml) (Berle, 1974; Braunstein *et al.*, 1989) (Figure 2). At term, hPL is also found in the fetal circulation and amniotic fluid, but at concentrations considerably less than in maternal serum (Keller *et al.*, 1970; Houghton *et al.*, 1984). However, at 12 to 19 weeks of gestation, concentrations of hPL in fetal circulation range from 4 to 500 ng/ml (mean = 82 ng/ml) (Hill *et al.*, 1988). *In situ* hybridization studies (McWilliams and Boime, 1980; Hoshina *et al.*, 1982; Boime *et al.*, 1982) have shown hPL mRNA to be localized exclusively to syncytiotrophoblast cells, and the concentration of hPL mRNA within individual syncytiotrophoblast cells do not change significantly throughout pregnancy (Hoshina *et al.*, 1982). However, the total mass of syncytiotrophoblast tissue increases as gestation progresses and is probably responsible for the fivefold increase in hPL mRNA between the first and last trimester (Boime *et al.*, 1982). Several growth factors, secretogogues, and metabolic regulators have been implicated in regulating hPL synthesis and secretion (Handwerger, 1991), but no definitive control mechanisms have so far emerged from these studies.

2. Rodent

Rodent PL-I and PL-II not only differ structurally, but show a different pattern of expression within trophoblast giant cells. PL-I synthesis, for example, is initiated shortly after implantation but terminates around midgestation in both the mouse (Nieder and Jennes, 1990; Faria *et al.*, 1991) and rat (Faria *et al.*, 1990). In both species, PL-II expression becomes detectable on Day 10 of gestation and continues until term. There is, therefore, a narrow window (Days 10–12) when the same trophoblast giant cells produce both forms of PL (Paria *et al.*, 1990, 1991). It has not been determined how this switch in production is carried out, but presumably it reflects differentiation events within the trophoblast. Initially rPL-II expression is confined to the trophoblast giant cells within the choriovitelline placenta (Faria *et al.*, 1990), but as pregnancy progresses, there is a shift to trophoblast giant cells within the chorioallantoic placenta, initially in the junctional zone and then later in the labyrinth zone (Soares, 1987; Campbell *et al.*, 1989; Deb *et al.*, 1991a). Rat PL-Iv, although structurally most similar to rPL-I, is expressed primarily by the cytotrophoblast within the junctional zone of the chorioallantoic placenta and exhibits a gestational time course similar to that of rPL-II (Deb *et al.*, 1991a; Robertson *et al.*, 1991).

3. Ruminants

Placental lactogen expression in cattle and sheep has been localized to the chorionic binucleate cells of the fetal cotyledon by both immunocytochemistry (Wooding, 1981; Duello *et al.*, 1986) and *in situ* hybridization (Milosavljec *et al.*, 1989; Kappes *et al.*, 1992). Migration of the chorionic binucleate cells out of the chorionic epithelium and their fusion with the uterine epithelium to form either trinucleated cells or a syncytium (see Section II) is thought to be the route by which ruminant PL is delivered to the maternal system. Although the pathways by which bPL and oPL enter the maternal blood circulation may be the same, the concentration of PL in circulation is considerably different between these two ruminants. In cattle, bPL is detectable in maternal serum by the fourth month of pregnancy (Beckers *et al.*, 1982) but remains low throughout gestation and generally does not exceed 1–2 ng/ml (Beckers *et al.*, 1982; Byatt *et al.*, 1987). Possibly it is cleared from the blood very rapidly. In contrast, oPL is detectable in pregnant ewes at Day 50 (Handwerger *et al.*, 1977; Martal and Djiane, 1977) and increases in concentration to peak values (0.5–2 μg/ml) about 10 days before term (Chan *et al.*, 1978; Butler *et al.*, 1981) (Figure 9). Maternal serum concentrations of PL in the goat follow a pattern similar to that in sheep (Currie *et al.*, 1990). In fetal circulation, both bPL (5–25 ng/ml) and oPL (~80 ng/ml) peak at midgestation, plateau, and/or gradually decline until term (Gluckman *et al.*, 1979; Beckers *et al.*, 1982; Byatt *et al.*, 1987). The amount of oPL that gets into maternal circulation reflects the amount of specific oPL mRNA found in cotyledonary tissue during mid- and late gestation (Figure 9) (Kappes *et al.*, 1992), and appears to result from an increase in the total number of chorionic binucleate cells within the cotyledon. The seemingly constant expression of oPL mRNA on a per cell basis is similar in this regard to hPL mRNA expression by syncytiotrophoblast cells (Hoshina *et al.*, 1982; see Section IV,C,1).

D. Gene Organization of the Placental Lactogens

The overall structural similarities between GH, PL, and PRL have led to the belief that they each diverged from a common ancestral gene (Niall *et al.*, 1971), which first duplicated approximately 350 million years ago (Barsch *et al.*, 1983; Miller and Eberhardt, 1983). This duplication event gave rise to GH and PRL precursor genes that then segregated onto two separate chromosomes resulting, for example, in the hGH and hPRL genes coming to reside on chromosomes 17 and 6, respectively (Owerbach *et al.*, 1980, 1981). In the mouse, the GH and PRL genes are located on chromo-

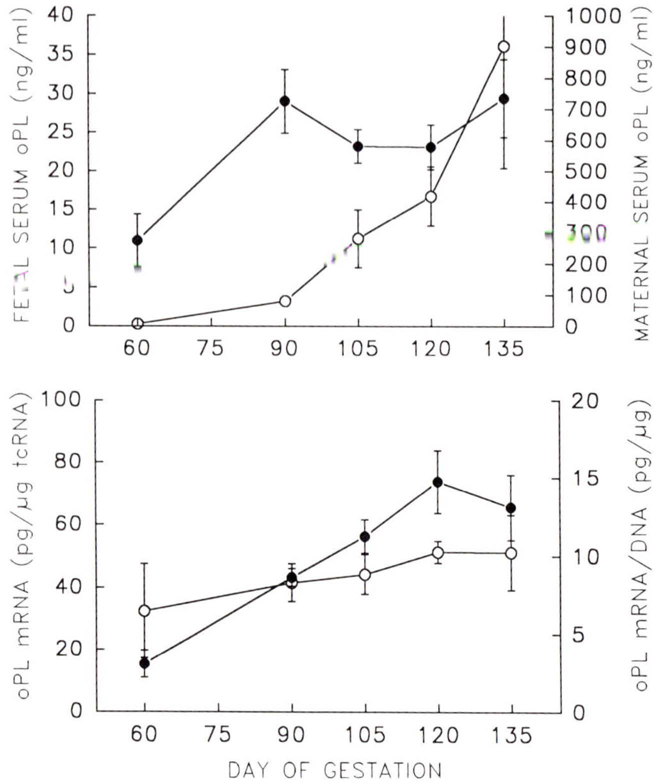

FIGURE 9 **(Top) Concentrations of oPL in maternal (open circles) and fetal (closed circles) serum from mid- to late gestation. (Bottom) Concentrations of oPL mRNA/μg total cellular RNA (closed circles) and oPL mRNA/μg cotyledonary tissue DNA (open circles) from mid- to late gestation. Data obtained from Kappes *et al.* (1992).**

somes 11 and 13, respectively (Jackson-Grusby *et al.*, 1988), while the rat genes are on chromosomes 10 and 17 (Cooke *et al.*, 1986). The gene(s) encoding PL is thought to have diverged after the separation of the main order of mammals, because the PL gene(s) is clearly derived from the GH gene in man and from the PRL gene in rodents. The hPL genes are located on chromosome 17 in close linkage with the hGH genes, and it has been suggested that the genes for hPL and hGH only diverged within the last 10–15 million years (Barsch *et al.*, 1983). Conversely, the mouse and rat PL genes are tightly linked with their respective PRL genes (Cooke *et al.*, 1986; Jackson-Grusby *et al.*, 1988; Levan *et al.*, 1991). By analogy, it seems reasonable to expect that the genes encoding the ruminant PLs will also be found on the same chromosomes as the PRL gene (see Section IV,B,3).

The genes encoding the two hGH genes (hGH-N and hGH-V) and the three PL genes (hPL_1, hPL_3, and hPL_4) are clustered within a 55-kb region on chromosome 17 (Barsch *et al.*, 1983; Chen *et al.*, 1989) (Figure 10). The confusing nomenclature for the hPL arises from initial reports that there were four rather than three genes. Included in the gene cluster are numerous 300-bp, highly repetitive Alu sequences that are believed to have contributed toward the recombinantorial events that allowed the genes to evolve (for discussion, see Hirt *et al.*, 1987; Chen *et al.*, 1989). Of the three hPL genes, only hPL_3 and hPL_4 have been unequivocally shown to be transcriptionally active (Barrera-Saldana *et al.*, 1983; Resendez-Perez *et al.*, 1990). The only difference between the polypeptides encoded by the two genes is at position −24 of the signal peptide (Barrera-Saldana *et al.*, 1983), where the alanine in hPL_3 is replaced with a proline in hPL_4. Thus, the fully processed polypeptides are identical. The hPL_1 gene has been considered a pseudogene since a G to A mutation at the donor splice site of the second intron (Hirt *et al.*, 1987; Chen *et al.*, 1989) has been assumed to render it nonfunctional. However, hPL_1 gene transcripts have been identified in a placental cDNA library (Chen *et al.*, 1989). One of these used an alternative splice site to overcome the mutant donor splice site, while the others resulted from incomplete processing of heterogeneous nuclear RNA. Recently Marcotty *et al.* (1992) detected four transcripts (0.9, 1.4, 1.6, and 3.8 kb) in placental polyadenylated-RNA with an oligonucleotide probe specific for hPL_1, but it remains unclear whether a functional hPL_1 protein is ever produced.

Because of the structural similarities between the ruminant PL and their respective PRL (Schuler *et al.*, 1988; Warren *et al.*, 1990b), it is assumed that they evolved from an ancestral PRL gene. The structure of the single bPL gene is known (Kessler and Schuler, 1991), although there is evidence for differential splicing and allelic variants (Yamakawa *et al.*, 1990; Kessler and Schuler, 1991).

E. Transcriptional Control of Placental Lactogen Gene Expression

1. Primates

As with the CGα-subunit, the PL present an interesting puzzle regarding tissue specificity of expression, and have been intensively studied. The genes evolved relatively recently from the GH gene (in the case of hPL) or the PRL gene (in the case of the roden or ruminant PL), yet are expressed in the trophoblast, including choriocarcinoma cells (Ringler and Strauss,

FIGURE 10 Schematic representation of the hGH/PL gene cluster as described by Barsch *et al.* (1983) and Chen *et al.* (1989). Arrows indicate the direction of transcription.

1990) rather than the pituitary. Indeed, the hPL_3 gene shows ~94% nucleotide sequence identity with the hGH-N gene over the 500-bp region immediately preceding their respective transcriptional start sites (Selby *et al.*, 1984) and possesses the pituitary-specific enhancer sequence that binds the transcription factor GHF-1/Pit-1 (Bodner *et al.*, 1988; Ingraham *et al.*, 1988; Karin *et al.*, 1990). However, despite the fact that the hPL_3 gene promoter region can bind GHF-1/Pit-1 (Lemaigre *et al.*, 1989b), this transcription factor is absent in the trophoblast and does not, therefore, contribute toward tissue-specific expression of hPL. Instead, basal expression is conferred by a region (−142 bp to −129 bp) in the proximal promoter (Fitzpatrick *et al.*, 1990), while a 138-bp fragment located 2.2 kb downstream of the hPL_3 gene has been identified as the likely tissue-specific enhancer (Walker *et al.*, 1990). Similar sequences have been located at the same relative position 3′ to the other hPL genes, but not to the hGH-N gene (Walker *et al.*, 1990). A 22-bp region within this enhancer region has been identified by DNase I protection and gel shift assays with nuclear extracts, and represents a motif that is known to bind the transcription factor, TEF-1 (Davidson *et al.*, 1988). Unfortunately, the latter is found in a variety of cell types and not just trophoblast. However, evidence has been presented that the trophoblast factor may be closely related but not identical to TEF-1 (Walker *et al.*, 1990). There are now many examples of extended families of DNA-binding proteins with subtly different DNA-binding properties but with often contrasting abilities to act as transactivators.

Progress has also been made recently on elucidating why the hPL genes are not expressed in pituitary cells. A likely explanation is the presence of a dominant repressor with specific ability to silence expression of the placental members of the hGH/hPL family within the pituitary (Nachtigal *et al.*, 1992). Thus, a combination of tissue-specific enhancers and repressors probably ensures trophoblast expression of hPL and related placental proteins. However, control is unlikely to be simple since the promoter region seems to be capable of binding a wide range of different transcription factors, including Sp1 (Lemaigre *et al.*, 1989b), which seems to be important for controlling initiation of transcription, AP-2 and NF-1 (Peritz *et al.*, 1988; Lemaigre *et al.*, 1989a; Courtois *et al.*, 1990). In addition, expression is modulated by cAMP, thyroid hormone, and glucocorticoids whose response elements have been mapped to either the 5′ flanking region of the gene or to the first intron (Slater *et al.*, 1985; Barlow *et al.*, 1986; Dana and Karin, 1989; Nickel and Cattini, 1991). Just as the specific biological activities of hPL remain to be defined, understanding how hPL synthesis is controlled during pregnancy remains a major challenge.

2. Rodents

The only rodent PL gene to be characterized functionally in any sort of detail is the one encoding mPL-II. Shida *et al.* (1992) have shown that the

5′-upstream region between −2700 and −569 relative to the transcriptional start site can confer expression on a reporter gene in trophoblast giant cells of a transgenic mouse. They also found a consensus cAMP response element or CRE between positions −255 and −248, but its functionality has not yet been tested. Extension of this type of transgenic analysis should soon provide considerably more information on the transcriptional control of the mouse PL genes. Similarly, gene ablation experiments, similar to those conducted with the gene for insulin-like growth factor-II (IGF-II) (DeChiara *et al.*, 1990) are likely to provide insight into the function of the PL genes in mice.

3. Ruminants

The structure of the single bovine PL gene has been reported (Kessler and Schuler, 1991), but little is known about control of its expression, except for the presence of certain motifs likely to bind known transcription factors such as AP-1, AP-2, and the thyroid hormone receptor. The oPL gene seems to have a fairly similar 5′ sequence within the region upstream of the likely transcription start site (Liang and Anthony, 1993), but since the two species evolved from each other so recently, such conservation is not unexpected. There is some evidence that the bPL gene is differentially spliced to give more than a single transcript (Yamakawa *et al.*, 1990; Kessler and Schuler, 1991), but whether these both provide functional proteins is unknown. The elucidation of transcriptional controls operating on ruminant PL expression is likely to remain difficult until cell lines of trophoblast origin from cattle or sheep become available for transfection studies (see Section VII,E for comparable discussion on IFNτ gene expression).

V. Prolactin-Related Proteins

Placentas from rodents and ruminants (and probably other species) produce several prolactin-related proteins that are distinct from the PL. In the rat, three prolactin-like proteins have been identified and their primary structures determined: PLP-A (Duckworth *et al.*, 1986b), PLP-B (Duckworth *et al.*, 1988), and PLP-C (Deb *et al.*, 1991b). The rat PLP amino acid sequences are only 25 to 32% identical to each other, and they share 25–31% identity with rPRL, 24–26% with rPL-I, and 27–30% with rPL-II (Deb *et al.*, 1991b) (Figure 8). All three of the rat PLP are glycoproteins, with PLP-A and C containing two consensus signals for N-linked glycosylation, and PLP-B containing one. PLP-A secreted into culture medium appears as two monomeric forms (M_r~29,000 and 33,000). Interestingly the 29,000 M_r form appears also to be localized to the nucleus (Deb and Soares, 1990). Secreted PLP-B is a 30,000 M_r glycoprotein (Ogilvie *et al.*, 1990) while PLP-C exists as a 25,000 M_r nonglycosylated form and a 29,000 M_r glycosylated form (Deb *et al.*, 1991c). The rat PLP genes are clustered closely with PRL and PL-I and -II on chromosome 17 (Cooke *et al.*, 1986; Deb *et al.*, 1991b; Levan *et al.*, 1991), suggesting that this whole family evolved from a common ancestral gene after PRL and GH diverged. Although the function and regulation of rat PLP are still unknown, their production has been localized to cytotrophoblast cells of the chorioallantoic placenta (PLP-A, Campbell *et al.*, 1989; Duckworth *et al.*, 1990; PLP-B, Duckworth *et al.*, 1990; Ogilvie *et al.*, 1990; PLP-C, Deb *et al.*, 1991b,c). Both are expressed in the second half of gestation in the trophoblast giant cells within the junctional zone of the chorioallantoic placenta; the developmental time frame is similar to that when rPL-II and rPL-Iv are also produced.

In the mouse, two prolactin-related proteins produced by the placenta have been identified—proliferin (PLF) (Linzer and Nathans, 1984; Linzer *et al.*, 1985) and proliferin-related protein (PRP) (Linzer and Nathans, 1985). Both PLF and PRP share approximately 30% amino acid sequence identity with mouse PRL, PL-I, and PL-II, and all of these genes are located on chromosome 13 (Jackson-Grusby *et al.*, 1988). Mouse PLF and PRP are both secreted glycoproteins that give multiple bands on electrophoretic gels (PLF, M_r~27,000 and 35,000; PRP, M_r~36–44,000) (Lee and Nathans, 1987; Colosi *et al.*, 1988). PLF is produced in trophoblast giant cells (Lee *et al.*, 1988) and reaches a peak concentration in maternal serum at Day 10 of pregnancy, i.e., at a time comparable to maximal mPL-I concentration. However, PLF remains detectable in maternal serum until term, whereas mPL-I does not. Mouse PRP can be measured in maternal serum from Days 10 to 18 of pregnancy, and is maximal at Days 12–14. In contrast to the other GH/PRL-related proteins, it is a product of the junctional zone cytotrophoblasts (Colosi *et al.*, 1988). No roles for either PLF or PRP during pregnancy have been defined.

Besides bPL, the placentas of cattle produce transcripts for six prolactin-related proteins (bPRP-I to VI; Schuler *et al.*, 1991). The products so encoded are likely to be glycoproteins with N-linked chains and are approximately 40% identical to bPRL and 30–34% identical to bPL in amino acid sequence (Kessler *et al.*, 1989). The synthesis and secretion of only one of these bPRP have been confirmed, however (bPRP-I; Zieler *et al.*, 1990), and its mRNA has been localized by *in situ* hybridization to the binucleate cells (Milosavljevic *et al.*, 1989), where it appears to be coexpressed with bPL. The 5′ region of the gene encoding bPRP-I has been sequenced (Ebbitt *et al.*, 1989) and is fairly similar to the corresponding region of the bPL gene (Kessler and Schuler, 1991). Both possibly contain similar or identical *cis*-acting sequences that account for their common pattern of expression. As with the rat PLP and mouse PLF and PRP, the function of the bovine PRP has not yet been defined, nor has the transcriptional control of their genes been investigated.

VI. Placental Growth Hormone

Recently the concentration of hGH was measured in pregnant women by using monoclonal antibodies that could distinguish the products of the hGH-N and hGH-V genes (Frankenne *et al.*, 1988). From approximately 22 weeks of pregnancy onward, the major circulatory form of hGH was found not to be pituitary-derived hGH-N, but rather placentally derived hGH-V. Placental hGH-V production has been localized to the syncytiotrophoblast (Liebhaber *et al.*, 1989) and is detectable in the maternal but not the fetal circulation (Frankenne *et al.*, 1988). Two transcripts for hGH-V appear to be generated from the hGH-V gene. One encodes 22,000 M_r secreted hGH-V gene product, while the second transcript retains intron 4 and is predicted to encode a 26,000 M_r polypeptide (hGH-V2) (Cooke *et al.*, 1988a,b). The hGH-V polypeptide differs from hGH-N at 13 amino acids and contains a site for Asn-linked glycosylation that appears to be used (Ray *et al.*, 1989; Frankenne *et al.*, 1990). The hGH-V2 product, arising through a differentially spliced transcript, is 230 amino acids in length, lacks an Asn-linked glycosylation site, but contains a hydrophobic tail that could potentially anchor it to the syncytiotrophoblast cell membrane (Cooke *et al.*, 1988b). Although the function of hGH-V is not known, it is able to bind to somatogen and lactogen receptors (Ray *et al.*, 1990). Its binding to the latter is significantly poorer than that of hGH-N and it is only weakly lactogenic in bioassays (MacLeod *et al.*, 1991), raising the possibility that hGH-V functions differently than hGH-N, possibly by binding to a variant form of the hGH receptor (Urbanek *et al.*, 1992).

VII. Trophoblast Interferons (IFNτ)

A. Function

In sheep and cattle there is no solid evidence for production of a gonadotropin homologous to hCG. Instead, the conceputs prevents the corpus luteum from regressing by inhibiting the release of luteolytic pulses of prostaglandin $F_{2\alpha}$ from the uterus. In the absence of induced luteolysis, pituitary LH is assumed to maintain the corpus luteum of pregnancy, although it has not been ruled out that other factors produced by the conceptus might assume a luteotropic role, particularly as pregnancy advances.

It is still not completely clear exactly how luteolysin release is inhibited in early pregnancy (see Bazer, 1992), but a search for the antiluteolytic substance led to the identification of a complex of proteins secreted by the early trophoblast. These proteins, originally named ovine and bovine trophoblast protein-1 (oTP-1 and bTP-1), unexpectedly turned out to be structurally related to type I interferons (Roberts *et al.*, 1990; reviewed by Roberts *et al.*, 1992a,b). Not only did these proteins show considerable sequence identity to type I IFN, such as IFNα, they also possessed the antiviral, antiproliferative, and immunomodulatory activities typical of such cytokines. When introduced into the uterine lumen of nonpregnant sheep or cattle, the purified proteins were able to inhibit pulsatile prostaglandin release and to extend the interestrous interval for several days and in some instances for weeks. Recombinant bovine IFNα, at a similar dosage, seemed not to be as potent as IFNτ at prolonging luteal lifespan. These observations have raised the possibility that, after breeding livestock, treatment with IFN during the period of maternal recognition of pregnancy might usefully supplement endogenous IFN production and rescue conceptuses that are otherwise lagging in their development (Schalue-Francis *et al.*, 1991).

Although there are numerous reports of constitutive IFN production by placenta or trophoblast tissue from a wide variety of species, including pigs, mice, and humans (see Roberts *et al.*, 1990, 1992a), there is no evidence that they serve an antiluteolytic role as they do in domestic ruminants. In general, production in such species is orders of magnitude less than in cattle and sheep, and in many cases the IFN themselves have been poorly characterized. However, these observations have led some investigators to propose a general role of IFN in pregnancy that might stem from their potential ability to modulate the maternal immune system (Chard, 1991). At present, there is little firm evidence for or against such a function.

B. General Structure

Although the IFN produced by conceptuses from domestic ruminants are undoubtedly related to other type I IFN, which include members of three known multigene families—the IFNα, $-\beta$, and $-\omega$ (formerly IFNα_{II}) (Table 1)—they constitute a distinct family themselves. They most closely resemble the 172-amino acid IFNω rather than the 166-residue IFNα, and Leaman and Roberts (1992) have proposed that they evolved from the former approximately 60 million years ago. However, the IFNτ are under different regulatory controls than the IFNω. They are not, for example, readily inducible by viruses (Cross and Roberts, 1991). These differences appear to be reflected in the organization of the promoter regions of their genes (see Section VI,E; Roberts *et al.*, 1992a).

All of the known IFNτ so far identified are 172 amino acids long (Table 1). There is, relative to the 166-residue IFNα, a 6-amino-acid extension at the carboxyl terminus of the IFNτ (Figure 11). All known bovine and some caprine IFNτ are glycosylated (Anthony *et al.*, 1988; Baumbach *et al.*, 1990), most probably at Asn^{78}. In contrast, the majority of ovine IFNτ appear not to be glycoproteins although some transcripts do code for a potential site

TABLE 1 Type I Interferons of Cattle

	Subtype			
	IFNα	IFNβ	IFNω	IFNτ
Number of genes	>15	~5	~4[a]	~4–5
Inducer	Virus	Virus	?[b]	Unknown
Cell of origin	Leukocytes and others	Fibroblasts and others	?	Trophectoderm
Genes	Intronless	Intronless	Intronless	Intronless
Length of polypeptide sequence	166	166	172[c]	172
Percent sequence identity to IFNα	100	~30	~75	~50
Antiviral activity	Yes	Yes	Yes	Yes
Antiproliferative activity	Yes	Yes	?	Yes

[a]Three to four genes are detected with a highly specific 3′ cDNA probe (Leaman and Roberts, 1992), but at least 15 with a less specific, full-length probe. Additional genes related to IFNω probably exist.
[b]IFNω are known to be virally inducible in humans.
[c]The human IFNω are either 172 or 174 amino acids in length.

for N-glycosylation at Asn[78] (Klemann *et al.*, 1990a). Recombinant bovine IFNτ, which lacks carbohydrate, seems to have properties identical to the native, glycosylated protein (Klemann *et al.*, 1990b). There is no evidence to suggest that the carbohydrate on any type I IFN has any function in relation to receptor interaction or signal transduction.

Despite the fact that the IFNτ show less than 30% amino acid sequence identity with the IFNβ and only ~50% and 75% identity with the IFNα and -ω (Table 1), they seem to interact with the same type I receptor system (see Roberts *et al.*, 1990, 1992a). Hydrophobicity/hydrophilicity plots and predictions of secondary structure suggest that all the type I IFN are relatively alike and probably fold similarly. In particular, all four cysteines that participate in disulfide bond formation in IFNα are conserved in IFNτ. Therefore, even though X-ray structural data are available only for murine IFNβ, it has been possible to construct a three-dimensional model for IFNτ, which is illustrated in Figure 11. There is reasonable but not irrefutable evidence to suggest that the regions of high amino acid conservation i.e., the upper end of helix E and the loop between helices A and B, probably interact with the receptor. As stated earlier, there is a common receptor system for the IFN, one component of which has been cloned from humans and mice (Uzé *et al.*, 1990, 1992). Signal transduction involves cytoplasmic activation of a tyrosine kinase, *tyk*-2, which phosphorylates a cytoplasmic transcription factor, ISGF-3 (Schindler *et al.*, 1992; Velazquez *et al.*,

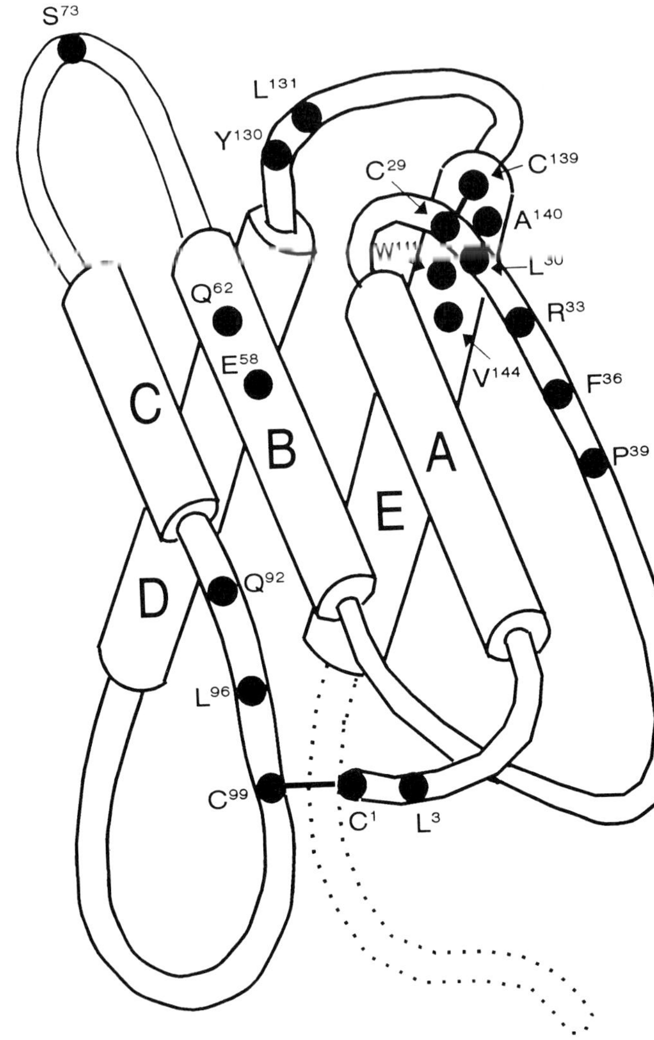

FIGURE 11 Diagram showing the predicted three-dimensional structure of an IFNτ modeled on the basis of predictions of secondary structure and a preliminary description of X-ray crystallography data on IFNβ (for details, see Roberts *et al.*, 1992b). The letters refer to the five α-helices. The solid circles represent amino acids highly conserved in all known type I IFN from a range of species. The broken line represents the carboxyl-terminus of the IFNτ whose placement is uncertain.

1992). The latter then migrates into the nucleus and is responsible for transactivating numerous IFN-responsive genes and hence controlling the major IFN responses of the cell. However, what remains unclear is whether there are secondary signal transduction pathways that can be exploited subtly by certain type I IFN relative to others. Such pathways might explain why the biological activities of each of these intriguing molecules do not always appear to be identical and do not necessarily reflect their respective K_d for the receptor.

C. Expression in Relation to Trophoblast Development

As discussed earlier, the domestic ruminants exhibit a diffuse synepitheliochorial type of placentation. However, the major phase of IFNτ production occurs prior to implantation, between about Days 13 to 17 in sheep and Days 15 to 21 in cattle. Early embryonic development in these species involves considerable growth of the trophoblast before it even attaches firmly to the uterine wall. After hatching, the blastocyst expands until it reaches a diameter greater than 1 mm (~Day 11 in sheep; ~Days 12–14 in cattle). This sphere then undergoes considerable growth and elongates to reach a length of several centimeters over the next 4 to 5 days, and only then begins to attach. Throughout, the embryonic disc remains relatively small and the predominant mass of tissue is trophoblast (see Roberts *et al.*, 1992a).

Figure 12 shows schematically the production of IFNτ by sheep conceptuses during this period of preimplantation development. It is important to note that peak production coincides with the period in which the CL would begin to regress if a conceptus were not present in the uterus. Although IFNτ expression is first detectable at about the time of blastocyst hatching (Days 8–9), production is relatively low (Ashworth and Bazer, 1989; Hernandez-Ledezma *et al.*, 1992). Massive secretion coincides with the period of elongation and can exceed 100 μg per day. On a per unit dry mass basis, synthesis is highest around Days 13 to 14 and then declines. The situation in the cow is very similar, except events are delayed by about 2 to 3 days (Roberts *et al.*, 1992a).

There is an excellent correlation between rates of IFNτ production and total amounts of mRNA present in the tissues. *In situ* hybridization procedures have shown a 5- to 10-fold rise in mRNA content per cell during the early elongation phase, and confirmed data obtained by immunocytochemistry that expression is confined to the mononucleate cells of the trophectoderm (Farin *et al.*, 1989). The binucleate cells, which are sparse prior to attachment and the beginnings of implantation, do not express IFNτ (Guillomot *et al.*, 1990). As the trophectoderm begins to attach to the uterine epithelium, there is a simultaneous fall in IFNτ expression. At no time does the inner cell mass of embryonic disk contain IFNτ transcripts.

The mechanisms that control IFN gene transcription during these few days of preimplantation development are unknown. It seems unlikely that genes are activated in response to environmental cues, such as the presence of cytokines in the reproductive tract (for a full discussion, see Roberts *et al.*, 1992a), and expression may therefore be genetically preprogrammed.

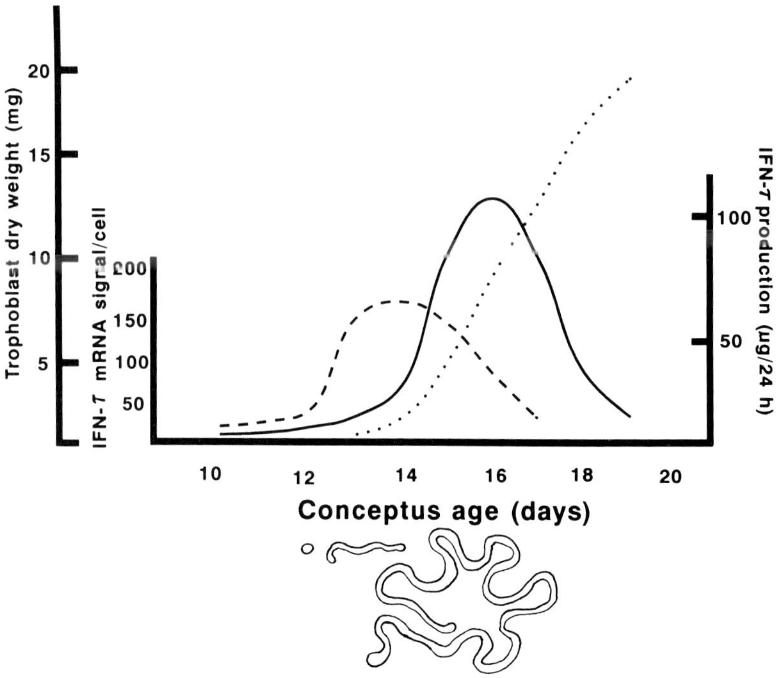

FIGURE 12 Diagram showing the relationship between day of ovine conceptus development and the amount of ovIFNτ produced (solid line); the dry weight of the trophoblast (dotted line); and the concentration of ovIFNτ mRNA per cell (broken line). The drawing below the graph depicts the morphology of the conceptus. Reproduced with permission from Roberts *et al.* (1993).

What is evident is that production is massive compared with IFNα production by leukocytes, and sustained for several days rather than hours. Expression is also not the result of viral induction and is confined to trophectoderm cells of preimplantation conceptuses. Consequently, control of IFNτ expression must differ significantly from that of other type I IFN.

D. Gene Organization of the IFNτ

There is a minimum of four to five IFNτ genes in cattle, sheep, and goats, and at least two in a related member of the *Bovidae*, the musk ox. Giraffes and deer also have the IFNτ gene. In cattle, the IFNτ genes are clustered on the same chromosome, close to the other type I IFN genes. All the genes so far cloned lack introns and are highly conserved across the four *Bovidae* species. Conservation is particularly high in the upstream promoter regions of the genes up to position −450 above the transcription start site, emphasizing that the unusual promoter organization noted below may be important in the control of tissue- and stage-specific expression. In this respect, they also differ from the IFNα and β genes, which start to diverge markedly from each other in sequence beyond −120. This observation suggests that these more distal regions, i.e., beyond −120, have a role in promoter regulation (see Leaman and Roberts, 1992; Roberts *et al.*, 1992a).

The mRNA transcribed from the IFNτ genes are about 1 kb in length and contain an open reading frame of 585 bases that encodes a 195-amino acid peptide from which a 23-residue signal peptide is removed. There is far greater nucleotide variability among genes within the open reading frame than there is in the nontranscribed promoter regions of the gene. As a consequence, different IFNτ genes within a species can differ as much as 10% in amino acid sequence, but vary very little in nucleotide sequences in regions of the gene likely to be involved in controlling expression.

E. Transcriptional Control of IFNτ Gene Expression

Since high levels of IFNτ synthesis are correlated with tissue concentrations of mRNA, it has generally been assumed that the controls operating on IFNτ expression are primarily transcriptional (Hansen *et al.*, 1991a). However, this assumption remains to be confirmed.

Sequence comparisons among all type I IFN genes, including the IFNτ, show that the proximal promoter regions share some superficial similarities. In Figure 13 we compare the IFNτ and IFNω genes. For example, each includes hexameric repeats of the sequence GAAANN (where N is any nucleotide) as well as AA ($^{G}/_{T}$) ($^{T}/_{G}$) GA hexamers that some consider to confer viral responsiveness to IFNα, -β, and -ω genes (see Roberts *et al.*, 1992b). However, these motifs are not clustered into structures resembling viral response elements analogous to the well-defined VRE on IFNα and -β genes (see Hansen *et al.*, 1991b). It is probably for this reason that the IFNτ are not activated by virus. In addition to the purine-rich sequences in the proximal promoter region, there are numerous A T T A-rich motifs in the more distal parts of the IFNτ promoter which resemble the core sequence recognized by many homeotic genes of the Antennapedia class (Roberts *et al.*, 1993). However, the functional significance of these sequences is not known.

The very narrow pattern of IFNτ expression strongly suggests that only trophoblast cells might be permissive for expression of transfected genes designed to test different promoter constructs. Unfortunately, no such trophoblast cell lines are available from ruminant species, and, as suspected, all nontrophoblast cell lines so far tested are nonpermissive. However, a bovine IFNτ gene transfected into a human choriocarcinoma cell line, e.g.,

```
                                                                         -400                      -380
oIFN-τ                                                                 AGTGACTGTGCATTCCTATGTGTAAGATAAGG
oIFN-ω                                                                          ATATATGGAATCCATAGAGCAGACA

                -360                      -340                      -320                      -300
oIFN-τ    AGGGAAAAATGCAGTTAAGAATCAAGGGAAAATTATGTTCCTGATATAAGATAAACAAAAGGAATGCTTATATGT
oIFN-ω    GAGCCTGGCTTCCTAAGGAACCTAAGAGGGAAAAATGGAGCTTCAGAGCAAGCCAAAA ATACACTCTTGGATGA

                      -280                      -260                      -240
oIFN-τ    ATTATACCTAAATTTGTGCCTAATAACTCTGTACACGTCTATAAGTCTTTCCATACGTACATCACTTCAGCCTTT
oIFN-ω    GAGGAGAGCAACACTTATGCATGAGAGATAAAGGGCAATAAAGTGCCTAATTACAGATGTCCATCAATTCTGAC

      -220                      -200                      -180                      -160
oIFN-τ    ATTTAGTTTCTCATTTAATTGATATACATTTATATTGACAAATCCAAATTTTATTGGGAAAAGTAAACTTCTACT
oIFN-ω    TTAAGAGAAAGA CTTTCTCATTTGATTGAAAAACATCCATTTGGGTGGG CACATCTGAAGTGACTAGGAAA A

            -140                      -120                      -100                      -80
oIFN-τ    ATAAAAATTAAGAGTTTAGATTGGCTACATTTTCTAGGTCAAG TAGAAAATATCTAAATGAAAA CAAACA GG
oIFN-ω    GGC........T...C....AG.AAG.ACA.....TT.A.TGTC.T..G....CA...CA......C...A.A..

                        -60                       -40                       -20
oIFN-τ    AAGTGAGG G AG GAATTT  TTGGATAATGAG   TACCATCTTCCCTATTTAAAG CCTTGCTTAGAAC CA
oIFN-ω    ....A.AA.T.C.TA...A.GA..A..A.....AAAT......G.....TATTTAAT.G...C........AG..

             +1
oIFN-τ    TCGTCATCAGAGAACCTACCTGAAGATTCCCCCTGACCCCATCTCAGCCAGTCCAGCAGCAGCCGCATCTTCCCC
oIFN-ω    .G. .................C...G....A..A...G..G..........C....T.................
```

FIGURE 13 Comparison of the nucleotide sequences within the promoter regions of the genes for an ovIFNτ and an ovIFNω. Dots indicate base identity. Note the relatively poor conservation of sequence above the TATA box and the complete loss of identity after −140. GAAAnn motifs are present in the proximal promoter regions of both genes. Data derived from Leaman and Roberts (1992).

Jar, is expressed (Cross and Roberts, 1991). Therefore, despite the fact that the human does not possess IFNτ genes, human trophoblast cells contain trophoblast-specific transcription factors capable of transactivating the IFNτ promoter. In contrast, an IFNω gene similarly transfected does not become constitutively active. This finding has allowed nested deletions of IFNτ promoters to be tested for their ability to drive transcription of a reporter gene in human choriocarcinoma cells. These deletion studies have indicated that a minimal promoter, 126 bp beyond the transcription start site, was about fivefold more active than the transfected, promoterless reporter gene, and that an additional enhancer element exists further upstream between −200 and −400 bp. However, beyond this very basic information, little is known about how transcriptional control of the IFNτ is mediated except that the method is strikingly different than that operating on other type I IFN. It should also be stressed that the promoter regions of the IFNτ do not contain a region comparable to the trophoblast-specific enhancer element of hCGα genes, nor do they possess any stretches of sequence comparable to regions within the CGβ or prolactin promoters.

Acknowledgment

This paper is a contribution from the Missouri Agricultural Experiment Station, Journal Series Number 11,790 and was supported by grants from the National Institutes of Health and the U.S. Department of Agriculture. The authors thank Drs. Michael Kubisch and Michel Villalta for critical reading of the chapter and Gail Foristal for preparing diagrams and assembling the manuscript.

References

Akerblom, I. E., Slater, P., Beato, M., Baxter, J. D., and Mellon, P. L. (1988). Negative regulation by glucocorticoids through interference with a cAMP responsive enhancer. *Science* 241, 350–353.

Albanese, C., Kay, T. W. H., Troccoli, N. M., and Jameson, J. L. (1991). Novel cyclic adenosine 3′,5′-monophosphate response element in the human chorionic gonadotropin β-subunit gene. *Mol. Endocrinol.* 5, 693–702.

Anthony, R. V., Helmer, S. D., Sharif, S. F., Roberts, R. M., Hansen, P. J., Thatcher, W. W., and Bazer, F. W. (1988). Synthesis and processing of ovine trophoblast protein-1 and bovine trophoblast protein-1, conceptus secretory proteins involved in the maternal recognition of pregnancy. *Endocrinology*, 123, 1274–1280.

Aplin, J. D. (1991). Implantation, trophoblast differentiation and haemochorial placentation: mechanistic evidence *in vivo* and *in vitro*. *J. Cell Sci.* 99, 681–692.

Arbiser, L. J., Arbiser, K. Z., and Majzoub, J. A. (1991). Regulation of gene expression in choriocarcinoma cells by methotrexate and hydroxyurea. *Endocrinology* 128, 972–978.

Arima, Y., and Bremel, R. D. (1983). Purification and characterization of bovine placental lactogen. *Endocrinology* 113, 2186–2194.

Ashitaka, Y., Nishimura, R., Futamura, K., Ohashi, M., and Tojo, S. (1974). Serum and chorionic tissue concentrations of human chorionic gonadotropin and its subunits during pregnancy. *Endocrinol. Jpn.* 21, 547–550.

Ashworth, C. J., and Bazer, F. W. (1989). Changes in ovine conceptus and endometrial function following asynchronous embryo transfer or administration of progesterone. *Biol. Reprod.* 40, 425–434

Auletta, F. J., and Flint, A. P. F. (1988). Mechanisms controlling corpus luteum function in sheep, cows, nonhuman primates and women, especially in relation to the time of luteolysis. *Endocr. Rev.* 9, 88–105.

Barlow, J. W., Voz, M. L. J., Eliard, P. H., Mathy-Hartert, M., De Nayer, P., Economidis, I. V., Belayew, A., Martial, J. A., and Rousseau, G. G. (1986). Thyroid hormone receptors bind to defined regions of the growth hormone and placental lactogen genes. *Proc. Natl. Acad. Sci. USA* 83, 9021–9025.

Barrera-Saldana, H. A., Seeburg, P. H., and Saunders, G. F. (1983). Two structurally different genes produce the same secreted human placental lactogen hormone. *J. Biol. Chem.* 258, 3787–3793.

Barsch, G. S., Seeburg, P. H., and Gelinas, R. E. (1983). The human growth hormone gene family: Structure and evolution of the chromosomal locus. *Nucleic Acids Res.* 11, 3939–3958.

Bartocci, A., Welker, R. D., Schlick, E., Chirigos, M. A., and Nisula, B. C. (1983). Immunosuppressive activity of human chorionic gonadotrophin preparations *in vivo:* evidence for gonadal dependence. *Cell. Immunol.* 82, 334–342.

Baumbach, G. A., Duby, R. T., and Godkin, J. D. (1990). N-glycosylated and unglycosylated forms of caprine trophoblast protein-1 are secreted by preimplantation goat conceptuses. *Biochem. Biophys. Res. Commun.* 172, 16–21.

Bazer, F. W. (1992). Mediators of maternal recognition of pregnancy in mammals. *Proc. Soc. Exp. Biol. Med.* 199, 373–384.

Beck, J. S. (1970). Time of appearance of human placental lactogen in the embryo. *N. Engl. J. Med.* 283, 189–190.

Beckers, J. F., deCoster, R., Wouters-Ballman, P., Fromont-Lienard, C., van der Zwalman, P., and Ectors, F. (1982). Dosage radioimmunologique de l'hormone placentaire somatotrope et mammotrope bovine. *Ann. Med. Vet.* 126, 9–21.

Berle, P. (1974). Pattern of the human chorionic somatomammotrophic (hCS) concentration ratio in maternal serum and amniotic fluid during normal pregnancy. *Acta Endocrinol. (Copenh.)* 76, 364–368.

Bewley, T. A., Dixon, J. S., and Li, C. H. (1972). Sequence comparison of human pituitary growth hormone, human chorionic somatomammotropin, and ovine pituitary growth and lactogenic hormones. *Int. J. Pept. Protein Res.* 4, 281–287.

Bielinska, M., Matzuk, M. M., and Boime, I. (1989). Site-specific processing of the N-linked oligosaccharides of the human chorionic gonadotropin α subunit. *J. Biol. Chem.* 264, 17113–17118.

Birken, S., and Canfield, R. E. (1980). Chemistry and immunochemistry of human chorionic gonadotropin. *In* "Chorionic Gonadotropin" (S. J. Segal, ed.), pp. 65–88. Plenum, New York.

Blithe, D. L., Richards, R. G., and Skarulis, M. C. (1991). Free alpha molecules from pregnancy stimulate secretion of prolactin from human decidual cells: a novel function for free alpha in pregnancy. *Endocrinology* 129, 2257–2259.

Bo, M., and Boime, I. (1992). Identification of the transcriptionally active genes of the chorionic gonadotropin β gene cluster *in vivo*. *J. Biol. Chem.* 267, 3179–3184.

Bodner, M., Castrillo, J. L., Theill, L. E. Deerinck, T., Ellisman, M., and Karin, M. (1988). The pituitary-specific transcription factor GHF-1 is a homeobox-containing protein. *Cell* 55, 505–518.

Boime, I., Boothby, M., Hoshina, M., Daniels-McQueen, S., and Darnell, R. (1982). Expression and structure of human placental hormone genes as a function of placental development. *Biol. Reprod.* 26, 73–91.

Boothby, M., Kukowska, J., and Boieme, I. (1983). Imbalanced synthesis of human chorionic gonadotropin α and β subunits reflects the steady state levels of the corresponding mRNAs. *J. Biol. Chem.* 258, 9250–9253.

Braunstein, G. D., Rasor, J. L., Engrall, E., and Wade, M. E. (1980). Interrelationships of human chorionic gonadotropin, human placental lactogen, and pregnancy-specific beta-1-glycoprotein throughout normal human gestation. *Am. J. Obstet. Gynecol.* 138, 1205–1213.

Butler, W. R., Fullenkamp, S. M., Cappiello, L. A., and Handwerger, S. (1981). The relationship between breed and litter size in sheep and maternal serum concentrations of placental lactogen, estradiol and progesterone. *J. Anim. Sci.* 53, 1077–1081.

Byatt, J. C., Shimomura, K., Duello, T. M., and Bremel, R. D. (1986). Isolation and characterization of multiple forms of bovine placental lactogen from secretory granules of the fetal cotyledon. *Endocrinology* 119, 1343–1350.

Byatt, J. C., Wallace, C. R., Bremel, R. D., Collier, R. J., and Bolt, D. J. (1987). The concentration of bovine placental lactogen and the incidence of different forms in fetal cotyledons and in fetal serum. *Don. Anim. Endo. Crinol.* 4, 231–241.

Byatt, J. C., Welply, J. K., Leimgruber, R. M., and Collier, R. J. (1990). Characterization of glycosylated bovine placental lactogen and the effect of enzymatic deglycosylation on receptor binding and biological activity. *Endocrinology* 127, 1041–1049.

Byatt, J. C., Warren, W. C., Eppard, P. J., Staten, N. R., Krivi, G. G., and Collier, R. J. (1992). Ruminant placental lactogens: structure and biology. *J. Anim. Sci.* 70, 2911–2923.

Campbell, W. J., Deb, S., Kwok, S. C. M., Joslin, J. A., and Soares, M. J. (1989). Differential expression of placental lactogen-II and prolactin-like protein-A in the rat chorioallantoic placenta. *Endocrinology* 125, 1565–1574.

Chan, J. S. D., Robertson, H. A., and Friesen, H. G. (1976). The purification and characterization of ovine placental lactogen. *Endocrinology* 98, 65–76.

Chan, J. S. D., Robertson, H. A., and Friesen, H. G. (1978). Maternal and fetal concentrations of ovine placental lactogen measured by radioimmunoassay. *Endocrinology* 102, 1606–1613.

Chard, T. (1991). Commentary: Interferon-α is a reproductive hormone. *J. Endocrinol.* 131, 337–338.

Chen, E. Y., Liao, Y. C., Smith, D. H., Barrera-Saldana, H. A., Gelinas, R. E., and Seeburgh, P. H. (1989). The human growth hormone locus nucleotide sequence, biology, evolution. *Genomics* 4, 479–497.

Colosi, P., Ogren, L., Thordarson, G., and Talamantes, F. (1987a). Purification and partial characterization of two prolactin-like glycoprotein hormone complexes from the midpregnant mouse conceptus. *Endocrinology* 120, 2500–2511.

Colosi, P., Talamantes, F., and Linzer, D. I. H. (1987b). Molecular cloning and expression of mouse placental lactogen I complementary deoxyribonucleic acid. *Mol. Endocrinol.* 1, 767–776.

Colosi, P., Swiergiel, J. J., Wilder, E. L., Oviedo, A., and Linzer, D. I. H. (1988). Characterization of proliferin-related protein. *Mol. Endocrinol.* 2, 579–586.

Colosi, P., Thordarson, G., Hellmiss, R., Singh, K., Forsyth, L. A., Gluckman, P. and Wood, W. I. (1989). Cloning and expression of ovine placental lactogen. *Mol. Endocrinol.* 3, 1462–1469.

Cooke, N. E., Coit, D., Shine, J., Baxter, J. D., and Martial, J. A. (1981). Human prolactin: cDNA structural analysis and evolutionary comparisons. *J. Biol. Chem.* 256, 4007–4016.

Cooke, N. E., Szpirer, C., and Levan, G. (1986). The related genes encoding growth hormone and prolactin have been dispersed to chromosomes 10 and 17 in the rat. *Endocrinology* 119, 2451–2454.

Cooke, N. E., Ray, J., Watson, M. A., Estes, P. A., Kuo, B. A., and Liebhaber, S. A. (1988a). Human growth hormone gene and the highly homologous growth hormone variant gene display different splicing patterns. *J. Clin. Invest.* 82, 270–275.

Cooke, N. E., Ray, J., Emery, J. G., and Liebhaber, S. A. (1988b). Two distinct species of human growth hormone-variant mRNA in the human placenta predict the expression of novel growth hormone proteins. *J. Biol. Chem.* 263, 9001–9006.

Courtois, S. J., Lafontaine, D. A., Lemaigre, F. P., Durviaux, S. M., and Rousseau, G. G. (1990). Nuclear factor-1 and activator protein-2 bind in a mutually exclusive way to overlapping promoter sequences and trans-activate the human growth hormone gene. *Nucleic Acids Res.* 18, 57–64.

Cross, J. C., and Roberts, R. M. (1991). Constitutive and trophoblast-specific expression of a class of bovine interferon genes. *Proc. Natl. Acad. Sci. USA* 88, 3817–3821.

Currie, W. B., Card, C. E., Michel, F. J., and Ignotz, G. (1990). Purification, partial characterization, and development of a specific radioimmunoassay for goat placental lactogen. *Reprod. Fertil.* 90, 25–36.

Dana, S., and Karin, M. (1989). Induction of human growth hormone promoter activity by the adenosine 3′,5′-monophosphate pathway involves a novel responsive element. *Mol. Endocrinol.* 3, 815–821.

Davidson, I., Xiao, J. H., Rosales, R., Staub, A. and Chambon, P. (1988). The HeLa cell protein TEF-1 binds specifically and cooperatively to two SV40 enhancer motifs of unrelated sequence. *Cell* 54, 931–942.

Davis, J. S., West, L. A., Weakland, L. L., and Farese, R. V. (1986). Human chorionic gonadotropin activates the inositol 1,4,5-triphosphate-Ca^{2+} intracellular signalling system in bovine luteal cells. *FEBS Lett.* 208, 287–291.

Deb, S., and Soares, M. J. (1990). Characterization of placental prolactin-like protein-A in intracellular and extracellular compartments. *Mol. Cell. Endocrinol.* 74, 163–172.

Deb, S., Faria, T. N., Roby, K. F., Larsen, D., Kwok, S. C. M., Talamantes, F., and Soares, M. J. (1991a). Identification and characterization of a new member of the prolactin family, placental lactogen-I variant *J. Biol. Chem.* 266, 1605–1610.

Deb, S., Roby, K. F., Faria, T. N., Szpirer, C., Levan, G., Kwok, S. C. M., and Soares, M. J. (1991b). Molecular cloning and characterization of prolactin-like protein C complementary deoxyribonucleic acid. *J. Biol. Chem.* 266, 23027–23032.

Deb, S., Roby, K. F., Faria, T. N., Larsen, D., and Soares, M. J. (1991c). Identification and immunochemical characterization of a major placental secretory protein related to the prolactin-growth hormone family, prolactin-like protein-C. *Endocrinology* 128, 3066–3072.

DeChiara, T. M., Efstratiadis, A., and Robertson, E. J. (1990). A growth-deficiency phenotype in heterozygous mice carrying an insulin-like growth factor II gene disrupted by targeting. *Nature* 345, 78–80.

Delegeane, A. M., Ferland, L. H., and Mellon, P. L. (1987). Tissue-specific enhancer of the human glycoprotein hormone α-subunit gene: dependence on cyclic AMP-inducible elements. *Mol. Cell. Biol.* 7, 3994–4002.

Duckworth, M. L., Kirk, K. L., and Friesen, H. G. (1986a). Isolation and identification of a cDNA clone of rat placental lactogen II. *J. Biol. Chem.* 261, 10871–10878.

Duckworth, M. L., Peden, L. M., and Friesen, H. G. (1986b). Isolation of a novel prolactin-like cDNA clone from developing rat placenta. *J. Biol. Chem.* 261, 10879–10884.

Duckworth, M. L., Peden, L. M., and Friesen, H. G. (1988). A third prolactin-like protein expressed by the developing rat placenta: complementary deoxyribonucleic acid sequence and partial structure of the gene. *Mol. Endocrinol.* 2, 912–920.

Duckworth, M. L., Schroedter, I. C., and Friesen, H. G. (1990). Cellular localization of rat placental lactogen II and rat prolactin-like proteins A and B by *in situ* hybridization. *Placenta* 11, 143–155.

Duello, T. M., Byatt, J. C., and Bremel, R. D. (1986). Immunohistochemical localization of placental lactogen in binucleate cells of bovine placentomes. *Endocrinology* 119, 1351–1355.

Ebbitt, D. M., Hurley, W. L., Kessler, M. A., McDonald, D. J., and Schuler, L. A. (1989). Characterization of the gene corresponding to bovine placental prolactin-related cDNA. I. Evolutionary implications. *DNA* 8, 161–169.

Enders, A. C., and Kim, I. K. M. (1991). Trophoblast-uterine interactions during equine chorionic girdle cell maturation, migration and transformation. *Am. J. Anat.* 192, 366–381.

Estes, P. A., Cooke, N. E., and Liebhaber, S. A. (1990). A difference in splicing patterns of the closely related normal and variant human growth hormone gene transcripts is determined by a minimal sequence divergence between two potential splice-acceptor sites. *J. Biol. Chem.* 265, 19863–19870.

Faria, T. N., Deb, S., Kwok, S. C. M., Talamantes, F., and Soares, M. J. (1990). Ontogeny of placental lactogen-I and placental lactogen-II expression in the developing rat placenta. *Dev. Biol.* 141, 279–291.

Faria, T. N., Ogren, L., Talamantes, F., Linzer, D. I. H., and Soares, M. J. (1991). Localization of placental lactogen-I in trophoblast giant cells of the mouse placenta. *Biol. Reprod.* 44, 327–331.

Farin, C. E., Imakawa, K., and Roberts, R. M. (1989). *In situ* localization of mRNA for the interferons, ovine trophoblast protein-1, during early embryonic development of the sheep. *Mol. Endocrinol.* 3, 1099–1107.

Farin, C. E., Cross, J. C., Tindle, N. A., Murphy, C. N., Farin, P. W., and Roberts, R. M. (1991). Induction of trophoblastic interferon expression in ovine blastocysts after treatment with double-stranded RNA. *J. Interferon Res.* 11, 151–157.

Fiddes, J. C. and Goodman, H. M. (1981). The gene encoding the common α subunit of the four human glycoprotein hormones. *J. Mol. Appl. Genet.* 1, 3–18.

Fiddes, J. C. and Talmadge, K. (1984). Structure, expression and evolution of the gene for the human glycoprotein hormones. *In* "Recent Progress in Hormone Research," Vol. 40, pp. 43–78. Academic Press, New York.

Fishel, S. B., Edwards, R. G., and Evans, C. J. (1984). Human chorionic gonadotropin secreted by preimplantation embryos cultured *in vitro*. *Science* 223, 816.

Fitzpatrick, S. L., Walker, W. H., and Saunders, G. F. (1990). DNA sequences involved in the transcriptional activation of a human placental lactogen gene. *Mol. Endocrinol.* 4, 1815–1826.

Forsyth, L. A. (1986). Variation among species in the endocrine control of mammary growth and function: the roles of prolactin, growth hormone, and placental lactogen. *J. Dairy. Sci.* 69, 886–903.

Frankenne, F., Closset, J., Gomez, F., Scippo, M. L., Smal, J., and Hennen, G. (1988). The physiology of growth hormones (GHs) in pregnant women and partial characterization of the placental GH variant. *J. Clin. Endocrinol. Metab.* 66, 1171–1180.

Frankenne, F., Scippo, M. L., VanBeeymen, J., Igout, A., and Hennen, G. (1990). Identification of placental human growth hormone as the growth hormone-V gene expression product. *J. Clin. Endocriol. Metab.* 71, 15–18.

Freemark, M. (1991). Immunoblot analysis of growth hormone and placental lactogen (PL) binding sites in fetal and maternal sheep liver. *In* "The Endocrine Society 73rd Annual Meeting Program & Abstracts," p. 418. [Abstract 1550]

Freemark, M., and Comer, M. (1989). Purification of a distinct placental lactogen receptor, a new member of the growth hormone/prolactin receptor family. *J. Clin. Invest.* 83, 833–889.

Freemark, M., Comer, M., and Handwerger, S. (1986). Placental lactogen and GH receptors in sheep liver: striking differences in ontogeny and function. *Am J. Physiol.* 251, E328–E333.

Freemark, M., Comer, M., Korner, G., and Handwerger, S. (1987). A unique placental lactogen receptor: Implications for fetal growth. *Endocrinology* 120, 1865–1872.

Freemark, M., Comer, M., and Korner, G. (1988). Differential solubilization of placental lactogen (PL)-and growth hormone-binding sites: further evidence for a unique PL receptor in fetal and maternal liver. *Endocrinology* 122, 2771–2779.

Galosy, S. S., Gertler, A., Elberg, G., and Laird, D. M. (1991). Distinct placental lactogen and prolactin (lactogen) receptors in bovine endometrium. *Mol. Cell. Endocrinol.* 78, 229–236.

Gaspard, U. J., Hustin, J., Reuter, A. M., Lambotte, R., and Franchimont, P. (1980). Immunofluorescent localization of placental lactogen, chorionic gonadotropin and its alpha and beta subunits in organ cultures of human placenta. *Placenta* 1, 135–144.

Gharib, S. D., Wierman, M. E., Shupnik, M. A., and Chin, W. W. (1990). Molecular biology of the pituitary gonadotropins. *Endocr. Rev.* 11, 177–199.

Gluckman, P. D., Kaplan, S. L., Rudolph, A. M., and Grumbach, M. M. (1979). Hormone ontogeny in the ovine fetus. II. Ovine chorionic somatomammotropin in mid- and late gestation in the fetal and maternal circulations. *Endocrinology* 104, 1828–1833.

Gluckman, P. D., Butler, J. H., and Elliott, T. B. (1983). The ontogeny of somatotropic binding sites in ovine hepatic membranes. *Endocrinology* 112, 1607–1612.

Guillomot, M., Michel, C., Gaye, P., Charlier, M., Trojan, J., and Martal, J. (1990). Cellular localization of an embryonic interferon, ovine trophoblastin, and its mRNA in sheep embryos during early pregnancy. *Biol. Cell* 68, 205–211.

Handwerger, S. (1991). Clinical counterpoint: The physiology of placental lactogen in human pregnancy. *Endocr. Rev.* 12, 329–336.

Handwerger, S., Crenshaw, C. Jr., Maurer, W. F., Barrett, J., Hurley, T. W., Golander, A., and Fellows, R. E. (1977). Studies on ovine placental lactogen secretion by homologous radioimmunoassay. *J. Endocr.* 72, 27–34.

Hansen, T. R., Cross, J. C., Farin, C. E., Imakawa, K., and Roberts, R. M. (1991a). Slowed transcription and rapid messenger RNA turnover contribute to a decline in synthesis of ovine trophoblast protein-1 during *in vitro* culture. *Biol. Reprod.* 45, 94–100.

Hansen, T. R., Leaman, D. W., Cross, J. C., Mathialagan, N., Bixby, J. A., and Roberts, R. M. (1991b). The genes for the trophoblast interferons and the related interferon-α_{II} possess distinct 5′-promoter and 3′-flanking sequences. *J. Biol. Chem.* 266, 3060–3067.

Harris, D. C., Machin, K. J., Evin, G. M., Morgan, F. J., and Isaacs, N. W. (1989). Preliminary x-ray diffraction analysis of human chorionic gonadotropin. *J. Biol. Chem.* 264, 6705–6706.

Hearn, J. P. (1986). The embryo-maternal dialogue during early pregnancy in primates. *J. Reprod. Fertil.* 76, 809–819.

Hearn, J. P., Webley, G. E., and Gidley-Baird, A. A. (1991). Chorionic gonadotrophin and embryo-maternal recognition during the peri-implantation period in primates. *J. Reprod. Fertil.* 92, 497–509.

Hernandez-Ledezma, J. J., Sikes, J. D., Murphy, C. N., Watson, A. J., Schultz, G. A., and Roberts, R. M. (1992). Expression of bovine trophoblast interferon in conceptuses derived by *in vitro* techniques. *Biol. Reprod.* 47, 374–380.

Hill, D. J., Freemark, M., Strain, A. J., Handwerger, S., and Milner, T. D. G. (1988). Placental lactogen and growth hormone receptors in human fetal tissues: relationship to fetal plasma human placental lactogen concentrations and fetal growth. *J. Clin. Endocrinol. Metab.* 66, 1283–1290.

Hirt, H., Kimelman, J., Birnbaum, M. J., Chen, E. Y., Seeburgh, P. H., Eberhardt, N. L., and Barta, A. (1987). The human growth hormone gene locus, structure, evolution, allelic variations. *DNA* 6, 59–70.

Hochberg, A., Rachmilewitz, J., Eldar-Geva, T., Salant, T., Schneider, T., and de Groot, N. (1992). Differentiation of choriocarcinoma cell line (JAr). *Cancer Res.* 52, 3713–3717.

Hodgen, G. D., and Itskovitz, J. (1988). Recognition and maintenance of pregnancy. *In* "The Physiology of Reproduction" (E. Knobil and J. Neill, eds.), pp. 1995–2021. Raven Press, New York.

Hoeffler, J. P., Deutsch, P. J., Lin, J., and Habener, J. F. (1989). Distinct adenosine 3′,5′-monophosphate and phorbol ester-responsive signal transduction pathways converge at the level of transcriptional activation by the interactions of DNA-binding proteins. *Mol. Endocrinol.* 3, 763–772.

Hollenberg, A. N., Albanese, C., Pestell, R. G., and Jameson, J. L. (1993). The CGβ gene cluster: analysis of functional promoter sequences. *In* "Trophoblast Cells: Pathways for Maternal-Embryonic Communication," (M. N. Soares, S. Handwerger, and F. Talamantes, eds.). Serono Symposia, Springer-Verlag, New York.

Hoshima, M., Boothby, M., and Boime, I. (1982). Cytological localization of chorionic gonadotropin I and placental lactogen mRNA during development of the human placenta. *J. Cell Biol.* 93, 190–198.

Houghton, D. J., Schackelton, P., Obieke, B. C., and Chard, T. (1984). Relationship of maternal and fetal levels of human placental lactogen to the weight and sex of the fetus. *Placenta* 5, 455–458.

Hurley, T. W., Handwerger, S., and Fellows, R. E. (1977). Isolation and structural characterization of ovine placental lactogen. *Biochemistry* 16, 5598–5604.

Ingraham, H. A., Chen, R., Mangalam, H. J., Elsholtz, H. P., Flynn, S. E., Lin, C. R., Simmons, D. M., Swanson, L., and Rosenfeld, M. G. (1988). A tissue-specific transcription factor containing a homeodomain specifies a pituitary phenotype. *Cell* 519–529.

Jackson, L. L., Colosi, P., Talamantes, F., and Linzer, D. I. H. (1986). Molecular cloning of mouse placental lactogen cDNA. *Proc. Natl. Acad. Sci. USA* 83, 8496–8500.

Jackson-Grusby, L. L., Pravtcheva, D., Ruddle, F. H., and Linzer, D. I. H. (1988). Chromosomal mapping of the prolactin/growth hormone gene family in the mouse. *Endocrinology* 122, 2462–2466.

Jameson, J. L., Jaffe, R. C., Gleason, S. L., and Habener, J. F. (1986a). Transcriptional regulation of chorionic gonadotropin α- and β-subunit gene expression by 8-bromoadenosin 3′-5′-monophosphate. *Endocrinology* 199, 2560–2567.

Jameson, J. L., Lindell, C. M., and Habener, J. F. (1986b). Evolution of different transcriptional start sites in the human luteinizing hormone and chorionic gonadotropin β subunit genes. *DNA* 5, 227–234.

Josimovich, J. B., Levitt, M. J., and Stevens, V. C. (1973). Comparison of baboon and human placental lactogen. *Ann. N.Y. Acad. Sci.* 148, 488–500.

Kappes, S. M., Warren, W. C., Pratt, S. L., Liang, R., and Anthony, R. V. (1992). Quantification and cellular localization of ovine placental lactogen messenger ribonucleic acid expression during mid- and late gestation. *Endocrinology* 131, 2829–2838.

Karin, M., Castrillo, J. L., and Theill, L. E. (1990). Growth hormone gene regulation: a paradigm for cell-type-specific gene activation. *Trends Genet.* 6, 92–96.

Keller, P. J., Gerber, C., Greub, H., and Schreiner, W. E. (1970). Studies on the metabolism of human placental lactogen in pregnancy. *Horm. Metab. Res.* 2, 265–267.

Kessler, M. A., and Schuler, L. A. (1991). Structure of the bovine placental lactogen gene and alternative splicing of transcripts. *DNA Cell Biol.* 10, 93–104.

Kessler, M. A., Milosavljevic, M., Zieler, C. G., and Schuler, L. A. (1989). A subfamily of bovine prolactin-related transcripts distinct from placental lactogen in the fetal placenta. *Biochemistry* 28, 5154–5161.

Kessler, M. A., Duello, T. M., and Schuler, L. A. (1991). Expression of prolactin-related hormones in the early bovine conceptus, and potential for paracrine effect on the endometrium. *Endocrinology* 129, 1885–1895.

Klemann, S. W., Imakawa, K., and Roberts, R. M. (1990a). Sequence variability among ovine trophoblast interferon mRNA. *Nucleic Acids Res.* 18, 6724.

Klemann, S. W., Li, J., Imakawa, K., Cross, J. C., Francis, H., and Roberts, R. M. (1990b). The production, purification and bioactivity of bovine trophoblast protein-1 (bovine trophoblast interferon). *Mol. Endocrinol.* 4, 1506–1514.

Kobata, A. (1988). Structures, function and transformational changes of the sugar chains of glycohormones. *J. Cell. Biochem.* 37, 79–90.

Kurman, R. J., Young, R. H., Norris, H. J., Main, C. S., Lawrence, W. D., and Scully, B. E. (1984). Immunocytochemical localization of placental lactogen and chorionic gonadotropin in normal placenta and trophoblastic tumors, with emphasis on intermediate trophoblast and the placental site trophoblastic tumor. *Int. J. Gynecol. Pathol.* 3, 101–121.

LaBonnardière, C., Martinat-Botte, F., Terqui, M., Lefèvre, F., Zourari, K., Martal, J., and Bazer, F. W. (1991). Production of two species of interferon by large white and Meischan pig conceptuses during the peri-attachment period. *J. Reprod. Fertil.* 91, 469–478.

Leaman, D. W., and Roberts, R. M. (1992). Genes for the trophoblast interferons in sheep, goat and musk ox, and distribution of related genes among mammals. *J. Interferon Res.* 12, 1–11.

Lee, C. S., Wooding, F. B. P., and Brandon, M. R. (1986). Immunogold co-localization of ovine placental lactogen and the antigen recognized by the SBU-3 monoclonal antibody in sheep placental granules. *J. Reprod. Fertil.* 78, 653–662.

Lee, S.-J., and Nathans, D. (1987). Secretion of proliferin. *Endocrinology* 120, 208–221.

Lee, S.-J., Talamantes, F., Wilder, E., Linzer, D. I. H., and Nathans, D. (1988). Trophoblastic giant cells of the mouse placenta as the site of proliferin synthesis. *Endocrinology* 122, 1761–1768.

Lefèvre, F., Mege, D., Martinat-Botte, F., Guillomot, M., Zouari, K., and LaBonnardière, C. (1990). Natural expression of gamma and alpha interferons by the porcine trophoblast in early gestation. *J. Interferon Res.* 10(Suppl. 1), S79.

Lemaigre, F. P., Courtois, S. J., Lafontaine, D. A., and Rousseau, G. G. (1989a). Evidence that the upstream stimulatory factor, the Sp1 transcription factor bind *in vitro* to the promoters of the human growth hormone gene. *Eur. J. Biochem.* 181, 555–561.

Lemaigre, F. P., Peers, B., Lafontaine, D. A., Mathy-Hartert, M., Rousseau, G. G., Belayew, A., and Martial, J. A. (1989b). Pituitary-specific factor binding to the human prolactin, growth hormone, placental lactogen genes. *DNA* 8, 149–159.

Levan, G., Szpirer, J., Szpirer, C., Klinga, K., Hanson, C., and Islam, M. Q. (1991). The gene map of the Norway rat (*Rattus novegicus*) and comparative mapping with mouse and man. *Genomics* 10, 699–718.

Li, C. H., Dixon, J. S., and Chung, D. (1973). Amino acid sequence of human chorionic somatomammotropin. *Arch. Biochem. Biophys.* 155, 95–110.

Liang, R., and Anthony, R. V. (1993). Isolation and characterization of the ovine placental lactogen gene. *In* "Trophoblast Cells: Pathways for Maternal Embryonic Communication" (M. N. Soares, S. Handwerger, and F. Talamantes, eds.) Serono Symposia, Springer-Verlag, New York.

Liebhaber, S. A., Urbanek, M., Ray, J., Tuan, R. S., and Cooke, N. E. (1989). Characterization and histological localization of human growth hormone-variant gene expression in the placenta. *J. Clin. Invest.* 83, 1985–1991.

Linzer, D. I. H., and Nathans, D. (1984). Nucleotide sequence of a growth-related mRNA encoding a member of the prolactin-growth hormone family. *Proc. Natl. Acad. Sci. USA* 81, 4255–4259.

Linzer, D. I. H., and Nathans, D. (1985). A new member of the prolactin-growth hormone gene family expressed in mouse placenta. *EMBO J.* 4, 1419–1423.

Linzer, D. I. H., Lee, S. J., Ogren, L., Talamantes, F., and Nathans, D. (1985). Identification of proliferin mRNA and protein in mouse placenta. *Proc. Natl. Acad. Sci. USA* 82, 4356–4359.

Lowman, H. B., Cunningham, B. C., and Wells, J. A. (1991). Mutational analysis and protein engineering of receptor-binding determinants in human placental lactogen. *J. Biol. Chem.* 266, 10982–10988.

Lusthader, J. W., Burken, S., Pileggi, N. P., Kolks, M. A., Pollak, S., Cuff, M. E., Yang, W., Hendrickson, W. A., and Canfield, R. E. (1989). Crystallization and characterization of human chorionic gonadotropin in chemically deglycosylated and enzymatically desialylated states. *Biochemistry* 28, 9239–9243.

MacLeod, J. N., Worsley, I., Ray, J., Friesen, H. G., Liebhaber, S. A., and Cooke, N. E. (1991). Human growth hormone-variant is a biologically active somatogen and lactogen. *Endocrinology* 128, 1298–1302.

Marcotty, C., Frankenne, F., and Scippo, M. L. (1992). Evidence for the development and dependent expression of CS-L gene in human placenta. *In* "The Endocrine Society 74th Annual Meeting Program & Abstracts," p. 407. [Abstract 1424]

Martal, J., and Djiane, J. (1975). Purification of lactogenic hormone in sheep placenta. *Biochem. Biophys. Res. Commun.* 65, 770–778.

Martal, J., and Djiane, J. (1977). The production of chorionic somatomammotrophin in sheep. *J. Reprod. Fertil.* 49, 285–289.

Maruo, T., Matsuo, H., Oishi, T., Hayashi, M., Nishino, R., and Mochizuki, M. (1987). Induction of differentiated trophoblast function by epidermal growth factor: relation of immunohistochemically detected cellular epidermal growth factor receptor levels. *J. Clin. Endocrinol. Dev.* 64, 744–750.

Masuhiro, K., Matsuzaki, N., Nishino, E., Taniguchi, T., Kameda, T., Li, Y., Saji, F., and Tanizawa, O. (1991). Trophoblast-derived interleukin-1 (IL-1) stimulates the release of human chorionic gonadotropin by activating IL-6 and IL-6-receptor system in first trimester human trophoblasts. *J. Clin. Endocrinol. Metab.* 72, 594–601.

Matzuk, M. M., Keene, J. L., and Boime, I. (1989). Site specificity of the chorionic gonadotropin N-linked oligosaccharides in signal transduction. *J. Biol. Chem.* 264, 2409–2414.

McWilliams, D., and Boime, I. (1980). Cytological localization of placental lactogen messenger ribonucleic acid in syncytiotrophoblast layers of human placenta. *Endocrinology* 107, 761–765.

Miller, W. L., and Eberhardt, N. L. (1983). Structure and evolution of the growth hormone gene family. *Endocr. Rev.* 4, 97–130.

Milosavljevic, M., Duello, T. M., and Schuler, L. A. (1989). *In situ* localization of two prolactin-related messenger ribonucleic acids to binucleate cells of bovine placentomes. *Endocrinology* 125, 883–889.

Morrish, D. W., Bhardwaj, D., Dabbagh, L. K., Marusy, R. H., and Siy, O. (1987). Epidermal growth factor induces differentiation and secretion of human chorionic gonadotropin and placental lactogen in normal human placenta. *J. Clin. Endocrinol. Dev.* 65, 1282–1290.

Morrish, D. W., Bhardwaj, D., and Parus, M. T. (1991). Transforming growth factor β1 inhibits placental differentiation and human chorionic gonadotropin and human placental lactogen secretion. *Endocrinology* 129, 22–26.

Moyle, W. R., Matzuk, M. M., Campbell, R. K., Logliani, E., Dean-Emig, D., Krichevsky, A., Barnett, R. W., and Boime, I. (1990). Localization of residues that confer antibody-binding specificity using human chorionic gonadotropin/luteinizing hormone β-subunit chimeras and mutants. *J. Biol. Chem.* 265, 8511–8518.

Murthy, G. S., Schellenberg, C., and Friesen, H. G. (1982). Purification and characterization of bovine placental lactogen. *Endocrinology* 111, 2117–2124.

Nachtigal, M. W., Nickel, B. E., and Cattini, P. A. (1992). Differential regulation of pituitary and placental members of the human growth hormone gene family. *In* "The Endocrine Society 74th Annual Meeting Program and Abstracts," p. 365. [Abstract 1256]

Nader, S., Berkowitz, A. S., Ochs, D., Held, B., and Winkel, C. A. (1988). Luteal phase support in stimulated cycles in an *in vitro* fertilization embryo transfer program: progesterone versus human chorionic gonadotropin. *In Vitro Fertil. Embryo Trans.* 5, 81–84.

Naylor, S. L., Chin, W. W., Goodman, H. M., Lasley, P. A., Grzechik, K. H., and Sakaguchi, A. Y. (1983). Chromosome assignment of the genes encoding the α and β-subunits of glycoprotein hormones in man and mouse. *Som. Cell Genet.* 9, 757–770.

Niall, H. D., Hogan, M. L., Sauer, R., Rosenblum, I. Y., and Greenwood, F. C. (1971). Sequences of pituitary and placental lactogenic and growth hormones: evolution from a primordial peptide by gene reduplication. *Proc. Natl. Acad. Sci. USA* 68, 866–869.

Nickel, B. E., and Cattini, P. A. (1991). Tissue specific expression and thyroid hormone regulation of the endogenous placental growth hormone variant and chorionic somatomammotropin genes in a human choriocarcinoma cell line. *Endocrinology* 128, 2353–2359.

Nieder, G. L., and Jennes, L. (1990). Production of mouse placental lactogen-1 by trophoblast giant cell *in utero* and *in vitro*. *Endocrinology* 126, 2809–2814.

Nilson, J. H., Bokar, J. A., Clay, C. M., Farmerie, T. A., Fenstermaker, R. A., Hamernik, D. L., and Keri, R. A. (1991). Different combinations of regulatory elements may explain why placenta-specific expression of the glycoprotein hormone α-subunit gene occurs only in primates and horses. *Biol. Reprod.* 44, 231–237.

Nisula, B. C., Blithe, D. L., Akur, A., Lefort, G., and Wehlman, R. E. (1989). Metabolic fate of human gonadotropin. *J. Steroid Biochem.* 33, 733–737.

Ogilvie, S., Duckworth, M. L., Larkin, L. H., Buhi, W. C., and Shiverick, K. T. (1990). *De novo* synthesis and secretion of prolactin-like protein-B by rat placental explants. *Endocrinology* 126, 2561–2566.

Ogren, L., and Talamantes, F. (1988). Prolactins of pregnancy and their cellular source. *Intl. Rev. Cytology* 112, 1–65.

Otani, F., Otani, T., and Boime, I. (1989). Effects of adenine nucleotides on choriogonadotropin α and β subunit synthesis. *Biochem. Biophys. Res. Commun.* 160, 6–11.

Owerbach, D., Rutter, W., Martial, J., and Baxter, J. (1980). Genes for growth hormone, chorionic somatomammotropin, and growth hormone-like gene are on chromosome 17 in humans. *Science* 209, 289–292.

Owerbach, D., Rutter, W. J., Cooke, N. E., Martial, J. A. and Shows, T. B. (1981). The prolactin gene is located on chromosome 6 in humans. *Science* 212, 815–816.

Peritz, L. N., Fodor, E. J. B., Silversides, D. W., Cattini, P. A., Baxter, J. D., and Eberhardt, N. L. (1988). The human growth hormone gene contains both positive and negative control elements. *J. Biol. Chem.* 263, 5005–5007.

Policastro, P. F., Daniels-McQueen, S., Carle, G., and Boime, I. (1986). A map of the hCGβ-[illegible]

Ray, J., Okamura, H., Kelly, P. A., Cooke, N. E., and Liebhaber, S. A. (1990). Human growth hormone-variant demonstrates a receptor binding profile distinct from that of normal pituitary growth hormone. *J. Biol. Chem.* 265, 7939–7944.

Ray, J., Jones, B. K., Liebhaber, S. A., and Cooke, N. E. (1989). Glycosylated human growth hormone variant. *Endocrinology* 125, 566–568.

Renfree, M. B. (1982). Implantation and placentation. *In* "Reproduction in Mammals. 2. Embryonic and Fetal Development" (C. R. Austin and R. V. Short, eds.), pp. 26–69. Cambridge University Press, Cambridge.

Resendez-Perez, D., Ramirezsolis, R., Varelaechavarria, A., Martinezrodriguez, H. G., and Barrerasaldana, H. A. (1990). Coding potential of transfected human placental lactogen genes. *Nucleic Acids Res.* 18, 4665–4670.

Reuter, A. M., Gaspard, U. J., Delville, J. L., Vrindts-Gevaert, Y., and Franchimont, P. (1980). Serum concentrations of human chorionic gonadotrophin and its alpha and beta subunits. *Clin. Endocrinol.* 13, 305–317.

Ringler, G. E., and Strauss, J. F. (1990). *In vitro* systems for the study of human placental endocrine function. *Endocr. Rev.* 11, 105–122.

Ringler, G. E., Kao, L-C., Ulloa-Aguirre, A., and Strauss, J. F. (1989). Control of the endocrine function of the human placenta by cyclic AMP. *In* "Blastocyst Implantation" (K. Yoshinaga, ed.), pp. 209–216. Serono Symposia, Adams Publishing, Boston.

Roberts, R. M., Farin, C. E., and Cross, J. C. (1990). Trophoblast proteins and maternal recognition of pregnancy. *In* "Oxford Reviews of Reproductive Biology" (S. R. Milligan, ed.), Vol. 12, pp. 147–180. Oxford University Press, Oxford.

Roberts, R. M., Cross, J. C., and Leaman, D. W. (1992a). Unique features of the trophoblast interferons. *Pharmacol. Ther.* 51, 329–345.

Roberts, R. M., Cross, J. C,. and Leaman, D. W. (1992b). Interferons as hormones of pregnancy. *Endocr. Rev.* 13, 432–452.

Roberts, R. M., Leaman, D. W., Hernandez-Ledezma, J. J., and Cosby, N. C. (1993). Trophoblast interferons: expression during development and gene organization. *In* "Trophoblast Cells: Pathways for Maternal-Embyronic Communication" (M. N. Soares, S. Handwerger and F. Talamantes, eds.). Serono Symposia, Springer-Verlag. New York.

Robertson, M. C., Gillespie, B., and Friesen, H. G. (1982). Characterization of the two forms of rat placental lactogen (rPL): rPL-I and rPL-II. *Endocrinology* 111, 1862–1866.

Robertson, M. C., Croze, F., Schroedter, I. C., and Friesen, H. G. (1990). Molecular cloning and expression of rat placental lactogen I complementary deoxyribonucleic acid. *Endocrinology* 127, 702–710.

Robertson, M. C., Schroedter, I. C., and Friesen, H. G. (1991). Molecular cloning and expression of rat placental lactogen-Iv, a variant of rPL-I present in late pregnancy rat placenta. *Endocrinology,* 129, 2746–2756.

Rosa, C., Amr, S., Birken, S., Wehlman, R., and Nisula, B. (1984). Effect of desialylation of human chorionic gonadotropin on its metabolic clearance rates in humans. *J. Clin. Endocrinol. Metab.* 59, 1215–1219.

Ryan, R. J., Keukmann, H. T., Charlesworth, M. C., McCormick, D. J., Milius, R. P., Calvo, F. O., and Vutyavanich, T. (1987). Structure-function relationships of gonadotropins. *Recent Prog. Horm. Res.* 43, 383–422.

Sairam, M. R. (1989). Role of carbohydrates in glycoprotein hormone signal transduction. *FASEB J.* 3, 1915–1926.

Schalue-Francis, T. K., Farin, P. W., Cross, J. C., Keisler, D., and Roberts, R. M. (1991). Effect of injected bovine interferon-α_I1 on oestrous cycle length and pregnancy success in sheep. *J. Reprod. Fertil.* 91, 347–356.

Schindler, C., Shuai, K., Prezioso, V. R., and Darnell, J. E. (1992). Interferon-dependent tyrosine phosphorylation of a latent cytoplasmic transcription factor. *Science* 257, 809–813.

Schuler, L. A., Shimomura, K., Kessler, M. A., Zieler, C. G., and Bremel, R. D. (1988). Bovine placental lactogen: molecular cloning and protein structure. *Biochemistry* 27, 8443–8448.

Schuler, L. A., Kessler, M. A., Tanaka, M., and Nakashima, K. (1991). Nomenclature clarification for the bovine placental prolactin-related hormones. *Endocrinology* 129, 2057.

Selby, M. J., Barta, A., Baxter, J. D., Bell, G. I., and Eberhardt, N. L. (1984). Analysis of a major human chorionic somatomammotropin gene. *J. Biol. Chem.* 259, 13131–13138.

Sherman, G. B., Wolfe, M. W., Farmerie, T. A., Clau, C. M., Threadgill, D. S., Sharp, D. C., and Nilson, J. H. (1992). A single gene encodes the β-subunits of equine luteinizing hormone and chorionic gonadotropin. *Mol. Endocrinol.* 6, 951–959.

Shida, M. M., Jackson-Grusby, L. L., Ross, S. R., and Linzer, D. I. H. (1992). Placental-specific expression from the mouse placental lactogen-II gene promoter. *Proc. Natl. Acad. Sci. USA* 89, 3864–3868.

Shimomura, K., and Bremel, R. D. (1988). Characterization of bovine placental lactogen as a glycoprotein with N-linked and O-linked carbohydrate side chains. *Mol. Endocrinol.* 2, 845–853.

Shome, B., and Friesen, H. G. (1971). Purification and characterization of monkey placental lactogen. *Endocrinology* 89, 631–641.

Shome, B., and Parlow, A. F. (1977). Human pituitary prolactin (hPRL): the entire linear amino acid sequence. *J. Clin. Endocrinol. Metab.* 45, 1112–1115.

Sibley, C. G., Hochberg, A., and Boime, I. (1991). Bromo-adenosine stimulates chorionic gonadotropin production in JAr and cytotrophoblast cells: evidence for effects on two stages of differnetiation. *Mol. Endocrinol.* 5, 582–586.

Slater, E. P., Rabenau, O., Karin, M., Baxter, J. D., and Beato, M. (1985). Glucocorticoid receptor binding and activation of a heterologous promoter by dexamethasone by the first intron of the human growth hormone gene. *Mol. Cell Biol.* 5, 2984–2992.

Soares, M. J. (1987). Developmental changes in the intraplacental distribution of placental lactogen and alkaline phosphatase in the rat. *J. Reprod. Fertil.* 79, 93–98.

Soares, M. J., Faria, T. N., Roby, K. F., and Deb, S. (1991). Pregnancy and the prolactin family of hormones: coordination of anterior pituitary, uterine, and placental expression. *Endocr. Rev.* 12, 402–423.

Southard, J. N., Do, L., Smith, W. C., and Talamantes, F. (1989). Hamster placental lactogen-II contains a structural feature unique among the growth hormone-prolactin placental lactogen family. *Mol. Endocrinol.* 3, 1710–1713.

Spellacy, W. N., Carlson, K. L., and Birk, S. A. (1966). Dynamics of human placental lactogen. *Am. J. Obstet. Gynecol.* 96, 1164–1173.

Steger, D. J., Altschmied, J., Buscher, M., and Mellon, P. L. (1991). Evolution of placenta-specific gene expression: comparison of the equine and human gonadotropin α-subunit genes. *Mol. Endocrinol.* 5, 243–255.

Steven, D. H. (1975). "Comparative Placentation." Academic Press, London.

Strauss, B. L., Pittman, R., Nilson, J. H., and Boime, I. (1993). Expression of the β subunit of chorionic gonadotropin in transgenic mice. *In* "Trophoblast Cells: Pathways for Maternal Embryonic Communication," (M. N. Soares, S. Handwerger, and F. Talamantes, eds.). Serono Symposia, Springer-Verlag, New York.

Sugino, H., Bousfield, G. R., Moor, W. T., and Ward, D. W. (1987). Structural studies on equine glycoprotein hormones: amino acid sequence of equine chorionic gonadotropin β subunit. *J. Biol. Chem.* 262, 8603–8609.

Talamantes, F., and Ogren, L. (1988). The placenta as an endocrine organ: Polypeptides. *In* "The Physiology of Reproduction" (E. Knobil and J. Neill, eds.), pp. 2093–2144. Raven Press, New York.

Talmadge, K., Vamvakopouics, W. C., and Fiddes, J. C. (1984). Evolution of the genes for the β subunits of human chorionic gonadotropin and luteinizing hormone. *Nature* 307, 37–40

Tuchinsky, D., and Hobel, C. J. (1973). Plasma hCG, estrone, estradiol, progesterone and 17-alphahydroxyprogesterone in human pregnancy. III. Normal human pregnancy. *Am. J. Obstet. Gynecol.* 117, 884–893.

Urbanek, M., MacLeod, J. N., Cooke, N. E., and Liebhaber, S. A. (1992). Expression of human growth hormone (hGH) receptor isoform is predicted by tissue-specific alternative splicing of exon 3 of the hGH receptor gene transcript. *Mol. Endocrinol.* 6, 279–287.

Uzé, G., Luffally, G., and Gresser, J. (1990). Genetic transfer of a functional human interferon α receptor into mouse cells: cloning and expression of its cDNA. *Cell* 60, 225–234.

Uzé, G., Luftalla, G., Bandu, M. T., Proudhon, D., and Mogensen, K. E. (1992). Behavior of a cloned murine interferon α/β receptor expressed in homospecific or heterospecific background. *Proc. Natl. Acad. Sci. USA* 89, 4774–4778.

Vandevoort, C. A., Molskness, T. A., Stouffer, R. L. (1988). Adenylate cyclase in the primate corpus luteum during chorionic gonadotropin treatment simulating early pregnancy: homologous versus heterologous desensitization. *Endocrinology* 122, 734–740.

Velazquez, L., Follous, M., Stark, G. R., and Pellegrini, S. (1992). A protein tyrosine kinase in the interferon α/β signalling pathway. *Cell* 70, 313–322.

Vinik, A. I., Kaplan, S. L., and Grumbach, M. M. (1973). Purification, characterization and comparison of immunological properties of monkey chorionic somatomammotropin with human and monkey growth hormone, human chorionic somatomammotropin and ovine prolactin. *Endocrinology* 92, 1051–1064.

Walker, W. H., Fitzpatrick, S. L., and Saunders, G. F. (1990). Human placental lactogen transcriptional enhancer. *J. Biol. Chem.* 265, 12940–12948.

Warren, W. C., Keisler, D. H., and Anthony, R. V. (1990a). Synthesis and secretion of ovine placental lactogen and its biochemical properties. *Dom. Anim. Endocrinol.* 7, 331–342.

Warren, W. C., Liang, R., Krivi, G. G., Siegel, N. R., and Anthony, R. V. (1990b). Purification and structural characterization of ovine placental lactogen. *J. Endocrinol.* 126, 141–149.

Wooding, F. B. P. (1981). Localization of ovine placental lactogen in sheep placentomes by electron microscope immunocytochemistry. *J. Reprod. Fertil.* 62, 15–19.

Wooding, F. B. P. (1992). Current topic: the synephitheliochorial placenta of ruminants: binucleate cell fusions and hormone production. *Placenta* 13, 101–113.

Xie, S., Low, B. G., Kramer, K. K., Nagel, R. J., Anthony, R. V., Zoli, A. P., Beckers, J.-F., and Roberts, R. M. (1991). Identification of the major pregnancy-specific antigens of cattle and sheep as inactive members of the aspartic proteinase family. *Proc. Natl. Acad. Sci. USA* 88, 10247–10251.

Yamakawa, M., Tanaka, M., Koyama, M., Kagesato, Y., Watahiki, M., Yamamoto, M., and Nakashima, K. (1990). Expression of new members of the prolactin growth hormone gene family in bovine placenta. *J. Biol. Chem.* 265, 8915–8920.

Zieler, C. G., Kessler, M. A., and Schuler, L. A. (1990). Characterization of a novel prolactin-related protein from bovine fetal placenta. *Endocrinology* 126, 2372–2382.

Index